Energy, Environment, and Sustainability

Series Editor

Avinash Kumar Agarwal, Department of Mechanical Engineering, Indian Institute of Technology Kanpur, Kanpur, Uttar Pradesh, India

AIMS AND SCOPE

This books series publishes cutting edge monographs and professional books focused on all aspects of energy and environmental sustainability, especially as it relates to energy concerns. The Series is published in partnership with the International Society for Energy, Environment, and Sustainability. The books in these series are edited or authored by top researchers and professional across the globe. The series aims at publishing state-of-the-art research and development in areas including, but not limited to:

- Renewable Energy
- Alternative Fuels
- Engines and Locomotives
- Combustion and Propulsion
- Fossil Fuels
- Carbon Capture
- Control and Automation for Energy
- Environmental Pollution
- Waste Management
- Transportation Sustainability

Review Process

The proposal for each volume is reviewed by the main editor and/or the advisory board. The chapters in each volume are individually reviewed single blind by expert reviewers (at least four reviews per chapter) and the main editor.

Ethics Statement for this series can be found in the Springer standard guidelines here https://www.springer.com/us/authors-editors/journal-author/journal-author-helpdesk/before-you-start/before-you-start/1330#c14214

More information about this series at http://www.springer.com/series/15901

Himanshu Tyagi · Prodyut R. Chakraborty ·
Satvasheel Powar · Avinash K. Agarwal
Editors

New Research Directions in Solar Energy Technologies

Editors
Himanshu Tyagi
Department of Mechanical Engineering
Indian Institute of Technology Ropar
Ropar, Punjab, India

Prodyut R. Chakraborty
Department of Mechanical Engineering
Indian Institute of Technology Jodhpur
Jodhpur, Rajasthan, India

Satvasheel Powar
School of Engineering
Indian Institute of Technology Mandi
Mandi, Himachal Pradesh, India

Avinash K. Agarwal
Department of Mechanical Engineering
Indian Institute of Technology Kanpur
Kanpur, Uttar Pradesh, India

ISSN 2522-8366 ISSN 2522-8374 (electronic)
Energy, Environment, and Sustainability
ISBN 978-981-16-0596-3 ISBN 978-981-16-0594-9 (eBook)
https://doi.org/10.1007/978-981-16-0594-9

This Springer imprint is published by the registered company Springer Nature Singapore Pte Ltd.
The registered company address is: 152 Beach Road, #21-01/04 Gateway East, Singapore 189721, Singapore

Preface

The International Society for Energy, Environment and Sustainability (ISEES) was found at Indian Institute of Technology Kanpur (IIT Kanpur), India, in January 2014 with an aim to spread knowledge/awareness and catalyze research activities in the fields of energy, environment, sustainability and combustion. The society's goal is to contribute to the development of clean, affordable and secure energy resources and a sustainable environment for the society and to spread knowledge in the above-mentioned areas and create awareness of the environmental challenges, which the world is facing today. The unique way adopted by the society was to break the conventional silos of specializations (engineering, science, environment, agriculture, biotechnology, materials, fuels, etc.) to tackle the problems related to energy, environment and sustainability in a holistic manner. This is quite evident by the participation of experts from all fields to resolve these issues. ISEES is involved in various activities such as conducting workshops, seminars and conferences in the domains of its interests. The society also recognizes the outstanding works done by the young scientists and engineers for their contributions in these fields by conferring them awards under various categories.

The Fourth International Conference on 'Sustainable Energy and Environmental Challenges' (IV-SEEC) was organized under the auspices of ISEES from 27–29 November 2019, at NEERI, Nagpur. This conference provided a platform for discussions between eminent scientists and engineers from various countries including India, USA, China, Italy, Mexico, South Korea, Japan, Sweden, Greece, Czech Republic, Germany, Netherland and Canada. In this conference, eminent speakers from all over the world presented their views related to different aspects of energy, combustion, emissions and alternative energy resource for sustainable development and cleaner environment. The conference presented one high-voltage plenary talk by Mrs. Rashmi Urdhwareshe, Director, Automotive Research Association of India (ARAI), Pune.

The conference included 28 technical sessions on topics related to energy and environmental sustainability including 1 plenary talk, 25 keynote talks and 54 invited talks from prominent scientists, in addition to 70+ contributed talks and 80+ poster presentations by students and researchers. The technical sessions in the conference included fuels, engine technology and emissions, coal and biomass combustion/gasification, atomization and sprays, combustion and modelling, alternative energy resources, water and water and wastewater treatment, automobile and other environmental applications, environmental challenges and sustainability, nuclear energy and other environmental challenges, clean fuels and other environmental challenges, water pollution and control, biomass and biotechnology, waste to wealth, microbiology, biotechnological and other environmental applications, waste and wastewater management, cleaner technology and environment, sustainable materials and processes, energy, environment and sustainability, technologies and approaches for clean, sensors and materials for environmental, biological processes and environmental sustainability. One of the highlights of the conference was the Rapid-Fire Poster Sessions in (i) engine/fuels/emissions, (ii) environment and (iii) biotechnology, where 50+ students participated with great enthusiasm and won many prizes in a fiercely competitive environment. Three hundred plus participants and speakers attended this three-day conference, where 12 ISEES books published by Springer, Singapore, under a special dedicated series 'Energy, environment and sustainability' were released. This was the third time in a row that such a significant and high-quality outcome has been achieved by any society in India. The conference concluded with a panel discussion on 'Balancing Energy Security, Environmental Impacts and Economic Considerations: Indian Perspective', where the panelists were Dr. Anjan Ray, CSIR-IIP Dehradun; Dr. R. R. Sonde, Thermax Ltd.; Prof. Avinash Kumar Agarwal, IIT Kanpur; Dr. R. Srikanth, National Institute of Advanced Studies, Bengaluru; and Dr. Rakesh Kumar, NEERI, Nagpur. The panel discussion was moderated by Prof. Ashok Pandey, Chairman, ISEES. This conference laid out the roadmap for technology development, opportunities and challenges in energy, environment and sustainability domain. All these topics are very relevant for the country and the world in present context. We acknowledge the support received from various funding agencies and organizations for the successful conduct of the Fourth ISEES Conference (IV-SEEC), where these books germinated. We would therefore like to acknowledge SERB, Government of India (special thanks to Dr. Sandeep Verma, Secretary); NEERI, Nagpur (special thanks to Dr. Rakesh Kumar, Director), CSIR, and our publishing partner Springer (special thanks to Swati Mehershi).

The editors would like to express their sincere gratitude to a large number of authors from all over the world for submitting their high-quality work in a timely manner and revising it appropriately at a short notice. We would like express our special thanks to all the reviewers who reviewed various chapters of this monograph and provided their valuable suggestions to improve the manuscripts.

This book is based on the solar energy technologies for various applications such as generation of electric power, heating and energy storage. We hope that the book would be of great interest to the professionals and postgraduate students involved in the field of solar energy.

Rupnagar, India Himanshu Tyagi
Jodhpur, India Prodyut R. Chakraborty
Mandi, India Satvasheel Powar
Kanpur, India Avinash K. Agarwal

Contents

Editors and Contributors

About the Editors

Dr. Himanshu Tyagi is an Associate Professor in the School of Mechanical, Materials and Energy Engineering, Indian Institute of Technology Ropar (IIT), India. He has previously worked at the Steam Turbine Design Division of Siemens (Germany and India) and at the Thermal and Fluids Core Competency Group of Intel Corp. (USA). He received his Ph.D. from Arizona State University, USA, in the field of heat transfer and specifically looked for the radiative and ignition properties of nanofluids. He obtained his Master's degree from University of Windsor, Canada, and his Bachelor's in Mechanical Engineering from IIT Delhi. At present, he is working to develop nanotechnology-based clean and sustainable energy sources with a team of several Ph.D. scholars, postgraduate, and undergraduate students. Among other awards, he has received Summer Undergraduate Research Award (SURA) from IIT Delhi, International Graduate Student Scholarship from University of Windsor Canada, Indo-US Science and Technology Forum (IUSSTF) grant awarded for organizing an Indo-US Workshop on 'Recent Advances in Micro/Nanoscale Heat Transfer and Applications in Clean Energy Technologies' at IIT Ropar.

Dr. Prodyut R. Chakraborty is an Associate Professor in the department of Mechanical Engineering, Indian Institute of Technology (IIT) Jodhpur, India. He received his Bachelor's in Mechanical Engineering from the North Bengal University in 2000, and his M.Sc. Engineering in 2004, and Ph.D. in 2011 from the Indian Institute of Science, Bangalore, India. He served two years at the Department of Material Physics in Space in German Aerospace Center (DLR), Cologne as a postdoctoral research fellow. He also worked as a Research Analyst at the Applied CFD Lab, G. E. Global Research Center Bangalore. His primary area of research is numerical modeling of alloy solidification, latent heat based energy storage systems for high temperature applications, thermal management and thermal comfort, and sorption cooling.

Dr. Satvasheel Powar is an Assistant Professor in the School of Engineering, Indian Institute of Technology (IIT) Mandi. He received a Bachelor of Production Engineering from Shivaji University in 2003, and M.Sc. (Mechanical Engineering) from Dalarna University, Sweden in 2005. He then served Greatcell Solar S.A., Switzerland, and G24i, the UK with a work focus on scalable process development. He received his Ph.D. in Chemistry/Materials Engineering from Monash University, Australia in 2013. Before joining IIT Mandi, he served at Nanyang Technological University, Singapore as a postdoctoral research fellow. His primary area of research is new generation solar photovoltaic and solar thermal utilization. Dr. Powar received a Bhaskara Advanced Solar Energy fellowship by Indo-US Science and Technology Forum (IUSSTF) to visit Lawrence Berkeley National Laboratory, University of California, Berkeley, USA for three months.

Prof. Avinash K. Agarwal joined the Indian Institute of Technology (IIT) Kanpur, India in 2001 after working as a post-doctoral fellow at the Engine Research Center, University of Wisconsin at Madison, USA. His interests are IC engines, combustion, alternate and conventional fuels, lubricating oil tribology, optical diagnostics, laser ignition, HCCI, emissions and particulate control, and large bore engines. Prof. Agarwal has published 290+ peer reviewed international journal and conference papers, 42 edited books, 78 books chapters and has 10000+ Scopus and 15300+ Google scholar citations. He is a Fellow of SAE (2012), Fellow of ASME (2013), Fellow of ISEES (2015), Fellow of INAE (2015), Fellow of NASI (2018), Fellow of Royal Society of Chemistry (2018), and a Fellow of American Association of Advancement in Science (2020). He is recipient of several prestigious awards such as Clarivate Analystics India Citation Award-2017 in Engineering and Technology, NASI-Reliance Industries Platinum Jubilee Award-2012; INAE Silver Jubilee Young Engineer Award-2012; Dr. C. V. Raman Young Teachers Award: 2011; SAE Ralph R. Teetor Educational Award - 2008; INSA Young Scientist Award-2007; UICT Young Scientist Award-2007; INAE Young Engineer Award-2005. Prof. Agarwal received Prestigious Shanti Swarup Bhatnagar Award-2016 in Engineering Sciences. For his outstanding contributions, Prof. Agarwal is conferred upon Sir J. C. Bose National Fellowship (2019) by SERB.

Contributors

Avinash K. Agarwal Department of Mechanical Engineering, Indian Institute of Technology Kanpur, Kanpur, Uttar Pradesh, India

Abhishek Agrawal The Energy and Resources Institute, New Delhi, India;
Indian Institute of Technology Delhi, New Delhi, India

Jyoti Bharj Department of Physics, Dr. B. R. Ambedkar National Institute of Technology Jalandhar, Jalandhar, Punjab, India

Rabinder Singh Bharj Department of Mechanical Engineering, Dr. B. R. Ambedkar National Institute of Technology Jalandhar, Jalandhar, Punjab, India

Anirban Bhattacharya School of Mechanical Sciences, IIT Bhubaneswar, Bhubaneswar, Odisha, India

Saikat Bhaumik Institute of Chemical Technology-IndianOil Odisha Campus, Bhubaneswar, India

M. Cabrera School of Engineering, Texas A&M International University, Laredo, TX, USA

Prodyut R. Chakraborty Department of Mechanical Engineering, Indian Institute of Technology Jodhpur, Jodhpur, Rajasthan, India

Prodyut Ranjan Chakraborty Department of Mechanical Engineering, Indian Institute of Technology Jodhpur, Jodhpur, India

Laltu Chandra Department of Mechanical Engineering, IIT BHU Varanasi, Varanasi, India

Atul Dhar School of Engineering, Indian Institute of Technology Mandi, Mandi, Himachal Pradesh, India

Shamini Dharmasena Energy, Power and Sustainability Group, Florida International University, Miami, USA

J. Figueroa School of Engineering, Texas A&M International University, Laredo, TX, USA

D. Ganta School of Engineering, Texas A&M International University, Laredo, TX, USA

Sairaj Gaunekar Department of Mechanical Engineering, Indian Institute of Technology Jodhpur, Jodhpur, India

Sampad Ghosh Interdisciplinary Graduate School of Engineering Sciences, Kyushu University, Fukuoka, Japan;
International Institute for Carbon-Neutral Energy Research (WPI-I^2CNER), Kyushu University, Fukuoka, Japan;
Department of Electrical and Electronic Engineering, Chittagong University of Engineering and Technology, Chittagong, Bangladesh

Shruti Goswami School of Engineering, Indian Institute of Technology Mandi, Mandi, Himachal Pradesh, India

Sivasankaran Harish International Institute for Carbon-Neutral Energy Research (WPI-I^2CNER), Kyushu University, Fukuoka, Japan

Alexandar Hernandez Energy, Power and Sustainability Group, Florida International University, Miami, USA

Rajan Kumar Department of Mechanical Engineering, Dr. B. R. Ambedkar National Institute of Technology Jalandhar, Jalandhar, Punjab, India

Vishwa Deepak Kumar Department of Mechanical Engineering, Indian Institute of Technology Jodhpur, Jodhpur, Rajasthan, India

Sudipto Mukhopadhyay Department of Mechanical Engineering, Indian Institute of Technology Jodhpur, Jodhpur, Rajasthan, India

Temitayo O. Olowu Energy, Power and Sustainability Group, Florida International University, Miami, USA

Marmik Pancholi School of Engineering and Applied Sciences, Ahmedabad University, Ahmedabad, Gujrat, India

Sushant Pandey School of Engineering, Indian Institute of Technology Mandi, Mandi, Himachal Pradesh, India

Parth Patil Department of Mechanical Engineering, Indian Institute of Technology Ropar, Rupnagar, Punjab, India

Satvasheel Powar School of Engineering, Indian Institute of Technology Mandi, Mandi, Himachal Pradesh, India

Dibakar Rakshit Centre for Energy Studies, Indian Institute of Technology Delhi, Hauz Khas, New Delhi, India

Arup K. Rath CSIR-National Chemical Laboratory, Pune, India

Prasenjit Rath School of Mechanical Sciences, IIT Bhubaneswar, Bhubaneswar, Odisha, India

K. R. Ravi Department of Metallurgical and Materials Engineering, Indian Institute of Technology Jodhpur, Jodhpur, India

Bidyut Baran Saha Interdisciplinary Graduate School of Engineering Sciences, Kyushu University, Fukuoka, Japan;
International Institute for Carbon-Neutral Energy Research (WPI-I^{2}CNER), Kyushu University, Fukuoka, Japan;
Department of Mechanical Engineering, Kyushu University, Fukuoka, Japan

Sudip K. Saha Diamond Harbour Women's University, Diamond Harbour, West Bengal, India

Pranaynil Saikia Centre for Energy Studies, Indian Institute of Technology Delhi, Hauz Khas, New Delhi, India

Prashant Saini School of Engineering, Indian Institute of Technology Mandi, Mandi, Himachal Pradesh, India

P. Salinas School of Engineering, Texas A&M International University, Laredo, TX, USA

Arif Sarwat Energy, Power and Sustainability Group, Florida International University, Miami, USA

Sutha Senthil Department of Electronics and Communication Engineering, Sri Venkateswara College of Engineering, Sriperumbudur, Chennai, India

Akshay Sharma School of Mechanical Sciences, IIT Bhubaneswar, Bhubaneswar, Odisha, India

Monia Sharma Department of Physics, Dr. B. R. Ambedkar National Institute of Technology Jalandhar, Jalandhar, Punjab, India

Amit Shrivastava Department of Mechanical Engineering, Indian Institute of Technology Jodhpur, Jodhpur, India

Gurkamal Nain Singh Department of Mechanical Engineering, Dr. B. R. Ambedkar National Institute of Technology Jalandhar, Jalandhar, Punjab, India

Uttam Singh School of Humanities and Social Sciences, Indian Institute of Technology Mandi, Mandi, Himachal Pradesh, India

Divyanshu Sood Centre for Energy Studies, Indian Institute of Technology Delhi, Hauz Khas, New Delhi, India

Yudhisther Surolia Department of Mechanical Engineering, Indian Institute of Technology Jodhpur, Jodhpur, Rajasthan, India

K. V. S. Teja Department of Mechanical Engineering, Indian Institute of Technology Ropar, Rupnagar, Punjab, India

Himanshu Tyagi Department of Mechanical Engineering, Indian Institute of Technology Ropar, Rupnagar, Punjab, India

Surya Prakash Upadhyay School of Humanities and Social Sciences, Indian Institute of Technology Mandi, Mandi, Himachal Pradesh, India

General

Introduction to New Research Directions in Solar Energy Technologies

Himanshu Tyagi, Prodyut R. Chakraborty, Satvasheel Powar, and Avinash K. Agarwal

The overall theme of this book is related to the topics of solar energy. This book has been divided into four parts. The first part relates to the general issues of clean and sustainable energy. The second part of this book deals with the topic of power generation using solar energy. It is followed by the applications of solar energy. The fourth part of the book deals with storage of energy, especially thermal energy storage (TES). In total, there are fifteen chapters in this book, including this one.

The first part of the book contains two chapters whose focuses are related to policy and security aspects of renewable energy sources. Chapter "Jawaharlal Nehru National Solar Mission: A Critical Analysis of Evolution and Challenges" provides a critical analysis of the origin of the Jawaharlal Nehru National Solar Mission (JNNSM) and its challenges, in an Indian context. India devised the JNNSM in the year 2008, in which the country has set very ambitious targets for installation of large units of rooftop and utility scale solar power plants. The cumulative target was to achieve installed capacity of more than 20 GW grid-connected solar power stations. Since India has the second highest populations in the world, along with some of the

H. Tyagi (✉)
Department of Mechanical Engineering, Indian Institute of Technology Ropar, Rupnagar, Punjab 140001, India
e-mail: himanshu.tyagi@iitrpr.ac.in

P. R. Chakraborty
Department of Mechanical Engineering, Indian Institute of Technology Jodhpur, Jodhpur, Rajasthan 342037, India

S. Powar
School of Engineering, Indian Institute of Technology Mandi, Mandi, Himachal Pradesh 175005, India

A. K. Agarwal
Department of Mechanical Engineering, Indian Institute of Technology Kanpur, Kanpur, Uttar Pradesh 208016, India

H. Tyagi et al. (eds.), *New Research Directions in Solar Energy Technologies*,
Energy, Environment, and Sustainability,
https://doi.org/10.1007/978-981-16-0594-9_1

highest population densities, the main genesis of this mission was to improve the environment quality and to reduce the carbon footprint of the country. This mission has been a huge success with the actual installations have surpassed the targets. This chapter also discusses the salient issues related to production of power from solar energy, for example the material waste which is generated over long periods. It also presents the comparison between large-scale and small-to-medium-scale solar power plants and highlights the issues and challenges in each category.

Chapter "Impact Analysis of Cyber Attacks on Smart Grid: A Review and Case Study" looks at the security aspects of a grid which connects electricity supplied by both conventional and non-conventional (renewable) energy sources. It presents a very comprehensive review of the articles published, which have dealt with the issue of cyber-attacks on smart distribution grids, and presents some solutions for such eventualities. It is expected that all major equipments, systems and utilities are increasing being connected together through the Internet, for ease of control and monitoring. While such systems (including smart grids) would benefit by such inter-connectivity, the same feature can present vulnerabilities due to them being exposed to cyber-attacks similar in scale and impact to the attacks on financial portals of banks and investments firms. With the presence of different types of subsystems (some having intermittent nature of power generation) and other may have storage capabilities, the real-time coordination is assumed to be very highly important. Moreover, the dynamics of this complicated system can have far reaching consequences, ranging from security threats as well as severe financial liabilities in the event of grid failure caused by cyber-attacks. In particular, one type of attack, the false data injection (FDI) on photovoltaic systems, has been covered in detail in this chapter.

The second section of the book includes four chapters. Of these, two chapters cover dye-sensitized solar cells and perovskite solar cells, the third-generation solar cell technologies. One chapter focuses on solar thermal-based building energy harvesting. The last chapter of this section focuses on self-cleaning coatings for solar photovoltaics to keep them dust-free. Chapter "A Perspective on Perovskite Solar Cells" discusses the advent of the latest class of organic–inorganic hybrid perovskite materials which are used in various applications. These applications include optoelectronic devices such as solar cells, light-emitting diodes, photodetectors and lasers. Synthetic controls of perovskite materials in the history of solar cell evolution have made perovskite solar cells (PSCs) the fastest growing technology. Through structure engineering, solvent chemistry, morphology and surface control, surface passivation and band engineering, the researchers are enhancing PSC performance. Perovskite materials have a number of unique characteristics, such as easily adjustable band-gap energy, solution processability, low-temperature long-range crystals and excellent charge transport. Perovskite-based solar cells are susceptible to humidity and temperature degradation. The efficiency of power conversion is comparable to monocrystalline silicon solar cells. In this chapter, the authors have addressed the recent developments in the strategy of synthesis and structural stability.

Chapter "Textile-Based Dye-Sensitized Solar Cells: Fabrication, Characterization, and Challenges" deals with the dye-sensitized solar cell technology (DSSC). There is a strong demand for wearable solar cells. Dye-sensitized solar cell technology is one of the promising alternatives; the use of natural plant-based dyes makes it environmentally friendly and cost-effective. However, a complete textile-based dye-sensitized solar cell (TDSSC) is still a major challenge. The chapter addresses the problem for TDSSC and also mentions all approaches to enhancing TiO_2 adhesion in order to make it ready for use in TDSSC. Different coating methods have been examined in Chap. "Textile-Based Dye-Sensitized Solar Cells: Fabrication, Characterization, and Challenges", applied in the manufacture of counter and working electrodes in textile-based DSSCs (TDSCs). Various forms of plant dye have been used to achieve excellent conversion efficiency in the TDSSC. A DSSC that is all-textile, with an efficiency rating comparable to or better than that of a DDSC glass, has yet to be produced. The challenges in the development of TDS SCs will be evaluated to help build a reliable and effective DSSS. The purpose of this chapter is to review and compare the different conductive coating methods reported in the DSSSC.

Chapter "Building Energy Harvesting Powered by Solar Thermal Energy" deals with the generation of electricity directly from buildings. It studies the latest techniques to convert thermal energy stored in the buildings to electricity using the thermoelectric generators (TEG). It discussed in detail the figure-of-merit and presents the results of electrical conductivity and Seebeck coefficient measurements. It uses the laser flash method to perform the thermal conductivity measurements. The cement composite itself is used as the thermoelectric material, which has lot of advantages in the context of the thermal energy stored within the buildings. This all is necessitated because due to the urban heat island (UHI) effect, the differential heat causes a significant increase in the temperature of buildings in urban concentrated localities. Such a temperature gradient allows the efficient utilization of the thermal energy and its effective conversion into electrical power. Once this is achieved at large scale, a huge power surplus can be achieved in almost all urban centres of the world where high solar insolation is experienced. Such a solution has a potential to solve both the issues—dissipation of thermal energy from the buildings (hence reducing the cooling load during the summers) and at the same time electrical power is produced. The addition of graphene nanoplatelets is shown to significantly improve the thermoelectric properties of cement.

Chapter "A Brief Review on Self-cleaning Coatings for Photovoltaic Systems" presents a brief review on self-cleaning coatings for photovoltaic systems: solar energy is the most efficient and inexpensive way to generate electricity. Recent reductions in energy costs have opened the door in developing countries in India to low-cost generation of energy. There are no moving parts in solar power plants; therefore, their maintenance is far lower. However, one of the biggest challenges is to clean up these installed solar panels. The particles of dust deposited on the solar modules reduce the transmission and therefore the generated energy. To ensure a good output, the surfaces of the PV panels have to be cleaned regularly. The new methods of cleaning are costly. One of the potential solutions is the self-cleaning nanostructured coatings. Chapter "A Brief Review on Self-cleaning Coatings for Photovoltaic Systems" offers a brief overview of the recent development of bio-imitative self-cleaning panel photovoltaic solar panels. A note outlines the considerations and developments in the production of robust photovoltaic self-cleaning coatings.

The third section of this book contains chapters which deal with applications of solar energy in various aspects of technological activities as well as in our daily lives. Chapter "Hybrid Electrical-Solar Oven: A New Perspective" deals with the applications of solar energy in the area of cooking, in particular by proposing the design of a new type of the hybrid oven, which works on both electrical and solar energies. This type of a hybrid design is very attractive for the end-user because it combines the two benefits of the two different types of energy source. A conventional electric oven is relatively fast and reliable (as long as uninterrupted electricity supply is available). On the other hand, a solar cooker may be slow and is dependent on the level of solar irradiation which may vary with time; however, it does not use high-grade energy for cooking purpose. Therefore, the design proposed by the authors in this chapter consists of a conventionally sized system so that it is user friendly and incorporates the benefits of both the systems. It has the reliability of the conventional electric oven, in case there is lack of solar irradiance. And simultaneously, it has the potential to utilize low-grade solar energy whenever available for cooking purposes. The authors have included a comprehensive literature review in this chapter and have highlighted the advantages of this system over prior published work.

Chapter "Performance Analysis of Vacuum Insulation Panels Using Real Gas Equation for Mitigating Solar Heat Gain in Buildings" deals with the usage of vacuum insulation panels (VIP) for energy conservation in buildings. India has a relatively very young and growing population. Additionally, there is large migration of population from rural to urban areas. Hence, it is expected that in the next few decades, huge amount of building construction will take place in India at a scale never seen before. It is observed that buildings consume huge amount of energy irrespective of their usage (whether it is used for residential, office, commercial or industrial purposes). Moreover, with rapidly increasing living standards indoor environment comfort is rapidly becoming the need of the hour. In such a scenario, it is imperative that energy efficient buildings be designed in the future. Under these conditions, the use of VIP is expected to improve the overall performance of buildings in various geographical and climatic regions of the country. This chapter considers a detailed analysis of the use of such VIP panels in a particular city (Jodhpur). Here the performance of the VIP-enabled building is studied in detail, and the benefits of its usage have been quantified. The presence of the vacuum region inside the panel leads to very good insulation properties of the building material and leads to higher efficiencies.

Another application of solar energy in the future could be to provide electricity for charging the electric vehicles. This topic has been studied in detail in Chap. "Solar-Based Electric Vehicle Charging Stations in India: A Perspective". Already several small versions of vehicles (ranging from two-wheelers, three-wheelers and even four-wheelers) have started to appear in various placed in the country. All these electric vehicles (EVs) require electricity to be charged (usually overnight). The Indian government is also heavily supporting the design and development and indigenously built EVs in India. The amount of change per vehicle determines to a large extend its range of operation as well as top speed. This mandates that suitable charging stations, which can provide quick charge to the batteries be made available. One issue that may need to be addressed is that the source of energy from where this electricity is produced becomes critical once the number of EVs drastically increase. The possibility of grid-connected charging stations or grid-independent charging stations needs to be studied in detail before large-scale operation of EVs become possible in India. In such a scenario, solar energy presents a very good alternative to conventional sources of energy. However, several challenges that are being faced by the intermittent nature of solar energy need to be also suitable addressed.

Chapter "Use of Phase Change Materials for Energy-Efficient Buildings in India" again presents a technical solution for reducing the heating and cooling load of buildings, by utilizing the phase change materials (PCM). These materials are abundantly available in the nature having wide variety of melting point temperature. Such materials can be utilized within the design of the buildings to provided additional thermal barrier to the leakage of heat (from inside the building to outside, during winters, and the reverse during summers). Incorporating PCM within building components enhances their thermal heat capacity as well as improves the energy efficiency of the buildings. Numerous researchers are experimenting with PCMs for their use in energy-efficient buildings. In this study, numerical modelling has been carried out for a PCM incorporated model that can be used depending on different climatic zones in India. The dimensions and boundary conditions used in numerical modelling are kept near the realistic weather conditions in various climate zones. The PCM selection has been carried out by taking into consideration the desirable thermophysical properties, operating temperature, availability and weather conditions in different locations. The results of this study show that the usage of PCM can significantly reduce the heating requirements in winter and the cooling loads during the summers. Large-scale adaption of this technology in building practices can lead to drastic reduction in the long-term energy used by bunds across the country.

The fourth section contains chapters within the theme of thermal energy storage associated with solar applications. There are four chapters in this section, and the salient feature of each of these four chapters is briefly described here. Chapter "Parabolic Dish Solar Cooker: An Alternative Design Approach Towards Achieving High Grade Thermal Energy Storage Solution" describes the thermal design aspects of parabolic dish solar cooker integrated with latent heat thermal energy storage. An alternative design approach for the solar cooker is described, where the double reflector-based assembly is used to locate the heavy storage container below the reflector primary reflector providing structural stability to the system. Energy requirement for domestic and commercial culinary is substantial in today's world with their carbon footprint rising with time. A well-known green tech available from decades is solar cooker. Conventional solar cooker, as the name applies, uses solar energy to cook food. One inherent drawback with such solar cooker is that it can cook only during the on-sun period. Integration of thermal energy storage allows such cookers to cook even during off-sun periods. Latent heat thermal energy storage is an obvious option for such application due to its high energy density, compactness and ability to supply heat consistently above 250 °C with suitable choice of phase change material. Sodium nitrate is chosen as the model phase change material (PCM), and the thermal storage container is filled with PCM-CEG (compressed expanded graphite) composite to enhance the thermal conductivity of PCM. Circumferentially distributed radially spread fins are also proposed to restrict the local hot spot temperature below the decomposition temperature of PCM. A detailed numerical model is presented. Results show that incorporation of CEG matrix and fins reduces the charging/discharging time significantly with large number of fins providing better uniformity of temperature field within the storage unit and restricting the local hotspot temperature conveniently below the decomposition limit.

Chapter "Experimental Investigation of a Sensible Thermal Energy Storage System" describes the experimental investigation of sensible heat thermal energy storage (SHTES) system. In this chapter, a SHTES system of in-house solar air tower simulator (SATS) is investigated. The SATS facility consists of subsystems such as an open volumetric receiver (OVAR), two TES, viz. primary and secondary, cross-flow heat exchanger, and solar convective furnace (SCF). In this system, hot air is generated in OVAR and transported to TES or SCF. A secondary SHTES is integrated to store the waste energy in magnesium silicate pebbles placed inside the thermal storage container once the hot air comes out of the SCF. An experimental investigation is carried out for charging and discharging of secondary SHTES. In the experiment, the ambient air with a mass flow rate of 0.0042 kg/s is heated by joule heating and supplied to secondary SHTES for its charging. For discharging, the ambient air with the same mass flow rate is supplied. K-type thermocouples are used to measure the temperature at different locations. The energy balance in the secondary SHTES is carried out with the help of measured temperature, and it is observed that approximately 60% of the inlet energy leaves the TES with the outgoing fluid, 12% is stored in the pebbles, 17% is stored in other components of the TES and 11% of energy is lost to ambient. The experiment on the secondary TES shows that the SHTES using pebbles performed well, and the system can be used to store waste heat after metal processing. However, the losses have to be minimized further, and improvements have to be carried out in the set-up to attain steady-state while charging. The presented work is the first demonstration of the use of locally available material to construct a SHTES system.

In Chap. "PCM Based Energy Storage Systems for Solar Water Heating", latent heat thermal energy storage (LHTES) systems for solar water heating are described. Numerical investigations are carried out for charging and discharging cycles. The numerical model based on enthalpy-porosity approach is described to apprehend phase change phenomena in the LHTES due to the flow of heat transfer fluid (HTF) through the thermal energy storage unit. The analysis is carried out for a two-channel storage configuration in two dimensions. The storage unit consists of an annular channel, where the PCM is placed within the outer annulus. The heat transfer fluid water flows through the inner channel. A thin wall of aluminium separates both the channels. Detailed parametric analyses are performed by varying the inlet conditions, storage dimensions and charging time on the melting and temperature evolution. It is reported that the outlet temperature of the water is highly dependent on the aspect ratio of the channel and on the discharged velocity. However, the initial temperature during charging and the aspect ratio of the PCM domain have a reasonable effect over outlet temperature. It is inferred from this investigation that the geometrical parameters and discharge velocity of the heat transfer fluid are vital to enhance the performance of the shell and tube-type heat exchanger.

In Chap. "Review on Thermal Performance Enhancement Techniques of Latent Heat Thermal Energy Storage (LHTES) System for Solar and Waste Heat Recovery Applications", a detailed review of latent heat thermal energy storage (LHTES) using phase change materials (PCMs) for solar and waste heat recovery (WHR) applications is presented. Temperature range of 40–200 °C is purposefully considered, as most of the heating and cooling applications for the domestic, commercial and public domain vary within this temperature range. The study is mainly directed towards the selection of suitable PCMs for solar air and water heating, solar stills, solar absorption cooling, waste heat recovery and solar thermal electricity generation. The importance is also given towards the thermal conductivity enhancement, selection of suitable heat exchangers and design parameters that aid in a prolonged energy storage interval. From the review, it is observed that the thermal conductivity of the PCMs could be enhanced up to 48 times by using different methods like introducing high thermal conductivity nanoparticles, the inclusion of metallic foams, using expanded graphite into the PCMs. Further enhancement in the heat transfer within PCMs can be improved by effectively designing the heat exchanger geometry and its components. The review presents comprehensive details of the possible extent of research existing in the field of latent heat thermal energy storage associated with solar and waste heat recovery applications.

This monograph consists of the following topics:

- Jawaharlal Nehru National Solar Mission: A Critical Analysis of Evolution and Challenges
- Impact Analysis of Cyber-Attacks on Smart Grid: A Review and Case Study
- A perspective on Perovskite Solar Cells
- Textile-Based Dye-Sensitized Solar Cells: Fabrication, Characterization, and Challenges
- Building Energy Harvesting Powered By Solar Thermal Energy
- A brief review on self-cleaning coatings for photovoltaic systems
- Hybrid Electrical-Solar Oven: A New Perspective
- Performance Analysis of Vacuum Insulation Panels using Real Gas Equation for Mitigating Solar Heat Gain in Buildings
- Solar-based Electric Vehicle Charging Stations in India: A Perspective
- Use of Phase Change Materials for Energy-efficient buildings in India
- Parabolic dish solar cooker: An alternative design approach towards achieving high-grade thermal energy storage solution
- Experimental Investigation of a Sensible Thermal Energy Storage System
- PCM-based Energy Storage Systems for Solar Water Heating

- Review on thermal performance enhancement techniques of latent heat thermal energy storage (LHTES) system for solar and waste heat recovery applications.

In this monograph, the topics are arranged in four sections: (i) General, (ii) Power Generation using Solar Energy, (iii) Applications of Solar Energy and (iv) Energy Storage.

Jawaharlal Nehru National Solar Mission: A Critical Analysis of Evolution and Challenges

Surya Prakash Upadhyay and Uttam Singh

1 Introduction

A judicious approach toward energy consumption is crucial not only for sustainable development, well-being and prosperity of any nation but also for the survival of human race and other creatures. The race for economic development, particularly since the Industrial Revolution in Europe, increased the demand for and consumption of energy which also led to extreme reliance on fossil fuels and biomass. Needless to say, developed nations grew rapidly as those nations made arrangements for energy requirements for industries. However, the development has been achieved at the expense of environment, climate and resource exploitation, particularly resources extracted from the colonies. The models of economic development, for example, modernization, made available in developed counties were also followed in post-colonial nations. The colonial experiences and underdevelopment required them to exploit available resources within the existing frameworks and methods. Globally, the horrors and limitations of economic models, available resources and technologies used for energy requirements become apparent since the late 1970s (no reference is required as it is framed by the authors). Simultaneously, rising awareness around global warming, the depletion of ozone layer and larger climatic change led scientists and policymakers look for alternative technologies for energy requirements.

This chapter takes the case of Jawaharlal Nehru National Solar Mission (JNNSM), a mission program by Indian government under National Action Plan for Climate Change (NAPCC), for the mitigation of climate change and commitment toward energy security, energy accessibility and sustainability. Interestingly, such a mission program is an outcome of overcoming internal compulsions and alignment with

S. P. Upadhyay (✉) · U. Singh
School of Humanities and Social Sciences, Indian Institute of Technology Mandi, Mandi, Himachal Pradesh 175005, India
e-mail: surya@iitmandi.ac.in

H. Tyagi et al. (eds.), *New Research Directions in Solar Energy Technologies*, Energy, Environment, and Sustainability,
https://doi.org/10.1007/978-981-16-0594-9_2

international conventions that began around the early 1980s. India began its journey toward renewable energy with the foundation of a new Ministry—Ministry of New and Renewable Energy. This chapter explores how the targets and goals for solar energy evolved with improvement in technologies during the last one decade. It explores the policy guidelines of JNNSM and discusses challenges of solar waste disposal. Further, the paper also analyzes methods of decentralized production and distribution and assesses the challenges of employing small-scale solar power plant. Lastly, India is setting-up several milestones for inclusive development; therefore, the availability of "energy for all" becomes cardinal for bringing energy justice to marginalized communities.

2 Background of Renewable Energy in India: Global and Local

India, as like other nations, compelled by the fluctuating oil prices in the global market and internal economic situations planned to switch toward alternative energy sources. The genesis of solar energy in India began with unstable oil prices in the 1970s; rising balance of payments in the 1980s; and, the realization of limitations of fossil fuels vis-a-vis to continue on the economic development and provisions of quality life to its citizens required India look for alternative sources of energy. The Ministry of New and Renewable Energy (MNRE), Government of India puts the situation as:

> The role of new and renewable energy has been assuming increasing significance in recent times with the growing concern for the country's energy security. Energy sufficiency was identified as the major driver for new and renewable energy in the country in the wake of the two oil shocks of the 1970s. The sudden increase in the price of oil, uncertainties associated with its supply and the adverse impact on the balance of payments position led to the establishment of the Commission for Additional Sources of Energy (CASE) in the Department of Science & Technology in March 1981. The Commission was charged with the responsibility of formulating policies and their implementation, programmes for development of new and renewable energy apart from coordinating and intensifying R&D in the sector. In 1982, a new department, i.e., Department of Non-conventional Energy Sources (DNES), that incorporated CASE, was created in the then Ministry of Energy. In 1992, DNES became the Ministry of Non-conventional Energy Sources. In October 2006, the Ministry was re-christened as the Ministry of New and Renewable Energy. (https://mnre.gov.in/history-background)

The Ministry's background history may suggest economic compulsions for the foundation of concerted efforts to look for new and renewable energy sources. Nevertheless, the economic burden is just one among several reasons for the creation of MNRE. What we need to underline is the global politics of climate change, international legal treaties and frameworks that developed throughout the 1980s and 1990s that led to the establishment of such ministries, not only in India but where else as well.

Much before the awareness of climate change developed across the globe, Massachusetts Institute of Technology (MIT) brought together US scientists through a program called Study of Man's Impact on Climate (SMIC), which made recommendations for global environmental problem (Howe 2017). An initial effort in the recent awareness on global environmental problem was the affirmation during United Nations Conference on the Human Environment in Stockholm 1972. Perhaps, India, as like many other developing economies, remained untouched by such awareness. The concerns for global warming and climate change brought developing economies in the frameworks of discussions and deliberations developed by Inter-governmental Panel on Climate Change (IPCC), which was created in 1988 by World Meteorological Organization (WMO) and United Nations Environment Program (UNEP). IPCC was created to assess available scientific information on climate change, social and economic impacts and suggest strategies to address climate change. Its precursor, a Commission, established by United Nations General Assembly through a Resolution in 1983 further led to the creation of World Commission on Environment and Development. Interestingly, the First Assessment Report of IPCC, released in 1990, differed significantly from the existing ideas and claims on global warming, particularly that of NASA scientist James Hansen's (Howe 2017).

The political atmosphere in the USA; the presidential campaign of George Bush (Senior); collapse of Soviet Union and thereby creation of unipolar world pursuing neoliberal market policies should be seen as the context for adoption of clean and green energy sources, more particularly solar and wind energy resources. The most remarkable event leading in the direction of sustainable development, climate change and global warming was the 1992 United Nations Conference on Environment and Development (UNCED) also called Rio Summit or Earth Summit. The Rio Summit introduced several mechanisms, legal instruments, including the foundation of United Nations Framework Convention on Climate Change (UNFCCC). The Second Assessment Report of the IPCC (1995) (Howe 2017) states that the science of climate change is directed toward: greenhouse gas concentrations have continued to increase since about 1750; anthropogenic aerosols tend to produce negative radiative forces; climate change has occurred over the past century; and, evidence suggests a discernible human influence on global context. Though the Berlin Mandate (1995) kept the developing economies out of the responsibilities of curbing effects of greenhouse gas, the Kyoto Protocol (1997) to the UNFCCC led to various debates, not only in the developed economies but also in developing world. The Kyoto Protocol envisaged several new mechanisms, such as, joint implementation by developed and developing countries and of efforts to control climate change; clean development mechanism (CDM) and carbon trading. Now, this is the context of global awareness on climate change, sustainable development and global warming that should form context for understanding what has happened in India around its efforts to harness potentials of renewable sources such as wind and solar energy.

Globally, around 80% of energy demands are fulfilled by non-renewable fossil fuels (Ummadisingu and Soni 2011). The depleting non-renewable energy sources and continuous demand for energy to drive the economy and improve quality of life, and the nations have devised plans and look for alternative sources of energy. As is evident from global discussions, excessive use of coal, natural gas and oil to fulfill the rising demand for energy has also caused greenhouse gas emissions leading to climate change and global warming. Though industrialization and use of fossil fuel are the main causes of climate change, other activities such as deforestation, construction and unsustainable agriculture practices are also seen responsible for climate change (Koneswaran and Nierenberg 2008; Steinfeld et al. 2006; Cole 1998; Houghton 2005). The risks and vulnerability of climate change vis-a-vis rising energy demands puts India under severe challenge and responsibility not only for the protection of fellow global citizens but also for the future generations who could live in clean and green environment. The need for sustainable growth and development have been argued, discussed and adopted in developmental models of most of the economies ensuring positive contribution toward sustainable environmental, economic and social development. In the coming decades, the non-renewable energy sources will further loose capacity to fulfill increased energy demands. The IPCC in its Fifth Appraisal Report has asserted that human activities during the last 50 years have caused 95% of warming up of the planet (https://www.ipcc.ch/site/assets/uploads/2018/02/ipcc_wg3_ar5_summary-for-policymakers.pdf). Renewable energy sources are the most remarkable option and the main answer for decrease of fossil fuel sources, greenhouse gas emission, geopolitical and other environmental issues (UNFCC 2015).

Therefore, challenged by above said contradictory situations and driven by responsibility toward sustainability, protection of environment and larger human welfare, we need to think much beyond the issues of climate change and sustainable models of development. These issues need to be tackled in multi-pronged methods, e.g., reduction in energy wastage, working for alternative sources of energy, designing appropriate technology that mitigate challenges of climate change and also secure people for their energy needs. It has also been demonstrated and argued that renewable energy has less adverse effect on environment. The promotions as well as harvesting of renewable energy sources are the key policy agenda of many countries (Owusu and Asumadu-Sarkodie 2016). In the UN General Assembly Summit, held in September 2015, 193 member states adopted 17 sustainable development goals (SDGs) and 169 targets that were brought in effect from January 1, 2016 (https://sustainabledevelopment.un.org/post2015/transformingourworld). Among 17 SDGs, emphasis has also been given on the issue of climate change focusing on renewable energy sources that have environmental, social and economic benefits. In light of the above facts, countries have begun focusing on issues of climate change and renewable energy through several policies. India has also adopted many such steps and one of the steps in this direction is to harvest renewable energy sources for consumption.

3 Potentials of Solar Power: JNNSM and India

Jawaharlal Nehru National Solar Mission (JNNSM) was formally launched on January 11, 2010, setting an aggressive focus of 22 GW solar power by 2022. The objective of JNNSUM is:

> To establish India as a global leader in solar energy, by creating the policy conditions for its diffusion across the country as quickly as possible (GoI, MNRE 2010). (https://mnre.gov.in/file-manager/UserFiles/draft-jnnsmpd-2.pdf)

The Ministry of New and Renewable Energy, Government of India, further states:

India is endowed with vast solar energy potential. About 5000 trillion kWh per year energy is incident over India's land area with most parts receiving 4–7 kWh per m^2 per day. Hence, both technology routes for conversion of solar radiation into heat and electricity, namely solar thermal and solar photovoltaics, can effectively be harnessed providing huge scalability for solar in India. Solar also provides the ability to generate power on a distributed basis and enables rapid capacity addition with short lead times. Off-grid decentralized and low-temperature applications will be advantageous from a rural electrification perspective and meeting other energy needs for power and heating and cooling in both rural and urban areas. From an energy security perspective, solar is the most secure of all sources, since it is abundantly available. Theoretically, a small fraction of the total incident solar energy (if captured effectively) can meet the entire country's power requirements. It is also clear that given the large proportion of poor and energy unserved population in the country, every effort needs to be made to exploit the relatively abundant sources of energy available to the country. While, today, domestic coal-based power generation is the cheapest electricity source, future scenarios suggest that this could well change (https://mnre.gov.in/solar).

This realization by the Ministry needs to be understood in the backdrop of international programs on climate change. What forces/circumstances/factors have shaped India's national solar mission? One of the objectives of UNFCC is stabilization of greenhouse gases and preventions of its effect on human and environmental health. As mentioned earlier, Kyoto Protocol alludes to the international legal system for environmental change process, wherein countries are asked to curb emissions. Further, the Cancun Agreement (2010) sets following terms and goals: GHG mitigation; transparency of actions; technology transfer; financial arrangements through Green Climate Fund; adaptation to impacts of climate change by vulnerable groups; concrete action plan for forest conservation; and, capability building of developing nations to achieve set targets (https://unfccc.int/process/conferences/the-big-picture/milestones/the-cancun-agreements). The Conference of the Parties (CoP) to the convention meets yearly to arrange and examine the global environmental change motivation and related duties from nations. In 2007, the IPPC in its Fifth Assessment Report raised serious concerns on changing weather in tropical and sub-tropical countries. IPPC report, an outcome of Bali Action Plan, advised individual countries to set up climate action plans and address issues of climate change through policy frameworks.

As a part of its commitment, India introduced National Action Plan for Climate Change (NAPCC) in 2008 which itself is an outcome of international negotiations and commitments. India prepared policy documents ahead of G-8 Summit in 2008 and Conference of Parties at Copenhagen in 2009. India established Prime Minister's Council on Climate Change (PMCCC) in 2007, which came up with NAPCC, which is a set of eight mission plans that proposed to address issues of climate change within the frameworks of sustainable approach and without compromising developmental needs of the country (Rattani 2018). NAPCC includes: National Mission on Sustainable Habitat; National Mission for Sustaining the Himalayan Ecosystem; National Mission for Sustainable Agriculture; National Solar Mission; National Mission for Enhanced Energy Efficiency; National Water Mission; National Mission on Strategic Knowledge for Climate Change; National Mission for Green India. JNNSM is one of the mission programs under NAPCC that provides broad objectives to mitigate challenges of climate change in varied fields. JNNSM, in 2010, targeted to generate 20 GW solar power by 2022 which was revised to 100 GW in 2015 (https://pib.nic.in/newsite/PrintRelease.aspx?relid%3D122566). The JNNSM policy document does not mention how these figures of 20 GW or its revised target of 100 GW is calculated. Perhaps, one has to simulate the larger situation as Phadke et al. (2016) have tried to calculate and arrive at the figures of renewable energy contribution. Utilizing various figures from the 13th Plan (Five Year Plan), Renewable Energy Missions documents and NAPCC plans, as Phadke et al. (2016) calculate that the government of India aims to achieve 15% contribution from the renewable sources in the total energy requirements of the country by the year 2020. Further, Phadke et al. (2016) argue that if the same trend of achievements of renewable energy sources continues, then the renewable energy could easily contribute up to around 20% of the total energy demand of the country (Phadke et al. 2016).

Most likely and intuitively, the calculated targets of production of solar energy have been arrived by considering the energy trilemma: energy access, energy security and sustainability. According to the Report of Expert Group of NITI Aayog (2015) (National Institution for Transforming India), around 400 million households in India do not have accessibility to electricity. As per the Central Electricity Authority, at the level of 2011–12, the demand for electricity will be more than double by 2021–22. The solar energy output in India has grown consistently from 4.59 billion units in 2014–2015 to 25.87 billion units in 2017–2018 (https://energy.economictimes.indiatimes.com/news/renewable/india-to-comfortably-achieve-100-gw-solar-energy-target-by-2022-government/65307602). Perhaps, understanding of energy trilemma alongside the improvement in estimation of solar potential in India; evolving methods in understanding intermittence, unpredictability and variability of solar energy; macroeconomic circumstances, e.g., increasing bill on imported coal, and decreasing prices of solar power inventories has led to re-estimation and revision of solar energy target from 20 to 100 GW by 2022. In June 2015, the Government of India argued that "the price of solar energy has come significantly from Rs. 17.90 per unit in 2010 to under Rs. 7 per unit, thereby reducing the need of viability gap funding (VGF)/generation-based incentives (GBI) per MW of solar power" (https://pib.nic.in/newsite/PrintRelease.aspx?relid%3D122566). It further asserts, "with technology advancement and market competition, this Green Power (solar energy) is expected to reach grid parity by 2017–2018. These developments would enable India to achieve its present target of 20,000 MW… But considering its international commitment toward green and climate-friendly growth trajectory, the Government of India has taken this path-breaking decision" (https://pib.nic.in/newsite/PrintRelease.aspx?relid%3D122566).

The JNNSM document provided broad framework that promotes solar energy through policy conditions, allows for adoption of indigenous technologies and initiation of researches for new technologies, and it also states provisions for subsidies and market support for cost reduction. Broadly, the policy document addresses energy needs, issues of climate change that reduces dependency on fossil fuel. The proposed structure for implementation of solar mission is embedded under MNRE which would report to the PMCCC with regard to its status and implementation. As per MNRE, the implementing structure of the mission comprises of four-level structure: Steering Group; Executive Committee; Solar Research Council; Mission Director. The JNNSM document expands the roles and responsibilities of various implementing structures that: the Steering Group comprising of the Minister of New and Renewable Energy and representatives of the states looks into the overall implementation of the JNNSM and approves various policies and financial norms; the Executive Committee headed by the secretary periodically reviews the implementation of projects that are approved by the Steering Group; the Solar Research Council comprises of eminent scientists who advise the mission on research and development, technologies and measures required for capacity building; while the Mission Director looks at how the targets of the mission are implemented. The initial JNNSM targets are divided into three-phase which is illustrated in Table 1.

In a resolution passed on July 1, 2015, the Government of India revised the cumulative target from 20 to 100 GW for the year 2022 and divided it into 40 GW through decentralized rooftop solar projects; 40 GW through utility-scale solar projects and 20 GW through Ultra Mega Solar Parks (UMSP) (https://web.archive.org/web/20170226140523/http://mnre.gov.in/file-manager/grid-solar/100000MW-Grid-Connected-Solar-Power-Projects-by-2021-22.pdf). The Government of India has also introduced the ideas of Solar Park and Ultra Mega Solar Power Project

Table 1 Phase-wise target for production (https://mnre.gov.in/file-manager/UserFiles/draft-jnnsmpd-2.pdf)

Component	Phase I (2010–2013)	Phase II (2013–17)	Phase III 2017–2022)	Cumulative target
Grid-connected solar (MW)	1000	3000	16,000	20,000
Off-grid solar (MW)	200	800	1000	2000
Solar thermal collector (million m^2)	7	8	5	20

(UMSPP) to speed-up production; tackle the issue of project cost per MW, and transmission cost. The government argues that it has observed limited capacity of individual projects which puts significant burden in terms of cost of establishment, grid connectivity, transmission line, water and other infrastructural requirements. In order to achieve 100 GW of solar energy by the year 2022, mission targets multiple schemes such as solar rooftop scheme, solar defense scheme, solar scheme for CPUs, solar PV power plant on the canal bank and channel top, solar pump. Under the frameworks of UNFCCC, India is also taking financial assistance from other countries as well. The overall investment that was required to achieve the set target is around 6 lakh crore with a rate of 6 crores per MW. In this context, the USA is investing $4 billion (24,400 crores) in terms of loan. India has also signed MoUs with US-based Sun Edison Inc. and Adani Enterprise to establish the largest solar photovoltaic (PV) manufacturing unit in India (https://archive.indiaspend.com/cover-story/why-the-us-backs-indias-great-solar-bet-45411). India signed the MoU with Germany and provide the soft loan of 1 billion Euros for the next five year to expand bi-lateral development cooperation in the solar energy field (https://www.pmindia.gov.in/en/news_updates/pm-undertakes-aerial-survey-of-cyclone-fani-affected-areas-in-odisha-reviews-the-situation/?comment=disable).

The second phase of the mission envisages involvement of multiple agencies including State Nodal Agencies and Departments implementing renewable energy programs, Solar Energy Corporation of India (SECI). For Solar Park and UMSP, and the land will be arranged by the respective State governments. The government also envisages the issues related with land acquisition and therefore has suggested the state governments to use fellow and non-agricultural land to avoid delay in project execution. The government also believes that establishing large solar parks will increase production, reduce costs, risk and delay. Solar Park and UMSP will have the capacity of 500 MW or above and MNRE sets the target to develop 25 solar parks in next five years (MNRE 2016) that would generate 20,000 MW. The government, in 2017, further revised the target from 20,000 to 40,000 MW and plans to build additional 25 solar parks (MNRE 2017) (https://mnre.gov.in/file-manager/UserFiles/Scheme-for-enhancement-of-capacity-to-40GW-Solar-Parks.pdf). The states can purpose solar park as per their capacity which will be established through open bid system. It will increase investments in the state through project developers. MNRE estimated total cost to establish these power plants as Rs. 4046.25 crore. The evaluated cost for development of a solar park would be around Rs. 0.95 crore per MW and capital cost of building a grid-connected solar power project is Rs. 6.91 crore per MW. Further, Solar Power Park Developer (SPPD) will build proposed site for solar park with appropriate and required infrastructure and amenities so risk of the task can be limited and lessen the number of required endorsement (https://mnre.gov.in/file-manager/UserFiles/Solar-Park-Guidelines.pdf). In first stance, the purposed solar parks are excessively good in the objective to accomplish the set targets and use of surplus land and minimize the cost of solar energy. India is paying significant attention to solar energy while also struggling to overcome problem of energy insecurity.

Howsoever benign these mission plans may be, JNNSM is constrained by unrevised institutional frameworks (Shrimali and Nekkalapudi 2014; Shrimali and Rohra

2012). Deshmukh et al. (2010) highlight shortcomings in JNNSM mission document, which, for them, is unclear with regard to production, technology, innovation, manufacturing of solar appliance, increased installation of off-grid solar plant (Deshmukh et al. 2010). Further, the mission document does not speak about costs to achieve targets, technology, strategy for promotion and inclusive approaches. As like with other programs it seems, in JNNSM as well, India has adopted frameworks and best practices from other countries without revisiting institutional compulsions, market structure and larger political economic situations. NAPCC document purposes to develop appropriate technologies for the success of the missions, but the action plan is mired by several lacunas, e.g., technologies to handle grid and off-grid connections and distribution. The success of any program depends on active involvement of citizen. The lack of frameworks for community involvement may not allow the Indian state to achieve the larger goals. It is more alarming that such as mission has acquired a character of market-driven program. Shrimali et al. (2014) mention two factors that motivate India to launch the solar mission: One, improve the energy security and reduce the dependency on the fossil fuel; two, the technological innovations for solar energy (Shrimali and Nekkalapudi 2014). As we have seen with regard to Cancun Agreement that there are provisions for technology transfers, but the market-driven approach may not allow developing countries to utilize global technologies. Similarly, the lack of trained personnel's and limited skilled human resources will add-up more crisis in handling and achieving mission goals. As indicated above and also discussed by other scholars, achieving this lofty target of 100 GW by the year 2022 within the atmosphere of poor financial and inadequate transmission infrastructure is a huge challenge. In this situation, India is investing huge amount in solar plant that provide subsidized solar electricity to the rich residential, commercial and industrial consumers (https://www.downtoearth.org.in/blog/who-needs-ultra-mega-solar-power-plants--43615).

The report on renewable energy and grid integration 2018 also identifies various challenges in establishing solar park such as land acquisition, lack of coordination between state agencies and SPPD, poor and inadequate infrastructure, non-availability of sufficient water, leveled land, drainage and roads, regulation of different state does not allow sub-leasing land (Kar et al. 2016). Rathore et al. (2018) also discuss barriers in the development of large solar project in India (Rathore et al. 2018). It has also been observed that without having any proper infrastructural, structural, technological and financial background it is not a rational approach to achieve the desired or set target within the time bound. In the case of Charanka Solar Park in Gujarat, Yenneti et al. (2016) highlight the inequalities created by large solar park. At regional and national level, solar parks have the positive benefits but at local community level, it has adverse effect. In distribution level, only upper caste and economically better-off get initial opportunities, while poor and marginal farmers loss their land and livelihood. Poverty and marginality are such problems that cannot be addressed by the large-scale project. Mohan (2017) studied land use of solar PV, wind and nuclear energy and argued that solar PV used high percentage of land as compared to wind and nuclear energy (Mohan 2017). Therefore, it is required to adopt innovative and alternative method to handle such issue through policy condition and involvement of community at large, which reduces wastage of land and resources. Perhaps, the canal bank and sideways on the highways, as proposed by the government, could be utilized effectively (Table 2).

4 Why JNNSM Should Be Important to India

India has around 17% of the world's population share (https://statisticstimes.com/demographics/population-of-india.php). As compared with the population and their energy needs, India does not have sufficient energy sources such as coal and oil. The available quality of coal is not very high and it mostly depends on the import of high-quality coal to meet the requirement. In the last couples of the year, the way Indian GDP is growing the import of coal has increased from 36.60 million tons during 2005–2006 to 199.88 million tons during 2015–2016 (https://coal.nic.in/content/production-supplies). The same story lies with oil based production of energy. As the solar energy sector grows, it ultimately reduces the import of fossil fuel from other countries. The price of fossil fuel is increasing which puts economic burden on India. With technological advancement and market rationalization, renewable energy may become cheap in coming years. As we have discussed in the previous sections, apart from energy need and depleting fossil fuel sources, climate change is a big concern to India. India is at fourth position in terms of CO_2 emissions in world producing 7% of global emission in 2017 (TERI 2018). The tropical situation of India, where the sun is available for longer hours per day and usual sunshine remains around 250–300 day per year with huge intensity created great potential for solar energy most part of the country (Basu et al. 2347). The JNNSM identified the development of solar energy, which hoped to mitigate the energy need of India. The question arises: why India, which has other socio-socio-economic problems, move to this expensive energy sources? It is a well-known fact that solar energy has environmental as well as socio-economic benefits by producing and consumption capacity. As we have discussed, the government has adopted a variety of strategies to promote solar energy with huge investment in JNNSM to produce clean energy. The government also encouraged private companies to produce solar energy by reducing customs duty. Indian Renewable Energy Status Report (2010) highlights India has the potential to become global leader in green economy (Arora et al. 2010). The way it has utilized its solar energy source through policy condition and technological innovation, it is a positive and remarkable step to become global leader. As the solar energy sector grows, it ultimately reduces the import of fossil fuel from other countries.

Looking at geographical location of India and its vast population, it may be difficult in financial and operation terms to connect all villages with a grid system and maintain the electricity line. One of the major ways to overcome several institutional and resource challenges is to envisage technologies and mechanisms that allow decentralized production at small scale. The small-scale projects may be devised which is

Table 2 Approved solar parks in various states in India (Parira 2016)

SI.no.	State	Capacity (MW)	Name of the solar power parks developer (SPPD)	Land identified at
1.	Andhra Pradesh	1500	AP Solar Power Corporation Pvt. Ltd., JVC of SECI, APGENCO and NREDCAP	NP Kunta of Ananthapuramu and Galiveedu of Kadapa Districts
2.		1000		Kurnool District
3.		1000		Galiveedu Madal, Kadapa district
4.		500		Talaricheruvu Village, Tadipatri Mandal, Anathapuramu District of Andhra Pradesh
5.	Arunachal Pradesh	100	Arunachal Pradesh Energy Development Agency (APEDA)	Tezu township in Lohit district
6.	Assam	69	JVC of APDCL and APGCL	Amguri in Sibsagar district
7.	Chhattisgarh	500	Chhattisgarh Renewable Energy Development Agency	Rajnandgaon, Janjgir Champa districts
8.	Gujarat	700	Gujarat Power Corporation Limited	Radhanesda, Vav, District Banaskantha
9.	Haryana	500	Saur Urja Nigam Haryana Ltd. (SUN Haryana)	Bugan in Hisar district, Baralu and Singhani in Bhiwani district and Daukhera in Mahindergarh District
10.	Himachal Pradesh	1000	HP State Electricity Board Ltd.	Spiti Valley of Lahaul and Spiti District
11.	Jammu & Kashmir	100	Jammu and Kashmir Energy Development Agency	Mohagarh and Badla Brahmana, District-Samba
12.	Karnataka	2000	Karnataka Solar Power Development Corporation Pvt. Ltd.	Pavagada Taluk, Tumkur district
13.	Kerala	200	Renewable Power Corporation of Kerala Limited	Paivalike, Meenja, Kinanoor, Kraindalam and Ambalathara villages of Kasargode district
14.	Madhya Pradesh	750	Rewa Ultra Mega Solar Limited	Gurh, Rewa, MP

(continued)

Table 2 (continued)

SI.no.	State	Capacity (MW)	Name of the solar power parks developer (SPPD)	Land identified at
15.		500	Rewa Ultra Mega Solar Limited	Neemuch and Mandsaur
16.		500	Rewa Ultra Mega Solar Limited	Agar and Shajapur
17.		500	Rewa Ultra Mega Solar Limited	Chhattarpur
18.		500	Rewa Ultra Mega Solar Limited	Rajgarh, Morena
19.	Maharashtra	500	M/s Sai Guru Mega Solar Park Pvt. Ltd. (formerly M/s Pragat Akshay Urja Ltd.)	Sakri, Dhule district of Maharashtra
20.		500	Maharashtra State Electricity Generating Company Ltd. (MAHAGENCO)	Dondaicha, district Dhule, Maharashtra
21.		500	M/s Paramount Solar Power Pvt. Ltd. (formerly M/s K P Power Pvt. Ltd.)	Taluka Patoda, district Beed, Maharashtra
22.	Meghalaya	20	Meghalaya Power Generation Corporation Ltd. (MePGCL)	West Jaintia Hills and East Jaintia Hills districts
23.	Nagaland	60	Directorate of New and Renewable Energy, Nagaland	Dimapur, Kohima and New Peren districts
24.	Odisha	1,000	Green Energy Development Corporation of Odisha Limited	Balasore, Keonjhar, Deogarh, Boudh, Kalahandi and Angul
25.	Rajasthan	680	Rajasthan Solar Park Development Company Ltd.	Bhadla Phase II, Bhadla, Rajasthan
26.		1000	Surya Urja Company of Rajasthan Ltd.	Bhadla Phase III, Bhadla, Rajasthan

(continued)

Table 2 (continued)

SI.no.	State	Capacity (MW)	Name of the solar power parks developer (SPPD)	Land identified at
27.		750	M/s Essel Surya Urja Company of Rajasthan Limited	Villages Ugraas, Nagnechinagar and Dandhu, tehsil Phalodi, district Jodhpur (450 MW) and villages Lavan and Purohitsar, tehsil Pokaran, district Jaisalmer (300 MW)
28.		500	M/s Adani Renewable Energy Park Rajasthan Limited	Bhadla Phase IV, Bhadla, Jodhpur Rajasthan
29.		421	M/s Adani Renewable Energy Park Rajasthan Limited	Fatehgarh and Pokaran, Jaisalmer, Rajasthan
30.	Tamil Nadu	500	To be finalized	Initially proposed in Ramanathapuram district. Site under revision
31.	Telangana	500	Telangana New & Renewable Energy Development Corporation Ltd. (TNREDC)	Gattu, Mehboob Nagar District
32.	Uttar Pradesh	600	Lucknow Solar Power Development Corporation Ltd.	Jalaun, Allahabad, Mirzapur, and Kanpur Dehat districts
33.	Uttarakhand	50	State Industrial Development Corporation Uttarakhand Limited (SIDCUL)	Industrial Area, Sitarganj (Phase I), Industrial Area, Sitarganj (Phase II) and Industrial Area, Kashipur
34.	West Bengal	500	West Bengal State Electricity Distribution Company Ltd.	East Mednipur, West Mednipur, Bankura
	Total	20,000		

less costly and can be managed by communities. Solar energy can be a viable option that can be installed on individual household and community level. The current production of energy is not enough to fulfill energy requirements and mostly energy needs fulfilled by fossil fuel (current energy production is demonstrated in Fig. 1, Table 3 show the renewable energy prospect in India and Table 4 show the achieved

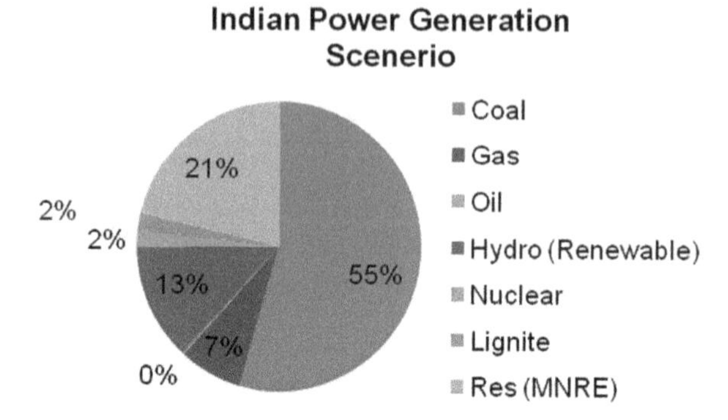

Fig. 1 Current scenario of energy production of India (TERI 2018)

status of renewable energy). Therefore, it is necessary to shift on renewal energy sources for clean energy and as well as to protect the environment. This is one of the aspects for launching largest solar energy program in mission mode. With the enabling policy condition, the government introduced different scheme to achieve the set targets such as the solar park, solar rooftop scheme, solar defense scheme, solar scheme for CPUs solar PV power plant, on the canal bank and channel top, solar pump and solar rooftop. To provide financial support and enabling environment government, it modifies the renewable energy (RE) framework such as RE policy, feed in tariff, renewable purchase obligation, grid connectivity and RE certificate mechanism (GoI, MNRE).

The government is focused to create market for solar energy through subsidies and tax incentives. Solar PV technologies are promoted to meet out the challenge of providing electricity for rural electricity for telecom and village electrification which are found to be very expensive technology (TERI 2018). Deshmukh et al. (2010) analyze and demonstrate how committed subsidy on off-grid solar plant and lighting system within JNNSM does not make its share for poor and rural areas beyond 7% of the total investment (Deshmukh et al. 2010).

5 Solar Energy: The Way Forward

In recent years, scholars and policymakers have pointed toward the need for mixed policies which can boost technological innovation and have an environmental solution (Jänicke and Lindemann 2010). The mixed policy may be promoted to achieve multiple goals. As Flanagan et al. (2011) also argue that this policy mix concept has the objective to fulfill one particular goal and same time it has other multiple goals, which goes simultaneously (Flanagan et al. 2011). In fact, the policy that has the objective to protect the environment, and at the same time, it fulfills multiple objective that are very much desired. In the case of JNNSM, it reduces the carbon emission in the environment by using solar energy, and at the same time, it secures energy

Table 3 Targets for solar energy (TERI 2018)

Component	Year-wise target (in MW)							
	2015–2016	2016–2017	2017–2018	2018–2019	2019–2020	2020–2021	2021–2022	Total
Rooftop grid-connected solar PV	200	4800	5000	6000	7000	8000	9000	40,000
Medium and large-scale grid-connected solar power	1800	7200	10,000	10,000	10,000	9800	8500	57,000

Table 4 Renewable energy prospect in India (TERI 2018)

Technology	Potential (MW)	Cumulative achievement so far
Wind	49,500 (50 m hub height) 102,800 (80 m hub height)	21,692.98
Solar	74,800	2970 MW
Small hydro (up to 25 MW)	19,700	3826.18
Biomass including bagasse cogeneration	22,500	4045.55

needs, creates jobs, avenues, technologies, accessibility and so on. It purposes the dominant socio-techno regime by using technological solution for multiple goals.

JNNSM is a prominent mission program which is a step forward in energy security. Currently, solar energy is comparatively more expensive than fossil fuel. The cost involved in the establishment of solar rooftop and grid-connected plant is of low capacity. The total population living under poverty in 2011 is 21.9% (https://www.adb.org/countries/india/poverty) and there is a significant decrease since 2004 which was 27.5% (https://data.gov.in/resources/percentage-people-below-poverty-line-india). Off-grid decentralization and low-temperature application are a favorable option for rural electrification. To find out the logic behind the ambitious solar mission, the deeply rooted problems such as poverty, unemployment, illiteracy, housing crisis, malnutrition and provision of clean drinking water come into the row. Deshmukh et al. (2010) argue that JNNSM is a non-participatory as public was not consulted and has no scope and assurance for public process (Deshmukh et al. 2010). In democratic process, policy formation is a subject related to larger public interest. In this process, public consultation is necessary to assure their participation and contribution. From its objective approach, it is not clear how the larger community is going to contribute in this mission to achieve the target.

It is disheartening but the JNNSM looks like a political mandate which stretched beyond the government itself. In the beginning, the involvement of different stakeholders is very minimal. Later on, the government takes the decision to do a consultation with different industry/technological/financial partners and implements the mission program (https://economictimes.indiatimes.com/industry/energy/power/mnre-to-hold-consultations-with-industry-on-solar-mission/articleshow/18402526.cms). Hence, it is difficult to access the achievability of the mission. The policy measurement represents a political challenge, which involved diverging interest of private sector where broader societal interest takes backseat. To find out the non-political and feasible method, active engagement of stakeholders becomes a necessary precondition for the mission. Further, it depends on the capacity and strategy of MNRE how it handles issues of stakeholders and larger interest of the public (Quitzow 2015). Not only MNRE but other departments too are not equipped with required capacity to implement programs. As Chibber (2003) argued, the government bodies depend on consultation of other expert agencies and public–private partnership (Chibber 2003). The mission policy measures are strongly focused on market support. In mission documents, there is an absence of consideration put on the other framework capacities, preparation or learning improvement and dispersion. In policy documents, it is mainly focused on to achieve the set target. There is no proper mechanism to address the issue of solar waste including environmentally hazardous materials from solar PV. According to the report of Bridge to India, the solar photovoltaic panel waste volume in India is estimated to grow to 2 million tons by 2030. India does not have any policy to deal with all these wastes. Another issue is related land used for solar panel installation. For solar energy production, a huge area is required, and most importantly, it depends on the topography of the land, ecosystem and biodiversity. The application of solar application can reduce the production capacity of cultivable land. The land effect can be reduced by establishing a solar project in the desert, mining area and brownfield. In brief, as Dawn et al. (2016) indicate (Dawn et al. 2016) following issues need to be considered seriously by policymakers:

- Availability of land for manufacturing and production.
- Weak relation between industry and government in term of guidelines, implementation, subsidy disbursal and land purchase.
- Less production in domestic manufacturing in solar appliance and technological innovation does not boost the local economy. Dependency on foreign-based solar technology and equipment due to low cost, resulted in local market faced the loss.
- Lack of public awareness regarding solar energy and its economic and environmental benefits.
- Involvement of different agencies such as MNRE, SECI, IREDA and electricity board in the implementation and disbursement of subsidy delay the implementation.
- Lack of trained manpower and appropriate technological orientation.

The case of Rewa Ultra Mega Solar Power Park (Madhya Pradesh, India) may serve as a template for future development of solar power. In Rewa UMSSP, instead of providing direct subsidies, the government concentrated on reducing final prices through systematic risk mitigation. Perhaps, as the Rewa UMSSP case shows, low tariff can be accomplished by avoiding subsidies and provisions of soft loans from the international financial bodies to support the construction may further mitigate the financial risk of projects. Further, the segregation of execution and management by separate teams of experts to avoid a variety of conflicts. The successful execution of the Rewa solar project suggests that barriers can be removed through the determined actions of the governments, establishing coordination among financial institutions and project developers, and the community could be mobilized and involved in the achievement of the targets and goals (https://www.ifc.org/wps/wcm/connect/b02cc29c-5105-4b32-9c71-ec723cf04367/scaling-infra-india-08.pdf?MOD=AJPERES&CVID=mSCUfFq).

6 Conclusion

Solar power is a national importance mission which has the potential to fulfill the requirement of energy needs in future. It becomes an alternative source of energy in place of fossil fuel. Though the benefits are not visible in short-run, but it has the capacity to contribute in the national growth. However, the way government policies and institutional structures support solar mission has raised several questions on governance and management of power sector. Reforms in power sector are also required along with enhanced understanding of structural realities of India. While India is struggling to build basic infrastructural facilities, creating amenities to strengthen economic growth, the government must take serious concern on the deep-rooted social problems and reachability of program goals to the poor and marginal sections and create avenues for inclusive growth. India needs inclusive growth where every section of society can take active part and get benefited. The technological mission need to be converted as social mission and indiscriminate use of redundant technology should be replaced with new technologies that are efficient and cost effective.

The launching of JNNSM provided a long-term strategic plan to achieve energy requirements for the future growth of economy, energy accessibility and India's commitment to contribute to green and clean development. The JNNSM defined objectives, targets, timeline; made arrangements for cooperation from various agencies. The JNNSM also seeks market support, support from the research and development that would also develop the human resources. As discussed, India introduced several schemes to achieve the target of 100 GW. The government opened electricity market for generation, transmission and distribution networks and coordination among various participating agencies (Sawhnet 2019). In particular, in the Phase II of JNNSM, the government worked through all possible options such as bundling, generation-based incentive (GBI), viability gap funding (VGF), transparent bidding, e-reverse auction and so on. Perhaps, these innovative methods would further help the government to achieve the current targets. Perhaps, the government can further work in these directions to achieve its new and under consideration targets of 4500 GW of renewable energy by the year 2030. Lastly, reforms in power sector are may be required along with enhanced understanding of structural realities of India. While India is struggling to build basic infrastructural facilities, creating amenities to strengthen economic growth, the government should take more concerted steps to avoid deep-rooted social problems and reachability of program goals to the poor and marginal sections and create avenues for inclusive growth.

References

Arora DS, Busche S, Cowlin S, Engelmeier T, Jaritz J, Milbrandt A, Wang S (2010) Indian renewable energy status report: background report for DIREC 2010. https://www.osti.gov/biblio/991558/

Basu SN, Karmakar A, Bhattacharya P (2015) Growth of solar energy in India: present status and future possibilities. Int J Electr Electron Comput Syst 3(5):2347–2820
Berlin Mandate (1995) https://unfccc.int/resource/docs/cop1/07a01.pdf. Accessed on 25 May 2019
Chibber V (2003) Locked in place: state-building and late industrialization in India. Princeton University Press, Princeton
Cole RJ (1998) Energy and greenhouse gas emissions associated with the construction of alternative structural systems. Build Environ 34(3):335–348. https://doi.org/10.1016/S0360-1323(98)00020-1
Dawn S, Tiwari PK, Goswami AK, Mishra MK (2016) Recent developments of solar energy in India: perspectives, strategies and future goals. Renew Sustain Energy Rev 62:215–235. https://doi.org/10.1016/j.rser.2016.04.040
Deshmukh R, Gambhir A, Sant G (2010) Need to realign India's national solar mission. Econ Polit Wkly 45(12):41–50
Flanagan K, Uyarra E, Laranja M (2011) Reconceptualising the 'policy mix' for innovation. Res Policy 40(5):702–713
Houghton RA (2005) Tropical deforestation as a source of greenhouse gas emissions, pp 13–21. https://www.edf.org/sites/default/files/4930_TropicalDeforestation_and_ClimateChange.pdf
Howe JP (2017) Making climate change history: documents from global warming's past. University of Washington Press, Seattle, pp 108–114, 205-215, 252-254
https://archive.indiaspend.com/cover-story/why-the-us-backs-indias-great-solar-bet-45411
https://coal.nic.in/content/production-supplies
https://data.gov.in/resources/percentage-people-below-poverty-line-india
https://economictimes.indiatimes.com/industry/energy/power/mnre-to-hold-consultations-with-industry-on-solar-mission/articleshow/18402526.cms
https://energy.economictimes.indiatimes.com/news/renewable/india-to-comfortably-achieve-100-gw-solar-energy-target-by-2022-government/65307602
https://mnre.gov.in/file-manager/UserFiles/draft-jnnsmpd-2.pdf
https://mnre.gov.in/file-manager/UserFiles/Scheme-for-enhancement-of-capacity-to-40GW-Solar-Parks.pdf
https://mnre.gov.in/file-manager/UserFiles/Solar-Park-Guidelines.pdf
https://mnre.gov.in/history-background. Accessed on 03 June 2019
https://mnre.gov.in/solar
https://pib.nic.in/newsite/PrintRelease.aspx?relid%3D122566
https://statisticstimes.com/demographics/population-of-india.php
https://sustainabledevelopment.un.org/post2015/transformingourworld
https://unfccc.int/process/conferences/the-big-picture/milestones/the-cancun-agreements
https://web.archive.org/web/20170226140523/http://mnre.gov.in/file-manager/grid-solar/100000MW-Grid-Connected-Solar-Power-Projects-by-2021-22.pdf
https://www.adb.org/countries/india/poverty
https://www.downtoearth.org.in/blog/who-needs-ultra-mega-solar-power-plants--43615
https://www.ipcc.ch/site/assets/uploads/2018/02/ipcc_wg3_ar5_summary-for-policymakers.pdf
https://www.pmindia.gov.in/en/news_updates/pm-undertakes-aerial-survey-of-cyclone-fani-affected-areas-in-odisha-reviews-the-situation/?comment=disable
Jänicke M, Lindemann S (2010) Governing environmental innovations. Environ Politics 19(1):127–141. https://doi.org/10.1080/09644010903396150
Kar SK, Sharma A, Roy B (2016) Solar energy market developments in India. Renew Sustain Energy Rev 62:121–133. https://doi.org/10.1016/j.rser.2016.04.043
Koneswaran G, Nierenberg D (2008) Global farm animal production and global warming: impacting and mitigating climate change. Environ Health Perspect 116(5):578–582. https://doi.org/10.1289/ehp.11034
Kyoto Protocol (1997) https://unfccc.int/sites/default/files/kpeng.pdf. Accessed on 25 May 2019
Mohan A (2017) Whose land is it anyway? Energy futures and land use in India. Energy Policy 110:257–262. https://doi.org/10.1016/j.enpol.2017.08.025

NITI Aayog (2015) Report of the expert group on 175 GW RE by 2022
Owusu PA, Asumadu-Sarkodie S (2016) A review of renewable energy sources, sustainability issues, and climate change mitigation. Cogent Eng. 3(1):1167990. https://doi.org/10.1080/23311916.2016.1167990
Parira AS (2016) Solar parks: accelerating the growth of solar power in India. Akshay Urja, Aug 2016, pp 15–16
Phadke A, Abhyankar N, Deshmukh R (2016) Techno-economic assessment of integrating 175 GW of renewable energy into the Indian grid by 2022. Ernest Orlando Lawrence Berkeley National Laboratory
Quitzow R (2015) Assessing policy strategies for the promotion of environmental technologies: a review of India's National Solar Mission. Res Policy 44(1):233–243. https://doi.org/10.1016/j.respol.2014.09.003
Rathore PKS, Rathore S, Singh RP, Agnihotri S (2018) Solar power utility sector in India: challenges and opportunities. Renew Sustain Energy Rev 81(P2):2703–2713
Rattani V (2018) Coping with climate change: an analysis of India's National Action Plan on Climate Change. Centre for Science and Environment, New Delhi. https://www.indiaenvironmentportal.org.in/files/file/coping-climate-change-NAPCC.pdf
Rewa Solar (India): Removing barriers to scale. https://www.ifc.org/wps/wcm/connect/b02cc29c-5105-4b32-9c71-ec723cf04367/scaling-infra-india-08.pdf?MOD=AJPERES&CVID=mSCUfFq
Sawhnet A (2019) Renewable energy sector: the way forward. https://energy.economictimes.indiatimes.com/energy-speak/renewable-energy-sector-the-way-forward/3952
Shrimali G, Nekkalapudi V (2014) How effective has India's solar mission been in reaching its deployment targets? Econ Polit Wkly 49(42):54–63
Shrimali G, Rohra S (2012) India's solar mission: a review. Renew Sustain Energy Rev 16(8):6317–6332
Steinfeld H, Gerber P, Wassenaar T, Castel V, Rosales M, de Haan C (2006) Livestock's long shadow: environmental issues and options. Food and Agriculture Organization of the United Nations, Rome
TERI (2018) Eight year of national solar mission: renewing and renewable targets. https://www.teriin.org/article/national-solar-mission-journey-toward-energy-security-and-sustainable-future
Ummadisingu A, Soni MS (2011) Concentrating solar power: technology, potential and policy in India. Renew Sustain Energy Rev 15(9):5169–5175
UNFCC (2015) Adoption of the Paris agreement. https://unfccc.int/resource/docs/2015/cop21/eng/l09.pdf
Yenneti K, Day R, Golubchikov O (2016) Spatial justice and the land politics of renewables: dispossessing vulnerable communities through solar energy mega-projects. Geoforum 76:90–99

Impact Analysis of Cyber Attacks on Smart Grid: A Review and Case Study

Temitayo O. Olowu, Shamini Dharmasena, Alexandar Hernandez, and Arif Sarwat

1 Introduction

Cyber physical systems is a term that applies to engineered systems that find their use in a variety of domains. Often these systems are a collection of sensors, actuators, and embedded devices that act as an interface with the real world. In addition, these devices communicate through short- or long-range communication channels to share data and create a seamlessly integrated network (Sundararajan et al. 2018, 2019). Cyber physical architecture can be used to improve existing traditional systems, as well as improve the quality of service provided by these systems. For example, one application of a CPS is wide-scale deployment of sensors and actuators that will be used to monitor key environmental changes in the world. The data can be aggregated to a database and used to make better decisions concerning the environment. This ties closely with disaster response, which is another area where the use of CPS can reduce the chaos caused by natural disasters or other large-scale emergencies. These systems can be implemented to manage evacuations and create scheduled departures that will reduce congestion and accidents that would further delay evacuations (Sundararajan et al. 2018; Gunes et al. 2014). Various other applications find promise in cyber physical systems. These include smart manufacturing, air transportation, robotics for service, and health care/medicine which includes anything from assistive devices to smart operating rooms (Gunes et al. 2014; Dharmasena et al. 2019). They create a

This work is funded by NSF under the grant numbers CNS-1553494 and CNS-1446570.

T. O. Olowu (✉) · S. Dharmasena · A. Hernandez · A. Sarwat
Energy, Power and Sustainability Group, Florida International University, Miami, USA
e-mail: tolow003@fiu.edu
URL: https://www.eps.fiu.edu

S. Dharmasena
e-mail: ikona001@fiu.edu

A. Hernandez
e-mail: ahern373@fiu.edu

A. Sarwat
e-mail: asarwat@fiu.edu

H. Tyagi et al. (eds.), *New Research Directions in Solar Energy Technologies*,
Energy, Environment, and Sustainability,
https://doi.org/10.1007/978-981-16-0594-9_3

highly monitored and controlled environments where human interaction is reduced. However, with any new technology, there are always a variety of challenges that must be overcome in order to realize widespread cyber physical system implementation. In order to ensure the systems are robust, several factors must be accounted for. These are inter-operability, predictability, security, reliability, dependability, and sustainability (Gunes et al. 2014). There is an ongoing drive by power utility companies to achieve smart distribution systems or the smart grid (Wadhawan et al. 2017; Chen et al. 2011; Sundararajan et al. 2019; Dharmasena et al. 2019). This typically involves the deployment of communication and control devices and integration of localized generations, distributions, and energy management systems to allow the physical grid become more autonomous, intelligent, and controllable (Sarwat et al. 2017; Hawrylak et al. 2012; Stefanov and Liu 2012; He and Yan 2016; Wei et al. 2014; Olowu et al. 2019a, b). As part of the smart grid architecture, the deployment of distributed energy resources (DERs) such as photovoltaic (PV) systems is becoming a good alternative to the conventional power generators (Olowu et al. 2018, 2019; Rahman et al. 2018; Jafari et al. 2018; Olowu et al. 2018, 2019; Debnath et al. 2020). The drive to achieve a smart distribution system has opened up new set of challenges for the utility companies (Zhaoyang 2014; Srivastava et al. 2013; Dagle 2012). Data communication and control between the physical systems and the cyber network have made the smart grid prone to cyber physical attacks (Parvez et al. 2016, 2017; Mekonnen et al. 2018; Odeyomi et al. 2020). In this paper, cyber physical attacks that occur in smart grid and their mitigation techniques are extensively reviewed under the three domains: device level, communication level, and application level in Sect. 2. In Sect. 3, a case study of fault data injection (FDI) in a production meter of a standard IEEE 34 test feeder with three PVs has is simulated, analyzed, and presented. Section 4 presents a proposed machine learning based protection architecture that can be used to mitigate the severe impact of FDI attack. Finally, Sect. 5 concludes the paper together with proposals for the future work.

2 Cyber Attacks on Smart Grid

This section discusses the features of the smart grid architecture and various level of cyber attacks that can be executed within it.

2.1 Smart Grid and Its Architecture

According to the definition by national institute of standard and technology (NIST), smart grid is a network that provides electricity efficiently, reliably, and securely. In other words, it is delivering electricity with brain (Smart grid 2019). Smart grid comprises of generation (including DERs), transmission, distribution, service providers, customers, and markets. Each of these components interacts with others which means that there is a bi-directional flow of power and communication in between. In order to facilitate the required functionalities, smart grid comprises of heterogeneous systems such as supervisory control and data acquisition (SCADA), advanced metering infrastructure (AMI), intelligent electronic devices (IEDs), human–machine interface (HMI), building management systems (BMS), DERs, and many more. Furthermore, there are different network protocols for the communication of these different systems. While these technologies make the grid smarter, it increases the system's vulnerability to cyber attacks. Smart grid and its cyber threats can be analyzed across three layers: device layer, communication layer, and application layer. In smart grid, there are many physical devices and their interfaces that falls into device layer. This includes smart meters, monitoring, and measuring units such as phasor measurement units (PMU), relays, and other protection devices. Communication layer incorporates the communication network between theses devices in generation, distribution, and consumer ends in the smart grid. It is very important to secure the data packet transfer between these devices. The processing and analytical platforms deliver high-end insights to analysts and operators which is listed as the application layer in this study (Saleem et al. 2019). Supervisory control and data acquisition (SCADA) systems and industrial control systems (ICS) are such examples for the application layer of smart grid.

2.2 *Layers of Cyber Attacks*

As discussed in Sect. 1, there is a growing concern as regards cyber attacks on the smart grid architecture. These attacks have wide ranging effects on the dynamics of the grid which include loss of generator synchronism, voltage collapse, frequency issues, prolong outages, and power quality issues among others. There are various possible attacks on a smart grid. These attacks can be categorized on the basis of what exactly is being compromised. As discussed in Sect. 2.1, smart grid can be classified into three domains, and so, the attacks can also categorize into these three levels. The attack happens in device level, communication level, and in application level. The authors of Li et al. (2012) proposed a different categorization for cyber attack, which are: device attack, data attack, privacy attack, and network availability attack. But the taxonomy of attacks used in this paper is simpler and shows a direct connection with the end target compared to the taxonomy proposed in Li et al. (2012). There are many other attacks that fall into above categories. In addition, there are many possible entry points for attackers. These include, but are not limited to, infected devices where an employee may inadvertently, or intentionally, plug in an infected USB. Attacking the network through vulnerabilities is another possibility if the IT infrastructure has holes or backdoors that can be accessed by hackers. Equipment preloaded with malware is another common entry point and is known as a supply chain attack. Phishing emails or social engineering also present a problem. Here, a hacker can obtain personal information and access the system as a valid user. This becomes difficult to detect as the intrusion does not appear as a threat to the system

(Conteh et al. 2016). Often, the addition of humans to the loop makes a system much more vulnerable to outside attack (Haack et al. 2009).

2.2.1 Device Layer

The device layer attack as its name connotes occurs when an attacker targets a grid device and seizes control of it. This could be used to wreak havoc by shutting off power or to gain control of communications, and mostly this level of attack can lead into another level. The puppet attack which is a variant of of denial of service (DoS) is a plausible attack type in AMI. AMI creates the bidirectional communication interface between smart meters (Wei et al. 2017) and utilities to share power consumption, outage, and electricity rate data. The attacker compromises several normal nodes in AMI and keep as puppet nodes to exhaust the system through flooding data packets. Attacks like puppet or time delay switch (TDS) attack only target one security parameter, availability, to affect AMI (El Mrabet et al. 2018).

2.2.2 Communications Layer

Attacks at communication level can be either data attacks or a network availability attack. The data attack involves either removing, adding, changing, or stealing the data being communicated. An example of this would be an attacker sending false price and meter information. This will lead to power shortages and overall cause a loss in the power companies' revenue (Mo et al. 2011). Privacy attacks can also be categorized under this type which involves stealing confidential information, which could be consumers electric bill or their daily energy usage. The network availability attack involves reducing or eliminating the functionality of the network that the devices are communicating on. A commonly seen example of this is a DOS attack. Based on previous statistics, momentary or prolonged shutdown of the grid can have devastating consequences. The distributed denial of service (DDOS) attacks that occur in networks overwhelm the Internet bandwidths and reduces the network performance through multiple compromised devices. There are several types of DDOS attacks: Slowloris, SYN flood, Ping of Death, ICMP flood, UDP flood, etc., which operates in different speeds (Ozgur et al. 2017). Slammer worm is another malware that attacked a nuclear power plant in Ohio in 2003. It disabled the plant's safety monitoring system for nearly 5 h. Slammer was one of the fastest worms at that period and had the capability to spread workwide in 15 min. Slammer sends a UDP datagram to the port 1434 of target computer, and it makes use of the buffer overflow vulnerability in the SQL server monitor for the execution.

2.2.3 Applications Layer

There are a variety of well-known attacks on the application level of the smart grid, and these include Stuxnet, Duqu, BlackEnergy3, etc. These attacks have become sophisticated and multifaceted making them harder to detect and prevent (Eder-Neuhauser et al. 2017). For example, Stuxnet compromised confidentiality, integrity, availability, and the accountability of a system, and it targeted SCADA and control devices (PLCs). Stuxnet was a multilayered attack that first infected a Windows computer through an infected USB and began replicating itself. Once it had integrated itself into the system, it found a certain program created by Siemens called Step7, and eventually found its way to the PLCs. Not only could the attacker spy on the systems, but they could also control the PLCs and in that way control the connected machinery (Kushner 2013). Stuxnet utilized four zero-day vulnerabilities of the system. It had a rootkit to hide the malicious files and processes from users and anti-malware software. There are several other Stuxnet related malware: Duqu, Flame, Triton. Similar to Stuxnet, Duqu uses a kernel driver to decrypt the dynamic load library (DLL) files, and it mainly targets the SCADA. The way the above attacks function vary greatly on what their specific target is. The Duqu and Flame attacks were slightly different than Stuxnet whose purpose was to cause physical damage to equipment. The Duqu attack was created to steal information about industrial control systems. DDOS type attacks occur in application level too. DDOS exhaust SCADA like systems by striking with simultaneous data requests and crash down the system. BlackEnergy-3 is a Trojan that is used for DDOS attacks. In 2015, a SCADA related industrial control system in a electricity distribution system in Ukraine was subjected to a BlackEnrgy-3 attack which caused power outages for around 225,000 customers for several hours. In this incident, the access to the network is gained through spear phishing and used a KillDisk to erase the master boot record and the logs of the impacted system (Sharing and Center 2016). A summary of reported cyber attacks in smart grid is given in Table 1.

2.3 *Cyber Attacks on DERs in a Smart Grid*

With the increase in utility- and small-scale DERs (particularly PV systems) in the grid, there is an increase in the vulnerability of the entire system. Figure 1 shows the different attack points in a DER integrated grid. There are two basic networks layers in a grid-tied DER system. These are the power layer (that allows the energy generated by the DER to be sent to the grid bidirectionally) and the communication and control layer (which allows remote monitoring, data logging, and remote controlling of the DERs).

Usually, DER devices have their individual energy management systems (EMSs) that control their power electronic converters such as smart inverters (SIs) (Qi et al. 2016). SIs typically have IP addresses that allow for a remote control of their operations. This makes them vulnerable to man-in-the-middle (MITM) attacks. The

Table 1 Reported cyber attacks in smart grid

Name of attack	Target of attack	Year	Attack details
Device layer			
Trojan.Laziok reconnaissance malware	Devices of energy companies	2015	Collected data from compromised devices, i.e., installed antivirus software, installed applications, CPU and GPU details, etc.
BlackEnergy	General Electric's HMI	2011	HMI of utility grid control systems
Communication layer			
Spear phishing, Havex malware for watering hole attack	ICS/ SCADA	2014	Espionage using OPC protocol to map devices on ICS network
Dragonfly 2.0	Western energy sector	2015–2017	Spear-phishing, Trojan-ware, watering hole attacks
Exploitation of vulnerabilities in firewall firmware	Power grid of Western US	2019	Outside party rebooting the company's firewalls to cause periodic "blind spots" for grid operators losing communication with multiple remote power generation sites for minutes at a time that lasted for around 10 h
Application layer			
Stuxnet worm	Programmable logic controllers of SCADA	2010	Travels via a USB stick. Exploits zero-day vulnerabilities of PLCs
Duqu worm	SCADA	2011	Designed to steal information about ICS (digital certificates, private keys)
Remote access Trojan; watering-hole attack	ICS/ SCADA	2014	Conducted by dragonfly, energetic bear
BlackEnergy3	Ukrainian grid control center	2015	Left 220000+ customers without power
Industroyer or crash override malware	Pivnichna substation ICS, Ukraine	2016	Power outage to one-fifth of Kiev

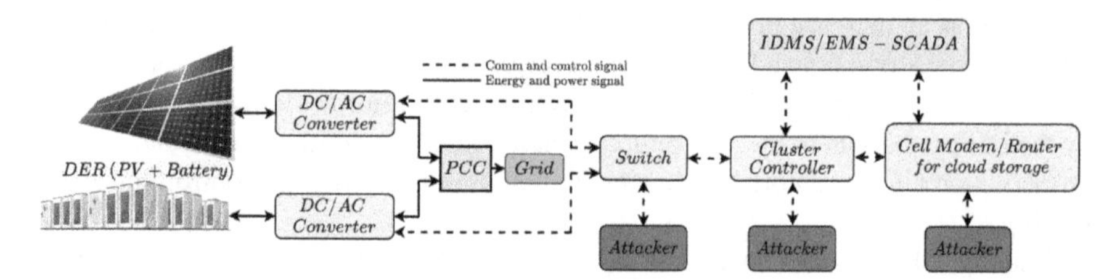

Fig. 1 Potential attack points on a grid-connected DER

knowledge of the SI's IP enables the attacker to gain a direct control of the SI and could potentially alter its SI settings. Depending on the level of DER penetration on the feeder where the DER is connected, altering the SI settings by an attacker could lead to severe changes in the grid's voltage and frequency (Teymouri et al. 2018) Another level of attack can be executed by compromising the utility's wide area network. This will allow the attacker send malicious and false commands to the DERs. This include sending false messages to enable the DERs make unnecessary control actions and operation. The false command messages to the DERs from the utility SCADA system could also be as a result of the attacker compromising the power systems data from the point of common coupling (PCC) being sent by the SIs. For example, if the voltage and frequency values coming from a DER's SI are compromised, this could cause the integrated distribution management system (IDMS) or energy management system (EMS) to send false control actions to other DER SIs.

2.4 *Detection and Mitigation Techniques for Cyber Attacks in Smart Grid*

Different ways in which the attacker infect and propagate through a system makes it difficult to detect and mitigate these attacks. These attacks all have different purposes depending on the goals of the hacker. They range from disrupting the normal operation of a system to stealing information from the local utility or its consumers. In addition, the way these systems breach the security of a smart grid vary from removable drives to client-to-server access. Nimda, an attack that occurred in 2003 for disrupting the smart grid, has several access points including email, client-to-server, server-to-client, and host-to-network sharing (Eder-Neuhauser et al. 2017; El Mrabet et al. 2018). All the different points of entry, attack methods, and different targets make attack detection, prevention, and elimination very difficult. Various detection and mitigation techniques are proposed in literature, and they are mostly specific to the type of the attack. And most of the time rather than using a single solution, several security measures are deployed together to mitigate attacks at every progression stage.

2.4.1 Detection Techniques

During pre-attack atmosphere, it is mainly monitoring and detection schemes applied at vulnerable locations. In this case, different detection techniques are introduced to get early warnings and prepare with proper counter measures. Intrusion detection system (IDS) is such major mechanism and can be found as anomaly-based detection, specification-based, and signature-based IDS. Currently, many anomaly-detection-based IDS are developed using machine learning techniques. The authors of (Ozay et al. 2016) present a review on different machine learning techniques to develop learning algorithms that can be employed to classify secure and attacked datasets. In Ozay et al. (2016), a statistical correlation-based scalable unsupervised anomaly detection engine for large-scale smart grids is proposed. The proposed scheme has reduced computational complexity by exploiting feature extraction through symbolic dynamic filtering. An IDS framework using blockchain for multimicrogrid (MMG) system is presented in Hu et al. (2019). This paper investigates that the vulnerability of MMGs for cyber attacks proposes a novel corroborative IDS that adopts a multipattern proposal generation method to reduce the false negative rate of intrusion detection.

False data injection is another common attack in smart grid. Therefore, many studies are carried out on false data injection detection (FDID). The paper (Wei et al. 2018) proposes a FDID technique that uses deep belief networks, and it uses unsupervised learning from the bottom of the restricted Boltzmann machine to have initial weights. Another recent study on FDID is presented in Ameli et al. (2020) which is focused on line current differential relays. It has been learnt that attacks on multiple relay can create catastrophic failures in the system, and therefore, this paper investigates coordinated attack scenario on line current differential relays and its consequences. Then, a FDID is proposed which uses the state space model of the faulty line together with positive and negative sequences of voltage to detect the attack. With specific reference to DER, some of the detection technique is proposed in literature. These include security information and event management (SIEM), data loss prevention (DLP) technology, and IDS (El Mrabet et al. 2018). The use of data-driven techniques can be used to implement real-time intrusion detection. This requires the use of machine learning algorithms to forecast accurate PV power generation and prediction of dynamic states of the network based of historical and extensive simulation data. Another detection approach specific to the power electronic converters is by regularly sampling the voltage and frequency at the PCC in other detect the rate of change of these parameters. A sudden change beyond the set tolerance could indicate a potential cyber attack on the SIs. The tolerance values of the detection algorithms are set based on the learning performance of the SIs.

2.4.2 Mitigation Techniques

The severity of cyber attacks on smart calls for the development of adequate and effective mitigation strategies. These techniques can be made proactively or reactively. This implies that steps can be taken to address a cyber physical threats before attack, during at or after an attack has been executed. Several approaches to addressing different types of attacks on cyber physical assets on smart grid has been proposed in literature. Authors of Srikantha and Kundur (2016), Farraj et al. (2016) proposed the use of game theoretic framework to mitigate cyber attacks on switching and control of physical assets in smart grid. In Srikantha and Kundur (2016), the authors demonstrated that power utility companies can devise a counter measure vectors against an attacker using the 2PZS (two-player zero sum) differential game formulation. A new iterative algorithm to solve the nonlinear 2PZS game was proposed. Their results showed that by applying the countermeasure vector, the utility company can successfully prevent the attack and keep the system stable. Also their formulation is able to determine the safety margin that will enable a proactive measures to be taken. In Farraj et al. (2016), a simplified model of a switching attacked is presented. The position and sign of the rotor speed is used to initiate a local control action (with resource constraints) to provide a counter measure against an attack n the generator. A game theoretic formulation is proposed to make this interaction between the attack switching action as well as the counterattack control mechanism. Their results shows that the proposed resource-constraint controller can effectively be called to action only when needed as well as meet the requirement of stabilizing the system. To prevent FDI attacks in smart grids, authors of Wang et al. (2017) proposed a data analytical technique to detect FDI attacks. The data-centric technique is based in margin-setting algorithm (MSA). MSA is a machine learning algorithm based on data analytics. The authors used a six-bus feeder for simulation and experimental validation of the proposed MSA algorithm on tho FDI attack scenarios. Their results showed that the proposed MSA algorithm performed better for FDI attack mitigation when compared with other machine learning algorithms such as support vector machine (SVM) and artificial neural network (ANN). Deep learning models have also been proposed to capture anomalies due to FDI attacks with validations showing high level of accuracy (He et al. 2017). One detection approach specific to the power electronic converters is by regularly updating the firmware of the SIs to minimize their vulnerabilities. Authors of McLaughlin et al. (2010) proposed the use of a firmware diversity approach that prevents or limits the possibility of a large-scale cyber attack on smart meters. This approach can also be deployed on SIs. The diversity in the SIs firmware of various DERs will make simultaneous large-scale attack difficult to achieve by the attacker since the vulnerabilities of these SI firmware will differ. An inverter internal anomaly (such as switch faults) detection and mitigation algorithm which adjusts the inverter voltage output using model predictive control technique was proposed by Fard et al. (2019). This control method prevents a complete shut down of the inverter which may lead to cascading shut down of other DERs in the network due to sudden increase in load seen by other DERs as a result of the sudden loss of power generation from the attacked DER's SI.

3 Case Study of FDI Attack on IEEE 34 Bus System

In order to visualize and quantify the potential impacts of an FDI attack on the grid, an IEEE 34 distribution feeder is developed by the IEEE PES test feeder working group. Its parameters are based on an actual distribution feeder located in Arizona in USA.

3.1 System Model and Simulation Setup

The IEEE 34 bus network (used as a case study) is integrated with three PV systems, a synchronous generator, and a battery energy storage system (BESS) as shown in Fig. 2. The nominal voltage rating of the feeder is 24.9 kV. The feeder has two voltage regulators between nodes 814–850 and 852–832. The substation transformer upstream of node 850 is a 2.5 MVA, 69/24.9 kV, $\Delta\ Y$. The combined rating of the load (modified) on the feeder is approximately 3.1 MW (active) and 0.689 MVAr (reactive). Node 838 is the farthest distance and its approximately $59km$ away from the substation transformer The specifications of the sources integrated into the feeder is as given in Table 2.

The IEEE 34 node test feeder, PVs, synchronous generator, and the battery energy storage are modeled using OpenDSS and MATLAB.

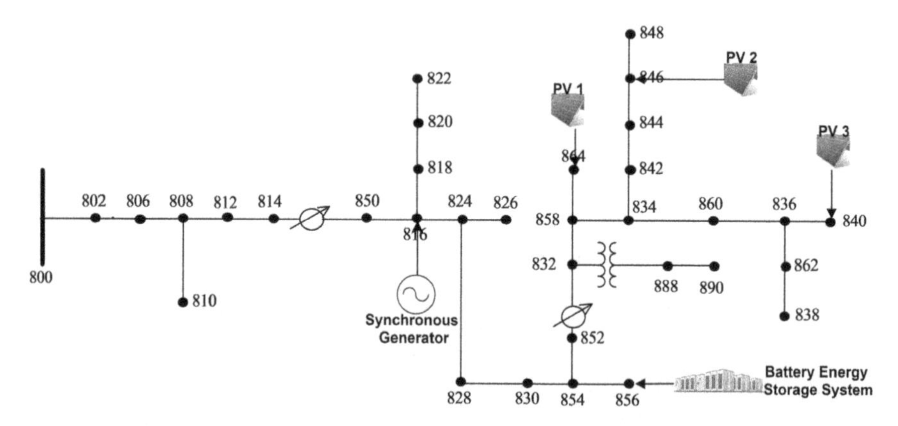

Fig. 2 IEEE 34 node distribution feeder with three PVs, a synchronous generator, and one battery energy storage system

Table 2 Generation specifications

Generation power (kW)	Maximum (kWh)	Inverter	Phases	Default PF	Energy
PV1	200	300	1	1	–
PV2	2050	3000	3	1	–
PV3	200	300	3	1	–
Synchronous generator	5000	5000	3	1	–
Battery energy storage	1000		1	1	12,000

3.2 Data Gathering, Cleaning, and Preprocessing

According to Sundararajan et al. (2019), the power generation from photovoltaic systems mainly depends the global horizontal irradiance (GHI) and the ambient temperature (consequently the module temperature).

$$P_{\mathrm{PV}^{\mathrm{gen}}}(t) = P_{\mathrm{DC}} \times \frac{\mathrm{GHI}(t)}{1000} \times G \times M \tag{1}$$

$$G = 1 + \frac{\%\mathrm{temp}_{-\mathrm{coeff}}}{100}[T(t) - 25] \tag{2}$$

$$M = b_{\mathrm{d}} \times b_{\mathrm{m}} \times b_{\mathrm{c}} \times b_{\mathrm{inv}} \tag{3}$$

where P_{dc} is the DC name plate capacity of the PV system, GHI(t) is the instantaneous value of the irradiance, b_d is the dirt/soil de-rate factor, b_m is the PV mismatch de-rate factor , b_c is the DC cable wiring de-rate factor, b_{inv} is the inverter plus transformer de-rate factor. The simulation software used to investigate this attack uses the instantaneous values of GHI, ambient temperature profiles, inverter, and PV efficiency-temperature de-rating factors in order to estimate the PV generation. A one-minute resolution (GHI) and temperature profile (as shown in Figs. 3 and 4) used for the PVs are actual data from the data acquisition system of a 1.4 MW PV plant located on the engineering Campus of Florida International University.

The GHI and temperature were acquired on the February 24, 2019 which was a typical cloudy day in Miami. Based on the method proposed by authors of Sundararajan et al. (2020), some of the missing data from measurements taken by the site data acquisition system were extrapolated. The production meter of the PV

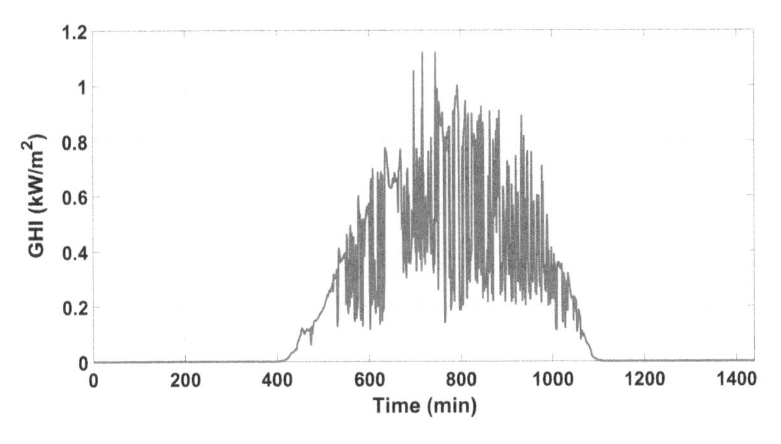

Fig. 3 One-minute resolution global horizontal irradiance profile of the location used for the simulation

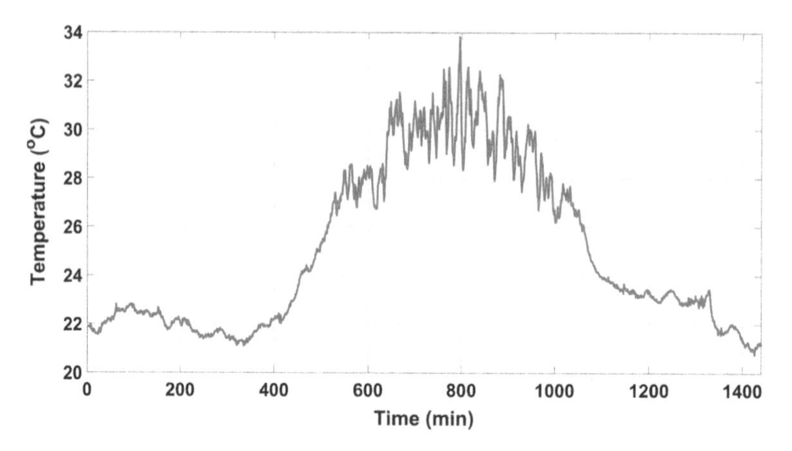

Fig. 4 PV ambient temperature profile

system is located the point of common coupling with the grid which measures the aggregation of the 46 string inverters installed on the PV site. Since the simulation software only takes the GHI measurements, temperature measurements, inverter, and PV efficiency-temperature de-rating factors as input, it is therefore imperative to verify the accuracy of the PV production generated by the software during the simulation with the actual data acquired from the production meter by the data acquisition system. The production meter values were correlated with the output if the simulation software for to verify the accuracy of the PV generation being simulated using the software. The correlation results shows a high level of accuracy between the actual PV generation and the values estimated by the software used for the simulation. For the power generation from the three PVs used in the simulation (200 kW, 2.05 MW, and 200 kW), the same weather parameters were used. Since the expression in (1) depends on the DC name plate rating of the PV, the individual power generation depends (or is directly proportional) to their respective name plate capacity.

3.3 Attack Scenario Construction

In order to implement the FDI attack in the production meter of the PV plant (PV 2), the power production data for a time window of 10 and 30 min is considered. The time resolution of the power generation data is one-minute. The attacks were synthetically generated by introducing error signals to the production estimations at time t. The attack can be modeled as expressed in Eq. 4, where P_{attack} and P_{actual} represent the tampered production measurements and real production measurements, respectively, for the attack scenario considered. $e(t)$is also a time series data which gives rise to the injected false data by the attacker.

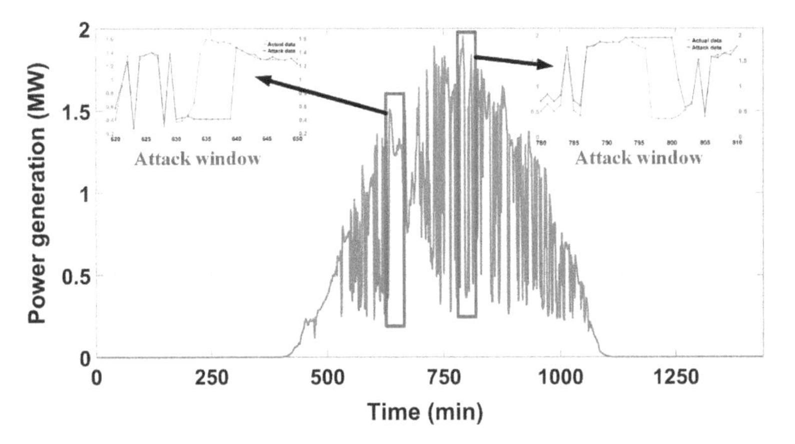

Fig. 5 Power generation profile showing the attack window

$$P_{\text{attack}}(t) = P_{\text{actual}}(t) + e(t) \tag{4}$$
$$P_{\text{attack}}\{P_1\},\ P_{\text{actual}}\{p_1\},\ e\{e_1\}$$

The FDI attack emulated within the time window between 620 and 650 min shows an error of 1.2 MW in the production meter measurement between 630 and 640 min. This drop as seen by the distribution command and control center will necessitate the need to ramp up the synchronous generator connected to bus 816 by the same amount. This will potentially cause excess generation in the network. Conversely, with time window of 780 and 810, the erroneous production meter measurements show an error of approximately 1.5 MW between 795 and 801 min. Consequently, this will lead to ramping down of the synchronous generator in order to maintain the stability of the system. The introduction of error values in the production meter measurement is as shown in Fig. 5. The impact of this attack on the power system network is analyzed and discussed in Sect. 3.4.

3.4 Simulation Results

Following the attack as described in Sect. 3.3, the synchronous generator was ramped up between 630 and 640 and ramped down 795 and 801 based on the false data received from the production meter of $PV2$. The nodal voltages of some buses in the network is as shown in Figs. 6 and 7. The result shows that there is a significant impact of the generator ramping on the voltage profile on almost all the buses in the network. Buses with close proximity to the synchronous generator were the most affected. For example, the voltage profile of nodes 850, 812, and 806 shows some spikes in their node voltages during the attack periods. The nodal voltage during the

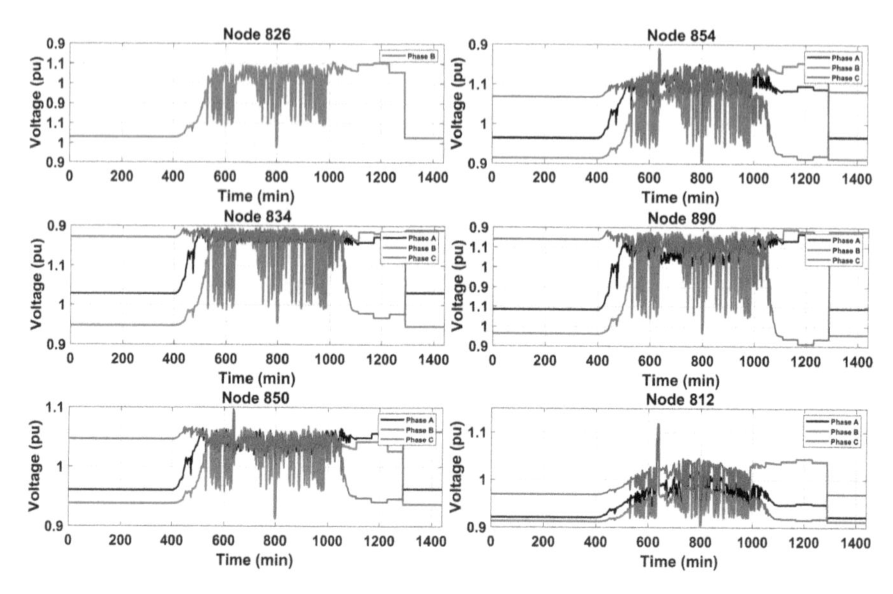

Fig. 6 Nodal voltage profile before, during, and after the attack

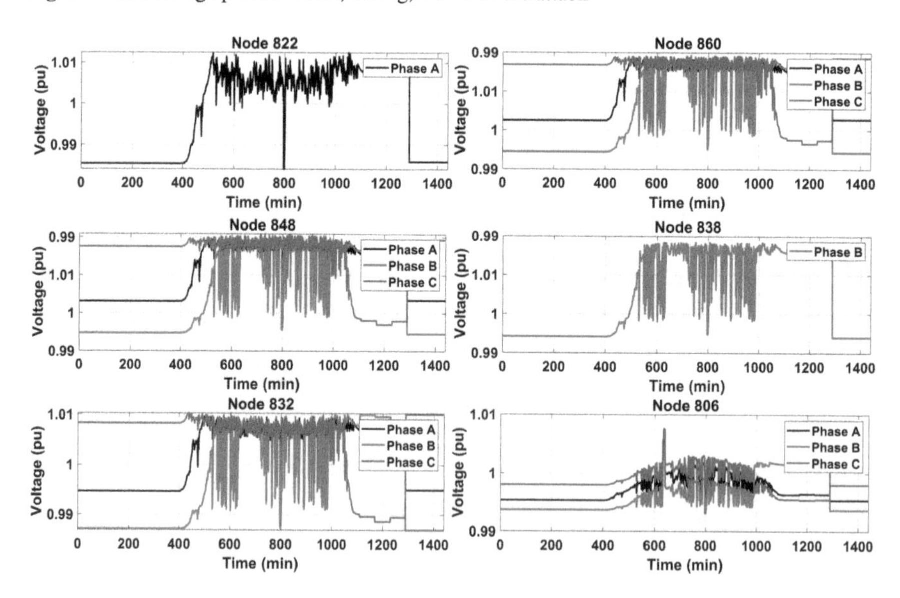

Fig. 7 Nodal voltage profile during before, during, and after the attack

first attack period went to 1.12 pu which is beyond the 1.05 pu threshold. This is an indication of a voltage collapse. For nodes close to the PVs, the fluctuations in the PV generation cause a severe variation in the nodal voltages.

It is worthy of note that the SI settings of the PVs is unity power factor. This means that the PVs does not carry out any voltage regulation in the network. For example, using SI settings of Volt-VAR will allow the PVs inject/absorb some reactive power which could potentially allow some voltage regulation in some of the nodes especially those closer to the location of the PVs. The impact of the BESS system can been seen in the nodal voltages at periods (beyond 1001 min time stamp) when the PV generation is no longer available. The current profile across some of the branches in the network is as shown in Figs. 8 and 9. Similar to the impact of the attack on the nodal voltage profiles, the branches close to the synchronous generator were the most impacted by the attack. Branches 810–808, 822–820, and 802–806 show significant spikes in their current profile. In practice, this could lead to erroneous tripping of the over-current relays and cause instability in the system. For branches close to the PVs, their current profiles is significantly impacted by the current injection by the PVs as seen in branches 832–852, 864–858, and 834–860. Branch 854–856 where the BESS is connected shows its current injection based on the charge and discharge profile attached to the BESS. This current injection allowed the some of the nodal voltages to be stabilized when the power generation from the PV ramps to zero (Fig. 10).

The total system's loss (which includes the line line loss and the transformer loss) is as shown in Fig. 11. The total network loss is also impacted by the attack. As it can be seen from Fig. 11, the second attack window caused a sharp increase in the network losses during this period which significantly affects the overall efficiency of the system

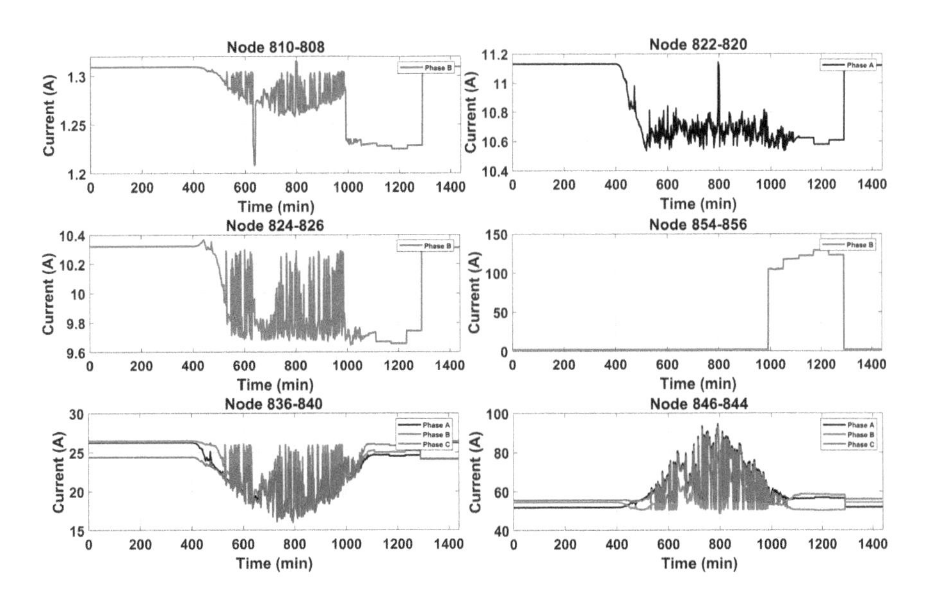

Fig. 8 Branch current profile before, during, and after the attack

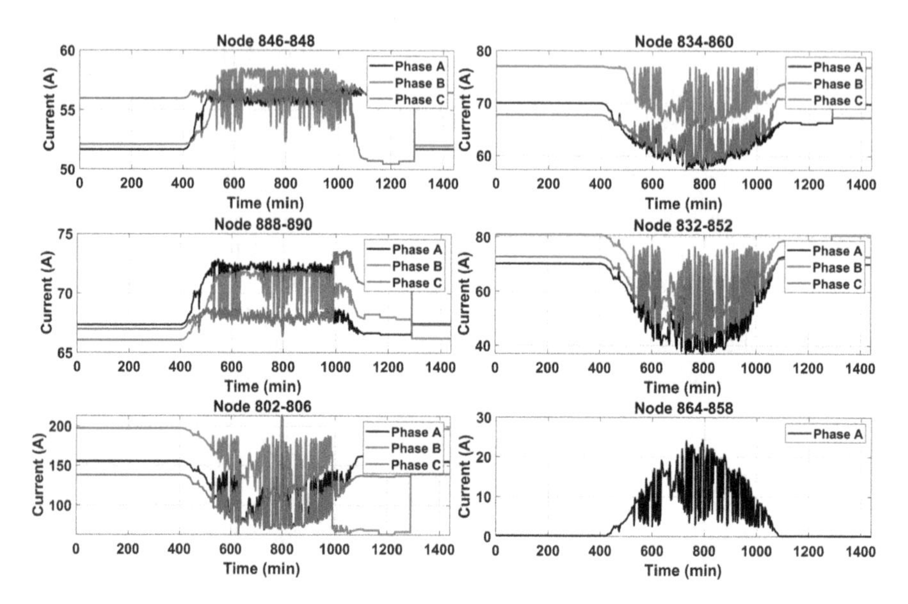

Fig. 9 Branch current profile before, during, and after the attack

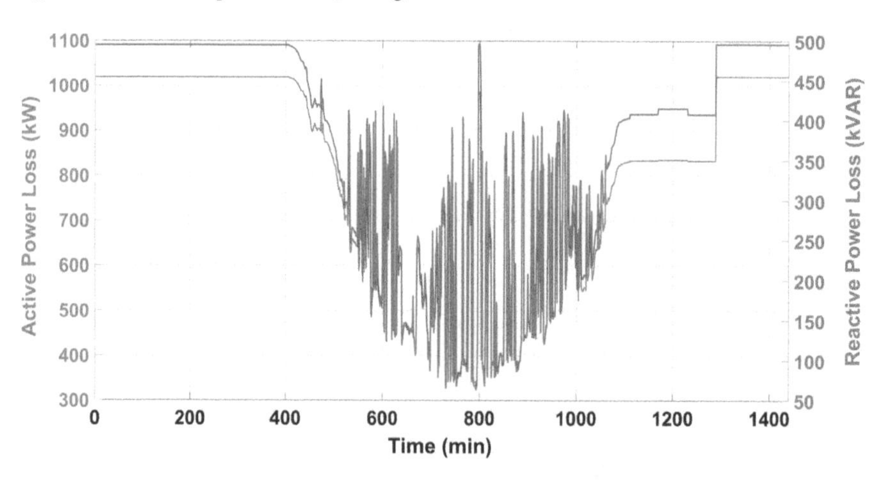

Fig. 10 Active and reactive power loss profile before, during, and after the attack

4 Machine Learning-Based Adaptive Protection Scheme

Figure 11 shows a proposed solution (under development) for smart grid protection with high PV penetration. This solution provides a multi-layer protection system using an machine learning-based protection device for isolation of fault currents from one cluster of PV systems to another. Each PV with SI is controlled by a PV control hub with an integrated WDAS. The weather forecast subsequent prediction

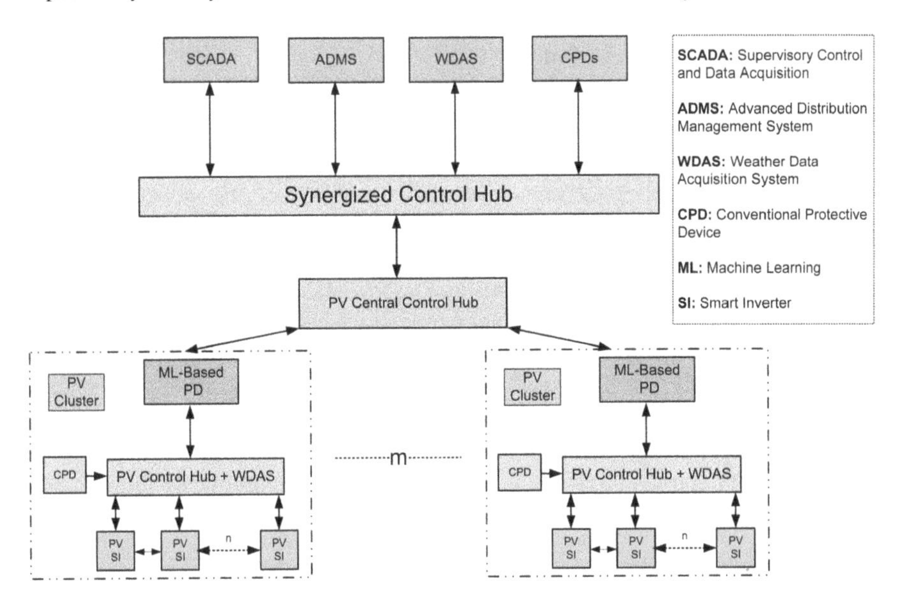

Fig. 11 Proposed machine learning adaptive protection scheme

of energy profile generation allows the PV control hub to determine the right settings for the PV SIs.

The settings of this SIs are dynamically controlled from the PV control hub. Within each PV cluster, the CPD continually sends in real-time current, voltage, and frequency parameters of the system to the PV control hub. This data is fed into the ML-based PD. The historical data of the network under normal and fault conditions are used to develop a classification model which is programmed into the ML-based PD. The fault parameters of the system are derided from extensive simulations carried out using the ADMS which allows various types of faults (such as single-to-ground, line-to-line-, line-to-line-to-ground, line-to-line-to-line, line-to-line-to-line-to-ground, and open circuit faults) to be simulated and their respective fault current values captured. The classification algorithm already programmed into the ML-based PD classifies if the system is system parameters normal, abnormal, or fault condition. Whenever the system is attacked and the conventional generator is wrongly ramped up, the new system parameters measured by the CPD will be classified as an abnormal and the ML-based PD disconnects this cluster of PV from the others. This would prevent other PV clusters from contributing to the abnormal system parameters. The PV central control hub is a wide area control that connects to the synergized control hub. The synergized control hub is located at the substation. This control hub is integrated with the SCADA, ADMS, and WDAS and controls the CPD at the substation level. The SCADA integrated ADMS has the fault location isolation and service restoration algorithm. This helps the systems to quickly located the abnormal section of the network and isolate it while restoring power to the normal section of the system as soon as possible. This is part of the self-healing process of the smart grid.

5 Conclusions and Future Work

With obvious increase in DER integration and the drive toward achieving smart distribution systems, there is an increase possibility of cyber attacks. As DERs continues to form a network with communications and control layers, the vulnerability and susceptibility to attacks consequently increase. This implies that smart grid systems can be simply regarded as a cyber physical network. This chapter presents a comprehensive review of vulnerabilities in smart grids and the impacts of cyber attacks. This chapter presented real-world case studies of successful attack on on multiple grid assets, including networks with high-penetration of distributed energy resources (DERs), and their impacts on the system. A specific case study of one of the prevalent attacks called FDI is presented. A real-world scenario of an FDI attack was done using an IEEE 34 bus system with three PVs, one synchronous generator, and one energy storage. A false command was sent to the synchronous generator based on false data received from the production meter by the command and control center. The eventual ramping up and down to dispatch the deficit or surplus of power from the PV lead to some severe impacts on the system's grid voltages, currents, and total system's power loss. The nodal voltage shows some voltage collapse with values going beyond and also below the $0.95 - 1.05pu$ thresholds . The current values also significantly increased in some branches in the network. The system power loss was also impacted by this attacks. This abnormal system parameters could potentially lead to erroneous tripping of the protective devices which will cause cascading failures and possible system collapse. A machine learning-based protection system was also proposed in this chapter which can be effective way of dealing wit FDI attacks by comparing the new system dynamic parameters using a classification model developed from historical data and fault simulation studies. The proposed holistic protection framework can help prevent a total system collapse during an FDI attack on grid assets especially at high DER penetration scenarios.

References

Ameli A, Hooshyar A, El-Saadany EF, Youssef AM (2020) An intrusion detection method for line current differential relays. IEEE Trans Inf Forens Secu 15:329–344

Chen TM, Sanchez-Aarnoutse JC, Buford J (2011) Petri net modeling of cyber-physical attacks on smart grid. IEEE Trans Smart Grid 2(4):741–749

Conteh NY, Schmick PJ (2016) Cybersecurity: risks, vulnerabilities and countermeasures to prevent social engineering attacks. Int J Adv Comput Res 6(23):31

Dagle JE (2012) Cyber-physical system security of smart grids. In: IEEE PES innovative smart grid technologies (ISGT) 2012, pp 1–2

Debnath A, Olowu TO, Parvez I, Dastgir MG, Sarwat A (2020) A novel module independent straight line-based fast maximum power point tracking algorithm for photovoltaic systems. Energies 13(12):3233

Dharmasena S, Choi S (2019) Model predictive control of five-phase permanent magnet assisted synchronous reluctance motor. In: IEEE Applied power electronics conference and exposition (APEC), 2019, pp 1885–1890

Dharmasena S, Olowu TO, Sarwat AI (2019) Bidirectional ac/dc converter topologies: a review. In. SoutheastCon 2019, pp 1–5
Eder-Neuhauser P, Zseby T, Fabini J, Vormayr G (2017) Cyber attack models for smart grid environments. Sustain Energy, Grids Networks 12:10–29
El Mrabet Z, Kaabouch N, El Ghazi H, El Ghazi H (2018) Cyber-security in smart grid: survey and challenges. Comput Electr Eng 67:469–482
Fard AY, Easley M, Amariucai GT, Shadmand MB, Abu-Rub H (2019) Cybersecurity analytics using smart inverters in power distribution system: Proactive intrusion detection and corrective control framework. In: IEEE International symposium on technologies for homeland security (HST). IEEE 2019, pp 1–6
Farraj A, Hammad E, Daoud AA, Kundur D (2016) A game-theoretic analysis of cyber switching attacks and mitigation in smart grid systems. IEEE Trans Smart Grid 7(4):1846–1855
Gunes V, Peter S, Givargis T, Vahid F (2014) A survey on concepts, applications, and challenges in cyber-physical systems. KSII Trans Internet Inf Syst 8(12):
Haack JN, Fink GA, Maiden WM, McKinnon D, Fulp EW (2009) Mixed-initiative cyber security: putting humans in the right loop. In: The first international workshop on mixed-initiative multiagent systems (MIMS) at AAMAS
Hawrylak PJ, Haney M, Papa M, Hale J (2012) Using hybrid attack graphs to model cyber-physical attacks in the smart grid. In: 2012 5th international symposium on resilient control systems, pp 161–164
He H, Yan J (2016) Cyber-physical attacks and defences in the smart grid: a survey. IET Cyber-Phys Syst: Theo Appl 1(1):13–27
He Y, Mendis GJ, Wei J (2017) Real-time detection of false data injection attacks in smart grid: a deep learning-based intelligent mechanism. IEEE Trans Smart Grid 8(5):2505–2516
Hu B, Zhou C, Tian Y, Qin Y, Junping X (2019) A collaborative intrusion detection approach using blockchain for multimicrogrid systems. IEEE Trans Syst Man, and Cybern Syst 49(8):1720–1730
Jafari M, Olowu TO, Sarwat AI (2018) Optimal smart inverters volt-var curve selection with a multi-objective volt-var optimization using evolutionary algorithm approach. In: North American Power Symposium (NAPS) 2018, pp 1–6
Kushner D (2013) The real story of stuxnet. IEEE Spect 3(50):48–53
Li X, Liang X, Lu R, Shen X, Lin X, Zhu H (2012) Securing smart grid: cyber attacks, countermeasures, and challenges. IEEE Commun Magaz 50(8):38–45
McLaughlin SE, Podkuiko D, Delozier A, Miadzvezhanka S, McDaniel PD (2010) Embedded firmware diversity for smart electric meters. In: HotSec
Mekonnen Y, Haque M, Parvez I, Moghadasi A, Sarwat A (2018) Lte and wifi coexistence in unlicensed spectrum with application to smart grid: a review. In: IEEE/PES Transmission and Distribution Conference and Exposition (T D) pp 1–5
Mo Y, Kim TH-J, Brancik K, Dickinson D, Lee H, Perrig A, Sinopoli B (2011) Cyber-physical security of a smart grid infrastructure. Proceedings of the IEEE 100(1):195–209
Odeyomi O, Kwon HM, Murrell DA (2020) Time-varying truth prediction in social networks using online learning. In: 2020 International Conference on Computing, Networking and Communications (ICNC), pp 1–5
Olowu TO, Jafari M, Sarwat AI (2018) A multi-objective optimization technique for volt-var control with high pv penetration using genetic algorithm. In: North American power symposium (NAPS) 2018, pp 1–6
Olowu TO, Sundararajan A, Moghaddami M, Sarwat A (2019) Fleet aggregation of photovoltaic systems: a survey and case study. In: 2019 IEEE Power Energy Society Innovative Smart Grid Technologies Conference (ISGT)
Olowu T, Sundararajan A, Moghaddami M, Sarwat A (2018) Future challenges and mitigation methods for high photovoltaic penetration: a survey. Energies 11(7):1782
Olowu TO, Jafari H, Moghaddami M, Sarwat AI (2019) Physics-based design optimization of high frequency transformers for solid state transformer applications. IEEE Ind Appl Soc Ann Meet 2019:1–6

Olowu T, Jafari H, Dharmasena S, Sarwat AI (2019) Photovoltaic fleet aggregation and high penetration: a feeder test case. SoutheastCon 2019:1–6
Olowu TO, Jafari M, Sarwat A (2019) A multi-objective voltage optimization technique in distribution feeders with high photovoltaic penetration. Adv Sci Tech Eng Syst J 4(6):377–385
Ozay M, Esnaola I, Yarman Vural FT, Kulkarni SR, Poor HV (2016) Machine learning methods for attack detection in the smart grid. IEEE Trans Neu Networks Learn Syst 27(8):1773–1786
Ozgur U, Nair HT, Sundararajan A, Akkaya K, Sarwat AI (2017) An efficient mqtt framework for control and protection of networked cyber-physical systems. In: IEEE Conference on Communications and Network Security (CNS) 2017, pp 421–426
Parvez I, Islam N, Rupasinghe N, Sarwat AI, Güvenç I (2016) LAA-based LTE and Zigbee coexistence for unlicensed-band smart grid communications. SoutheastCon 2016:1–6
Parvez I, Sarwat AI, Pinto J, Parvez Z, Khandaker MA (2017) A gossip algorithm based clock synchronization scheme for smart grid applications. In: North American Power Symposium (NAPS) 2017, pp 1–6
Qi J, Hahn A, Lu X, Wang J, Liu C-C (2016) Cybersecurity for distributed energy resources and smart inverters. IET Cyber-Phys Syst Theo Appl 1(1):28–39
Rahman S, Moghaddami M, Sarwat AI, Olowu T, Jafaritalarposhti M (2018) Flicker estimation associated with pv integrated distribution network. SoutheastCon 2018:1–6
Saleem D, Sundararajan A, Sanghvi A, Rivera J, Sarwat AI, Kroposki B (2019) A multidimensional holistic framework for the security of distributed energy and control systems. IEEE Syst J 1–11
Sarwat AI, Sundararajan A, Parvez I (2017) Trends and future directions of research for smart grid iot sensor networks. In: International symposium on sensor networks, systems and security. Springer, pp 45–61
Sharing EI, Center A (2016) Analysis of the cyber attack on the Ukrainian power grid: defense use case pp 1–5
Smart grid 2019. [Online]. Available: https://www.nist.gov/el/smart-grid
Srikantha P, Kundur D (2016) A der attack-mitigation differential game for smart grid security analysis. IEEE Trans Smart Grid 7(3):1476–1485
Srivastava A, Morris T, Ernster T, Vellaithurai C, Pan S, Adhikari U (2013) Modeling cyber-physical vulnerability of the smart grid with incomplete information. IEEE Trans Smart Grid 4(1):235–244
Stefanov A, Liu C (2012) Cyber-power system security in a smart grid environment. In: IEEE PES Innovative Smart Grid Technologies (ISGT) 2012, pp 1–3
Sundararajan A, Chavan A, Saleem D, Sarwat AI (2018) A survey of protocol-level challenges and solutions for distributed energy resource cyber-physical security. MDPI Energ 9:2360
Sundararajan A, Sarwat AI, Pons A (2019) A survey on modality characteristics, performance evaluation metrics, and security for traditional and wearable biometric systems. ACM Comput Surv 52(2):1–35
Sundararajan A, Hernandez AS, Sarwat A (2020) Adapting big data standards, maturity models to smart grid distributed generation: critical review. IET Smart Grid (2020)
Sundararajan A, Khan T, Moghadasi A, Sarwat AI (2018) Survey on synchrophasor data quality and cybersecurity challenges, and evaluation of their interdependencies. J Mod Power Syst Clean Energy 1–19
Sundararajan A, Olowu TO, Wei L, Rahman S, Sarwat AI (2019) Case study on the effects of partial solar eclipse on distributed pv systems and management areas. IET Smart Grid (2019)
Teymouri A, Mehrizi-Sani A, Liu C-C (2018) Cyber security risk assessment of solar pv units with reactive power capability. In: IECON 2018–44th Annual Conference of the IEEE Industrial Electronics Society. IEEE, pp 2872–2877
Wadhawan Y, Neuman C, AlMajali A (2017) Analyzing cyber-physical attacks on smart grid systems. In: Workshop on modeling and simulation of cyber-physical energy systems (MSCPES) 2017, pp 1–6
Wang Y, Amin MM, Fu J, Moussa HB (2017) A novel data analytical approach for false data injection cyber-physical attack mitigation in smart grids. IEEE Access 5, 26-022–26-033

Wei J, Kundur D, Zourntos T, Butler-Purry KL (2014) A flocking-based paradigm for hierarchical cyber-physical smart grid modeling and control. IEEE Trans Smart Grid 5(6):2687–2700
Wei L, Sundararajan A, Sarwat AI, Biswas S, Ibrahim E (2017) A distributed intelligent framework for electricity theft detection using Benford's law and Stackelberg game. Resilience week (RWS) 2017:5–11
Wei L, Gao D, Luo C (2018) False data injection attacks detection with deep belief networks in smart grid. In: Chinese Automation Congress (CAC) 2018, pp 2621–2625
Zhaoyang D (2014) Smart grid cyber security. In: 2014 13th International Conference on Control Automation Robotics Vision (ICARCV), pp 1–2 (2014)

Power Generation Using Solar Energy

A Perspective on Perovskite Solar Cells

Saikat Bhaumik, Sudip K. Saha, and Arup K. Rath

1 Introduction

1.1 Global Energy Crisis

Global energy demand rises by 1.3% each year, and the empirical evidence points toward a steady increase in energy consumption until the year 2040 (International Energy Agency 2019). Almost one billion of the global population still do not have access to electricity, highlighting the need for additional energy. Electricity drives the modern civilization, and energy security is paramount to the sustainable development and prosperity of the human civilization. The over-reliance on fossil fuel to meet the growing energy demand has contributed in a major way to aggravate the current energy problem. Rapid consumption of fossil fuel severely depleted the global energy reserves and their combustion destroying our environment by pouring the harmful greenhouse gasses into it. To contain the global average temperature to increase below 2 °C rapid cut in greenhouse gasses has been suggested by multiple scientific and environmental forums.

S. Bhaumik (✉)
Institute of Chemical Technology-IndianOil Odisha Campus, Gajapati Nagar, Bhubaneswar 751013, India
e-mail: s.bhaumik@iocb.ictmumbai.edu.in

S. K. Saha
Diamond Harbour Women's University, South 24 Paraganas, Diamond Harbour, West Bengal 743368, India
e-mail: sudipsaha.dhwu@gmail.com

A. K. Rath
CSIR-National Chemical Laboratory, Pune 411008, India
e-mail: ak.rath@ncl.res.in

H. Tyagi et al. (eds.), *New Research Directions in Solar Energy Technologies*,
Energy, Environment, and Sustainability,
https://doi.org/10.1007/978-981-16-0594-9_4

1.2 Renewable Clean Energy Alternatives

The generation of energy from renewable energy sources with a minimum carbon footprint could be a possible way forward to the present energy crisis. Several renewable energy technologies have been explored over the years, among which most common technologies with large-scale production capabilities are wind, hydro, geothermal, biomass, and solar. These energy resources are intermittent, in terms of geographical positioning and duration of availability of the energy sources during a day; however, they can be complimentary to develop smart electricity grid to meet all the energy needs. For example, wind speed varies during a day, or the hydropower resource may not be available in all the places; similarly, solar light is inaccessible during the night time. A smart combination of renewable energy sources together with storage capacity (batteries, fuel production, and pumped-storage hydro) would meet most of the energy demands of our daily life.

Among all renewable energy sources, solar energy is particularly interesting as it provides an inexhaustible and universal source of energy. Annually earth receives around 1×10^9 TWh of solar energy with a typical intensity of around 1000 W/m^2 on the ground. Cumulative global annual energy consumption is merely 0.012% (124,290 TWh) (Morton 2006; https://www.iea.org/reports/world-energy-outlook-2019) of the annual solar irradiance. This is to say that the sun provides the earth with as much energy every hour as human civilization consumes every year. Though solar energy is ubiquitous, the conversion of solar energy to useful electrical energy is not as economical when compared to fossil fuel-based alternatives.

1.3 Solar Cell Technology

A solar cell device absorbs the incident light and converts it to the usable electrical power, known as the photovoltaic effect, discovered by French physicist Edmond Becquerel in 1839. Typically, a semiconductor of appropriate band-gap (E_G) is used in solar cells to absorb solar radiation. Photons having energies higher than E_G are absorbed by the semiconductor to create excited electrons in the conduction band and vacancy of electrons in the valance band. These negative and positive charge carriers are separated and extracted from the solar cells to achieve electrical power. Carrier selective contacts, on either side of the absorber layer, are used for the preferential collection of electron and hole at the terminal electrodes of a solar cell. Solar radiation is a panchromatic one and extends within the broad energy range of 3.5–0.5 eV. The wide bandwidth of the solar spectrum makes it challenging to harness in the solar cell devices. Semiconductor having a specific band-gap E_G is transparent to the photons having energy lower than E_G, which is accounted for transmittance loss in solar cells. Similarly, absorption of high energy photons ($>E_G$) excites the valance band electrons to deep inside the conduction band. The high energy photon releases their excess energy quickly (~10^{-12} s) through the emission of phonon to reach the conduction band minimum. The loss of the excess energy for the high energy photons (>E_G) is known as thermalization loss for the solar cell. Additional losses in a solar cell are emission loss due to spontaneous emission of the photoexcited electron from the conduction band to valance band (Hirst and Ekins-Daukes 2011). Those losses are qualified as intrinsic loss, as they cannot be overcome by device and material optimization in the high-performing solar cells. The fundamental efficiency limit for single-junction solar cells is limited to 33.7%, formulated in 1961 by Shockley and Queisser (Shockley and Queisser 1961). The first actual solar cell was developed in 1954 at Bell laboratories using the silicon semiconductor to show the photoconversion efficiency of 6%. Over the last 60 years, research and development have pushed the PCE of silicon solar cells to record 26.7% (Yoshikawa et al. 2017), very close to its theoretical limit of 29% (Andreani et al. 2018). The silicon solar cell technology is already a proven technology with more than 90% accumulated share in the PV market (Andreani et al. 2018).

1.4 *The Emergence of Perovskite Solar Cells (PSCs)*

In last ten years, the organic–inorganic hybrid perovskite solar cells (PSCs) have emerged as a potential alternative to the existing photovoltaic technologies as their efficiency has improved from 3.8% in 2009 to 25.1% in 2019 as shown in Fig. 1 (Kojima et al. 2009; NREL solar energy chart: https://www.nrel.gov/pv/assets/pdfs/pv-efficiency-chart 2019). The unprecedented growth of the PSCs is associated with the fact that the perovskite materials can be synthesized from low-cost solution

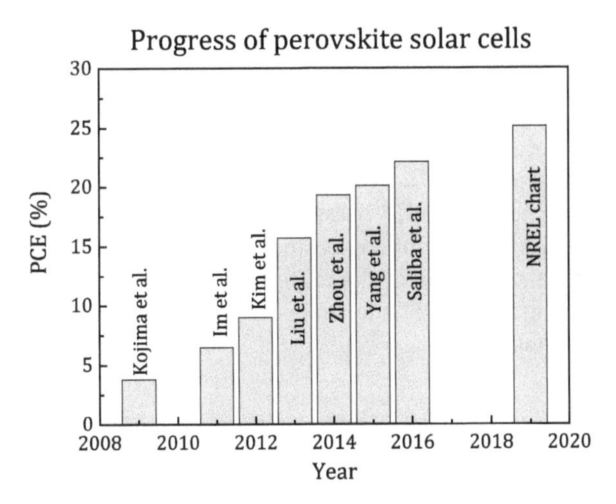

Fig. 1 Graphical representation of progress of PSCs (Kim et al. 2012; Yang et al. 2015a; Kojima et al. 2009; NREL solar energy chart: https://www.nrel.gov/pv/assets/pdfs/pv-efficiency-chart 2019; Im et al. 2013; Zhou et al. 2014b; Liu and Kelly 2014; Saliba et al. 2016a)

processing and can be made crystalline below 200 °C, promising for the cost competitiveness of the disruptive solar cell technology (Green et al. 2014; Zuo et al. 2016). The band-gap of the perovskite semiconductors can be tunable for a range of 2.2–1.2 eV through composition engineering. Hybrid perovskite systems possess the direct band-gap with the strong absorption coefficient (>10^4 cm^{-1}), which requires only 1-μm-thick perovskite layer to absorb the full solar radiation above their band-gap. Long-range crystallinity of solution-processed perovskite layer contributes significantly to achieve high carrier mobility (1–30 cm^2 V^{-1} S^{-1}) and long carrier lifetime (~100 ns) in their solid films (Johnston and Herz 2016). The carrier diffusion length in PSCs exceeds 10 μm as a result (Tainter et al. 2019), allowing efficient extraction of photogenerated carriers in solar cells. The aforementioned traits make hybrid perovskites unique for the solar cells (Grätzel 2014) as well as other optoelectronic applications, like light-emitting diodes (Tan et al. 2014; Zhang et al. 2017a), lasers (Chen et al. 2016a; Stylianakis et al. 2019), and photodetectors, (Hu et al. 2014a; Wang and Kim 2017).

Despite several advancements, there are concerns over the stability of the perovskite materials (Wang et al. 2019), which hinders their prospect for commercialization. The organic–inorganic hybrid perovskite materials suffer from poor stability when exposed to heat, oxygen, moisture, and even illumination (Lee et al. 2015a; Smecca et al. 2016; Aristidou et al. 2015). The instability of perovskite materials lies with the fact that the perovskite crystals are ionic, and there exits significant empty space within the perovskite unit cells which makes them soft crystals and the volatile nature of the organic component in the perovskite crystal. Significant progress has been made to improve the stability of perovskite semiconductor through compositional engineering to minimize the crystal strain, tuning the unit cell toward cubic structure, and replacement of the volatile component (Wang et al. 2019; Asghar et al. 2017). Development of the two-dimensionally confined perovskite layer, interlinked by long-chain organic molecules has been investigated to reduce the moisture induce degradation of the perovskite layer (Grancini et al. 2017; Tsai et al. 2016). However, the special confinement and insulating ligands have a detrimental effect on charge transport. Nevertheless, it is an interesting approach to improve the stability of perovskite semiconductors.

In this book chapter, a detailed overview of PSCs will be discussed. Progress in synthesis strategies in terms of composition engineering and structure–property correlation to attain the high photovoltaic efficiency will be explained. The evolution of device engineering for the PSCs to attain high photovoltaic efficiency and better stability will be summarized.

2 Intrinsic Properties of Perovskites

2.1 *Structural Properties*

Metal halide perovskites are known as the common cluster of compounds with general chemical formula of AMX_3, where A is organic or inorganic cations (typically $MA = CH_3NH_3^+$, $FA = HC(NH_2)_2^+$, Cs^+, K^+, Rb^+, etc.), M is metal cations (Pb^{2+}, Sn^{2+}, Eu^{2+}, Ge^+, etc.), and X is halide anions (Cl^-, Br^-, I^-) (Kojima et al. 2009; NREL solar energy chart: https://www.nrel.gov/pv/assets/pdfs/pv-efficiency-chart 2019; Sum and Mathews 2014). In a perovskite unit cell, A cation is situated at the eight corners of the cube, while M-cation is placed at the body center and six numbers of X anions are located at the face centers. AMX_3 belongs to an extended large family of organic–inorganic metal halide perovskites where the $[MX_6]^{4-}$ octahedra can grow three-dimensional (3D), two-dimensional (2D), one-dimensional (1D), or zero-dimensional (0D) crystal structures having the same unit cell as shown in Fig. 2. For example, in $MAPbI_3$ each $[PbI_6]^{4-}$ octahedra is connected with six

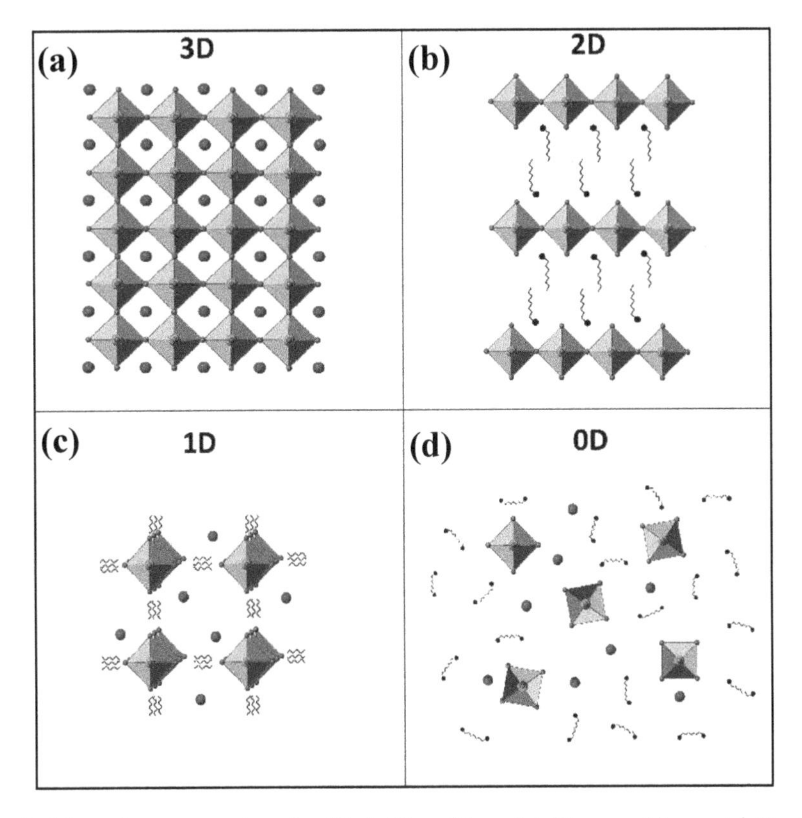

Fig. 2 Schematic representation of **a** 3D, **b** 2D, **c** 1D, and **d** 0D perovskite crystal structure, respectively

neighbors of iodide forming a 3D network while MA^+ is located at the void of the network (see Fig. 2a). In other words, the $[PbI_6]^{4-}$ octahedras are connected three dimensionally in the crystal structure. For the 2D case (see Fig. 2b), longer organic cations like $CH_3(CH_2)_nNH_3^+$ situated at A-site where each $[PbI_6]^{4-}$ octahedron is connected with four neighbors iodide anions, forming a 2D network layer that is sandwiched between two $CH_3(CH_2)_nNH_3^+$ layers and the chemical formula becomes A_2PbI_4. It results in multiple quantum well structures from the stacking of these sandwiched layers via van der Waals interaction with the $CH_3(CH_2)_nNH_3^+$ layer as the barrier. In 1D case (see Fig. 2c), each octahedra is attached to two opposite corners with neighboring octrahedras and forming parallel infinite chains (e.g., $(C_{10}H_{21}NH_3)_2PbI_4$). Lastly, for 0D structure (e.g., Cs_4PbI_6), each $[PbI_6]^{4-}$ octrahedra is separated by four Cs^+ ions to form an isolated molecule resembles to a quantum dot array (see Fig. 2d).

The lead halide perovskite structure (AMX_3), in which A-site cation, plays a very important role for formation of stable perovskite crystal structure (Park and Seok 2019; Correa-Baena et al. 2017). The stability parameter is characterized the Goldschmidt tolerance factor (t),

$$t = \frac{(r_A + r_X)}{\sqrt{2} \times (r_M + r_X)}$$

where r_A, r_M, and r_X are the ionic radii of the A-site cation, metal cation, and halide, anions, respectively. For an example, the ionic radii of I^- and Pb^{2+} ions are 2.03 Å (r_X) and 1.33 Å (r_M), respectively, as represented in Fig. 3, and the radii of A-site cation is in the range of 2.3–2.8 Å (r_A). When the perovskite is formed using Cs^+, MA^+, or FA^+ as the A-site cations, the optical properties of the perovskites change with the cation result in a red-shifting absorbance onset. In other words, band-gap of the perovskite material changes in this order, MA^+ (1.55 eV) < Cs^+ (1.5 eV) < FA^+ (1.45 eV). However, the volume per $APbI_3$ unit changes from 222, 248, and 256 $Å^3$ for Cs^+, MA^+, and FA^+ cations, respectively. Thus, Cs^+ and MA^+ differ significantly in radial size, but the band-gap changes a little compared to difference between the MA^+ and FA^+ cations.

Halide substitution. The advantage of these metal halide lead halide perovskites is the capability to tune their optoelectronic properties by substitution the halide ions. For an example, the iodine ions in $MAPbI_3$ perovskite structure can be substituted with both Cl^- and Br^- anions (Correa-Baena et al. 2017). Same substitution of the ions can be possible for $MAPbBr_3$ and $MAPbCl_3$ perovskite structures. While halide substitution, the band-gap of the perovskites changes 2.97, 2.24, and 1.53 eV for the $MAPbCl_3$, $MAPbBr_3$, and $MAPbI_3$ perovskite, respectively. At room temperature, $MAPbCl_3$ and $MAPbBr_3$ perovskites are found to be in a cubic structure while the phase changes to a tetragonal structure at lower temperatures. Moreover, $MAPbI_3$ crystallizes to tetragonal crystal structure, whereas $FAPbI_3$ crystallizes to hexagonal δ-phase or cubic α-phase at room temperature as shown in Fig. 4.

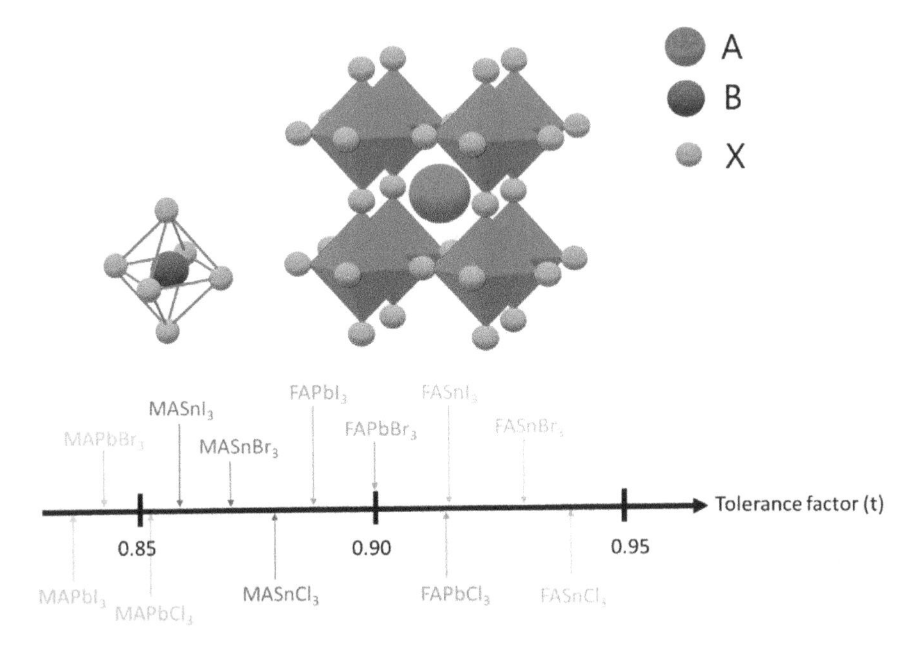

Fig. 3 Tolerance factor of ABX_3 perovskites with the various compositions of A, B, and X-sites

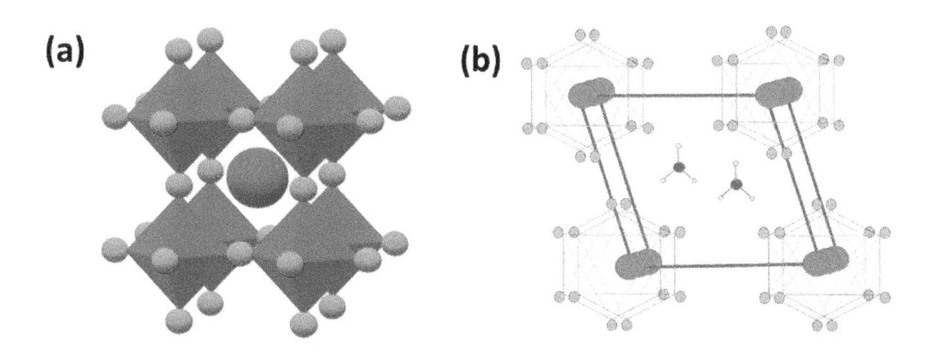

Fig. 4 Crystal structure of **a** cubic α-phase and **b** hexagonal δ-phase of perovskites

Organic cation substitution. Similar to halide substitution, organic cations can also be replaced in the perovskites (Correa-Baena et al. 2017). For an example, MA^+ cations can be exchanged with slightly bigger sized FA^+ cations. The cation exchange in perovskites has very small impact on the optical band-gap or very little change in band-gap observed. DFT computations demonstrate that organic cations do not contribute to the electronic states close to the band edges. However, with cation exchange the crystal lattices changes which results in a slight change in the band-gap.

Organic/inorganic ion mixing. Simultaneous exchange of both organic cations and anions has been done as well. For an example, the performance of $MAPbI_3$ perovskite-based solar cells is not highly efficient (Correa-Baena et al. 2017). On the other hand, $FAPbI_3$ and $CsPbI_3$ perovskite structures (cubic phase) are not stable at room temperature. However, the compositional mixing of MA/FA/Cs/Br/I perovskites has been studied enormously. $FAPbI_3$ perovskite-based photovoltaics appears to give better device performance over $MAPbI_3$, but some MA^+ cations in $FAPbI_3$ perovskites stabilizes the perovskite structure. Introducing Br^- anions in $FAPbI_3$ perovskites allows to tune the band-gap, enhance structural stability, and improve the device performance. Therefore, it is very important to design principle to mix cations and halides to achieve final perovskite compositions that is advantageous while evading their disadvantages.

2.2 *Electronic Structure*

Perovskite is known for the common cluster of compounds with general chemical formula. The electronic band structures of organic–inorganic perovskites can be calculated by using a semiempirical technique based on the extended Huckel theory and an ab initio method based on the Hartree–Fock theory (Sum and Mathews 2014). Another approach is using ultraviolet photoelectron spectroscopy and first principles density functional theory (DFT) band calculations at the room temperature. DFT calculations for 3D $MAPbI_3$ perovskite crystals reveal that the valence band maxima contain Pb 6p–*I* 5p σ antibonding orbital, while the conduction band minima comprise of Pb 6p–*I* 5s σ antibonding and Pb 6p–*I* 5p π antibonding orbitals as represented in Fig. 5. Nevertheless, DFT calculations also show that the A-site cation has a very little influence on the band-gap energy, of which is mainly determined by the $[PbI_4]^{6-}$ network.

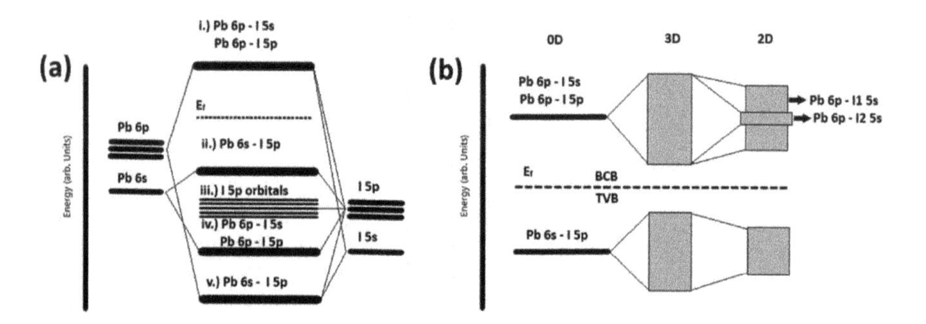

Fig. 5 Bonding energy diagram of **a** $[PbI_4]^{6-}$ cluster, **b** 0D, 3D, 2D band structure of $MAPbI_3$ perovskites

Absorption coefficient. The absorption coefficient of materials is described as the amount of a given color of light is absorbed by the material for a given thickness. Thats means that more light absorbs by a material then its absorption coefficient will be higher. The absorption coefficient is represented by the Greek letter "α". It has units of cm^{-1} because it defines the amount of light absorbed per unit thickness of the material. Since the material absorbance varies with the wavelength of the light, so the absorption coefficient is a function of wavelength/color. For an example, the absorption coefficient of $MAPbI_3$ thin films is around 1.5×10^4 cm^{-1} at 550 nm, that gives the penetration depth is only 0.66 mm for 550 nm light. However, for 700 nm light, the absorption coefficient of $MAPbI_3$ thin films is around 0.5×10^4 cm^{-1} and corresponding penetration depth is around 2 mm (Park 2015). Higher penetration depths led to more incoming light can be absorbed by the perovskite films, which is essential for high-efficiency PSCs.

Balanced charge transport behaviors. The charge transport properties of $MAPbX_3$ perovskites were reported by Xing et al. (2013) and Stranks et al. (2013). Transient spectroscopic analysis reveals that upon absorbing the light perovskites exhibited balanced electron- and hole-transporting behavior. The calculated electron diffusion length for $MAPbI_3$ thin film is around 130 nm while the hole diffusion length is calculated to 100 nm (Xing et al. 2013). However, by doping Cl^- ions in $MAPbI_3$ perovskite, the electron and hole diffusion length enhanced to 1069 nm and 1213 nm, respectively (Stranks et al. 2013). The longer and balanced charge diffusion lengths results in improved solar cells device performance.

3 Perovskite Structure Formation Techniques

3.1 Single Crystals (SCs)

Solution temperature-lowering (STL) method. In this method, the solubility of the lead halide perovskites in acid halide solvents (e.g., HI, HBr, HCl) plays an important role for perovskite crystal growth. The perovskite materials solubility changes significantly with temperature. This mechanism is generally used for perovskite SCs growth. At first perovskite, seed crystals are dipped into an acid halide solvent at certain temperature (see Fig. 6a). Upon lowering the temperature, the saturation of the solute in the solvent takes place and corresponding crystal growth start around the perovskite seed crystals. High-quality $MAPbI_3$ SCs can be grown by this temperature-lowering process. In a glass beaker, $MAPbI_3$ seed crystals are spanned by a stirrer in HI solvent at 65 °C (Dang et al. 2015). By lowering the temperature to 40 °C, saturation of perovskites solute in the HI solvent expedite the crystal formation and finally 10 mm × 10 mm × 8 mm-sized $MAPbI_3$ SCs. These as-synthesized SCs exhibit two natural facets in the directions of (100) and (112) crystal planes. The advantages of this method are that the crystal growths are easily controlled with temperature and high-quality large-size SCs can be obtained.

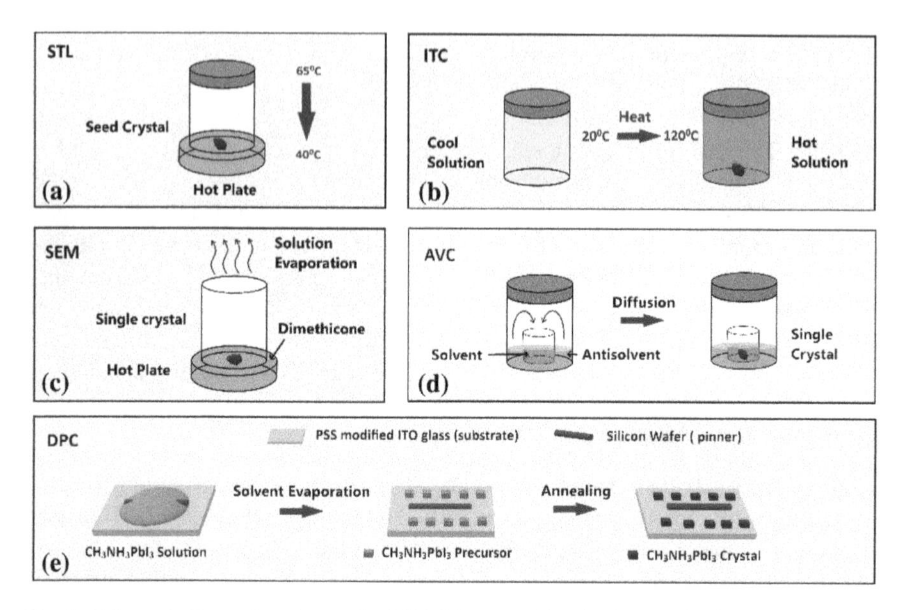

Fig. 6 Schematic illustration of growth of $MAPbI_3$ perovskite single crystals via various synthetic methods as **a** STL, **b** ITC, **c** AVC, **d** SEM, and **e** DPC method, respectively

Inverse temperature crystallization (ITC) method. The ITC crystal growth mechanism is totally opposite to the temperature-lowering method. The perovskite materials whose solubility in a particular solvent are decreasing with increasing the temperature. Several research groups investigated the lead halide perovskites solubility in *N*, *N*-dimethylformamide (DMF), dimethylsulfoxide (DMSO), and γ-butyrolactone (GBL). Interestingly, in these solvents the perovskite crystal structure formation was observed with increase of solution temperature. By this ITC method, mm-sized $MAPbX_3$ and $FAPbX_3$ ($X = Cl^-$, Br^-, I^-) SCs were obtained via using different organic solvents (Saidaminov et al. 2015; Liu et al. 2015a).

Basically, in the solvent mixture organic solvents such as DMF and DMSO are connected with lead halides and form intermediate adducts. The perovskite SCs can be developed by removing the organic solvents at higher temperature (see Fig. 6b). For example, when MAI and PbI_2 are mixed in DMF, an intermediate $MAPbI_3$-DMF adduct phase is formed due to a strong interaction of DMF–MA bonding. In a similar way, $MAPbI_3$-DMSO adduct phase is formed due to the interaction of DMSO–PbI_2 bonding, when MAI and PbI_2 are mixed in DMSO. The $MAPbI_3$ single crystal was obtained by removing the DMF or DMSO solvent via annealing.

Anti-solvent vapor-assisted crystallization (AVC) method. The AVC method is used to grow the perovskite crystals that are highly soluble in a solvent but have very poor solubility in other solvent (Shi et al. 2015). By this method, mm-sized $MAPbX_3$ SCs were obtained by using the anti-solvent dichloromethane (DCM), that is slowly diffused into the solution containing MAX and PbX_2 ($X = Br^-$, I^-) dissolved in DMF or GBA solvents (see Fig. 6c). This growth method for preparation of the hybrid halide perovskites SCs are highly efficient and applicable. However, it is difficult to develop large-size SCs that is important for large-area optoelectronic device applications.

Slow evaporation method (SEM). This slow evaporation method is a traditional and easy solution-based process for growth of SCs. Liao et al. (2015) have prepared the SCs of hybrid perovskite analogue (benzylammonium)$_2$PbX$_4$ (X = Cl^-, Br^-). In this process, stoichiometric mixture of benzylammonium chloride and $PbCl_2$ was mixed in concentrated HCl aqueous solution. Bulk (benzylammonium)$_2$PbX$_4$ crystals with the dimensions of 5 mm × 10 mm × 2 mm were obtained via the slow evaporation of DMF solution at 90 °C (see Fig. 6d). The preferred growth of single crystalline planes is extended along the [001] direction. Although this process is highly efficient, sometimes this method is hard to control precisely, that limits the industrial applications.

Droplet-pinned crystallization (DPC) method. Micrometer-sized $MAPbI_3$ single crystalline arrays can be formed using this DPC method (Jiang and Kloc 2013). At first, $MAPbI_3$ precursor solution was drop-casted on PEDOT:PSS-coated indium tin oxide (ITO) glass substrate, on which smaller wafers were placed (see Fig. 6e). Upon annealing, the precursor solvent was evaporated and rectangular-shaped $MAPbI_3$ SCs arrays were formed within some minutes. This method is very useful for growth of micrometer-sized SCs. It also provides a platform to grow single crystalline thin films.

3.2 Thin Films

Perovskites thin films are usually grown using solution process and vapor-phase deposition techniques. Very careful control on several processing parameters, such as the perovskite film thickness, crystallinity, perovskite phase purity, and perovskite film morphology, plays a significant role in achieving high-quality perovskite thin films and corresponding final device performance. The optimized perovskite thin-film processing steps can lead to desired perovskite thin-film thickness, highly crystalline films, uniform morphology, bigger crystal sizes, and less defect states. For solution-processed perovskite thin films, the processing parameters are types of perovskite precursors and solvent mixtures, precursor solubility, spin-coating speed, solvent engineering steps, types of anti-solvents, volume of the anti-solvent, time of anti-solvent injection, post-film thermal annealing temperature and time.

Single-step solution deposition. It is the simplest way to prepare a perovskite thin film via solution-processed spin-coating method. This perovskite thin films formation depends on various components, like substrate on which perovskite film will be deposited, precursors, solvents/mixed solvent and followed by spin-coating parameters. After spin coating, the semiconducting thin film is further annealed for faster crystallization process. The final film crystallinity, thickness, and morphology depend

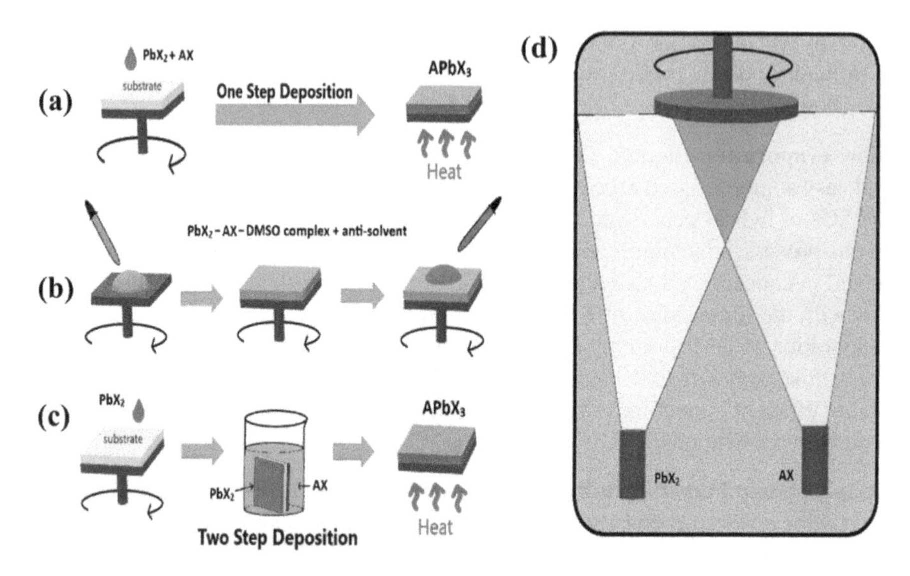

Fig. 7 Schematic illustration of formation of $MAPbI_3$ perovskite thin films via **a** one-step solution deposition process, **b** solvent engineering process, **c** two-step deposition process, and **d** vacuum processing technique, as mentioned in the diagram

on various processing parameters. Single-step perovskite films formation was first introduced by Im et al. in 2011 (Im et al. 2013). In this process, they prepared a precursor solution by mixing an equimolar MAI and PbI_2 powders in γ-butyrolactone (GBL) solvent at 60 °C for 12 h under vigorous stirring. Then the mixed solution was filtered through a 0.45 mm size PVDF filter for final thin-film formation via spin coating (see Fig. 7a). Then they spin-coated the precursor concentrations are in various concentrations of 10.05, 20.13, 30.18, 40.26 to 41.22 wt%. They observed that mesoporous TiO_2 films are better than compact films of TiO_2 for growth of thick uniform perovskite films. On the top of mesoporous TiO_2 film, the precursor solution was dropped and waits for one min to penetrate the solution into a mesoporous TiO_2 layer, which was then spin-coated at 2000 rpm for 40 s in an ambient atmosphere. The spin-coated $MAPbI_3$ film was annealed at different temperatures, and tetragonal crystal structure was formed. With increasing the concentration of the perovskite precursor, the film formation abruptly changes. A yellow-colored perovskite film was obtained for 10.05 wt% of precursor concentration, while it transformed to black color when the wt% concentration increased to 40.26 wt%. The color change of perovskite thin films at different precursor concentrations is ascribed due to higher perovskite precursor concentrations led to enhanced precursor interaction upon annealing and corresponding formation of the black perovskite phase. The UV-vis absorption spectra of films prepared from different concentrations of the perovskite precursor showed an increase in light absorption with higher precursor concentrations. It is also observed that up on annealing from temperature 40–100 °C, the perovskite film led to an increase in the absorption intensity, however, beyond 100 °C the absorption intensity decreases.

Selection of mixed solvents to dissolve perovskite precursors is also very important for formation of smooth perovskite surface with uniform crystal domains (Kim et al. 2014a). A mixed precursor solution of DMF and GBL was used to dissolve MAI and PbI_2 at 60 °C inside a nitrogen glovebox. The morphology of spin-coated perovskite films hugely differs when the films when spin-coated from different solvents like only DMF, only GBL, and mixed DMF and GBL solvents. The perovskite films formed from GBL solvents show formation of larger crystal grains with poor surface coverage. However, the perovskite films prepared from DMF show an improved morphology but non-uniform crystal dimensions. When the perovskite films developed from mixed solvent of DMF:GBL, the films display a smooth surface morphology with denser packing having uniform crystal dimension of 100 nm. The root-mean-square (RMS) roughness of the perovskite thin film reduces to 6.6 nm (for DMF:GBL) from 24.53 nm (for GBL) and 8.88 nm (for DMF). The crystallization process of perovskites varies from different solvents to solvents due to different evaporation rate of each solvent during spin-coating process. A higher evaporation rate of a solvent could lead to irregular thin-film surface morphology while mixture of different solvents may increase precursor solubility and controls the evaporation rate and resulted in an uniform and compact perovskite film.

Solvent engineering approach. This method is slightly modified compared to conventional single-step solution deposition process where some volume of an anti-solvent is drop-casted on top of perovskite film during spin coating of the perovskite precursor (see Fig. 7b). This solvent engineering process for perovskite preparation was first introduced by Jeon et al. in 2014 (Jeon et al. 2014). Usually, perovskite precursors are dissolved in a mixed solvent of GBL and DMSO. An anti-solvent that does not dissolve the perovskite but is miscible with GBL and DMSO, such as toluene, chlorobenzene, and chloroform, was dropped during the spin coating to facilitate an intermediate complex (MAI–PbI_2–DMSO) film and reduces the growth kinetic for perovskite crystallization process. The intermediate phase is confirmed by X-ray diffraction (XRD) and Fourier-transform infrared spectroscopy (FTIR) analysis (Beckmann 2010). This intermediate thin film was finally annealed at 130 °C and fully converted into perovskite phase. The resulted films are very uniform, dense, and smooth over larger active area.

Two-step deposition: Two-step deposition or sequential deposition technique, in which individual precursor layers, is deposited separately and interacts together to develop a final perovskite thin film (Burschka et al. 2013). At first, lead halide films are grown on the substrate via spin coating and later this film is dipped in MAI in IPA solution, leading to formation of $MAPbI_3$ perovskite thin films (see Fig. 7c). The precursor PbI_2 dissolved in DMF solvent and the solution was spin-coated on top of a mesoporous TiO_2 layer, followed by annealing at 70 °C to form yellow-colored PbI_2 film. This film was then dipped in a 5–10 mg/ml MAI precursor in IPA for several seconds of time, followed by rinsing in IPA and annealing at 70 °C. The yellow PbI_2 films transformed into black-colored films confirmed the formation of the perovskite phase. The thin-film morphology greatly depends on MAI solvent concentration and dipping time. This procedure is very useful for large-area device fabrication with reproducible high-quality crystalline perovskite film and excellent photovoltaic device performance.

Vacuum processing technique. In vapor deposited technique, the perovskite precursor powders MAI and $PbCl_2$ are placed separately in a thermal boat and simultaneously thermally evaporated on a substrate to form a perovskite film (Liu et al. 2013). Usually, the MAI and $PbCl_2$ precursors were evaporated with a molar ratio of 4:1 in a vacuum of 10^{-5} mbar and resulted in a dark reddish-brown-colored perovskite ($MAPbI_{3-x}Cl_x$) film (see Fig. 7d). The resulted perovskite films show a complete, crystalline, uniform coverage, and larger grain size compared to the conventional solution-processed perovskite. This procedure is also very advantageous for large-area device fabrication.

4 Basic Principle of PSCs

Mainly three key parameters play crucial role for device performance of the PSCs. First, the active perovskite material, the material needs to be designated in such a way that it has optimized band-gap and high-phase stability. However, the band-gap of the perovskites can be tuned in the visible to infrared spectral range by mixing/exchanging the halide anions ($X = Cl^-$, Br^-, I^-) or replacing the cations (MA^+, $C_2H_5NH_3^+$, FA^+, Cs^+, K^+, Rb^+). Second, the surface morphology of the perovskite thin films, uniform top layer, and large crystals formation is very essential for efficient charge transport and good diode characteristics; this is widely governed by precursor/solvent selections and post-processing conditions. Third, selection of proper charge transporting materials and optimizing the energy levels alignment is between the perovskite and the neighboring transport layers in order to enable effective charge transport throughout the device. In this regards, detailed study on the position of conduction band (CB) minimum and valence band (VB) maximum is required for perovskites and transporting materials.

The basic principle of solar cells is to convert the solar energy into electrical energy. The solar energy originates from the Sun, considered as a blackbody with a light spectrum at the temperature of about 5800 K. As the Sunlight passes through the Earth's atmosphere, it is attenuated by light scattering and some part is absorbed via chemical interactions with the atmospheric molecules. The atmosphere absorbs certain wavelengths of light and changing the amount of light reaching the Earth's surface. The water vapor mostly contributes for absorption of Sunlight spectra while molecular nitrogen, oxygen, and carbon dioxide also absorb some parts. The final solar spectrum that reaches to the Earth's surface varies with the light path length covered through the atmosphere. AM 1.5 spectrum with light intensity of 100 mW cm^{-2} are standardized as the measuring conditions for characterization of solar cells (see Fig. 8).

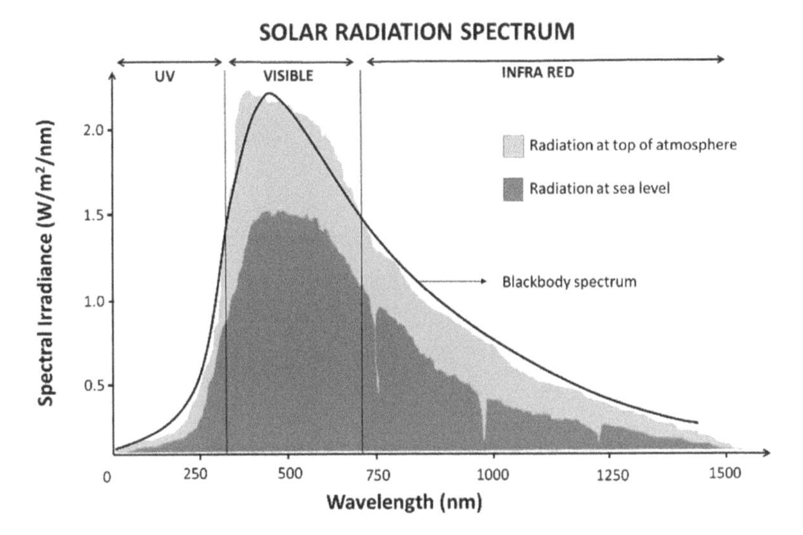

Fig. 8 Spectral power of the AM 1.5 global solar spectrum

4.1 Charge Generation and Transport in Perovskite Materials

For the semiconductor-based photovoltaics, the photons from Sunlight with enough energy can excite the electrons from the valence band to the conduction band across the band-gap and that results in generation of charge carriers. The photogenerated charge carriers can be separated out with proper band alignment of transporting layers such that electrons and holes are moving in opposite directions and store at counter electrodes. This is how solar cells are working under Sunlight. When Sunlight is falling on any absorbing active material, there are two types of loss mechanisms happened that hinder the solar cell efficiency. In a single-junction solar cell, the absorbing material cannot absorb photons having energy less than the band-gap of the active material and the light directly transmit through the material (see Fig. 9a). These low energy photons do not contribute to solar cells device performance. However, the photons having energy higher than the band-gap, absorbed by the active material, and the charge carriers are excited to higher energy states. After a very short time period, these excited charge carries relaxes to band edges of the active material through non-radiative thermalization process (see Fig. 9b). Here the excess energy of the incident photons is losses via this non-radiative process. Finally, the band-gap energy contributes to the device electricity. These two loss mechanisms reduce half of incident solar energy conversion to electrical energy.

Generation of charge carriers. In an ideal semiconductor, the valence and conduction bands are not flat. The band alignment is depending on the k-vector in the phase space that describes the momentum of an electron in the semiconductor. So, the energy of an electron is dependent on its momentum because of the periodic crystal structure of the semiconductor. If the maximum energy of the valence band and the

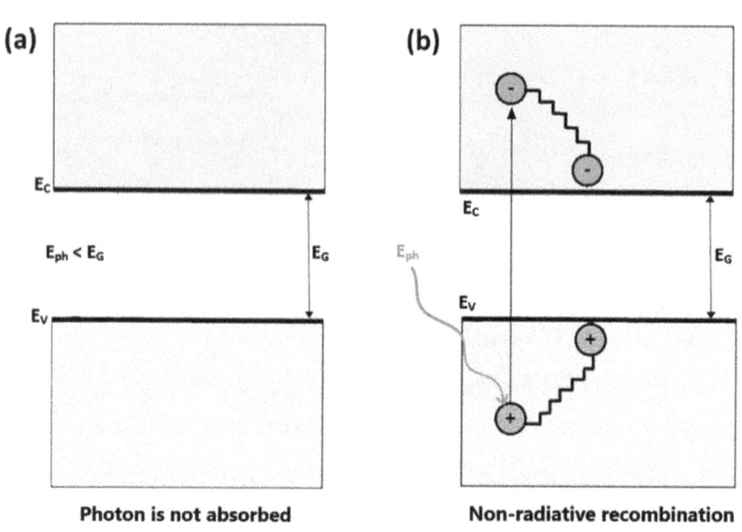

Fig. 9 Schematic representation of two types of energy losses in a solar cell when photon incident on the device. **a** If $E_{ph} < E_G$, the incident photon is not absorbed by the semiconductor. **b** If $E_{ph} > E_G$, a part of photon energy relaxes through thermalization process

minimum energy of the conduction band matches at the same *k*-vector in the *E*–*k* space, an electron can be excited from the valence band to the conduction band without a change in the momentum. Such a semiconductor is called a direct band-gap material. If the maximum energy of the valence band and the minimum energy of the conduction band does not match at same *k*-vector, so the electron cannot be excited without changing its momentum, such materials are called an indirect band-gap material. The absorption coefficient of a direct band-gap material is much higher than an indirect band-gap material; thus, the absorbing semiconductor can be much thinner for a direct band-gap semiconductor.

In a semiconductor, electrons can only stay at energy levels below the valence band edge (E_v) and above the conduction band edge (E_c). Between these two energy levels no allowed energy states exist for the electrons. Hence, the band-gap energy difference is, $E_g = E_c - E_v$. When the Sunlight with incident photons has energy higher than the band-gap of the semi-conducting material, which are absorbed and subsequently excite electrons from an initial energy level E_i to a higher energy level E_f as shown in Fig. 10a. So, the photons with an energy smaller than E_g incident on the ideal semiconductor, it will not be absorbed by the material. If an electron is excited from E_i energy level to E_f energy level, a void is created at E_i energy level and the void acts like a positive charged particle and is so-called a hole. Therefore, the absorption of a photon leads to formation of an electron–hole pair where electron stays at conduction band and hole stays are at valence band.

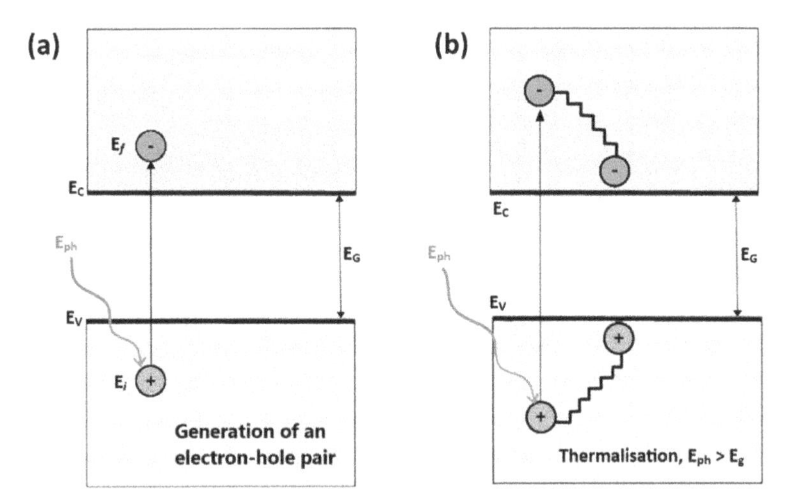

Fig. 10 Schematic representation of **a** absorption of a photon in a semiconductor with band-gap of E_G. The incident photon with energy $E_{ph} = h\gamma$ excites an electron from E_i to E_f in the semiconductor and at E_i a hole is created. **b** If $E_{ph} > E_G$, a part of the energy is thermalized

Separation of the photogenerated charge carriers. The excited electron can stay at the conduction energy level for a very short period of time, which is so-called the lifetime of the electron. The time period is generally in millisecond to microsecond scale. After the lifetime, the electrons will return back to the valence energy level from the conduction energy level and recombine with the holes. The electron–hole recombination energy will be released either as a photon (radiative recombination) or relaxes via lattice vibrations (non-radiative recombination) as shown in Fig. 10b. One can store the energy for further use before the electron–hole recombination process by separating out the electrons and holes via an external circuit and it is so-called solar cells circuit. A solar cell has to be designed such that the electrons and holes can transfer to opposite directions in presence of favorable adjacent energy levels of transporting materials in a very short span of time, i.e., less than the lifetime of the excited electrons. This requirement limits the thickness of the absorber materials.

Collection of the photo-generated charge carriers. Finally, the opposite charge carriers are extracted from the solar cells with electrical contacts and the energy is stored in a battery. Here, the chemical energy of the electron–hole pairs is converted into electrical energy.

4.2 *Characterization Processes of PSCs*

Power conversion efficiency (PCE). The most important parameter to characterize the solar cells is the PCE, which can be obtained from the current density–voltage

(J–V) characterization. As demonstrated in Fig. 11, under illumination and without external bias, a negative current is flowing through the circuit. The current density under these conditions through the solar cells device is called short-circuit current density (J_{sc}). When a positive bias is applied to the circuit, up to an open-circuit voltage (V_{oc}) a negative current is transporting though the circuit, indicating the power generation by the solar cell. The PCE of a solar cell is defined as the ratio of the maximum power output (P_{out}) to the incident light power (P_{in}), and it is represented by the formula,

$$\mathrm{PCE} = \frac{P_{\mathrm{out}}}{P_{\mathrm{in}}} = \frac{I_{\mathrm{sc}} \times V_{\mathrm{oc}} \times \mathrm{FF}}{P_{\mathrm{in}}}$$

where the fill factor (FF) can be thought of as the biggest rectangular area (blue area; see Fig. 11) covered under the current voltage curve. It is the quotient of maximum solar power output and the product of V_{oc} and I_{sc}.

External quantum efficiency. Another important parameter to characterize solar cells is the external quantum efficiency (EQE). It is the conversion ratio of the incident photon into electron in a solar cell, which is also known as the incident photon to converted electron (IPCE). The EQE is defined as the ratio of the number of output electrons to the number of incident photons at different wavelengths. EQE of a solar cell is depended on many factors, such as light absorption-co-efficient of the active layer, exciton generation efficiency, exciton dissociation efficiency, and carrier extraction efficiency.

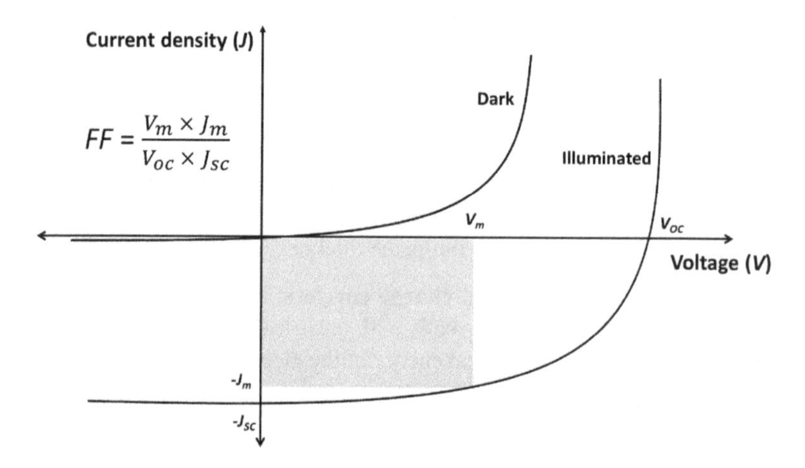

Fig. 11 Schematic representation of J–V characteristics of a solar cell

5 Working Principle of PSCs

5.1 Device Architectures

Selection of the transporting materials with proper band alignment is a crucial part. In PSCs, perovskite layer acted as an absorber layer. Depending upon the electron and hole transport toward light entering electrode, PSCs can be classified into two parts *n-i-p* structure and *p-i-n* structure. One can easily understand this classification by checking the position of transporting layer [*n-i-p*: electron-transporting materials (ETMs) layer placed on top of TCO; and *p-i-n*: hole-transporting materials (HTMs) layer placed on top of TCO]. These two types of solar cells device structures are subclassified into two other parts mainly mesoscopic and planer structure. The mesoscopic structure incorporates a mesoporous layer (in front of light entering window), whereas the planar structure consists of all planar layers of transporting layers. Sometimes PSCs are classified without transporting layers.

Regular *n-i-p* structure. PSCs are often called as solid-state dye sanitized solar cells (SS-DSSCs). PSCs are modified form of conventional DSSC. Here the light absorption was governed by the solid-state perovskite material in DSSCs, and it was fabricated from dyes. Conventional PSCs structure is *n-i-p* (see Fig. 12) type, where the *n*-type metal oxide layer (transporting layer) was deposited on top of the conducting oxide layer. On the top of this *n*-type transporting layer, intrinsic absorber perovskite layer was grown and finally the *p*-type transporting layer and top electrode was deposited in succession. The efficiency of the PSCs mostly depends on the proper selection of transporting materials. Mesoscopic layer (like, TiO_2) was deposited as a transporting layer in front of the light entering window to enhance the charge collection ability from perovskite layer and as a result the PSCs device performance enhanced.

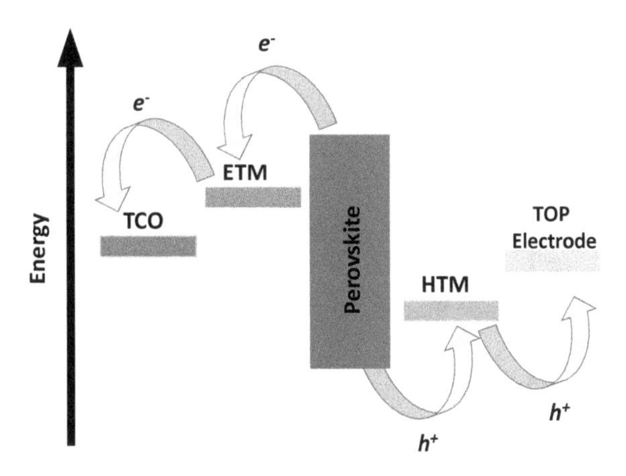

Fig. 12 Schematic energy band diagram of direct *n-i-p* structure

Inverted *p-i-n* structure. Inverted *p-i-n* structure was imported from the organic solar cells. Where the *p*-type hole-transporting layer (HTL) was deposited on top of the TCO and the *n*-type electron-transporting layer (ETL) was deposited under the top electrode. Here the holes and electrons are forced to move in two different directions and collected at different electrodes (see Fig. 13). In this device configuration, mesoscopic and planer PSCs were fabricated in the presence of different perovskite compositions.

Electron-transporting layer-free structure. A compact *p*-type transporting layer was deposited on top of transparent conductive oxide (see Fig. 14a), or on top of perovskite layer (see Fig. 14b) to fabricate electron-transporting layer-free PSCs. On top suitable electrodes are deposited. This type of PSCs is fabricated on planar structure and helps to achieve high PCE by increasing the open-circuit voltage (V_{oc}). Recently, scientists have developed a suitable method for fabricating PSCs without one transporting layer and achieved higher PCE.

The essential requirements of HTM layers in PSCs are due to several advantages such as,

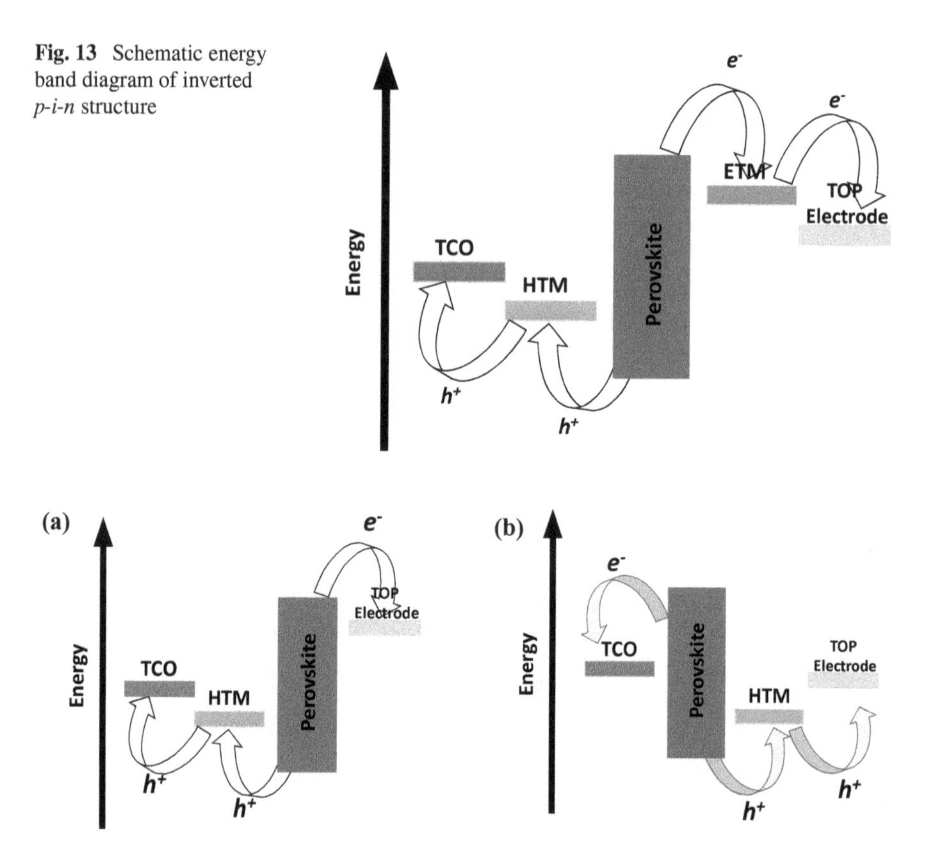

Fig. 13 Schematic energy band diagram of inverted *p-i-n* structure

Fig. 14 Schematic energy band diagram of PSC without ETL **a** direct and **b** inverted structure

(1) Suitable band matching with perovskite layer helps to attain high V_{oc} values.
(2) Enhancement of the fill factor (FF) by reducing the series resistance governed by the high hole mobility in the device.
(3) The optical loss in the devices can be minimized by proper selection of low optical absorption of the transporting material in the visible region.

5.2 PSCs with Various HTM Layers and Their Device Performance

The thickness of the transporting layer should be maintained at an optimum value to overcome the resistive losses and allowed better crystalline growth of the perovskite. Some of the PSCs device structures are shown in Fig. 15.

Nanoparticles as HTMs

Nanomaterials are a suitable candidate for large-area optoelectronic industry due to their low production costs and ease synthesis helps. Superior optoelectronic properties such as easy processability, low synthesis cost, tunable band-gap, high mobility serve the nanomaterials as an alternative choice as a transporting layer material for PSCs. Tunable optoelectronic properties of nanomaterials help easy band matching with perovskite. As a result, the use of nanomaterials in the PSCs expected to improve the performance the device indeed. In the previous study, on colloidal QDs have been used in DSSC to get a device performance of 7% as an alternative to molecular dyes (Ip et al. 2012). Here, in this section we will discuss the device performance of in PSCs with various hole-transporting nanomaterials, such as Cu_2ZnSnS_4, WO_3, NiO, CuS nanocrystals (NCs) together with $CuInS_2$- and PbS-based QDs.

Cu_2ZnSnS_4 Nanoparticles (NPs). Earth-abundant Cu_2ZnSnS_4 (CZTS) NPs display some interesting properties such as high hole mobility (6–30 cm^2 V^{-1} s^{-1}), ideal band-gap (1.5 eV) and permit them to use as a HTL in PSCs (Chen et al. 2013; Zhou et al. 2013; Walsh et al. 2012; Vanalakar et al. 2015). Wu et al. introduced the CZTS NPs as a transporting material due to the band matching with absorber perovskite layer in which the PCE was reached to 13% (Wu et al. 2015a). In this report, the NPs were synthesized with different reaction times of 20, 30, and 40 min to achieve different sizes such as CZTS-20, CZTS-30, and CZTS-40, and optical absorption was verified with mostly used organic HTM spiro-OMeTAD. After analyzing the SEM images, it was revealed that smooth, uniform, and homogeneous perovskite layer was formed on top of CZTS NPs. When compared with the commonly used spiro-OMeTAD HTM layer, it was found that the CZTS layer improved the interfacial contact between the perovskite layer and the top electrode to reduce the non-radiative recombination at the interfaces. The results suggested that the CZTS NPs can be effectively acts as a low-cost HTM layer as its able to transport holes effectively in PSCs (Fig. 16).

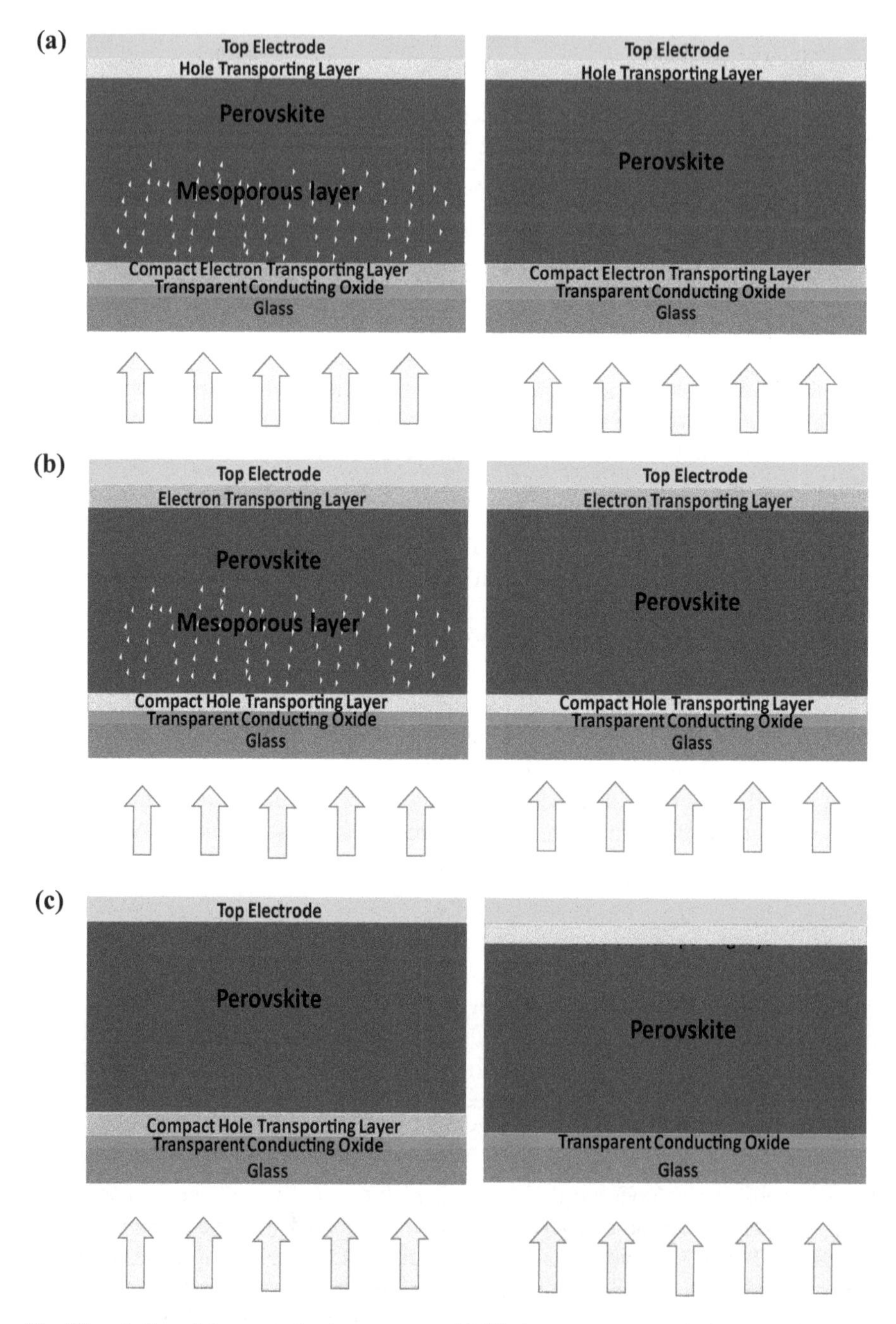

Fig. 15 **a** Left to right, normal *n-i-p* structure of PSCs in mesoporous and planar configuration of ETL. **b** Left to right, normal *p-i-n* structure of PSCs in mesoporous and planar configuration of HTL. **c** Structure of PSC without ETL (left) and without HTL (right)

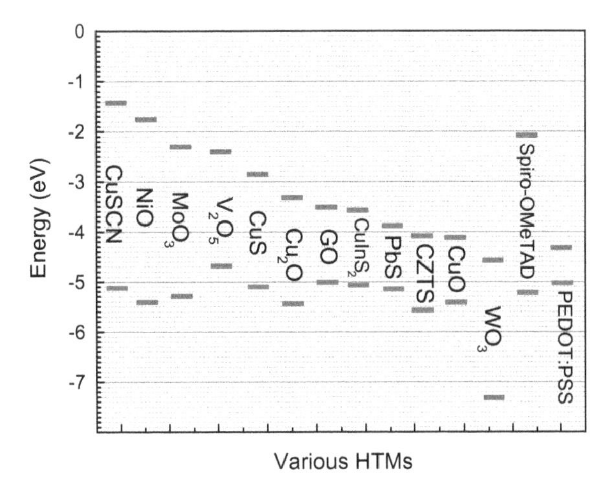

Fig. 16 Schematic energy band position of various HTMs used in PSCs

WO_3 NCs. In 2015, Li et al. had shown the successful use of WO_3 NCs as an efficient inorganic HTL in PSCs (Li 2015). Highly transparent WO_3 NCs layer (87% transparency) allows the incoming photons to be absorbed by the perovskite layer, high work function, and deep energy levels of WO_3 makes them efficient HTM in PSCs. It was also observed that the WO_3-based PSCs were more stable compared to the other organic and nanocrystalline HTLs-based solar cells. It was found that the device performance was considerably well with WO_3 as HTL and a PCE of 7.68% observed (Table 1).

NiO NCs. In 2014, Zhu et al. had successfully demonstrated the use of NiO NCs as an efficient HTL in PSCs. The NiO layer (30–40 nm thick) was grown by conventional sol–gel method and the corresponding solar cells device exhibited a PCE of 9.11% (Zhu et al. 2014). From the photoluminescence (PL) measurements, it was found that the NiO thin films have superior hole extraction ability from the perovskite layer compared to PEDOT:PSS layer.

CuS NPs. Rao et al. also discovered the use of CuS NPs as a hole-transporting material in an inverted perovskite solar cell (Rao et al. 2016). CuS NPs-based solar cells exhibit some interesting features, such as reproducibility, low charge recombination, effective hole extraction, higher stability, and ability to modify work function of ITO layer, found perfect band matching with the perovskite layer. From the SEM images, it was confirmed that the CuS NPs provide good surface coverage on ITO substrate. AFM studies of root-mean square (RMS) roughness suggest that no pinholes were formed in the CuS layer. From the J–V curves, it was discovered that the solar cell device exhibited PCE value of 16.2% and stability of the device boosted up to 250 h.

Copper indium disulfides ($CuInS_2$) colloidal Quantum Dots (QDs). Less toxic suitable band-gap (1.45 eV) with high extinction coefficient in the visible region of $CuInS_2$ QDs enables them to use as an HTM layer in PSCs. Lv et al. introduced $CuInS_2$ as an HTM in PSCs by replacing organic HTMs (Kolny-Olesaik and Weller

Table 1 Performance of perovskite (PVK) solar cells based on various oxides and NCs-based HTMs

Architecture	Cell structure for ***various oxides***	PCE (%)	J_{sc} (mA cm^{-2})	V_{oc} (V)	FF	References	Year
MoS_2 (planar *p-i-n*)	ITO/MoS_2/PVK/TiO_2/Ag	20.53	26.25	0.93	0.84	Kohnehpoushi et al. (2018)	2018
MoS_2 (mesoporous *n-i-p*)	FTO/TiO_2/PVK/MoS_2-QD/Au	20.12	22.81	1.11	0.79	Najafi et al. (2018)	2018
MoS_2 (mesoporous *n-i-p*)	FTO/TiO_2/PVK/MoS_2/spiro-OMeTAD/Au	13.09	21.09	0.93	0.66	Capasso et al. (2016)	2016
CZTS NCs	$MAPbI_3$	13.0	20.5	1.06	0.58	Wu et al. (2015a)	2015
VO_x (planar *p-i-n*)	ITO/VO_x/PVK/PCBM/C60/BCP/Al	11.7	20.8	1.02	0.56	Xiao et al. (2016)	2016
WO_3 (mesoporous *n-i-p*)	FTO/WO_3/TiO_2/PVK/spiro-OMeTAD/Ag	11.24	17.00	0.87	0.76	Mahmood et al. (2015a)	2015
WO_3 (mesoporous *n-i-p*)	FTO/WO_3/Cs_2CO_3/PCBM/PVK/P3HT/Au	10.49	20.4	0.84	0.61	Chen et al. (2016)	2016
NiO NCs	$MAPbI_3$	9.11	16.2	0.88	0.63	Zhu et al. (2014)	2014
WO_x (mesoporous *n-i-p*)	FTO/WO_3/TiO_2/PVK/spiro-OMeTAD/Ag	8.99	21.77	0.71	0.58	Wang et al. (2015a)	2015
$CuInS_2$ quantum dots	$MAPbI_3$	8.38	18.6	0.92	0.48	Lv et al. (2015)	2015
PbS quantum dots	$MAPbI_3$	7.88	12.1	0.86	0.72	Hu et al. (2015)	2015
WO_3 NCs	$MAPbI_3$	7.68	18.1	0.92	0.64	Li (2015)	2015
WO_3 (planar *p-i-n*)	ITO/WO_3/PVK/PCBM/Al	7.68	18.10	0.92	0.64	Li (2015)	2015

(continued)

Table 1 (continued)

Architecture	Cell structure for ***various oxides***	PCE (%)	J_{sc} (mA cm^{-2})	V_{oc} (V)	FF	References	Year
WO_3(mesoporous *n-i-p*)	FTO/WO_3/PVK/spiro-OMeTAD/Au	7.04	18.00	0.66	0.61	Zhang et al. (2016a)	2016
MoS_2 (planar *p-i-n*)	ITO/MoS_2/PVK/PCBM/Al	6.01	12.60	0.84	0.57	Dasgupta et al. (2017)	2017
PbS QDs	$MAPbI_3$	4.88	18.6	0.87	0.49	Li et al. (2015a)	2015
Carbon QDs	$MAPbI_3$	3.0	7.8	0.51	0.74	Paulo et al. (2016)	2016

2013). The use of $CuInS_2$ QDs (VB: 5.00–5.05 eV) improves the hole conduction from perovskite layer (VB: −5.43 eV) due to suitable band matching (Pan et al. 2014). The devices were fabricated by depositing colloidal $CuInS_2$ QDs on top of the $MAPbI_3$-coated TiO_2 films, during testing under suitable light intensity these cells delivered a PCE of 6.57%. The device performance was improved with a PCE of 8.38%, which was achieved by using modified $CuInS_2$ QDs with ZnS shell layer. The ZnS layer helps to reduce the non-radiative carrier recombination at the interface of TiO_2 and HTM layer (Santra et al. 2013).

Lead sulfide (PbS) QDs. Excellent optoelectronic properties of PbS QDs, such as high absorption coefficient, low band-gap, tunable band-gap, large exciton Bohr radius (~18 nm), motivate to implement them as a *p*-type HTM layer in PSCs (Hodes 2013; Snaith 2013). Solution-processed PbS QDs as an inorganic HTM layer in PSCs was successfully demonstrated and attained a PCE of 7.5% (Hu et al. 2015). By varying the band-gap of PbS QDs, the energy level alignments between the perovskite and PbS QDs can easily be tuned to achieve optimized device performance (Tang et al. 2011). Dai et al. have reached device efficiency nearly 8% with a good device stability by using colloidal PbS QDs as an inorganic HTM layer (Li et al. 2015a). This article investigated the solar cells performance by optimizing the PbS QDs layer thickness. Here two-step spin-coating method provides a higher PCE of 7.8%, whereas one-step method contributed lower PCE 4.73%. Enhanced solar cells device performance was achieved due to the low recombination in two-step processed perovskite thin films. From the SEM images of the two-step processed $MAPbI_3$, thin films deposited on TiO_2 layer showed more uniform surface coverage over the one-step process. These results discovered the efficient use of PbS QDs as a low-cost HTM in perovskite/QD hybrid solar cells.

Copper-based HTMs

Copper Iodide (CuI). Inexpensive, stable, wide band-gap, with high conductivity make CuI as a favorable candidate for HTM layer in PSCs. Christians et al., used CuI using the drop casting method and recorded a PCE of 6% (Christians et al. 2014). When compared with the conventional spiro-OMeTED HTM layer, it was found that the electrical conductivity of CuI solar cells was two orders higher and resulted in significant improvement in device performance. The stability test of these solar cell devices was carried out under constant illuminations of 100 mW/cm^2 AM 1.5G for a period of 2 h without encapsulation under ambient conditions. It was found that J_{sc} value remained constant for CuI based solar cells, while there is a decrement of J_{sc} of 10% from initial value was observed for spiro-OMeTAD-based solar cells. CuI HTM-based PSCs exhibited higher V_{oc} value with lower device efficiency compared to the conventional spiro-OMeTAD HTM-based solar cells (Table 2).

Cuprous oxide (Cu_2O) and Copper oxide (CuO). Earth-abundant and easy processable Cu_2O and CuO are typically *p*-type semiconductors, with suitable band alignment matches well with the perovskite ($MAPbI_3$) energy levels. Copper-based oxide materials have shown high hole mobility of 100 cm^2 V^{-1} s^{-1}, which makes them suitable as an HTM material (Shao et al. 2010; Bao et al. 2009). Zuo et al. had

Table 2 Performance of various PSCs where CuI was used as HTM layer

Architecture	Cell structure for ***CuI***	PCE (%)	J_{sc} (mA cm^{-2})	V_{oc} (V)	FF	References	Year
Planar *p-i-n* (spin coating)	ITO/CuI/PVK/PCBM/BCP/Ag	16.8	22.6	0.99	0.71	Sun et al. (2016)	2016
Mesoscopic *n-i-p* (spin coating)	FTO/TiO_2/PVK/spiro-OMeTAD:CuI/Ag	16.67	21.52	1.06	0.73	Li et al. (2016a)	2016
Mesoscopic *n-i-p* (thermally evaporated)	FTO/CuI/PVK/PCBM/PEI/Ag	14.3	20.3	1.04	0.68	Wang et al. (2017a)	2017
Planar *p-i-n* (spin coating)	FTO/CuI/PVK/PCBM/Al	13.58	21.06	1.04	0.62	Chen et al. (2015b)	2015
Mesoscopic *n-i-p* (thermally evaporated)	FTO/TiO_2/PVK/CuI/Cu	9.24	22.99	0.85	0.47	Nazari et al. (2017)	2017
Mesoscopic *n-i-p* (doctor blading)	FTO/bl-TiO_2/PVK/CuI/graphite/Cu	7.5	16.7	0.78	0.57	Sepalage et al. (2015)	2015
Mesoscopic *n-i-p* (doctor blading)	FTO/bl-TiO_2/mp-TiO_2/PVK/CuI/Au	6.0	17.8	0.55	0.62	Christians et al. (2014)	2014
Mesoscopic *n-i-p* (spray Coating)	FTO/bl-TiO_2/mp-TiO_2/PVK/CuI/Au	5.6	22.6	0.64	0.39	Huangfu et al. (2015)	2015
Mesoscopic *n-i-p* (thermally evaporated)	FTO/TiO_2/PVK/CuI/Au	4.38	22.8	0.66	0.29	Gharibzadeh et al. (2016)	2016

successfully introduced the copper-based oxide materials as HTMs in PSCs. They have achieved maximum PCE of 13.35% with Cu_2O and 12.16% with CuO considered as inorganic HTMs (Zuo and Ding 2015). Cu_2O and CuO thin films showed smooth surface morphology that was analyzed from AFM study. The RMS values were obtained as of 2.81 nm and 3.32 nm, respectively. Highly crystalline $MAPbI_3$ films were observed on Cu_2O and CuO thin films compared with PEDOT:PSS films. These oxide transporting layers facilitate the charge transport and as a result increase in the device performance (Dong et al. 2015a). Cu_2O solar cells were found more stable compared to PEDOT:PSS-based solar cells. The PCE value of Cu_2O HTM-based solar cells reduced from 11.02 to 9.96%, whereas for PEDOT:PSS HTM-based solar cells declined from 10.11 to 6.79%, when the devices were kept for 70 days in the nitrogen filled glove box. In another study, Yu et al. have reported Cu_2O HTM-based PSCs with maximum PCE of 11.0% (Yu et al. 2016) (Table 3).

Sun et al. had reported a PCE of 17.1% by using CuO_x as an inorganic HTM layer in PSCs and exhibited short-circuited current density of 23.2 mA/cm^2, open-circuit voltage of 0.99 V, and fill factor of 0.74 (Sun et al. 2016). The devices were found to be stable for approximately 200 h. The better device performance of CuO_x-based PSC devices was achieved due to efficient hole transport from perovskite layer to HTM layer. Low contact resistance of CuO_x layer was beneficial for such enhanced device performance. From AFM study, it was discovered that the surface roughness or RMS value of ITO surface was 4.7 nm and the RMS value was decreased to 4.2 nm with addition of CuO_x HTM layer on top of ITO surface that improves the overall surface morphology and prevents short-circuited current leakage inside the device. Rao et al., have reported an output power efficiency of 19.0% where $MAPbI_{3-x}Cl_x$ have been used as an absorber layer and CuO_x used as a HTM layer (Rao et al. 2016). Doping of Cl offers the better surface morphology compared to undoped perovskites thin film and improved the hole mobility which in turn enhanced the device performance. The SEM morphology of $MAPbI_3$ films revealed that the average perovskite particle size was very small and form several grain boundaries, which trigger the non-radiative trap assisted recombination in turn reduced the overall device performance.

Copper thiocyanate (CuSCN). CuSCN is immensely used as an inorganic HTM layer in PSCs due to their some promising characteristics, such as their high optical transparency, high hole mobility of 0.01–0.1 $cm^2\ V^{-1}\ s^{-1}$ and good chemical stability (O'Regan et al. 2000; Tsujimoto et al. 2012; Pattanasattayavong et al. 2013a, b). Qin et al. had reported the fabrication of PSCs with copper thiocyanate (CuSCN) used as an HTM and achieved device PCE of 12.4% (Qin et al. 2014a). Ye et al. found in inverted PSCs with CuSCN HTM that exhibited an average PCE of 16.6% which was better compared to other conventional organic HTM layers used in PSCs (Ye et al. 2015). It was found that the device fabricated from perovskite layer deposited on top of a CuSCN via one-step deposition method was much efficient compared to two-step deposition. In one-step deposition process, the perovskites crystallized slowly and resulted in low surface roughness. The high device performance also signifies the smaller interface contact resistance between the perovskite layer and the CuSCN layer (Table 4).

Table 3 Performance of various copper oxide HTM layer-based PSCs

Architecture	Cell structure for ***copper oxide***	PCE (%)	J_{sc} (mA cm^{-2})	V_{oc} (V)	FF	References	Year
CuO_x planar *p-i-n* (spin coating)	ITO/CuO_x/PVK/PCBM/BCP/Ag	16.8	22.2	1.0	0.76	Sun et al. (2016)	2016
Cu_2O planar *p-i-n* (spin coating)	ITO/Cu_2O/PVK/PCBM/Ca–Al	13.35	16.52	1.07	0.76	Zuo and Ding (2015)	2015
CuO planar *p-i-n* (spin coating)	ITO/CuO/PVK/PCBM/Ca–Al	12.61	15.82	1.06	0.73	Zuo and Ding (2015)	2015
Cu:CrO_x planar *p-i-n* (spin coating)	FTO/Cu-CrO_x/PVK/PCBM/Ag	10.99	16.02	0.98	0.70	Qin et al. (2016)	2016
Cu_2O Mesoscopic *n-i-p* (spin coating)	FTO/TiO_2/PVK/Cu_2O/Au	8.93	15.8	0.96	0.59	Nejand et al. (2016)	2016
Cu_2O planar *p-i-n* (spin coating)	ITO/Cu_2O/PVK/PCBM/Al	8.30	15.60	0.92	0.58	Chatterjee and Pal (2016)	2016
Cu_xO planar *p-i-n* (electro spray)	ITO/Cu_xO/PVK/C60/BCP/Al	5.83	17.22	0.78	0.48	Bu et al. (2017)	2017

Table 4 Performance of various CuSCN HTM layer-based PSCs

Architecture	Cell structure for ***CuSCN***	PCE (%)	J_{sc} (mA cm^{-2})	V_{oc} (V)	FF	References	Year
Mesoscopic *n-i-p* (spin coating)	FTO/TiO_2/PVK:CuSCN/rGO/Au	20.8	23.35	1.13	0.77	Arora et al. (2017)	2017
Mesoscopic *n-i-p* (spin coating)	ITO/TiO_2/PVK/spiro-OMeTAD:CuSCN/Ag	18.02	22.01	1.06	0.77	Li et al. (2016a)	2016
Planar *n-i-p* (spin coating)	ITO/CuSCN/PVK/C60/Ag	17.2	22.3	1.09	0.71	Wijeyasinghe et al. (2017)	2017
Planar *n-i-p* (electro deposition)	ITO/CuSCN/PVK/C60/BCP/Ag	16.6	21.9	1.00	0.76	Ye et al. (2015)	2015
Mesoscopic *n-i-p* (doctor blading)	FTO/bl-TiO_2/mp-TiO_2/PVK/CuSCN/Au	16.6	21.8	1.1	0.69	Madhavan et al. (2016)	2016
Planar *p-i-n* (spin coating)	ITO/CuSCN/PVK/PCBM/PDINO/Al	16.4	21.4	1.05	0.73	Wang et al. (2016a)	2016
Mesoscopic *n-i-p* (spin coating)	FTO/bl-TiO_2/mp-TiO_2/PVK/CuSCN/Au	15.43	20.8	1.06	0.70	Madhavan et al. (2016)	2016
Mesoscopic *n-i-p* spray-coated	FTO/Al_2O_3/CuSCN/Au/PMMA	13.3	21.07	0.98	0.64	Liu et al. (2016a)	2016

(continued)

Table 4 (continued)

Architecture	Cell structure for ***CuSCN***	PCE (%)	J_{sc} (mA cm^{-2})	V_{oc} (V)	FF	References	Year
Mesoscopic *n-i-p* (doctor blading)	FTO/bl-TiO_2/mp-TiO_2/PVK/CuSCN/Au	12.4	19.7	1.026	0.62	Qin et al. (2014a)	2014
Planar *p-i-n* (spin coating)	ITO/CuSCN/PVK/PCBM/Al	12.4	19.1	1.0	0.65	Wang et al. (2016a)	2016
Planar *p-i-n* (spin coating)	ITO/CuSCN/PVK/PCBM/LiF-Ag	10.5	15.76	1.06	0.63	Zhao et al. (2015)	2015
Planar *p-i-n* (spin coating)	ITO/CuSCN/PVK/PCBM/C60/Ag	10.22	12.2	1.07	0.76	Jung et al. (2015a)	2015
Mesoscopic *n-i-p* (spin coating)	FTO/SnO_2/PVK/CuSCN/Au	8.38	18.99	0.96	0.45	Murugadoss et al. (2016)	2016
Mesoscopic *n-i-p* (doctor blading)	FTO/bl-TiO_2/PVK/CuSCN/Au	7.19	18.42	0.97	0.40	Ito et al. (2015)	2015
Mesoscopic *n-i-p* (drop casting)	FTO/bl-TiO_2/PVK/CuSCN/Au	6.4	18.53	0.727	0.617	Chavhan et al. (2014)	2014
Mesoscopic *n-i-p* (spin coating)	FTO/cp-TiO_2/mp-TiO_2/Sb_2S_3/PVK/CuSCN/Au	5.12	17.2	0.57	–	Ito et al. (2014b)	2014
Mesoscopic *n-i-p* (doctor blading)	FTO/bl-TiO_2/mp-TiO_2/PVK/CuSCN/Au	4.85	14.5	0.63	0.53	Ito et al. (2014a)	2014
Mesoscopic *n-i-p* (electro deposition)	FTO/CuSCN/PVK/PCBM/Ag	3.8	8.8	0.67	–	Subbiah et al. (2014)	2014

Nickel oxide (NiO_x). NiO_x is also ambient stable having larger band-gap (5.4 eV), and suitable energy levels match well with energy band to perovskite and make them as a potential candidate for HTM layer in PSCs. Yin et al. observed a higher PCE of 16.47% on a ITO-coated glass substrate compared to those of ITO-PEN (polyethylene naphthalate) substrate solar cells (PCE of 13.43%). Here solution-processed NiO_x was used as a HTM layer in the inverted planar heterojunction (Yin et al. 2016). From the J–V curve and EQE spectra, it was exposed that NiO_x-based device show low recombination than that of PEDOT:PSS which is commonly used HTM in perovskite solar cell. The use of pristine and copper-doped NiO_x as HTMs was represented by Kim et al., publicized the use of pristine and copper-doped NiO_x as HTMs for high performing and stable planar PSCs (Kim et al. 2014b). After careful analysis of J–V curves of Cu:NiO_x HTM-based PSC showed the maximum PCE of 15.40% compared to pristine NiO_x HTM exhibited PCE of 8.94%.

Wei et al. fabricated PSCs using Li–Mg-doped NiO as HTM and Ti(Nb)O_x used as an ETL, showing a highest PCE of 18.3% with a J_{sc}= 20.4 mA/cm^2, V_{oc} = 1.08 V, FF = 0.83 (Chen et al. 2015a). Introduction of Li^+ and Mg^{2+} into the NiO lattice increased the conductivity and avoid the undesirable shift of valence band (Chen et al. 2015a; Alidoust et al. 2014; Huang et al. 2014; Deng et al. 2012). For the stability test, the devices were kept under dark condition for 1000 h and it was found that the efficiency was reduced to its 97% from the initial value. Same experiment was carried out under 1000 h constant illumination condition, and the device PCE value was reduced to 90% from its initial value (Table 5).

Carbon Materials-based HTMs

Advantageous optoelectronic properties of carbon (*C*) materials, such as carbon nanotubes, graphene, and graphene oxide, as a transporting layer have gained significant attention in the field of organic electronics. It was observed that the carbon material-based solar cells had achieved a maximum PCE of 15.5% with better stability (Aitola et al. 2016). Zheng et al. had reported to achieve a PCE of 12.8% by using graphene sheet doped functionalized thiolated nanographene (TSHBC) as a HTM layer in PSCs and the device improved to an efficiency of 14% (Cao et al. 2015a) (Fig. 17; Table 6).

NiO_x/PEDOT:PSS. Use of hybrid PEDOT:PSS/NiO_x HTL was reported by Park et al. in an inverted planar device architecture and the device exhibited a PCE of 15.1% (Park et al. 2015). The hybrid PEDOT:PSS/NiO_x transporting layer was deposited by spin-coating different concentrations of (0.1, 1.0 and 5.0%) PEDOT:PSS solution on top of NiO_x layer. The device fabricated with 1.0% PEDOT:PSS/NiO_x as HTL displayed highest PCE of 13.9% compared to other compositions. Device with pure PEDOT:PSS (11.8%) and bare NiO_x (12.7%) achieved a lower PCE. From the impedance spectroscopy studies, it was observed that there was a subsequent reduction of internal resistance for PEDOT:PSS/NiO_x HTL based PSCs.

Spiro-OMeTAD. Due to several advantageous features of spiro-OMeTAD HTM layer, such as favorable glass transition temperature (T_g = 120 °C), easy solubility, ideal ionization potential, suitable absorption spectrum, and solid-state morphology

Table 5 Performance of various NiO HTM layer-based PSCs

Architecture	Cell structure for ***NiO***	PCE (%)	J_{sc} (mA cm^{-2})	V_{oc} (V)	FF	References	Year
Planar *p-i-n* (spraying)	FTO/NiO_x/$FAPbI_3$/PCBM/TiO_x/Ag	20.65	23.09	1.10	0.81	Xie et al. (2017)	2017
Planar *p-i-n* (spin coating)	ITO/NiO_x/PVK/PCBM/ZrAcac/Al	20.5	–	–	–	Yue et al. (2017)	2017
Planar *p-i-n* spray pyrolysis	FTO/NiO_x/PVK/PCBM/Ag	19.58	22.68	1.12	0.77	Wu et al. (2017)	2017
Planar *p-i-n* (spin coating)	FTO/Cs:NiO_x/PVK/PCBM/ZrAcac/Ag	19.35	21.77	1.12	0.79	Chen et al. (2017a)	2017
Planar *p-i-n* (spin coating)	ITO/NiO_x/PVK/PCBM/ZrAcac/Al	18.69	22.17	1.08	0.78	Chen et al. (2017b)	2017
Planar *p-i-n* (spin coating)	ITO/*Cu*:NiO_x/PVK/PCBM/BCP/Ag	18.66	20.76	1.11	0.81	He et al. (2017)	2017
Planar *p-i-n* (spin coating)	FTO/NiO_x/PVK/PCBM/Ag	18.6	22.8	1.09	0.75	Hu et al. (2017a)	2017
Planar *p-i-n* (spin coating)	ITO/NiO_x/PVK/PCBM/c-HATNA/Bis-C60/Ag	18.21	21.25	1.09	0.79	Zhu et al. (2018)	2018
Planar *p-i-n* (spin coating)	ITO/NiO_x/PVK/PCBM/Al	18.0	21.79	1.12	0.74	Nie et al. (2018)	2108
Planar *p-i-n* (combustion)	ITO/Cu:NiO/PVK/Bis-C60/C60/Ag	17.46	21.60	1.05	0.77	Jung et al. (2015b)	2105
Mesoscopic *p-i-n* (pulsed laser deposition)	ITO/PLD-NiO/PVK/PCBM/LiF-Al	17.3	20.2	1.06	0.81	Park et al. (2015)	2015
Planar *p-i-n* (spin coating)	ITO/NiO_x/PVK/PCBM/Ag	17.2	21.4	1.03	0.78	Kim et al. (2017)	2017

(continued)

Table 5 (continued)

Architecture	Cell structure for ***NiO***	PCE (%)	J_{sc} (mA cm^{-2})	V_{oc} (V)	FF	References	Year
Planar *p-i-n* (electrodeposition)	ITO/NiO_x/PVK/PCBM/Ag	17.1	22.6	1.05	0.72	Park et al. (2017)	2017
Planar *p-i-n* (spin coating)	ITO/NiO_x/PVK/PCBM/Ag	16.55	21.22	1.04	0.75	Ciro et al. (2017)	2017
Planar *p-i-n* (spin coating)	ITO/NiO_x/PVK/PCBM/Ag	16.47	20.58	1.07	0.75	Yin et al. (2016)	2016
Planar *p-i-n* (spin coating)	ITO/NiO_x/PVK/ZnO/Al	16.1	21.01	1.01	0.76	You et al. (2016)	2016
Planar *p-i-n* (spin coating)	FTO/Cu:NiO/PVK/PCBM/Ag	15.40	18.75	1.11	0.72	Kim et al. (2015)	2015
Planar *p-i-n* (vacuum deposition)	ITO/NiO_x/PVK/PCBM/BCP/Ag	15.4	18.6	1.06	0.78	Pae et al. (2018)	2018
Mesoscopic *n-i-p* (drop casting)	FTO/c-TiO_2/m-TiO_2/PVK:NiO-MWCNTs	15.38	22.38	0.90	0.76	Yang et al. (2017)	2017
Planar *p-i-n* (spin coating)	ITO/NiO-PEDOT/PVK/PCBM/Ag	15.1	20.1	1.04	0.72	Park et al. (2015)	2015
Multi-layered *n-i-p* (screen printing)	FTO/bl-TiO_2/mp-TiO_2/PVK/mp-Al_2O_3:PVK/mp-NiO:PVK/C	15.03	21.62	0.92	0.76	Cao et al. (2015b)	2015
Planar *p-i-n* (spray pyrolysis deposition)	FTO/NiMgLiO/PVK/PCBM/Ti(Nb)O_x/Ag	15.00	20.96	1.09	0.67	Chen et al. (2015a)	2015
Multi-layered *n-i-p* (screen printing)	FTO/bl-TiO_2/mp-TiO_2: PVK/mp-ZrO_2: PVK/mp-NiO: PVK/C	14.9	21.36	0.92	0.76	Xu et al. (2015a)	2015
Planar *p-i-n* (spin coating)	FTO/NiO_x/PVK/PCBM/Ag	14.42	17.93	1.09	0.74	Yin et al. (2015)	2015

(continued)

Table 5 (continued)

Architecture	Cell structure for ***NiO***	PCE (%)	J_{sc} (mA cm^{-2})	V_{oc} (V)	FF	References	Year
Mesoscopic *n-i-p* (dip coating)	FTO/c-TiO_2/m-TiO_2/m-ZrO_2: PVK/NiO-NS	14.2	20.4	0.97	0.72	Liu et al. (2015b)	2015
Mesoscopic *p-i-n* (spray pyrolysis deposition)	FTO/NiO/meso-Al_2O_3:PVK/PCBM/BCP/Ag	13.5	18.0	1.04	0.72	Chen et al. (2015c)	2015
Planar *p-i-n* (spin coating)	ITO-PEN/NiO_x-based flexible device/PVK/PCBM/Ag	13.43	18.74	1.04	0.69	Yin et al. (2016)	2016
Planar *p-i-n* (spin coating)	ITO/NiO_x/PVK/PCBM/LiF/Al	13.4	19	1.03	0.69	Kwon et al. (2016)	2016
Mesoscopic *n-i-p* (dip coating)	FTO/c-TiO_2/m-TiO_2/m-ZrO_2:PVK/NiO-NP	12.4	19.5	0.93	0.68	Liu et al. (2015b)	2015
Mesoscopic *p-i-n* (sputtering)	ITO/bl-NiO_x/nc-NiO/PVK/PCBM/BCP/Al	11.6	19.8	0.96	0.61	Wang et al. (2014b)	2014
Multi-layered *n-i-p* (screen printing)	FTO/bl-TiO_2/mp-TiO_2:PVK/mp-NiO:PVK/*C*	11.4	18.2	0.89	0.71	Liu et al. (2015c)	2015
Planar *p-i-n* (spin coating)	FTO/NiO NCs/PVK/PCBM (1.5 wt% PS)/*Al*	10.68	15.62	1.07	0.64	Bai et al. (2015)	2015
Planar *p-i-n* (vacuum thermal evaporation)	ITO/NiO_x/PVK/PCBM/C60/BCP/*Al*	10.6	18	1.06	0.56	Xiao et al. (2016)	2016
Planar *p-i-n* (sputtering)	FTO/NiO/PVK/PCBM/BCP/Au	9.83	15.17	1.10	0.59	Mamun et al. (2017)	2017
Mesoscopic *p-i-n* (spin coating)	ITO/NiO_x/nc-NiO/PVK/PCBM/BCP/Al	9.51	13.24	1.04	0.69	Wang et al. (2014a)	2014
Planar *p-i-n* (spin coating)	FTO/NiO/PVK/PCBM/Au	9.11	16.27	0.88	0.64	Zhu et al. (2014)	2014

(continued)

Table 5 (continued)

Architecture	Cell structure for ***NiO***	PCE (%)	J_{sc} (mA cm^{-2})	V_{oc} (V)	FF	References	Year
Planar *p-i-n* (spin coating)	ITO/NiO/PVK/PCBM/Ag	8.94	14.13	1.08	0.58	Kim et al. (2015)	2015
Planar *p-i-n* (spin coating)	ITO/NiO/PVK/PCBM/BCP/Al	7.8	12.43	0.92	0.68	Jeng et al. (2014)	2014
Planar *p-i-n* e-beam evaporator	ITO/NiO_x/PVK/PCBM/BCP/Al	7.75	13.16	0.90	0.65	Wei-Chih et al. (2015)	2015
Planar *p-i-n* (spin coating)	ITO/NiO/PVK/PCBM/Al	7.6	15.4	1.05	0.48	Hu et al. (2014b)	2014
Planar *p-i-n* (electro deposition)	FTO/NiO/PVK/PCBM/Ag	7.26	14.2	0.79	0.65	Subbiah et al. (2014)	2014
Mesoscopic *n-i-p* (screen printing)	FTO/bl-NiO/nc-NiO/PVK/PCBM/Al	1.50	4.94	0.83	0.35	Tian et al. (2014)	2014
Mesoscopic *n-i-p* (screen printing)	FTO/NiO_x/PVK/Pt/FTO	0.71	9.47	0.21	0.36	Wang et al. (2014c)	2014

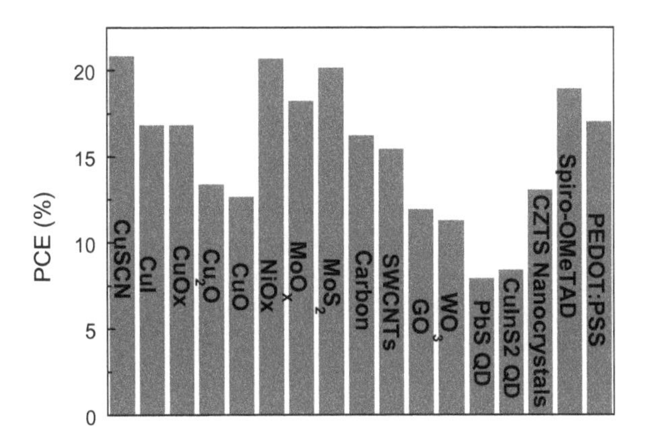

Fig. 17 Graphical representation of PCE value associated, different HTM materials used in perovskite solar cells

introduced them the most commonly used organic HTM layer in PSCs which can provide the high output (Huang et al. 2016a). Spiro-OMeTAD was introduced by Bach et al. in 1998 (Bach et al. 1998) in DSSCs to use them as an efficient heterojunction layer formed with dye absorbers achieved a good PCE. Recently, many researchers extensively utilized the spiro-OMeTAD as HTM layer in solid-state DSSCs. Using spiro-OMeTAD, the PSCs boosted the device PCE up to 22%. The devices fabricated with spiro-OMeTAD suffered low stability toward water, light, and heat due to amorphous nature and the chemical structure spiro-OMeTAD. Pristine and various-doped spiro have been used as a HTM layer in PSCs to overcome the low stability of the devices. These various dopants in the spiro-OMeTAD enhance the electrical conductivity, stability of the device. The first doping effect was employed in spiro-OMeTAD by lithium and antimony-based salts $Li[(CF_3SO_2)_2N]$ (Li-TFSI) and achieved the highest PCE of 7.2% (Burschka et al. 2013). Doping of antimony in spiro-OMeTAD resulted in generation of free charge carriers via oxidization. Generated Li^+ ions inside the system by the ionic lithium improve the device performance, but due to extreme hygroscopic nature of lithium accelerates the decomposition of perovskite and degrades the device performance quickly. The presence of pinhole channels from the bottom to the top across the organic transporting layer spiro-OMeTAD accelerates the degradation processes, the pinholes are generated from the migration of Li-TFSI film, Hawash et al. (2015). Hua et al. (2016) had successfully demonstrated the use of fluorine-doped spiro-OMeTAD and found enhanced stability of PSCs. Improvement of the device performance was observed when non-hygroscopic materials (tetrafluoro-tetra cyanoquinodimethane [F4-TCNQ])-doped spiro-OMeTAD used instead of pristine spiro-OMeTAD. The energy offset between the perovskite and the spiro-OMeTAD has to be small enough in order to achieve a high V_{oc} (Ou et al. 2017). In 2018, Hawash et al. observed that spiro-OMeTAD with various new additives and dopants excels pristine spiro-OMeTAD both in terms of device performance and stability (Hawash et al. 2018) (Fig. 18; Table 7).

PEDOT:PSS mixed polymer-based HTMs. The HOMO energy level of PEDOT film at −5.0 eV facilities the hole extraction from the perovskite layer due to band

Table 6 Performance of various carbon and carbon CNT HTM layer-based PSCs

Architecture	Cell structure for ***carbon***	PCE (%)	J_{sc} (mA cm^{-2})	V_{oc} (V)	FF	References	Year
Carbon mesoporous *p-i-n* (e-beam)	FTO/PEDOT:PSS/$MAPbI_{3-x}Cl_x$/PCBM/C	16.2	23.69	0.96	0.71	Wang et al. (2016b)	2016
Carbon mesoporous *n-i-p* (doctor blading)	FTO/TiO_2/mp-TiO_2/PVK/CuPc nanorods/C	16.1	20.8	1.05	0.74	Zhang et al. (2016c)	2016
Carbon Multi-layered *n-i-p* (screen printing)	FTO/TiO_2/mp-TiO_2/ZrO_2/PVK/C	15.60	21.45	0.94	0.77	Rong et al. (2017)	2017
Carbon mesoporous *n-i-p* (doctor blading)	FTO/TiO_2/mp-TiO_2/PVK/doped-TPDI/C	15.5	20.1	1.03	0.75	Zhang et al. (2015b)	2015
Carbon mesoporous *n-i-p* (doctor blading)	FTO/TiO_2/mp-TiO_2/PVK/Spiro-OMeTAD/C	15.29	20.42	1.12	0.67	Gholipour et al. (2016)	2016
Carbon/graphite multi-layered *n-i-p* (screen printing)	FTO/c-TiO_2/TiO_2/Al_2O_3/NiO/C/PVK	15.03	21.62	0.91	0.76	Cao et al. (2015b)	2015
Carbon mesoporous *n-i-p* (screen printing)	FTO/TiO_2/mp-TiO_2/PVK/Al_2O_3/C	15.0	22.43	0.89	0.75	Tsai et al. (2017)	2017
Carbon multi-layered *n-i-p* (blade coating)	FTO/TiO_2/mp-TiO_2/PVK/ZrO_2/NiO/C/graphite	14.9	21.36	0.92	0.76	Cao et al. (2015b)	2015
Carbon mesoporous *n-i-p* (paint)	FTO/TiO_2/mp-TiO_2/PVK/C	14.58	21.83	1.0	0.67	Chang et al. (2016b)	2016
Carbon mesoporous *n-i-p* (screen printing)	FTO/TiO_2/mp-TiO_2/PVK/C	14.38	21.27	1.04	0.65	Chen et al. (2016f)	2016
Carbon multi-layered *n-i-p* (screen printing)	FTO/TiO_2/mp-TiO_2/ZrO_2/PVK/C	14.35	19.31	1.00	0.74	Hou et al. (2017)	2017

(continued)

Table 6 (continued)

Architecture	Cell structure for ***carbon***	PCE (%)	J_{sc} (mA cm^{-2})	V_{oc} (V)	FF	References	Year
Carbon mesoporous *n-i-p* (screen printing)	FTO/TiO_2/mp-TiO_2/PVK/NiO nanosheets/C	14.2	20.4	0.97	0.72	Liu et al. (2015c)	2015
Carbon Multi-layered *n-i-p* (screen printing)	FTO/m-TiO_2/ZrO_2/*C* (perovskite infiltration)	14.3	21.5	0.86	0.77	Hashmi et al. (2017)	2017
C60 mesoporous *p-i-n* (e-beam)	FTO/PEDOT:PSS/PVK/PCBM/C60	14.0	22.47	0.97	0.64	Mamun et al. (2017)	2017
Carbon multi-layered *n-i-p* (screen printing)	FTO/TiO_2/mp-TiO_2/ZrO_2/PVK/C	13.89	19.21	0.92	0.78	Chen et al. (2016c)	2016
Carbon mesoporous *n-i-p* (hot press)	FTO/TiO_2/PVK/Carbon/Al	13.53	21.30	1.00	0.63	Wei et al. (2015a)	2015
Carbon multi-layered *n-i-p* (screen printing)	FTO/TiO_2/mp-TiO_2/ZrO_2/PVK/C	13.41	22.93	0.87	0.67	Yang et al. (2015b)	2015
Carbon mesoporous *n-i-p* (screen printing)	FTO/TiO_2/mp-TiO_2/PVK/ZrO_2/C	13.24	18.15	0.96	0.76	Chen et al. (2016d)	2016
Carbon multi-layered *n-i-p* (screen printing)	FTO/bl-TiO_2/mp-TiO_2: PVK/mp-ZrO_2: PVK/C	13.14	20.45	0.90	0.72	Liu et al. (2015d)	2015
Carbon multi-layered *n-i-p* (screen printing)	FTO/bl-TiO_2/mp-TiO_2: PVK/mp-ZrO_2:PVK/C	12.84	22.8	0.86	0.66	Liu et al. (2015d)	2015
Carbon mesoporous *n-i-p* (screen printing)	FTO/TiO_2/mp-TiO_2/PVK/Al_2O_3/C	12.3	20.04	0.85	0.72	Chan et al. (2016)	2016
Carbon multi-layered *n-i-p* (printed)	FTO/m-TiO_2/Al_2O_3/NiO_x/C (dip coating PVK)	12.12	17.22	0.95	0.69	Behrouznejad et al. (2017)	2017
Carbon mesoporous *n-i-p* (doctor blading)	FTO/TiO_2/mp-TiO_2/PVK/C	12.02	22.67	1.00	0.53	Nouri et al. (2017)	2017

(continued)

Table 6 (continued)

Architecture	Cell structure for ***carbon***	PCE (%)	J_{sc} (mA cm^{-2})	V_{oc} (V)	FF	References	Year
Carbon multi-layered *n-i-p* (printing)	FTO/m-TiO_2/2D-3D PVK/ZrO_2/C	11.9	23.60	0.86	0.59	Grancini et al. (2017)	2017
Carbon multi-layered *n-i-p* (screen printing)	FTO/bl-TiO_2/mp-TiO_2: PVK/mp-ZrO_2:PVK/C	11.63	18.06	0.89	0.72	Zhang et al. (2015a)	2015
Carbon planar *p-i-n* (inkjet printing)	FTO/TiO_2/PVK/C	11.60	17.20	0.95	0.71	Wei et al. (2014)	2014
Carbon mesoporous *n-i-p* (spin coating)	FTO/TiO_2/PVK/C	11.44	21.43	0.89	0.60	Zhang et al. (2016d)	2016
Carbon multi-layered *n-i-p* (screen printing)	FTO/TiO_2/NiO/PVK/*C*	11.4	18.2	0.89	0.71	Liu et al. (2015c)	2015
Carbon mesoporous *n-i-p*	FTO/TiO_2/mp-TiO2/PVK/mp-ZrO_2/C	11.07	–	–	–	Xu et al. (2017)	2017
Carbon/graphite multi-layered *n-i-p* (screen printing)	FTO/TiO_2/mp-TiO2/Al_2O_3/PVK/C	11.03	15.37	1.04	0.69	Cao et al. (2015c)	2015
Carbon mesoporous *n-i-p* (rolling transfer)	FTO/TiO_2/PVK/C (cathode soot)	11.02	17.00	0.90	0.72	Sepalage et al. (2015)	2015
Carbon mesoporous *n-i-p* (doctor blading)	FTO/TiO_2/mp-TiO_2/PVK/*C*/PDMS	10.8	23.5	0.97	0.47	Liu et al. (2016b)	2016
Carbon/graphite multi-layered *n-i-p* (screen printing)	FTO/d-TiO_2/mp-TiO_2/PVK/ZrO_2/C/graphite	10.64	20.1	0.87	0.61	Rong et al. (2014)	2014
Carbon mesoporous *n-i-p* (blade coating)	FTO/d-TiO_2/W-doped mp-TiO_2/PVK/C	10.53	20.79	0.86	0.59	Xiao et al. (2017)	2017

(continued)

Table 6 (continued)

Architecture	Cell structure for ***carbon***	PCE (%)	J_{sc} (mA cm^{-2})	V_{oc} (V)	FF	References	Year
Carbon mesoporous *n-i-p* (doctor blading)	FTO/TiO_2/mp-TiO_2/PVK/C	10.4	20.1	0.80	0.64	Li et al. (2017b)	2017
Carbon Mesoporous *n-i-p* (screen printing)	FTO/TiO_2/mp-TiO_2/PVK/C	10.19	19.1	0.97	0.55	Chen et al. (2015d)	2015
Carbon mesoporous *n-i-p* (hot press)	FTO/TiO_2/PVK/C/graphite sheet	10.20	18.73	0.95	0.57	Yang et al. (2014)	2014
Carbon mesoporous *n-i-p* (screen printing)	FTO/c-TiO_2/silver contact/mp-TiO_2/PVK/ZrO_2/C	9.53	16.57	0.86	0.67	Wang et al. (2016b)	2016
Carbon mesoporous *n-i-p* (spin coating)	FTO/TiO_2/PVK/C	9.35	15.98	0.90	0.65	Wei et al. (2015b)	2015
Carbon multi-layered *n-i-p* (screen printing)	FTO/TiO_2/mp-TiO_2/ZrO_2/PVK/C	9.10	15.10	0.88	0.68	Hou et al. (2017)	2017
Carbon mesoporous *n-i-p* (doctor blading)	FTO/TiO_2/PVK/C	9.08	21.02	0.80	0.54	Zhou et al. (2014a)	2014
Carbon mesoporous *n-i-p* (doctor blading)	FTO/C-ZnO/PVK/C	8.73	19.98	0.81	0.54	Zhou et al. (2015)	2015
Carbon mesoporous *n-i-p* (doctor blading)	FTO/TiO_2/mp-TiO_2/PVK/Spiro-OMeTAD/C	8.70	18.42	1.08	0.43	Gholipour et al. (2016)	2016
Carbon mesoporous *n-i-p* (spin coating)	FTO/TiO_2/PVK/*C*	8.61	14.20	1.01	0.60	Chen et al. (2014)	2014
Carbon mesoporous *n-i-p* (doctor blading)	FTO/TiO_2/PVK/C	8.31	16.78	0.90	0.55	Zhang et al. (2014)	2014
Carbon multi-layered *n-i-p* (screen printing)	FTO/d-ZnO/PVK/ZnO NR layer/ZrO_2/C	8.23	14.82	0.96	0.58	Wang et al. (2016b)	2016

(continued)

Table 6 (continued)

Architecture	Cell structure for ***carbon***	PCE (%)	J_{sc} (mA cm^{-2})	V_{oc} (V)	FF	References	Year
Carbon mesoporous *n-i-p* printing	FTO/TiO_2/mp-TiO_2/PVK/C	8.09	8.35	1.35	0.72	Chen et al. (2016e)	2016
Carbon multi-layered *n-i-p* (doctor blading)	FTO/TiO_2/mp-TiO_2/porous Al_2O_3/PVK/*C*	8.0	15.1	0.78	0.68	Xu et al. (2014)	2014
Carbon Planar *p-i-n* (doctor blading)	FTO/TiO_2/PVK/*C*	8.07	18.56	0.77	0.56	Zhou et al. (2015)	2015
Carbon mesoporous *n-i-p* (doctor blading)	FTO/TiO_2/mp-TiO_2/PVK/*C*	7.29	18.40	0.81	0.50	Yue et al. (2016)	2016
Carbon multi-layered *n-i-p* (screen printing)	FTO/bl-TiO_2/mp-TiO_2: PVK/mp-ZrO2:PVK/*C*	7.08	15.24	0.87	0.54	Wang et al. (2015b)	2015
Carbon mesoporous *n-i-p* (doctor blading)	FTO/TiO_2/mp-TiO_2/PVK/*C*	6.88	18.3	0.85	0.44	Liu et al. (2016c)	2016
Carbon mesoporous *n-i-p* (doctor blading)	FTO/TiO_2/mp-TiO_2/PVK/boron and phosphorus co doped *C*/Al	6.78	12.35	0.90	0.61	Chen et al. (2017c)	2017
Carbon multi-layered *n-i-p* (screen printing)	FTO/bl-TiO_2/mp-TiO_2: PVK/mp-ZrO_2:PVK/*C*	6.64	12.4	0.88	0.61	Ku et al. (2013)	2013
Carbon multi-layered *n-i-p* (screen printing)	FTO/d-ZnO/PVK/(ZnO/TiO_2 NR layer)/ZrO_2/*C*	5.56	14.19	0.84	0.46	Wang et al. (2016b)	2016
Carbon mesoporous *n-i-p* (screen printing)	FTO/TiO_2/mp-TiO_2/PVK/Al_2O_3/*C*	5.13	19.63	0.46	0.57	Tsai et al. (2016)	2016
Carbon/graphite mesoporous *n-i-p* (printing)	FTO/TiO_2/mp-TiO_2/PVK/C	5.0	5.7	1.29	0.68	Chang et al. (2016a)	2016

(continued)

Table 6 (continued)

Architecture	Cell structure for ***carbon***	PCE (%)	J_{sc} (mA cm^{-2})	V_{oc} (V)	FF	References	Year
Carbon planar *p-i-n* (doctor blading)	ITO/PEN/ZnO/PVK/*C*	4.29	13.38	0.76	0.42	Zhou et al. (2015)	2015
Carbon mesoporous *n-i-p* (soot of burning candle)	FTO/TiO_2/mp-TiO_2/PVK/spiro-OMeTAD/*C*/FTO	4.24	12.30	0.82	0.42	Zhang et al. (2016b)	2016
SWCNTs:spiro-OMeTAD mesoporous *n-i-p* (CVD process)	FTO/TiO_2/PVK/SWCNTs: Spiro-OMeTAD	15.5	20.3	1.1	0.61	Im et al. (2013)	2011
SWCNTs mesoporous *n-i-p* (spin coating)	FTO/TiO_2/mp-Al_2O_3:PVK/SWCNTs/spiro-OMeTAD/Ag	15.4	21.4	1.02	0.71	Habisreutinger et al. (2014b)	2014
SWCNTs Mesoporous *n-i-p* (spin coating)	FTO/TiO_2/mp-Al_2O_3:PVK/SWCNTs:P3HT-PMMA/Ag	15.3	22.71	1.02	0.66	Habisreutinger et al. (2014a)	2014
MWCNTs mesoporous *n-i-p* (drop casting)	FTO/TiO_2/mp-TiO_2/PVK/Al_2O_3/*B*-doped MWNTs	15.23	–	0.92	0.77	(Zheng et al. 2017)	2017
GO mesoporous *n-i-p* (spin coating)	FTO/TiO_2/PVK/graphene oxide/spiro-OMeTAD/*Au*	15.1	20.2	1.04	0.73	Li et al. (2014a)	2014
MWCNTs planar *n-i-p* (spin coating)	FTO/TiO_2/PVK/MWCNT :spiro-OMeTAD/*Au*	15.1	–	–	–	Lee et al. (2015b)	2015
SWCNTs-carbon mesoporous *n-i-p* (spin coating)	FTO/TiO_2/Al_2O_3/PVK/SWCNTs-*C*	14.7	21.26	1.01	0.69	Li et al. (2016b)	2016
SWCNTs mesoporous *n-i-p* Press transfer	FTO/TiO_2/mp-TiO_2/PVK/spiro-OMeTAD/SWCNTs	13.6	20.3	1.1	0.61	Im et al. (2013)	2011

(continued)

Table 6 (continued)

Architecture	Cell structure for ***carbon***	PCE (%)	J_{sc} (mA cm^{-2})	V_{oc} (V)	FF	References	Year
CNTs Mesoporous *n-i-p* (dripping/screen printing)	FTO/b-TiO_2/mp-TiO_2/PVK/CNT/*C*	13.57	18.97	1.00	0.71	Ryu et al. (2017)	2017
SWCNTs/GO/PMMA mesoporous *n-i-p* (spin coating)	FTO/TiO_2/PVK/SWCNTs/GO/PMMA	13.3	19.4	0.95	0.72	Wang et al. (2016c)	2016
Thiolatednano graphene mesoporous *n-i-p* (spin coating)	FTO/TiO_2/PVK/thiolated nanographene	12.81	20.56	0.95	0.66	Kim et al. (2014a)	2014
SWCNTs mesoporous *n-i-p* (screen printing)	FTO/TiO_2/mp-TiO_2/Al_2O_3/NiO/SWNCTs	12.7	20.7	0.95	0.64	Liu et al. (2017b)	2017
MWCNTs mesoporous *n-i-p* (drop cast)	FTO/TiO_2/PVK/MWCNTs	12.67	18.00	0.88	0.80	Wei et al. (2015b)	2015
Graphene planar *p-i-n* (chemical vapor deposition)	FTO/TiO_2/PVK/spiro-OMeTAD/PEDOT:PSS/graphene	12.37	19.17	0.96	19.17	You et al. (2015)	2015
GO planar *n-i-p* (spin coating)	–	11.90	19.18	0.88	0.71	Liu et al. (2015e)	2015
MWCNTs multi-layered *n-i-p* (doctor blading)	FTO/d-TiO_2/(TiO_2/SiO_2)/PVK/*C*	11.6	21.3	0.93	0.59	Cheng et al. (2016)	2016
Multi-layered graphene mesoporous *n-i-p* (spin coating)	FTO/TiO_2/PVK/multi-layered graphene	11.5	16.7	0.94	0.73	Yan et al. (2015)	2015

(continued)

Table 6 (continued)

Architecture	Cell structure for ***carbon***	PCE (%)	J_{sc} (mA cm^{-2})	V_{oc} (V)	FF	References	Year
CNTs planer *p-i-n* (PDMS stamp)	ITO/PEDOT:PSS/PVK/PCBM/CNTs	11.16	19.5	0.97	0.59	Mielczarek and Zakhidov (2014)	2014
GO planar *p-i-n* (spin coating)	ITO/graphene oxide/PVK/PCBM/ZnO/*Al*	11.11	15.59	0.99	0.72	Wu et al. (2014)	2014
rGO planar *p-i-n* (spin coating)	ITO/reduced graphene oxide/PCBM/PCB/*Ag*	10.8	15.4	0.98	0.72	Yeo et al. (2015)	2015
CNTs mesoporous *n-i-p* (transfer)	FTO/TiO_2/mp-TiO_2/PVK/Al_2O_3/CSCNTs/PMMA	10.54	17.22	0.85	0.71	Luo et al. (2016)	2016
CNTs *p-i-n*	ITO/PEDOT:PSS/PVK/PCBM/CNTs	10.5	18.1	0.79	0.73	Jeon et al. (2017)	2017
MWCNTs mesoporous *n-i-p* (spin coating)	FTO/TiO_2/PVK/MWCNTs	10.30	15.60	0.88	0.75	Wei et al. (2015b)	2015
Graphene mesoporous *n-i-p* (doctor blading)	FTO/TiO_2/(PVK/3DHG)/3DHG	10.06	0.89	0.63	18.11	Wei et al. (2017)	2017
SWCNTs/GO mesoporous *n-i-p* (spin coating)	FTO/TiO_2/PVK/SWCNTs/GO	9.8	19.1	0.89	0.57	Liu et al. (2017b)	2017
CNTs film Mesoporous *n-i-p* (attaching)	FTO/TiO_2/PVK/CNT film:spiro-OMeTAD	9.90	18.1	1.00	0.55	Li et al. (2014b)	2014
CNTs mesoporous *n-i-p* (transfer)	FTO/TiO_2/mp-TiO_2/PVK/Al_2O_3/CSCNTs/PMMA	9.37	16.21	0.84	0.69	Luo et al. (2016)	2016
Iodide-r-GO mesoporous *n-i-p* (spin coating)	FTO/TiO_2/PVK/iodide-reduced graphene oxide: spiro-OMeTAD/Au	9.31	16.73	0.91	0.61	Luo et al. (2015)	2015

(continued)

Table 6 (continued)

Architecture	Cell structure for ***carbon***	PCE (%)	J_{sc} (mA cm^{-2})	V_{oc} (V)	FF	References	Year
SWCNTs mesoporous *n-i-p* (chemical vapor deposition)	FTO/TiO_2/PVK/SWCNTs	9.1	20.3	0.97	0.46	Im et al. (2013)	2011
CNTs mesoporous *n-i-p* (transfer)	FTO/TiO_2/mp-TiO_2/PVK/Al_2O_3/CSCNTs/PMMA	8.60	14.91	0.85	0.68	Luo et al. (2016)	2016
CNTs mesoporous *n-i-p* (transfer)	FTO/TiO_2/mp-TiO_2/PVK/Al_2O_3/CSCNTs/PMMA	8.35	15.81	0.82	0.64	Luo et al. (2016)	2016
CNTs mesoporous *n-i-p* (chemical vapor deposition)	*Ti* foil/TiO_2 nanotube/PVK/spiro-OMeTAD: CNTs	8.31	14.36	0.99	0.68	Wang et al. (2015c)	2015
Bamboo-structured CNTs mesoporous *n-i-p* (spin coating)	FTO/TiO_2/PVK/bamboo-structured CNTs:P3HT/Au	8.3	18.75	0.86	0.52	Cai et al. (2015)	2015
CNTs film mesoporous *n-i-p* (attaching)	FTO/TiO_2/PVK/CNTs film	6.87	15.46	0.88	0.51	Li et al. (2014b)	2014
CNTs mesoporous *n-i-p* (transfer)	FTO/TiO_2/mp-TiO_2/PVK/Al_2O_3/CSCNTs/PMMA	6.81	14.43	0.88	0.53	Luo et al. (2016)	2016
Graphene mesoporous *n-i-p* (spin coating)	FTO/TiO_2/PVK bilayer/SG	6.7	14.2	0.54	0.878	Luo et al. (2016)	2016
Single-walled CNTs planar *n-i-p* (chemical vapor deposition)	Single-walled-CNTs/PEDOT:PSS/PVK/PCBM/*Al*	6.32	14.9	0.79	0.54	Jeon et al. (2015a)	2015

(continued)

Table 6 (continued)

Architecture	Cell structure for ***carbon***	PCE (%)	J_{sc} (mA cm^{-2})	V_{oc} (V)	FF	References	Year
Graphene planar *n-i-p* (chemical vapor deposition)	FTO/TiO_2/PVK/spiro-OMeTAD/graphene	6.2	12.56	0.90	0.55	Lang et al. (2015)	2015
Graphite mesoporous *n-i-p* (spin coating)	FTO/TiO_2/PVK/Graphite	6.13	10.30	0.93	0.64	Wei et al. (2015b)	2015
SWCNTs mesoporous *n-i-p* (spin coating)	FTO/TiO_2/PVK/SWCNTs	4.9	10.47	0.73	0.64	Wang et al. (2016c)	2016
CNTs sheet mesoporous *n-i-p* (dip-coated)	Stainless Steel/TiO_2/PVK/spiro-OMeTAD/CNTs sheet	3.3	10.2	0.66	0.48	Qiu et al. (2014)	2014

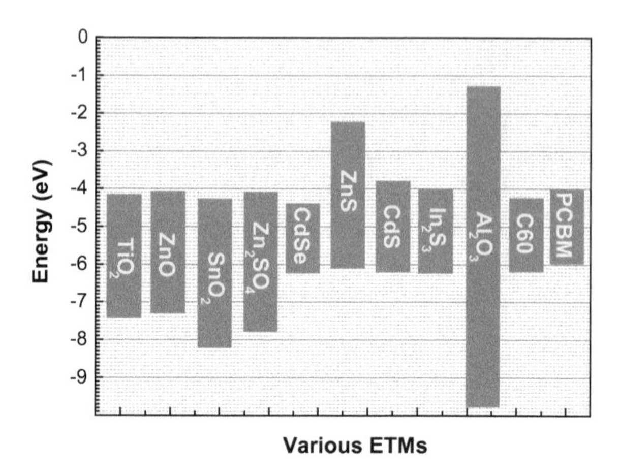

Fig. 18 Schematic energy band position of various ETMs used in PSCs

matching. In 2014, You et al. (2014) had successfully demonstrated the use of poly (3,4-ethylene dioxythiophene) polystyrene sulfonate polymer (PEDOT:PSS) as an HTM to fabricate high-efficiency PSCs at a low temperature, where PEDOT:PSS and PCBM were used as hole and electron transport layers, respectively. The device exhibited a PCE of 11.5% on glass/ITO substrate, while a 9.2% PCE was achieved in a flexible substrate (polyethylene terephthalate/ITO). The successfully grown of conducting poly (3,4-ethylene dioxythiophene) polystyrene sulfonate polymer (PEDOT:PSS) and applys them as a potential HTM in PSCs was demonstrated by Jiang et al. in 2017, (Jiang et al. 2017). They had deposited PEDOT:PSS layer on top of perovskite and showed an excellent PCE of 17.0%. One of the major drawback of PSCs fabricated by depositing PEDOT:PSS on top of perovskite layer is their limited stability in ambient atmosphere, modified by the inverted structure of PSCs, where perovskite layer was deposited on top of PEDOT: PSS layer. In 2017, Luo et al. (2017a) achieved a PCE of 15.34% by using GO-modified PEDOT:PSS. They had found that the device was less effective with the application of PEDOT:PSS (11.90%) layer in PSCs. It was also found that the devices fabricated with GO-modified PEDOT:PSS were much stable as compared to unmodified one and maintained PCE up to 83.5% of the initial PCE value after aging for 39 days. During the spin coating of the GO solution (ethanol in the GO solution), the hydrophilic PSS material can be partially removed from the surface. Energy levels of various HTMs used in PSCs and the highest achieved efficiency presented in Figs. 16 and 17, respectively.

Table 7 Performance of various spiro-OMeTAD HTM layer-based PSCs

Structure type	Device	PCE (%)	J_{sc} (mA cm^{-2})	V_{oc} (V)	FF	References	Year
Planar *p-i-n*	FTO/SnO_2/$FA_{0.83}MA_{0.17}Pb(I_{0.83}Br_{0.17})_3$/SWNTs:spiro-OMeTAD/Ag	18.9	22.10	1.14	0.75	Habisreutinger et al. (2014a)	2014
Planar *p-i-n*	FTO/TiO_2/PSK/Pristine- spiro-OMeTAD/Ag	18.7	22.90	1.12	0.73	Habisreutinger et al. (2014a)	2014
Planar *n-i-p*	FTO/TiO_2/PSK/HT1:HT2:spiro-OMeTAD/Al	18.04	21.40	1.11	0.73	Hua et al. (2016)	2016
Planar *p-i-n*	ITO/TiO_2/PSK/CuSCN: spiro-OMeTAD/Ag	18.02	21.01	1.06	0.77	Li et al. (2016a)	2016
Planar *p-i-n*	FTO/TiO_2/$FA_{0.85}Cs_{0.15}PbI_3$/spiro-OMeTAD with H_2SO_4 acid/Ag	17.7	21.6	1.06	0.78	Li et al. (2017a)	2017
Planar *p-i-n*	FTO/TiO_2/$FA_{0.85}Cs_{0.15}PbI_3$/spiro with H_3PO_4 acid/Ag	17.6	21.90	1.06	0.76	Li et al. (2017a)	2017
Planar *p-i-n*	FTO/TiO_2/$FA_{0.85}Cs_{0.15}PbI_3$/spiro-OMeTAD :CF3PA/Ag	17.5	21.6	1.04	0.78	Li et al. (2017a)	2017
Planar *p-i-n*	ITO/TiO_2/PSK/CuI: spiro-OMeTAD/Ag	16.67	21.52	1.06	0.73	Li et al. (2016a)	2016
Planar *p-i-n*	FTO/TiO_2/PSK/Ni-nanobelts: spiro-OMeTAD/Ag	16.18	21.64	1.02	0.73	Liu et al. (2017a)	2017
Planar *n-i-p*	FTO/TiO_2/PSK/Li-TFSI:tBP: spiro-OMeTAD/Ag	12.66	19.70	0.97	0.64	Di Giacomo et al. (2014)	2014
Planar *n-i-p*	FTO/TiO_2/PSK/F4-TCNQ:spiro-OMeTAD/Ag	10.59	18.72	0.94	0.48	Huang et al. (2016a)	2016

5.3 PSCs with Various ETM Layers and Their Device Performance

In the PSCs, the conventional compact ETL layer was used due to two different reasons, primarily to extract the photogenerated electrons from perovskite layer, and secondly, working as a hole blocking layer. This also indicates that the compact ETL layer should able to hinder the reverse movement of the electrons from the FTO substrate to the perovskite layer. It is important for the ETL layer to have continuous, uniform, and high transparent and thin structure for the better performance of the solar cells. The $MAPbBr_3$ and $MAPbI_3$-based PSCs efficiency were limited to 3.8% while using TiO_2 as an ETM layer. Later scientists have devoted themselves to improve the device performance and achieved an efficiency of more than 15%. Recently, Yang et al. have reported 19.3% efficient PSC using a polyethyleneimine (PEI) thin layer on TiO_2 as ETM fabricated in air (Zhou et al. 2014b) (Table 8).

Recently, metal oxides have gained much attention as a suitable ETL layer due to their good stability, high electron mobility, easy processability, and high transparency. Among various metal oxide materials, TiO_2, Al_2O_3, ZnO, SnO_2, SrO_2, etc., have shown potential in the perovskite cell device performance. TiO_2 is widely used in PSCs as an ETL layer. Ultra-thin TiO_2 could smooth the surface and keep the uniformity where the mesoscopic TiO_2 has better light scattering effect prolonging the incident light path. Park et al. (Lee et al. 2014) studied the effect of the crystal phase and morphology of the TiO_2 for the device performance of PSCs. They have reported that the rutile TiO_2 film was better than the anatase TiO_2 film because of the smooth surface of perovskite capping layer and lower conduction band position of the rutile TiO_2 film than that of the perovskite layer. Mali et al. have reported the use of transporting layer made by two TiO_2 layers, atom layer deposited (ALD) ultra-thin TiO_2 was deposited on top of the surface of one-dimensional TiO_2 nanorod arrays (Mali et al. 2015a). The device structure with 4.8 nm ALD passivated TiO_2 nanorod achieved the output PCE of 13.45%. In this method, it helped the light absorption and avoided the high-temperature processed $TiCl_4$ treatment. It is conventional that the nanorods, nanotubes, and nanofibers have shown better electron transport ability than nanoparticle films because of their directional charge transport properties. The PCE-based on TiO_2 nanorods grown in different method, with water-HCl solution was found to be higher than that with ethanol-HCl solution, which could be attributed to their special orientation, good optical properties, high conductivity, fast charge transfer, and reduced charge recombination (Wu et al. 2015b). TiO_2 is not suitable for flexible devices because it needs to be annealed at high temperature to get better crystallite (Table 9).

Al_2O_3 mesoscopic scaffold was used as an ETL layer and $MAPbCl_{3-x}I_x$ as the light absorber in PSCs (Lee et al. 2012). An optimal thick Al_2O_3 layer was fabricated at low temperature by atomic layer deposition (ALD) method. Introduction of Al_2O_3 layer effectively blocks the electron recombination between the perovskite and fluorine-doped tin oxide (FTO) layer and enhances the electron transport between the junction (Zhang et al. 2017b). It was observed that the perovskite cells with a

Table 8 Performance of various metal oxides used as ETM layer in PSCs

Fabrication method	ETL thickness (nm)	Device structure	PCE (%)	J_{sc} (mA-cm^{-2})	V_{oc} (V)	FF	References	Year
SC/SG	10/50	FTO/TiO_2/ZnO/$MAPbI_3$/spiro-OMeTAD/Au	17.2	20.8	1.08	0.71	Xu et al. (2015b)	2015
PA	40	FTO/c-TiO_2/$MAPbI_3$/spiro-OMeTAD/Au	15.2	20.5	1.06	0.70	Choi et al. (2016)	2016
TO	15	FTO/c-TiO_2/mp-TiO_2: $MAPbI_3$/spiro-OMeTAD/Au	15.1	22.0	1.09	0.63	Ke et al. (2014)	2014
TO	15	FTO/c-TiO_2/mp-TiO_2: $MAPbI_3$/spiro-OMeTAD/Au	15.1	22.0	1.09	0.63	Ke et al. (2014)	2014
TO	20	FTO/c-TiO_2/mp-TiO_2: $MAPbI_3$/spiro-OMeTAD/Au	14.8	21.3	1.09	0.64	Ke et al. (2014)	2014
TO	10	FTO/c-TiO_2/mp-TiO_2: $MAPbI_3$/spiro-OMeTAD/Au	14.6	22.2	1.09	0.60	Ke et al. (2014)	2014
TO	25	FTO/c-TiO_2/mp-TiO_2: $MAPbI_3$/spiro-OMeTAD/Au	14.2	21.1	1.09	0.62	Ke et al. (2014)	2014
ALD	10	FTO/c-TiO_2/$MAPbI_{3-x}Cl_x$/P3HT/Ag	13.6	24.3	0.98	0.57	Lu et al. (2015)	2015
ALD	10	FTO/c-TiO_2/$MAPbI_{3-x}Cl_x$/P3HT/Ag	13.6	24.3	0.98	0.57	Lu et al. (2015)	2015
ED	29	FTO/c-TiO_2/mp-TiO_2: $MAPbI_3$/spiro-OMeTAD/Au	13.6	20.0	1.00	0.68	Su et al. (2015)	2015
SC	60	FTO/c-TiO_2/mp-TiO_2: $MAPbI_3$/spiro-OMeTAD/Au	13.5	21.2	1.09	0.58	Ke et al. (2014)	2014

(continued)

Table 8 (continued)

Fabrication method	ETL thickness (nm)	Device structure	PCE (%)	J_{sc} (mA-cm^{-2})	V_{oc} (V)	FF	References	Year
DC	55	FTO/c-TiO_2/mp-TiO_2: $MAPbI_3$/spiro-OMeTAD/Au	12.8	20.1	0.88	0.73	Hong et al. (2015)	2015
ALD	50	FTO/c-TiO_2/mp-TiO_2: $MAPbI_3$/spiro-OMeTAD/Au	12.6	18.7	0.93	0.72	Wu et al. (2014)	2014
ALD	5	FTO/c-TiO_2/$MAPbI_{3-x}Cl_x$/P3HT/Ag	12.5	23.6	0.95	0.56	Lee et al. (2012)	2012
SC	40	FTO/c-TiO_2/$MAPbI_3$/spiro-OMeTAD/Ag	12.5	18.1	1.03	0.67	Choi et al. (2016)	2016
TO	5	FTO/c-TiO_2/mp-TiO_2: $MAPbI_3$/spiro-OMeTAD/Au	12.3	20.7	1.07	0.56	Ke et al. (2014)	2014
RFMS	60	FTO/c-TiO_2/$MAPbI_3$/spiro-OMeTAD/Ag	12.1	20.6	1.09	0.54	Chen et al. (2015e)	2015
SC	99	FTO/c-TiO2/mp-TiO_2: $MAPbI_3$/spiro-OMeTAD/Au	11.0	18.9	0.83	0.70	Hong et al. (2015)	2015
ED	14	FTO/NS-c-TiO_2/mp-TiO_2 :$MAPbI_3$/spiro-OMeTAD/Au	10.7	17.2	0.97	0.64	Choi et al. (2016)	2016
ALD	20	FTO/c-TiO_2/$MAPbI_{3-x}Cl_x$/P3HT/Ag	10.6	23.2	0.98	0.47	Lu et al. (2015)	2015
SC	60	FTO/c-TiO_2/mp-TiO_2: $MAPbI_3$/spiro-OMeTAD/Au	10.4	17.1	1.02	0.60	Su et al. (2015)	2015
SP	50	FTO/c-TiO_2/mp-TiO_2: $MAPbI_3$/spiro-OMeTAD/Au	8.8	17.4	0.87	0.58	Wu et al. (2014)	2014

(continued)

Table 8 (continued)

Fabrication method	ETL thickness (nm)	Device structure	PCE (%)	J_{sc} (mA-cm^{-2})	V_{oc} (V)	FF	References	Year
SC	n/a	FTO/c-TiO_2/$MAPbI_{3-x}Cl_x$/P3HT/Ag	8.7	18.4	0.92	0.51	Lu et al. (2015)	2015
SG	65	FTO/c-TiO2/$MAPbI_3$/spiro-OMeTAD/Ag	8.4	19.5	0.90	0.48	Wu et al. (2015c)	2015
SG	35	FTO/c-TiO_2/$MAPbI_3$/spiro-OMeTAD/Ag	8.1	14.6	0.93	0.60	Wu et al. (2015c)	
SC	50	FTO/c-TiO_2/mp-TiO_2: $MAPbI_3$/spiro-OMeTAD/Au	6.5	15.3	0.79	0.54	Wu et al. (2014)	2014
SG	25	FTO/c-TiO_2/$MAPbI_3$/spiro-OMeTAD/Ag	4.9	9.5	0.96	0.53	Wu et al. (2015c)	2015
SP/SC	80/22	FTO/TiO_2/SnO_2/$MAPbI_3$/spiro-OMeTAD/Au	2.2	11.2	0.62	0.32	Wang et al. (2015e)	2015
SG	95	FTO/c-TiO_2/$MAPbI_3$/spiro-OMeTAD/Ag	1.5	9.9	0.83	0.19	Wu et al. (2015c)	2015

SG—sol–gel; ALD—atomic layer deposition; TO—thermal oxidation, ALD—atomic layer deposition; ED—electrodeposition; DC—dip coating; NS—nanosheet; RFMS—radio frequency magnetron sputtering; PA—potentiostatic anodization; SC—spin coating; SG—sol–gel; SP—spray pyrolysis; TO—thermal oxidation

Table 9 Performance of PSCs based on TiO_2 as ETM layer in various device structures

Device architecture	Particles size (nm)	PCE (%)	J_{sc} (mA- cm^{-2})	V_{oc} (V)	FF	References	Year
FTO/c-TiO_2/mp-TiO_2:$MAPbI_3$/spiro-OMeTAD/Au	100a	18.4	22.7	1.05	0.77	Yang et al. (2016a)	2016
FTO/c-TiO_2/mp-TiO_2:$MAPbI_3$/spiro-OMeTAD/Au	50	17.2	21.6	1.05	0.76	Do Sung et al. (2015)	2015
FTO/c-TiO_2/mp-TiO_2:$MAPbI_3$/spiro-OMeTAD/Au	50	16.9	21.9	1.02	0.76	Yang et al. (2016a)	2016
FTO/c-TiO_2/mp-TiO_2:$MAPbI_3$/spiro-OMeTAD/Au	65	15.5	21.3	1.01	0.72	Do Sung et al. (2015)	2015
FTO/c-TiO_2/mp-TiO_2:$MAPbI_3$/spiro-OMeTAD/Au	40	15.2	21.5	1.03	0.70	Do Sung et al. (2015)	2015
FTO/c-TiO_2/mp-TiO_2:$MAPbI_3$/spiro-OMeTAD/Au	250a	15.0	19.4	1.05	0.74	Huang et al. (2016b)	2016
FTO/c-TiO_2/mp-TiO_2:$MAPbI_3$/spiro-OMeTAD/Au	30	14.7	20.1	0.99	0.74	Yang et al. (2016a)	2016
FTO/c-TiO_2/mp-TiO_2:$MAPbI_3$/spiro-OMeTAD/Au	30	14.3	21.6	1.00	0.67	Do Sung et al. (2015)	2015
FTO/c-TiO_2/mp-TiO_2:$MAPbI_{2.7}Br_{0.3}$/PTAA/Au	15	12.8	18.8	1.04	0.66	Sarkar et al. (2014)	2014
FTO/c-TiO_2/mp-TiO_2 :$MAPbI_{2.7}Br_{0.3}$/PTAA/Au	10	11.7	18.6	1.00	0.63	Sarkar et al. (2014)	2014

5 nm Al_2O_3 layer revealed a PCE of 16.2%, which is much higher than the device fabricated without Al_2O_3 layer (PCE ~ 11.0%) (Table 10).

In an inverted PSCs, ZnO and SnO_2 are usually used as an ETL layer because of their excellent electron mobility property. Low-temperature synthesized ZnO layer was used as a transporting layer in PSCs and the device exhibited a PCE value of 15.7% (Liu and Kelly 2014). SnO_2 has higher electron mobility and deeper conduction band than TiO_2. It was reported that SnO_2 shows better environmental stability with beneficial for charge transportation from perovskite layer to electron transport layer. SnO_2 layer is generally fabricated at low temperature, and the corresponding solar cells had achieved a PCE of 13%. Such devise showed a good device stability in ambient environment (Song et al. 2015a). A higher device efficiency of 17.2% was achieved by preparing the SnO_2 as a transporting layer. This layer was fabricated by spin coating of $SnCl_2{\cdot}H_2O$ precursor and annealing at 180 °C in the air (Ke et al. 2015a). A high efficiency of 19.9% was achieved by adopting solution-processed SnO_2 that was fabricated at the temperature of 150 °C (Jiang et al. 2016). WO_3 transporting layer are also exhibiting a good stability and higher mobility compared to TiO_2 layer. It was observed that WO_3-based devices are more sensitive to ambient moisture compared to TiO_2-based devices. The WO_3 ETM layer-based PSCs also degrade faster (Gheno et al. 2017). These devices showed better photovoltaic device performance when TiO_2 NPs were covered on WO_3 thin-film surface. Amorphous WO_x:TiO_x composites were fabricated at a relatively low temperature, and they were very effective as a ETL layer in PSCs. The addition of TiO_x and WO_x could raise the Fermi level and simultaneously suppress the non-radiative charge recombination at the perovskite interfaces (Wang et al. 2016d) (Table 11).

Doping in the metal oxides reduces the surface vacancies and other defects and reduces the charge recombination probability, which could exhibit better device performance due to their better electron mobility and suitable energy levels. The oxygen vacancies on TiO_2 surface are not beneficial for charge transport and also detrimental for efficient device performance (Leijtens et al. 2013). A 15% enhancement in short-circuit current density was observed when the device fabricated with Y^{3+}-doped TiO_2 layer. TiO_2 doped with Y^{3+} was used to modify the morphology of the perovskite active layer and improved the electron transfer properties (Qin et al. 2014b). Al-doped TiO_2 shown enhanced conductivity as the doping might remove the oxygen defects from the TiO_2 lattice causes the device stability (Pathak et al. 2014). Mg-doped TiO_2 demonstrated better optical properties and better energy level alignment with the perovskite active layer which provided better electron transportation (Wang et al. 2015). Nitrogen-doped ZnO (N:ZnO) nanorods also enhance the electron mobility and corresponding PSC devices exposed higher PCE (Mahmood et al. 2015b) (Table 12).

Fullerenes and their derivatives. Fullerene and its derivatives are commonly used as an ETL layer in inverted PSCs due to their band matching with perovskite layer and better electron transport ability. C60, phenyl-C60-butyric acid methyl ester (PC61BM), and indene-C60 bisadduct (ICBA) were firstly employed as ETL layers in $MAPbI_3$-based PSCs (Jeng et al. 2013). The open-circuit voltage of corresponding

Table 10 Performance of PSCs based on ZnO as ETM layer in various solar cells

Structure of ZnO ETL	Device structure	PCE (%)	J_{sc} (mA cm^{-2})	V_{oc} (V)	FF	References	Year
Compact	ITO/c-ZnO/$MAPbI_3$/PTAA/Au	17.7	20.1	1.13	0.78	Heo et al. (2016)	2016
Compact	ITO/c-ZnO/$MAPbI_3$/spiro-OMeTAD/Ag	15.7	20.4	1.03	0.75	Wang et al. (2015c)	2015
Compact	PEN/ITO/c-ZnO/$MAPbI_3$/PTAA/Au	15.6	18.7	1.10	0.76	Heo et al. (2016)	2016
Compact	FTO/c-ZnO/$MAPbI_3$/spiro-OMeTAD/Ag	14.2	19.5	1.03	0.71	Hadouchi et al. (2016)	2016
Compact	ITO/c-ZnO/$MAPbI_3$/spiro-OMeTAD/Ag	13.9	19.9	1.07	0.65	Song et al. (2015b)	2015
Nanowall	ITO/mp-ZnO: $MAPbI_3$/spiro-OMeTAD/Ag	13.6	18.9	1.00	0.72	Tang et al. (2016)	2016
Compact	ITO/c-ZnO/$MAPbI_3$/spiro-OMeTAD/MoO_3/Ag	13.4	22.4	1.04	0.57	Liang et al. (2014)	2014
Compact	FTO/c-ZnO/mp-Al_2O_3: $MAPbI_3$/spiro-OMeTAD/Ag	13.1	20.4	0.98	0.66	Dong et al. (2014)	2014
Compact	ITO/c-ZnO/PCBM/$MAPbI_3$/spiro-OMeTAD/MoO_3/Ag	12.2	18.2	1.00	0.67	Kim et al. (2014c)	2014
Nanorod	FTO/c-ZnO/mp-ZnO:$MAPbI_3$/spiro-OMeTAD/Au	11.1	20.1	0.99	0.56	Son et al. (2014)	2014
Compact	ITO/c-ZnO/$MAPbI_3$/spiro-OMeTAD/Ag	10.91	22.6	0.91	0.53	Zhang et al. (2015)	2015
Nanosheet + nanorod	FTO/c-ZnO/mp-ZnO:$MAPbI_3$/spiro-OMeTAD/Ag	10.4	18.0	0.93	0.62	Mahmood et al. (2014)	2014

(continued)

Table 10 (continued)

Structure of ZnO ETL	Device structure	PCE (%)	J_{sc} (mA cm^{-2})	V_{oc} (V)	FF	References	Year
Compact	ITO/c-ZnO/PEI/$MAPbI_3$/spiro-OMeTAD/Au	10.2	16.8	0.88	0.69	Cheng et al. (2015)	2015
Compact	PET/ITO/c-ZnO/$MAPbI_3$/spiro-OMeTAD/Ag	10.2	13.4	1.03	0.74	Wang et al. (2015c)	2015
Nanorod	FTO/c-ZnO/mp-ZnO:$MAPbI_3$/spiro-OMeTAD/Au	8.9	17.0	1.02	0.51	Kumar et al. (2013)	2013
Compact	ITO/c-ZnO/$MAPbI_3$/PTB7-Th/MoO_3/Ag	8.4	14.3	0.86	0.68	Kim et al. (2014c)	2014
Compact	ITO/c-ZnO/PCBM/$MAPbI_3$/spiro-OMeTAD/Au	6.4	16.0	0.88	0.46	Cheng et al. (2015)	2015
Nanorod	PET/ITO/c-ZnO/mp-ZnO:$MAPbI_3$/spiro-OMeTAD/Au	2.6	7.5	0.80	0.43	Kumar et al. (2013)	2013

Table 11 Performance of PSCs based on SnO_2 as ETM layer in various device structures

Material of c-ETL/mp-ETL	Device structure	PCE (%)	J_{sc} (mA cm^{-2})	V_{oc} (V)	FF	References	Year
SnO_2/—	FTO/c-SnO_2/C60-SAM/$MAPbI_3$/spiro-OMeTAD/Au	19.0	21.4	1.13	0.79	Wang et al. (2016e)	2016
SnO_2/—	FTO/c-SnO_2/$MAPbI_3$/spiro-OMeTAD/Au	18.2	22.8	1.10	0.73	Ma et al. (2017)	2017
SnO_2/—	FTO/c-SnO_2/PCBM: $MAPbI_3$/spiro-OMeTAD/Au	17.3	21.5	1.11	0.73	Chen et al. (2017d)	2017
SnO_2/—	FTO/c-SnO_2/$MAPbI_3$/spiro-OMeTAD/Au	17.2	23.3	1.11	0.67	Ke et al. (2015a)	2015
SnO_2/—	FTO/c-SnO_2/$MAPbI_3$/spiro-OMeTAD/Au	16.1	19.5	1.10	0.74	Ke et al. (2015b)	2015
TiO_2/—	FTO/c-TiO_2/$MAPbI_3$/spiro-OMeTAD/Au	15.2	22.5	1.06	0.64	Ke et al. (2015a)	2015
SnO_2/—	FTO/c-SnO_2($TiCl_4$)/$MAPbI_3$/spiro-OMeTAD/Au	14.7	20.0	1.00	0.67	Rao et al. (2015)	2015
—/SnO_2	FTO/mp-SnO_2-TiO_2:$MAPbI_3$/spiro-OMeTAD/Ag	14.2	21.2	1.02	0.66	Han et al. (2015)	2015

(continued)

Table 11 (continued)

Material of c-ETL/mp-ETL	Device structure	PCE (%)	J_{sc} (mA cm^{-2})	V_{oc} (V)	FF	References	Year
TiO_2/TiO_2	FTO/c-TiO_2/mp-TiO_2:$MAPbI_3$/spiro-OMeTAD/Ag	13.9	20.5	0.97	0.72	Dong et al. (2015b)	2015
TiO_2/—	FTO/c-TiO_2($TiCl_4$)/$MAPbI_3$/spiro-OMeTAD/Au	13.4	19.2	1.01	0.69	Rao et al. (2015)	2015
SnO_2/—	ITO/c-SnO_2/$MAPbI_3$/spiro-OMeTAD/Ag	13.0	19.5	1.08	0.62	Song et al. (2015a)	2015
TiO_2/—	FTO/c-TiO_2/$MAPbI_3$/spiro-OMeTAD/Ag	12.3	19.9	1.00	0.62	Song et al. (2015a)	2015
SnO_2/TiO_2	FTO/c-SnO_2/mp-TiO_2:$MAPbI_3$/spiro-OMeTAD/Ag	10.3	31.0	0.86	0.39	Dong et al. (2015b)	2015
SnO_2/SnO_2	FTO/c-SnO_2/mp-SnO_2($TiCl_4$): $MAPbI_3$/spiro-OMeTAD/Au	10.2	17.4	0.93	0.63	Li et al. (2015b)	2015
—/SnO_2	FTO/mp-SnO_2($TiCl_4$) + $MAPbI_3$/spiro-OMeTAD/Au	8.5	17.2	0.80	0.62	Zhu et al. (2015)	2015
SnO_2/—	FTO/c-SnO_2/$MAPbI_3$/CuSCN/Au	8.4	19.0	0.96	0.45	Murugadoss et al. (2016)	2016

Table 12 Performance of PSCs based on WO_3 as ETM layer in various device structures

Material	Device structure	PCE (%)	J_{sc} ($mA\text{-}cm^{-2}$)	V_{oc} (V)	FF	References	Year
Fe_2O_3	FTO/c-Fe_2O_3/mp-Fe_2O_3/$MAPbI_3$/spiro-OMeTAD/Au	18.2	22.7	1.01	0.79	Luo et al. (2017b)	2017
Nb_2O_5	FTO/c-Nb_2O_5/$MAPbI_3$/spiro-OMeTAD/Au	17.1	22.9	1.04	0.72	Ling et al. (2015)	2015
CeO_2	FTO/CeO_2/PCBM/MAPbI3/spiro-OMeTAD/Ag	17.0	23.3	1.06	0.69	Wang et al. (2017b)	2017
SiO_2	FTO/c-TiO_2/mp-SiO_2: $MAPbI_{3-x}Cl_x$/spiro-OMeTAD/Ag	16.6	22.2	1.08	0.69	Yu et al. (2016)	2015
WO_3	ITO/c-WO_3/SAMs/$MAPbI_3$/spiro-OMeTAD/Ag	14.9	21.9	1.02	0.67	Hou et al. (2015)	2015
In_2O_3	FTO/c-In_2O_3/PCBM/$MAPbI_3$/spiro-OMeTAD/Au	14.8	20.1	1.08	0.69	Hu et al. (2017b)	2017
SiO_2	FTO/c-TiO_2/mp-SiO_2: $MAPbI_{3-x}Cl_x$/spiro-OMeTAD/Au	13.9	18.3	1.02	0.69	Lee et al. (2016)	2016
Al_2O_3	FTO/c-TiO_2/mp-Al_2O_3 :$MAPbI_2Cl$: spiro-OMeTAD/Ag	12.3	18.0	1.02	0.67	Ball et al. (2013)	2013
CdSe	ITO/CdSe/$MAPbI_3$/spiro-OMeTAD/Ag	11.7	17.4	0.99	0.68	Wang et al. (2014d)	2014
SiO_2	FTO/c-TiO_2/mp-SiO_2: $MAPbI_{3-x}Cl_x$/spiro-OMeTAD/Au	11.5	16.4	1.05	0.66	Hwang et al. (2014)	2014
WO_3	FTO/c-WO_3/NS-mp-WO_3($TiCl_4$): $MAPbI_3$/spiro-OMeTAD/Ag	11.2	17.0	0.87	0.76	Mahmood et al. (2015a)	2015
Fe_2O_3	FTO/c-Fe_2O_3/$MAPbI_3$/spiro-OMeTAD/Au	11.2	20.6	0.90	0.61	Hu et al. (2017b)	2017
CdS	ITO/CdS/$MAPbI_3$/spiro-OMeTAD/Au	11.2	16.1	1.05	0.66	Liu et al. (2015f)	2015
Al_2O_3	FTO/c-TiO_2/mp-Al_2O_3: $MAPbI_2Cl$:spiro-OMeTAD/Ag	10.9	17.8	0.98	0.63	Lee et al. (2012)	2015
ZrO_2	FTO/c-TiO_2/mp-ZrO_2: $MAPbI_3$/spiro-OMeTAD/Ag	10.8	17.3	1.07	0.59	Bi et al. (2013)	2013
Nb_2O_5	FTO/c-Nb_2O_5/mp-Al_2O_3: $MAPbI_{3-x}Cl_x$/spiro-OMeTAD/Au	10.3	12.8	1.13	0.72	Kogo et al. (2015)	2015
In2O_3	FTO/c-In_2O_3/$MAPbI_3$/spiro-OMeTAD/MoO_3/Al	9.9	19.7	0.95	0.53	Dong et al. (2016)	2016

(continued)

Table 12 (continued)

Material	Device structure	PCE (%)	J_{sc} (mA-cm^{-2})	V_{oc} (V)	FF	References	Year
WO_3	FTO/c-WO_3/MAPbIxCl3-x/spiro-OMeTAD/Ag	9.4	22.8	0.74	0.60	Wang et al. (2015a)	2015
WO_3	FTO/c-WO_3/NR-mp-WO_3($TiCl_4$) :$MAPbI_3$/spiro-OMeTAD/Ag	9.1	15.0	0.86	0.70	Mahmood et al. (2015b)	2015
Zn_2SnO_4	FTO/c-Zn_2SnO_4/mp-Zn_2SnO_4: MAPbI3/spiro-OMeTAD/Au	7.7	13.0	0.99	0.59	Mali et al. (2015b)	2015
$SrTiO_3$	FTO/c-TiO_2/mp-$SrTiO_3$: $MAPbI_{3-x}Cl_x$/spiro-OMeTAD/Au	7.6	14.9	0.93	0.55	Bera et al. (2015)	2015
Al_2O_3	FTO/c-TiO_2/mp-Al_2O_3: $MAPbI_2Cl$:spiro-OMeTAD/Au	7.2	12.8	0.93	0.61	Carnie et al. (2013)	2013
Zn_2SnO_4	FTO/c-Zn_2SnO_4/mp-Zn_2SnO_4: $MAPbI_3$/spiro-OMeTAD/Au	7.0	13.8	0.83	0.61	Oh et al. (2015)	2015
WO_3	FTO/c-WO_3/NP-mp-WO_3($TiCl_4$): $MAPbI_3$/spiro-OMeTAD/Ag	6.1	11.0.	0.83	0.67	Mahmood et al. (2015a)	2015
ZnS	ITO/ZnS/$MAPbI_3$/; spiro-OMeTAD/Au	1.0	2.25	0.98	0.44	Liu et al. (2015f)	2015

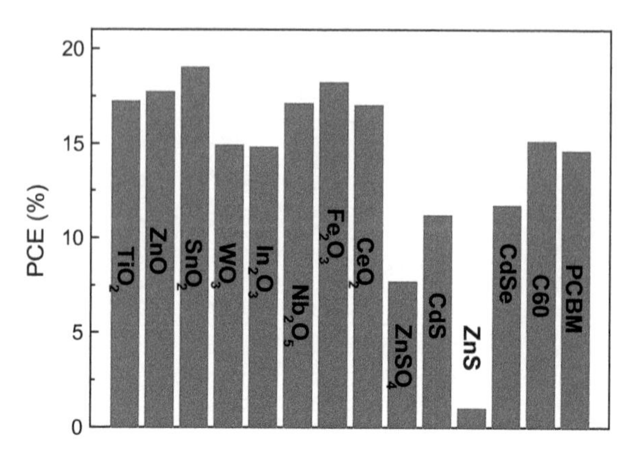

Fig. 19 Graphical representation of PCE values associated with various ETM materials used in perovskite solar cells

PSCs was achieved as 0.55, 0.65, and 0.75 V, respectively. The higher open-circuit voltage was achieved for ICBA because of its higher LUMO level compared to C60 and PC61BM. The PCE of fullerene-based PSCs is mostly related to electron mobility of these transport layers (C60, ICBA, PC61BM) (Liang et al. 2015). High mobility fullerene will beneficial for the charge transportation. Fullerene materials can also passivate the interfacial trap states of perovskite layer and reduce the energy barrier between the electrode and perovskite layer.

The new fullerene derivative C70-DPM-OE was synthesized by phenyl groups of diphenyl methano fullerene (DPM) moiety with oligoether chain (Xing et al. 2016). This oligoether chain can reduce the work function of the metal cathode and passivate the trap states of the perovskite layer. The device fabricated with C70-DPM-OE as an ETL have achieved the PCE of 16% which was much higher than the conventional PC61BM as electron transport layer. Another fullerene pyrrolidine derivative, *N*-methyl-2-pentyl-(Santra et al. 2013) fullerene pyrrolidine (NMPFP) was prepared by a simple solution process (Chen et al. 2019). The NMPFP thin film revealed a higher conductivity than the PC61BM thin film. The device fabricated with NMPFP as an ETL layer exhibited a PCE of 13.83% (Fig. 19; Table 13).

Doping in the fullerenes is a good way to improve the mobility and the device efficiency of PSCs. MAI-doped fullerene C60, exhibited dramatically increased conductivity by over 100 times (Bai et al. 2016). The iodide as a Lewis base anion in the MAI-dopant acted as an electron donor. When the iodide contacts fullerene after solvent drying, the electron density would redistribute from iodide to fullerene. Enhanced free electron density in fullerene presented a higher conductivity. The improved conductivity plays a significant role in boosting the device performance; the perovskite solar cell delivered an efficiency of 19.5% with a high fill factor of 80.6%.

Low-dimensional lead halide perovskites

In recent years, quasi-2D halide perovskites have intensively investigated as active layers in solar cells comparing with the conventional 3D perovskites structures. Smith et al. reported the solar cell from quasi-2D halide perovskites with a device structure

Table 13 Performance of PSCs based on fullerenes as ETM layer in various device structures

Material	Device structure	PCE (%)	J_{sc} (mA-cm^{-2})	V_{oc} (V)	FF	References	Year
TiO_2/IL	ITO/c-TiO_2/c-[BMIM]BF_4/$MAPbI_3$/PTAA/Au	19.6	22.8	1.12	0.77	Yang et al. (2016c)	2016
SnO_2/PCBM	FTO/c-SnO_2/PCBM/$MAPbI_3$/spiro-OMeTAD/Au	19.1	22.61	1.12	0.76	(Ke et al. 2016)	2016
TiO2/IL	FTO/c-TiO_2/c-[EMIM]PF_6/$MAPbI_3$/spiro-OMeTAD/Au	18.5	22.9	1.10	0.74	Wu et al. (2016)	2016
TiO_2/PCBM	ITO/c-TiO_2/PCBM/$MAPbI_3$/spiro-OMeTAD/Au	18.4	22.2	1.11	0.75	Kegelmann et al. (2017)	2017
TiO_2/PCBM	FTO/c-TiO_2/PCBM/$MAPbI_3$/spiro-OMeTAD/Au	17.9	21.0	1.11	0.77	Tao et al. (2015)	2015
TiO_2/graphene	FTO/c-TiO_2/mp-Gr:$MAPbI_3$/spiro-OMeTAD/Au	17.2	22.8	1.05	0.72	Tavakoli et al. (2016)	2016
TiO_2/IL	FTO/c-TiO_2/c-[EMIM]I/$MAPbI_3$/spiro-OMeTAD/Au	17.2	23.7	1.10	0.66	Wu et al. (2016)	2016
Graphene/TiO_2	FTO/c-Gr: TiO_2/Al_2O_3: $MAPbI_{3-x}Cl_x$/spiro-OMeTAD/Au	15.6	21.9	1.04	0.73	Wang et al. (2014e)	2014
PCBM	FTO/PEI/c-PCBM/$MAPbI_3$/PTAA/Au	15.3	21.8	0.98	0.72	Ryu et al. (2015)	2015
C60	ITO/c-C60/$MAPbI_3$/spiro-OMeTAD/Au	15.1	18.9	1.08	0.75	Ke et al. (2015c)	2015
IL	PET/ITO/c-BenMeIM-Cl/$MAPbI_3$/spiro-OMeTAD/Au	15.0	20.6	1.00	0.73	Yang et al. (2016b)	2016

(continued)

Table 13 (continued)

Material	Device structure	PCE (%)	J_{sc} (mA-cm^{-2})	V_{oc} (V)	FF	References	Year
C60	FTO/c-C60/$MAPbI_3$/spiro-OMeTAD/Au	14.5	19.6	1.07	0.69	Wojciechowski et al. (2015)	2015
PCBM	FTO/PCBM/$MAPbI_3$/spiro-OMeTAD/Au	14.6	20.4	1.03	0.69	Ke et al. (2016)	2016
IL	FTO/c-[EMIM]PF_6/$MAPbI_3$/spiro-OMeTAD/Au	14.2	21.6	1.05	0.63	Wu et al. (2016)	2016
PCBM	ITO/c-PC71BM/$MAPbI_3$/spiro-OMeTAD/MoO_3/Ag	13.9	20.5	1.08	0.63	Upama et al. (2017)	2017
PCBM	ITO/c-PC61BM/$MAPbI_3$/spiro-OMeTAD/MoO_3/Ag	12.7	23.9	0.84	0.63	Upama et al. (2017)	2017
IL	FTO/c-[EMIM]I/$MAPbI_3$/spiro-OMeTAD/Au	9.2	15.2	1.03	0.59	Wu et al. (2016)	2016
Graphene	FTO/c-Gr/Al_2O_3: $MAPbI_{3-x}Cl_x$/spiro-OMeTAD/Au	5.9	14.6	0.90	0.48	Wang et al. (2014e)	2014

of $FTO/TiO_2/(PEA)_2(MA)_2[Pb_3I_{10}]$/spiro-OMeTAD/Au and the device showed a PCE of 4.71% (Quan et al. 2016). In this work, the quasi-2D halide perovskite was obtained by mixing $C_6H_5(CH_3)_2NH_3I$ (PEAI), MAI and PbI_2 with a molar ratio of 2:2:3 and form a quasi-2D perovskite structure of $(PEA)_2(MA)_2[Pb_3I_{10}]$. The number of layers ($n$) was determined to be 3. It was observed that the decrease in dimension resulting an increase in the band-gap and exciton binding energy. It was shown that the devices were more stable when fabricated with quasi-2D halide perovskites and remained stable after 46 days, whereas the 3D halide perovskite started to decompose after 4–5 days, as it was evident from the XRD spectra. Improved stability of the perovskites was observed using all-inorganic halide perovskite ($CsPbX_3$), reported by many groups. $CsPbI_3$ having band-gap of 1.73 eV was the suitable for solar cells. Swarnakar et al. had grown α-$CsPbI_3$ perovskite QDs which was found stable for several months in ambient air (Fang et al. 2015). These QDs were also used as active layer in PSCs and deposited on TiO_2/FTO substrates. The $CsPbI_3$ QDs-coated substrates were then immersed in saturated methyl acetate (MeOAc) solution for several times to yield a desired thick perovskite thin film. Such QDs-based solar cells revealed a PCE of 10.77% with a V_{oc} of 1.23 eV and fill factor of 0.65. More recently, Sanehira et al. boosted the PCE up to 13.43% by using $CsPbI_3$ QDs in the device and resulted in enhancement of charge carrier mobility of QD films (Sanehira et al. 2017).

Mixed cations

FA/MA-Cs mixed-cation perovskite

The α-$Cs_xFA_{1-x}PbI_3$ perovskite layer was fabricated via a regular one-step solvent engineering method with different Cs/FA ratios. Such thin films showed comparatively more stable and do not transform from α-phase to δ_H-phase (Li et al. 2016b). $FA_{0.9}Cs_{0.1}PbI_3$ thin-film-based PSCs displayed both the superior stability and efficiency (16.5%) in solar cell devices compared to pure $FAPbI_3$ perovskite (Lee et al. 2015c). Later, Yi et al. had achieved the improved PCE to 18% by replacing both a small fraction of the iodide and bromide anions as $Cs_{0.2}FA_{0.8}PbI_{2.84}Br_{0.16}$. McMeekin et al. also reported a FA/Cs mixed cation perovskite of $FA_{0.83}Cs_{0.17}PbI_3$ with some added bromide achieve a band-gap of 1.75 eV (Yi et al. 2016). The structure of perovskite $FA_{0.83}Cs_{0.17}Pb(I_{0.6}Br_{0.4})_3$ was used in a solar cell and achieved an open-circuit voltage 1.2 volts with PCE of over 17%. The mixed perovskites of $FA_{0.83}Cs_{0.17}Pb(I_{0.6}Br_{0.4})_3$ absorber layer was used in a *p-i-n* solar cell structure where *n*-doped C60 used as an electron collecting layer where 80% of the original efficiency sustained after 650 h under ambient air without encapsulation and stable over 3400 h with encapsulation (Wu et al. 2016). MA cation had also been alloyed with Cs^+ to form the MA/Cs mixed perovskites. Chio et al. represented the use of perovskite containing Cs^+ and MA^+ with [6,6]-phenyl-C60 butyric acid methyl ester (PCBM) as an electron acceptor (Choi et al. 2014). 10% of Cs^+ ions doping in $MAPbI_3$ perovskite improved the efficiency from 5.51 to 7.68%.

Ternary cation perovskite

The successful exhibition of binary cation perovskites with enhanced performance and stability might open a new way to use of ternary cation perovskite using MA^+, FA^+, and Cs^+ cations-based perovskite in solar cells. Saliba et al. first reported the triple cation perovskite with FA/MA/Cs where Cs^+ ions could improve the film quality for FA/MA mixture (Saliba et al. 2016a). 5% CsI was incorporated into the mixed-cation perovskite which was known as $(FAPbI_3)_{0.83}(MAPbBr_3)_{0.17}$ to suppress non-perovskite phase and enhanced the crystallization process with a highly stabilized PCE (21.1%) and stability was reported. The solar cells still maintained an efficiency at 18% after 250 h. Saliba et al. investigated several alkali cations and found the radii of Rb^+ was only slightly smaller than the favorable cations of Cs^+, MA^+, and FA^+. They have successfully alloyed a small amount of RbI (about 5–10%) into Cs/MA/FA mixed-cation perovskite to achieve a record efficiency of 21.6% on small areas and 19.0% on large area (0.5 cm^2) under AM 1.5G. The addition of Rb^+ ions suppresses the formation of unwanted yellow phase of Cs- or FA-based perovskite structures and improves the stability of solar cells. These devices based on Rb/Cs/MA/FA perovskite were found 95% efficient from its initial value after 500 h at 85 °C under continuous illumination, which could meet the industrial standards for reliable solar cells (Saliba et al. 2016b). However, more than 10% RbI addition into the mixed-cation perovskite resulted in a Rb-rich phase and was destructive to the solar cells.

Mixed Cation to stabilize the Pb-Sn alloy metal halide perovskite

Due to the similar electronic structure and the similar ionic radii of Sn^{2+} (1.35 Å) to Pb^{2+} (1.49 Å), tin is a useful candidate for the application of lead-free perovskites. Hao et al. had reported the optical band-gap 1.3 eV in $MASnI_3$ with high absorption at 950 nm, which is ideal for solar cell application (Hao et al. 2014a). It was found that the presence of Pb and Sn in methylammonium iodide perovskite showed a narrower band-gap of 1.3 eV which extend the absorption onset to the near-IR region (Hao et al. 2014b; Ogomi et al. 2014; Im et al. 2015). According to Shockley–Queisser theory, the PCE of approximately 30% can be achieved by replacing 15% Pb^{+2} by Sn^{+2} ions in the perovskite structure (Anaya et al. 2016). Pure Sn-based PSCs had shown relatively low PCE apart from that their tendency of oxidizing to Sn^{+2} to Sn^{+4} state damage the perovskite structure. Zhao and co-workers had found the lesser oxidation probability observed for Sn atom in $FASnI_3$ than in $MASnI_3$ due to the stronger hydrogen bond in $FASnI_3$ (Wang et al. 2016f). Liao et al. had used the mixed cation (FA/MA) and (Sn/Pb) perovskite materials in the inverted device structure by using $FASnI_3$ and $MAPbI_3$ precursor solutions (Liao et al. 2016). The as-fabricated device structure with this perovskite ITO/PEDOT:PSS/$(FASnI_3)_{0.6}(MAPbI_3)_{0.4}$/C60/BCP/Ag showed a PCE of 15.08%. The $MA_{0.5}FA_{0.5}Pb_{0.75}Sn_{0.25}I_3$ PSCs were fabricated using one-step spin-coating method and resulted in a PCE of 14.19% with a great stability (Yang et al. 2016d). The fabricated solar cells maintained PCE of 94% of its initial value after 30 days when kept in an inert atmosphere and retained 80% of initial PCE value after 12 days when exposed to an ambient atmosphere (30–40% R.H.) (Table 14).

Table 14 Performance of PSCs based on various perovskite active layer

Device architecture	Thickness (nm)/E_g (eV)	PCE (%)	J_{sc} (mA cm^{-2})	V_{oc} (V)	FF	References	Year
FTO/c-TiO_2/mp-TiO_2/$Rb_{0.05}Cs_{0.05}(FA_{0.83}MA_{0.17})_{0.90}Pb(I_{0.83}Br_{0.17})_3$/PTAA/Au	~500/1.63	21.8	22.8	1.18	0. 81	Duong et al. (2017)	2017
FTO/c-TiO_2/mp-TiO_2/(FA/MA)Pb(I/Br)$_3$/spiro-OMeTAD/Au	~400c/1.6	20.8	24.6	1.16	73	Bi et al. (2016)	2016
FTO/c-TiO_2/mp-TiO_2/$FA_{4/6}MA_{2/6}Pb(I_{5/6}Br_{1/6})$/spiro-OMeTAD/Au	~400c/1.64	20.67	23.7	1.14	0.76	Jacobsson et al. (2016)	2016
FTO/c-TiO_2/mp-TiO_2/$Rb_{0.05}FA_{0.8}MA_{0.15}PbI_{2.55}Br_{0.45}$/spiro-OMeTAD/Au		19.6	22.5	1.17	0.75	Zhang et al. (2017c)	2017
FTO/c-TiO_2/mp-TiO_2/$Rb_{0.05}Cs_{0.10}FA_{0.85}Pb(I_{0.83}Br_{0.17})_3$/spiro-OMeTAD/Au		19.3	22.3	1.16	0.75	Saliba et al. (2016b)	2016
FTO/c-TiO_2/mp-TiO_2/$Cs_{0.05}(FA_{0.83}MA_{0.17})_{0.95}Pb(I_{0.83}Br_{0.17})_3$/spiro-OMeTAD/Au	500/1.6	19.2 ± 0.91	22.69 ± 0.75	1.132 ± 0.025	0.75 ± 0.18	Saliba et al. (2016a)	2016
FTO/SnO_2/C_{60}/$Cs_{0.17}FA_{0.83}Pb(I_{0.6}Br_{0.4})$/spiro-OMeTAD/Au		18.3	23	1.06	0. 75	Wang et al. (2017c)	2017
ITO/PTAA/$MA_{0.6}FA_{0.4}PbI_3$/ICBA/C60/BCP/Cu	1.55	18.3	23	1.03	0.77	Deng et al. (2016)	2016
FTO/c-TiO_2/mp-TiO_2/$MA_{0.91}Cs_{0.09}PbI_3$/spiro-OMeTAD/Au	~400	18.1	22.57	1.06	0.76	Niu et al. (2017)	2017
FTO/c-TiO_2/mp-TiO_2/$Cs_{0.1}FA_{0.8}MA_{0.1}Pb(I_{0.83}Br_{0.17})_3$/spiro-OMeTAD/Au	~500	18.1	21.5	1.16	0.73	Matsui et al. (2017)	2017
FTO/c-TiO_2/mp-TiO_2/$Rb_x(FAPbI_3)_{1-y}(MAPbBr_3)_y$/PTAA/Au		17.68 ± 0.91	22.86 ± 0.37	1.12 ± 0.01	0.70± 0.19	Duong et al. (2016)	2016

(continued)

Table 14 (continued)

Device architecture	Thickness (nm)/E_g (eV)	PCE (%)	J_{sc} (mA cm^{-2})	V_{oc} (V)	FF	References	Year
FTO/c-TiO_2/mp-TiO_2/$(FAPbI_3)_{0.85}Pb(MABr_3)_{0.15}$/spiro-OMeTAD/Au	mp-TiO_2 with PVSK infiltrared(∼200 nm) + capping layer (∼650 nm)/1.57−1.58	17.6	22.5	1.13	0.69	Sveinbjornsson et al. (2016)	2016
FTO/Nb-TiO_2/$FA_{0.84}MA_{0.16}Pb(I_{0.84}Br_{0.16})$/spiro-OMeTAD/Au	∼500	17.6	23.4	1.00	0.74	Chen et al. (2017f)	2017
ITO/PEDOT:PSS/$MA_{0.80}FA_{0.20}PbI_{3-y}Cl_y$/PC61BM/C60/LiF/Ag	∼280/1.58	17.45	21.55 ± 0.55	1.10 ± 0.01	0.75	Isikgor et al. (2016)	2016
FTO/c-TiO_2/mp-TiO_2/Rb(5%) in$FA_{0.75}(MA_{0.6}Cs_{0.4})_{0.25}PbI_2Br$/PTAA/Au	mp-TiO_2 with PVSK infiltrated (∼180 nm) + capping layer (∼400 nm)/1.73	17.4	18.3	>1.1		Duong et al. (2017)	2017
FTO/c-TiO_2/mp-TiO_2/$Cs_{0.2}FA_{0.8}PbI_{2.84}Br_{0.16}$/spiro-OMeTAD/Au	1.49	17.35	21.9	1.07	0.74	Yi et al. (2016)	2016
FTO/c-TiO_2/mp-TiO_2/$Cs_{0.2}FA_{0.8}PbI_{2.84}Br_{0.16}$/spiro-OMeTAD/Au	1.58	17.35	21.9	1.073	0.74	Yi et al. (2016)	2016
FTO/SnO_2/$Cs_{0.1}FA_{0.8}MA_{0.0.1}Pb(I_{0.83}Br_{0.17})_3$/spiro-OMeTAD/Au		17.3	22.4	1.13	0.68	Matsui et al. (2017)	2017
FTO/c-TiO_2/mp-TiO_2/$Rb_{0.05}FA_{0.95}PbI_3$/spiro-OMeTAD/Au	∼480 (capping layer)/1.53	17.16	23.93	1.07	0.67	Park et al. (2017)	2017
FTO/SnO_2/$PC_{60}BM$/$FA_{0.83}Cs_{0.17}Pb(I_{0.6}Br_{0.4})_3$/spiro-OMeTAD/Ag	1.74	17.1	19.4	1.2	0.75	McMeekin et al. (2016)	2016
FTO/c-TiO_2/$MA_{0.7}FA_{0.3}PbI_3$/spiro-OMeTAD/Ag	38 ± 9 1.54	17.02	22.03	1.03	0.75	Li et al. (2016d)	2016

(continued)

Table 14 (continued)

Device architecture	Thickness (nm)/E_g (eV)	PCE (%)	J_{sc} (mA cm^{-2})	V_{oc} (V)	FF	References	Year
FTO/c-TiO_2/$Cs_{0.1}FA_{0.9}PbI_3$/spiro-OMeTAD/Ag (Au)	1.55	16.5	23.5	1.06	0.66	Lee et al. (2015a)	2015
FTO/c-TiO_2/C60/$MA_{1-x}FA_xPbI_3$/spiro-OMeTAD/Au	340	16.48	22.51	1.00	0.74	Chen et al. (2017e)	2017
FTO/c-TiO_2/$Cs_xFA_{1-x}PbI_3$/spiro-OMeTAD/Ag		16.4	20.4	1.09	0.74	Liu et al. (2017c)	2017
FTO/c-TiO_2/mp-TiO_2/$Rb_{0.05}FA_{0.95}PbI_3$/spiro-OMeTAD/Au		16.2	23.8	1.03	0.66	Zhang et al. (2017c)	2017
FTO/c-SnO_2/C60-SAM/$Cs_{0.2}FA_{0.8}PbI_3$/spiro-OMeTAD/Au	435	16.18 ± 0.50	21.73 ± 0.51	1.03 ± 0.02	0.72 ± 0.12	Yu et al. (2016)	2016
FTO/c-TiO_2/$Cs_{0.15}FA_{0.85}PbI_3$/spiro-OMeTAD/Ag	1.52	16.1	20.39	1.06	0.74	Li et al. (2016c)	2016
FTO/c-TiO_2/mp-TiO_2/$FAPbI_3$-MABr/spiro-OMeTAD/Au	~450	15.98	22.17	1.07	0.67	Zheng et al. (2016)	2016
FTO/c-TiO_2/mp-TiO_2/$(MAPbI_3)_{1-x}(CsPbBr_3)_x$/spiro-OMeTAD/Au	360	15.9 ± 0.52	20.9 ± 0.42	1.07 ± 0.01	0.71 ± 0.02	Niu et al. (2016)	2016
FTO/c-TiO_2/mp-TiO_2/$MA_xFA_{1-x}PbI_3$/WO_3/Ag	1.57	15.86	20.85	1.04	0.73	Kim et al. (2016)	2016
FTO/c-TiO_2/mp-TiO_2/$Cs_{0.2}FA_{0.8}PbI_3$/spiro-OMeTAD/Au	1.56	15.69	21.5	1.02	70.1	Yi et al. (2016)	2016
FTO/c-TiO_2/mp-TiO_2/$MA_{0.25}FA_{0.75}PbI_3$/spiro-OMeTAD/Ag	300	15.51	21.10	1.03	0.71	Qin et al. (2014b)	2014
FTO/NiO/$Cs_{0.15}FA_{0.85}PbI_3$/PC61BM/PFNBr/Ag		15.38	20.81	1.04	0.71	Huang et al. (2017)	2017

(continued)

Table 14 (continued)

Device architecture	Thickness (nm)/E_g (eV)	PCE (%)	J_{sc} (mA cm^{-2})	V_{oc} (V)	FF	References	Year
ITO/NiO/$FA_{0.83}Cs_{0.17}Pb(I_{0.6}Br_{0.4})_3$/LiF/$PC_{60}BM$/$SnO_2$/ZTO/ITO/LiF/Ag	1.63	14.5	18.7	0.98	0.79	Bush et al. (2017)	2017
FTO/c-TiO_2/mp-TiO_2/(FA/MA)Pb(I/Br)$_3$/spiro-OMeTAD/Au	mp-TiO_2 with PVSK infiltrated (~150 nm) + capping layer (~400 nm)	14.5	21	1.01	0.69	Reyna et al. (2016)	2016
FTO/c-TiO_2/mp-TiO_2/$MA_{0.6}FA_{0.4}PbI_3$/spiro-OMeTAD/Au		14.23	20.87	0.98	69.97	Pellet et al. (2014)	2014
FTO/ZnO/$MA_{0.6}FA_{0.4}PbI_3$/spiro-OMeTAD/Ag	315	13.4	22.39	0.98	0.61	Mahmud et al. (2016)	2016
FTO/c-TiO_2/mp-TiO_2/$MA_{0.6}FA_{0.4}PbI_3$/spiro-OMeTAD/Au	~300/1.53	13.4	18.15	1.03	0.72	Pellet et al. (2014)	2014
ITO/PEDOT:PSS/$MA_{0.6}FA_{0.4}PbI_3$/PC61BM/Ag		13.28 ± 0.37	20.96 ± 0.14	0.98 ± 0.01	0.65 ± 0.12	Chen et al. (2016g)	2016
ITO/PEDOT:PSS/$MA_{0.6}FA_{0.4}PbI_3$/PC61BM/Ca/Ag	~300	13.0	18.95	0.94	0.73	Liu et al. (2015g)	2015
ITO/PEDOT:PSS/$(FAPbI_3)_{0.8}(MAPbBr_3)_{0.2}$/C60/BCP/Ag	280	12	20.6	0.88	0.66	Chen et al. (2016h)	2016
ITO/PEDOT:PSS/$(FAPbI_3)_{0.8}(MAPbBr_3)_{0.2}$/C60/BCP/Ag		11.8 ± 0.20	20.1 ± 0.5	0.87 ± 0.01	0.66 ± 0.12	Matsui et al. (2017)	2017
FTO/c-TiO_2/$MA_{0.13}FA_{0.87}PbI_3$/spiro-OMeTAD/Au	1.52 ± 0.02	8.73	15.7	1	0.56	Wang et al. (2014c)	2014
ITO/PEDOT:PSS/$MA_{0.9}Cs_{0.1}PbI_3$/PC60BM/Al		7.68	10.10	1.05	0.73	Choi et al. (2014)	2014
FTO/c-TiO_2/$(MAPbI_3)(FAPbI_3)$/Au	1.53–1.54	4.0	9.58	0.77	0.54	Aharon et al. (2015)	2015

6 Challenges of PSCs

6.1 Hysteresis

The current–voltage hysteresis behavior in PSCs appeared due to several internal factors, such that (a) hysteresis is mostly depending on the perovskite material, (b) selective contact materials, i.e., HTM, ETM, play crucial roles in this behavior, including the material and the morphology, and (c) the typical charge generation/recombination processes ($\approx$ns) in PSCs (Snaith et al. 2014). These effects cause non-uniform current flow in forward and reverse biasing condition. The shape of measured J–V curves and corresponding device efficiency hugely depends on scan direction, light source, delay in measurement time, and voltage bias conditions before measurement. In last few years, different mechanisms have been proposed to be responsible such hysteresis effects, like, ferroelectricity, charge trapping/detrapping, and ionic migration. Out of them ionic migration seems to appear as dominative factor.

Ferroelectricity. Ferroelectricity of a material depends on the hysteretic swapping of ferroelectric domains in an external electric field. Even though the $MAPbI_3$ perovskite crystal structure possesses centrosymmetric tetragonal space group, however the reorientation of the organic cations and the distortion of the $[PbI_6]^{4-}$ cages can result in a polarization (Sherkar and Jan Anton Koster 2016). Secondly, the ability of the perovskite crystal structure to switch this polarization by an external field can originate the hysteresis effect. Several research groups have confirmed the polarization switching in both amplitude and phase by using piezoelectric force microscopy (PFM) in the perovskite materials (Chen et al. 2015f; Coll et al. 2015).

Charge trapping/detrapping. Generally, the perovskite thin films are grown via low-temperature solution process that creates defects in the perovskite crystal structure, which impact the charge separation/recombination and charge transport through the device (Ono and Qi 2016). Spectroscopic characterization confirms that these traps are mostly accumulated at the interfaces or surface, where the periodic crystalline structure is not liable (Wu et al. 2015d; D'Innocenzo et al. 2014). Such defects can be passivated by several methods and can significantly improve the device performance and decreases the hysteresis. Luminescence characteristics confirmed that oxygen exposure led to a significant reduce the density of the trap state in the perovskite material (Motti et al. 2016). Lewis bases treatment to the perovskite thin films also can reduce the recombination centers and enhance the photoluminescence intensity and lifetime (deQuilettes et al. 2015). It has been observed that by using PCBM or other organic molecules either in grain boundary or an interlayer or mixed as bulk heterojunction structure can able to passivate defect states and correspondingly reduce the hysteresis effect (Meng et al. 2016).

Ionic migration. Several experimental observations point out that the ionic migration through the perovskite thin films is a dominant factor for the origin of hysteresis.

Mobile charge carriers not only impact on the hysteresis effect in J–V curves but also influence the emission properties of the perovskites (e.g., photoluminescence, electroluminescence, blinking), and they induce capacitive effects (Richardson et al. 2016; deQuilettes et al. 2016). In an external electric field, charge carriers move toward the opposite interfaces (perovskite/ETL or perovskite/HTL) (De Bastiani et al. 2016). These accumulations of charge carriers at the interfaces result in change of the internal field and a modulation of interfacial barriers which ultimately results in the hysteretic behavior. For an example, the migrating ions in $MAPbI_3$ are MA^+, I^-, H^+ ions. These ions originate during growth of perovskites at low-temperature fabrication and include vacancies like Schottky defects, interstitial defects, and Frenkel defects (Azpiroz et al. 2015; Yuan and Huang 2016).

Suppression of hysteresis. It is very important to reduce the hysteresis effects in PSCs. It will be very beneficial to not only reduce the hysteresis but also to increase the long-term stability of PSC devices. There are several ways to reduce defect states as: (1) reducing the concentration of defects/ions in the precursor solution, (2) hindering the motion of these ions from the perovskite octahedral crystal structure, and (3) reducing the interfacial barriers and accelerating the interfacial charge-transfer process (Gangishetty et al. 2016). Larger crystal grains possess very few defects that are the source of mobile ions. It has been observed that if the surface recombination in the material is low and diffusion length is long, then the hysteresis is weak. Improved crystallinity, better fabrication process, and improved contacts are going to reduce hysteresis effects. If the perovskite films deposited on mesoporous TiO_2 layers, forms a uniform, compact and dense perovskite grains, leading to negligible hysteresis (Yang et al. 2016e).

6.2 Stability

Moisture. Water molecules are strongly interacting with perovskite molecules as the perovskite structure itself is soluble in water. In the presence of limited humidity at the atmosphere during perovskite thin-film processing time can improve the thin-film morphology and can lead to improve solar cell device performance. However, when the PSCs are exposed to the atmosphere with a relatively high humidity, a detrimental device performance was occurred. So, it is very important to understand the effect of moisture on perovskite materials to achieve highly stable PSCs that can last over a decade. It has been observed that if the $MAPbI_3$ perovskite film is exposed to a warm humid air at room temperature, the perovskite film decomposes into hydroiodic acid (soluble in water), solid PbI_2, and CH_3NH_2 (either released as a gas or dissolved in water) (Frost et al. 2014). This process is irreversible and in results the device degrades. When the films are exposed to cool humid air, then water is slowly incorporated into the crystal and results in homogeneous uniform films throughout the sample. In this case, the process is reversible in films and enhances the performance of solar cells (Hao et al. 2014c). When spiro-OMeTAD is used as

HTM for PSCs, these results in many pinholes that poorly protect the perovskite against the atmospheric water and forms PbIOH as a degradation product (Ono et al. 2015; Chen et al. 2017g). Synthesis of hydrophobic HTMs forms pinhole-free films and improves device performance and stability (Kwon et al. 2014). To improve the moisture stability of PSCs, the encapsulation of the devices is a key for long-term stability.

Heat. Heating in the perovskite thin films can result in the internal crystal structural changes. For an example, $MAPbI_3$ perovskites undergo a reversible crystal-phase transition between tetragonal and cubic symmetry in the temperature range of 54–57 °C (Baikie et al. 2013). Changes to the electronic band structure can modify the band alignment and potentially reduce the photovoltaic performance. Additionally, cycling between the two crystal phases during the day and night cycles is likely to lead to material fatigue and shorter device lifetimes. $CsPbI_3$ is a large band-gap of 2.8 eV at room temperature and has a cubic phase with a band-gap of 1.7 eV at high temperatures (Giesbrecht et al. 2016). $FAPbI_3$ has a band-gap of 2.2 eV also which possess hexagonal structure at room temperature, similar to the PbI_2 lattice, so-called δ-phase. However, depending on the growth temperature, $FAPbI_3$ can also be crystallized in the α-phase with trigonal symmetry (Binek et al. 2015). It is very difficult to find a suitable synthesis route such that the solar cells remain stable at normal operating conditions. To overcome this issue, Cs^+, MA^+, FA^+ and halides were mixed to form cubic structure at room temperature and allow the band-gap tuning, which enhance the device stability of the device.

Mixed-Halide, mixed-cation perovskites. The intrinsic stability of the perovskite crystal structures has been studied immensely with the DFT calculations and experimental observations. It has been observed that $MAPbI_3$ is very unstable in the normal atmospheric conditions while $MAPbCl_3$ is the most stable (Buin et al. 2015). Partially substituting the I^- ions with Br^- or Cl^- ions can improve the stability of the perovskite structure compared to the pure $MAPbI_3$ perovskite material. However, under illumination condition $MAPb(Br_xI_{1-x})_3$ mixed-halide perovskites undergo a reversible phase separation into I^- and Br^- ion-rich domains (Hoke et al. 2015). This creates recombination centers inside the film, which limits the open-circuit voltage. This also leads to a poor photovoltaic device performance and poor long-term stability. However, in a mixed-cation, mixed-halide perovskite ($Cs_yFA_{1-y}Pb(Br_xI_{1-x})_3$), we suppress unwanted halide segregation. Small content between $0.10 < y < 0.30$ of Cs^+ cations inside the perovskite structure shows a high crystallinity and good optoelectronic properties (Rehman et al. 2017). Upon adding, FA^+ cations significantly improve the solar cells performance and good long-term stability under continuous illumination condition. Rb^+ cations are also an alternative cation that further increasing the photovoltaic performance (Saliba et al. 2016c).

Defect states: Another main reason of perovskite instability is the defect chemistry of perovskite structure. Since perovskite materials are ionic in nature, their solution-processed growth process enables formation of defect states. That defect states generally form at the thin-film surfaces or the grain boundaries of perovskite films. Point defects, such as organic cation (MA/FA) vacancies, and halide anion (I/Br) vacancies, can easily develop in the perovskite structure due to their low formation energies. These defect states create shallow electronic energy levels close to the band edges and resulted in lower device performance in PSCs. Such defects also play a significant role in chemical degradation of the perovskite materials and leads to instability in PSCs (Wang et al. 2016g). Ion vacancies can diffuse into the perovskite crystals and also trigger cations and anions to diffuse at the surfaces and grain boundaries. Therefore, systematic control of defect states is essential during perovskite thin-film growth process that can enhance PSCs device efficient and stability.

There are several strategies to reduce these unwanted defect states in perovskite films (Noel et al. 2014; Abdi-Jalebi et al. 2018; Saidaminov et al. 2018; Yang et al. 2020). Firstly, use of excess MA/FA organic cations during film growth or after the thin-film annealing process can compensate with thermally evaporated organic cations. Secondly, introduction of larger organic molecules: such as phenethylamine, polyethylenimine, and trifluoroethylamine, which are difficult to evaporate. Thirdly, guanidinium, an organic cation that connect with perovskite structure via forming hydrogen bonds, has been used to suppress the formation of iodide vacancies. Fourthly, addition of KI into perovskites passivating iodine vacancies and small ions (Cl and Cd) doped into the perovskite lattice also suppress the formation of halide vacancies via lattice strain relaxation. Such perovskite films exhibited better device efficiency and stability compared to pristine PSCs.

7 Future Outlooks

In the last decade, organometal halide perovskites (OHPs) have emerged as a promising alternative to the existing commercially available solar cells. High photovoltaic efficiency, low materials cost, solution-phase deposition, low-temperature processing steps, and long-range crystallinity sets the OHPs apart from the contemporary emerging solar cell technologies (Petrus et al. 2017). Their performance has rapidly increased from 3.8% in 2009 to 23.7% in 2019, thanks to their high absorption coefficient, high carrier mobility, and long carrier lifetime. The commercial success of the PSCs, however, depends on their long-term stability, high-efficiency large-area module formation, and the toxicity issue related to the use of lead in the perovskite semiconductors (Petrus et al. 2017; Asghar et al. 2017).

Compositional engineering has been applied to a great effect to minimize the crystal strain in the perovskite crystals, which has improved their thermal stability significantly (Jeon et al. 2015b). The use of hydrophobic surfactants in the development of 2D/3D perovskite crystal has increased its moisture resistance considerably in solid films (Grancini et al. 2017). In PSCs, efficiency, stability, and J–V hysteresis are closely correlated. Interfacial defect and bulk recombination through intermediate trap states are adversely affecting the solar cell properties. Interfacial engineering at both ETL/perovskite and HTL/perovskite and grain boundary engineering are believed to be the right research direction toward stable and high-efficiency PSCs

(Grancini et al. 2017). Further, hydrophobic ETL and HTL layer development would act as a deterrent toward moisture ingress to the sensitive perovskite layer. Better encapsulation may also be critical to improve the PSCs device stability and it is very essential to resolve this stability issue for further industrial commercialization (Jena et al. 2019).

Large-area compatible roll-to-roll processing needs to be developed for smooth, reproducible, and pinhole-free deposition of perovskite and other solution-processed layers (ETL/HTL) for large-scale production of PSCs. A significant amount of research efforts have been made in this direction in recent years, resulting in a mini-module with a certified efficiency of 12.1% with an aperture area >36 cm^2 (Chen et al. 2017h). The efficiencies of perovskite solar cell modules are still notably lagging behind the small-area cells. A loss in solar cell performance in large-area modules is observed due to several issues like higher series resistance, lower shunt resistance, non-uniform surface morphology, pinholes, and unavoidable dead area. However, the efficiency gap between small-area cell and large-area modules for PSCs is much larger than the well-established PV technologies, as of today. Challenges in scaling up PSCs involve development of deposition methods for growth of uniform films over larger area and need reliable module design process (laser scribing process, interconnection properties, optimal width of subcells, etc.).

As of date, only lead-based perovskite systems show efficiency exceeding 20%, however, the toxicity of lead is a serious concern for their commercialization. Efforts have been made to replace Pb with nontoxic and earth-abundant alternatives (Shockley and Queisser 1961). Among all the metals, tin (Sn) and germanium (Ge) can form genuine perovskite structure since they both fulfill the coordination, ionic size, and charge balance prerequisites (Ke et al. 2019). Sn-based perovskites are potential candidate to replace Pb-atoms as they exhibit very similar optoelectronic characteristics, such as suitable optical band-gaps, high absorption coefficient, and reasonable charge carrier mobilities. However, Sn has two main oxidation states, +2 and the slightly more stable +4, which makes them even more unstable than Pb-based perovskite systems. Sn-based perovskite systems show higher *p*-doping and higher conductivity arising from the oxidation of Sn^{2+}/Sn^{4+}. The photovoltaic efficiency of Sn-based perovskites has reached 9%; however, the solar cells suffer significantly from low open-circuit voltage and low fill factor.

As the PCE of PSCs approaching their theoretical limits set by the Shockley–Queisser equation for single-junction solar cells, alternative architectural platforms are being explored in the form of the multifunction tandem solar cell. Composition engineering has been utilized to achieve the band-gap tunability of perovskite systems to develop perovskite (high-E_g)/perovskite (low-E_g) tandem solar cells (Zhao et al. 2017). High band-gap perovskite systems are also being explored to develop perovskite/crystalline-Silicon tandem solar cells (Sahli et al. 2018). Four-terminal and monolithic two-terminal tandem architectures have been successfully realized on both perovskite/perovskite and perovskite/crystalline-Silicon tandem solar cells. In the future, a lot more works are expected in this direction to boost the PCE over 30%.

References

Abdi-Jalebi M, Andaji-Garmaroudi Z, Cacovich S, Stavrakas C, Philippe B, Richter JM, Alsari M, Booker EP, Hutter EM, Pearson AP, Lilliu S, Savenije TJ, Rensmo H, Divitini G, Ducati C, Friend RH, Stranks SD (2018) Maximizing and stabilizing luminescence from halide perovskites with potassium passivation. Nature 555(7697):497–501

Aharon S, Dymshits A, Rotem A, Etgar L (2015) Temperature dependence of hole conductor free formamidinium lead iodide perovskite based solar cells. J Mater Chem A 3:9171–9178

Aitola K, Sveinbjoärnssona K, Correa-Baenab JP, Kaskelac A, Abated A, Tianc Y, Johanssona MJE, Graätzel M, Kauppinenc IE, Hagfeldtb A, Boschloo G (2016) Carbon nanotube-based hybrid hole-transporting material and selective contact for high efficiency perovskite solar cells. Energy Environ Sci 9:461–466

Alidoust N, Toroker MC, Keith JA, Carter EA (2014) Significant reduction in NiO band gap upon formation of $Li_xNi_{1-x}O$ alloys: applications to solar energy conversion. Chem Sus Chem 7:195–201

Anaya M, Correa-Baena JP, Lozano G, Saliba M, Anguita P, Roose B, Abate A, Steiner U, Grätzel M, Calvo ME, Hagfeldt A, Miguez H (2016) Optical analysis of $CH_3NH_3Sn_xPb_{1-x}I_3$ absorbers: a roadmap for perovskite-on-perovskite tandem solar cells. J Mater Chem A 4:11214–11221

Andreani LC, Bozzola A, Kowalczewski P, Liscidini M, Redorici L (2018) Silicon solar cells: toward the efficiency limits. Adv Phys-X 4(1):1548305

Aristidou N, Sanchez-Molina I, Chotchuangchutchaval T, Brown M, Martinez L, Rath T, Haque SA (2015) The Role of oxygen in the degradation of methylammonium lead trihalide perovskite photoactive layers. Angew Chem Int Ed 54(28):8208–8212

Arora N, Dar MI, Hinderhofer A, Pellet N, Schreiber F, Zakeeruddin SM, Grätzel M (2017) Perovskite solar cells with CuSCN hole extraction layers yield stabilized efficiencies greater than 20. Science 358:768–771

Asghar MI, Zhang J, Wang H, Lund PD (2017) Device stability of perovskite solar cells—a review. Renew Sustain Energy Rev 77:131–146

Azpiroz JM, Mosconi E, Bisquert J, De Angelis F (2015) Defect migration in methylammonium lead iodide and its role in perovskite solar cell operation. Energy Environ Sci 8(7):2118–2127

Bach U, Lupo D, Comte P, Moser J, Weissortel F, Salbeck J, Spreitzer H, Gratzel M (1998) Solid-state dye-sensitized mesoporous TiO_2 solar cells with high photon-to-electron conversion efficiencies. Nature 395:583

Bai Y, Yu H, Zhu ZL, Jiang K, Zhang T, Zhao N, Yang SH, Yan H (2015) High performance inverted structure perovskite solar cells based on a PCBM:polystyrene blend electron transport layer. J Mater Chem A 3:9098–9102

Bai Y, Dong Q, Shao Y, Deng Y, Wang Q, Shen L, Wang D, Wei W, Huang J (2016) Enhancing stability and efficiency of perovskite solar cells with crosslinkable silane-functionalized and doped fullerene. Nat Commun 7:12806

Baikie T, Fang Y, Kadro JM, Schreyer M, Wei F, Mhaisalkar SG, Graetzel M, White TJ (2013) Synthesis and crystal chemistry of the hybrid perovskite $(CH_3NH_3)PbI_3$ for solid-state sensitized solar cell applications. J Mater Chem A 1(18):5628–5641

Ball JM, Lee MM, Hey A, Snaith HJ (2013) Low-temperature processed meso-superstructured to thin-film perovskite solar cells. Energy Environ Sci 6:1739–1743

Bao Q, Li CM, Liao L, Yang H, Wang W, Ke C, Song Q, Bao H, Yu T, Loh KP, Guo J (2009) Electrical transport and photovoltaic effects of core–shell CuO/C60 nanowire heterostructure. Nanotechnology 20:065203

Beckmann PA (2010) A review of polytypism in lead iodide. Cryst Res Tech 45(5):455–460

Behrouznejad F, Tsai CM, Narra S, Diau EW, Taghavinia N (2017) Interfacial investigation on printable carbon-based mesoscopic perovskite solar cells with NiO_x/C back electrode. ACS Appl Mater Interfaces 9:25204–25215

Bera A, Wu K, Sheikh A, Alarousu E, Mohammed OF, Wu T (2015) Perovskite oxide $SrTiO_3$ as an efficient electron transporter for hybrid perovskite solar cells. J Phys Chem C 118:28494–28501

Bi D, Moon S-J, Haäggman L, Boschloo G, Yang L, Johansson EMJ, Nazeeruddin MK, Graätzel M, Hagfeldt A (2013) Using a two-step deposition technique to prepare perovskite ($CH_3NH_3PbI_3$) for thin film solar cells based on ZrO_2 and TiO_2 mesostructures. RSC Adv 3:18762–18766
Bi D, Tress W, Dar MI, Gao P, Luo J, Renevier C, Schenk K, Abate A, Giordano F, Correa Baena J-P (2016) Efficient luminescent solar cells based on tailored mixed-cation perovskites. Sci Adv 2:e1501170
Binek A, Hanusch FC, Docampo P, Bein T (2015) Stabilization of the trigonal high-temperature phase of formamidinium lead iodide. J Phys Chem Lett 6:1249–1253
Bu IYY, Fu YS, Li JF, Guo TF (2017) Large-area electrospray-deposited nanocrystalline Cu_XO hole transport layer for perovskite solar cells. RSC Adv 7:46651–46656
Buin A, Comin R, Xu J, Ip AH, Sargent EH (2015) Halide-dependent electronic structure of organolead perovskite materials. Chem Mater 27(12):4405–4412
Burschka J, Pellet N, Moon S-J, Humphry-Baker R, Gao P, Nazeeruddin MK, Grätzel M (2013) Sequential deposition as a route to high-performance perovskite-sensitized solar cells. Nature 499(7458):316–319
Bush KA, Palmstrom AF, Yu ZJ, Boccard M, Cheacharoen R, Mailoa JP, McMeekin DP, Hoye RLZ, Bailie CD, Leijtens T (2017) 23.6%-efficient monolithic perovskite/silicon tandem solar cells with improved stability. Nat Energy 2:17009
Cai ML, Tiong VT, Hreid T, Bell J, Wang HX (2015) An efficient hole transport material composite based on poly (3-hexylthiophene) and bamboo-structured carbon nanotubes for high performance perovskite solar cells. J Mater Chem A 3:2784–2793
Cao J, Liu Y-M, Jing X, Yin J, Li J, Xu B, Tan Y-Z, Zheng N (2015a) Well-defined thiolated nanographene as hole-transporting material for efficient and stable perovskite solar cells. J Am Chem Soc 137:10914–10917
Cao K, Zuo Z, Cui J, Shen Y, Moehl T, Zakeeruddin SM, Grätzel M, Wang M (2015b) Efficient screen printed perovskite solar cells based on mesoscopic$TiO_2/Al_2O_3/NiO$/carbon architecture. Nano Energy 17:171–179
Cao K, Cui J, Zhang H, Li H, Song JK, Shen Y, Cheng YB, Wang MK (2015c) Efficient mesoscopic perovskite solar cells based on the $CH_3NH_3PbI_2Br$ light absorber. J Mater Chem A 3:9116–9122
Capasso A, Matteocci F, Najafi L, Prato M, Buha J, Cinà L, Pellegrini V, Carlo AD, Bonaccorso F (2016) Few-layer: MoS_2 flakes as active buffer layer for stable perovskite solar cells. Adv Energy Mater 6:1600920
Carnie MJ, Charbonneau C, Davies ML, Troughton J, Watson TM, Wojciechowski K, Snaith H, Worsley DA (2013) A one-step low temperature processing route for organolead halide perovskite solar cells. Chem Commun 49:7893–7895
Chan CY, Wang YY, Wu GW, Diau EWG (2016) Solvent-extraction crystal growth for highly efficient carbon-based mesoscopic perovskite solar cells free of hole conductors. J Mater Chem A 4:3872–3878
Chang X, Li W, Zhu L, Liu H, Geng H, Xiang S, Liu J, Chen H (2016a) Carbon-based CsPbBr 3 perovskite solar cells: all-ambient processes and high thermal stability. ACS Appl Mater Interfaces 8:33649–33655
Chang X, Li W, Chen H, Zhu L, Liu H, Geng H, Xiang S, Liu J, Zheng X, Yang Y, Yang S (2016b) Colloidal precursor-induced growth of ultra-even $CH_3NH_3PbI_3$ for high-performance paintable carbon-based perovskite solar cells. ACS Appl Mater Interfaces 8:30184–30192
Chatterjee S, Pal AJ (2016) Introducing Cu_2O thin films as a hole-transport layer in efficient planar perovskite solar cell structures. J Phys Chem C 120:1428–1437
Chavhan S, Miguel O, Grande HJ, Gonzalez-Pedro V, Sanchez RS, Barea EM, Mora-Sero I, Tena-Zaera R (2014) Organo-metal halide perovskite-based solar cells with CuSCN as the inorganic hole selective contact. J Mater Chem A 2:12754–12760
Chen SY, Walsh A, Gong XG, Wei SH (2013) Classification of lattice defects in the Kesterite Cu_2ZnSnS_4 and $Cu_2ZnSnSe_4$ Earth-abundant solar cell absorbers. Adv Mater 25:1522–1539

Chen H, Wei Z, Yan K, Yi Y, Wang J, Yang S (2014) Liquid phase deposition of TiO_2 nanolayer affords $CH_3NH_3PbI_3$/nanocarbon solar cells with high open-circuit voltage. Faraday Discuss 176:271–286

Chen W, Wu Y, Yue Y, Liu J, Zhang W, Yang X, Chen H, Bi E, Ashraful I, Graätzel M, Han L (2015a) Efficient and stable large-area perovskite solar cells with inorganic charge extraction layers. Science 350:944–948

Chen WY, Deng LL, Dai SM, Wang X, Tian CB, Zhan XX, Xie SY, Huang RB, Zheng LS (2015b) Low-cost solution-processed copper iodide as an alternative to PEDOT:PSS hole transport layer for efficient and stable inverted planar heterojunction perovskite solar cells. J Mater Chem A 3:19353–19359

Chen W, Wu YZ, Liu J, Qin CJ, Yang XD, Islam A, Cheng YB, Han LY (2015c) Hybrid interfacial layer leads to solid performance improvement of inverted perovskite solar cells. Energy Environ Sci 8:629–640

Chen HN, Wei ZH, Zheng XL, Yang SH (2015d) A scalable electrodeposition route to the low-cost, versatile and controllable fabrication of perovskite solar cells. Nano Energy 15:216–226

Chen C, Cheng Y, Dai Q, Song H (2015e) Radio frequency magnetron sputtering deposition of TiO_2 thin films and their perovskite solar cell applications. Sci Rep 5:17684

Chen B, Shi J, Zheng X, Zhou Y, Zhu K, Priya S (2015f) Ferroelectric solar cells based on inorganic–organic hybrid perovskites. J Mater Chem A 3(15):7699–7705

Chen ST, Roh K, Lee J, Chong WK, Lu Y, Mathews N, Sum TC, Nurmikko A (2016a) A photonic crystal laser from solution based organo-lead iodide perovskite thin films. ACS Nano 10(4):3959–3967

Chen CM, Lin ZK, Huang WJ, Yang SH (2016b) WO_3 nanoparticles or nanorods incorporating Cs_2CO_3/PCBM buffer bilayer as carriers transporting materials for perovskite solar cells. Nanoscale Res Lett 11:464

Chen JZ, Xiong YL, Rong YG, Mei AY, Sheng YS, Jiang P, Hu Y, Li X, Han HW (2016c) Solvent effect on the hole-conductor-free fully printable perovskite solar cells. Nano Energy 27:130–137

Chen J, Rong Y, Mei A, Xiong Y, Liu T, Sheng Y, Jiang P, Hong L, Guan Y, Zhu X, Hou X, Duan M, Zhao J, Li X, Han H (2016d) Hole-conductor-free fully printable mesoscopic solar cell with mixed-anion perovskite $CH_3NH_3PbI_{(3-x)}(BF_4)_x$. Adv. Energy Mater 6:1502009

Chen HN, Zheng XL, Li Q, Yang YL, Xiao S, Hu C, Bai Y, Zhang T, Wong KS, Yang SH (2016e) An amorphous precursor route to the conformable oriented crystallization of CH3NH3PbBr 3 in mesoporous scaffolds: toward efficient and thermally stable carbon-based perovskite solar cells. J Mater Chem A 4:12897–12912

Chen HN, Wei ZH, He HX, Zheng XL, Wong KS, Yang SH (2016f) Solvent engineering boosts the efficiency of paintable carbon-based perovskite solar cells to beyond 14%. Adv Energy Mater 6:1502087

Chen C-C, Chang SH, Chen L-C, Kao F-S, Cheng H-M, Yeh S-C, Chen C-T, Wu W-T, Tseng Z-L, Chuang CL (2016g) Improving the efficiency of inverted mixed-organic-cation perovskite absorber based photovoltaics by tailing the surface roughness of PEDOT: PSS thin film. Sol Energy 134:445–451

Chen LC, Tseng ZL, Huang JK (2016h) A study of inverted-type perovskite solar cells with various composition ratios of $(FAPbI3)_{(1-x)}(MAPbBr\ 3)_x$. Nanomaterials 6:183

Chen W, Liu FZ, Feng XY, Djurisic AB, Chan WK, He ZB (2017a) Cesium doped NiO_x as an efficient hole extraction layer for inverted planar perovskite solar cells. Adv Energy Mater 7:1700722

Chen W, Xu L, Feng X, Jie J, He Z (2017b) Metal acetylacetonate series in interface engineering for full low-temperature-processed, high-performance, and stable planar perovskite solar cells with conversion efficiency over 16% on 1 cm^2 scale. Adv Mater 29:1603923

Chen M, Zha RH, Yuan ZY, Jing QS, Huang ZY, Yang XK, Yang SM, Zhao XH, Xu DL, Zou GDL (2017c) Boron and phosphorus co-doped carbon counter electrode for efficient hole-conductor-free perovskite solar cell. Chem Eng J 313:791–800

Chen Z, Yang G, Zheng X, Lei H, Chen C, Ma J, Wang H, Fang G (2017d) Bulk heterojunction perovskite solar cells based on room temperature deposited hole-blocking layer: suppressed hysteresis and flexible photovoltaic application. J Power Sources 351:123–129

Chen J, Xu J, Xiao L, Zhang B, Dai S, Yao J (2017e) Mixed-Organic-Cation $(FA)_x(MA)_{1-x}PbI_3$ planar perovskite solar cells with 16.48% efficiency via a low-pressure vapor-assisted solution processl. ACS Appl Mater Interfaces 9:2449–2458

Chen B-X, Li W-G, Rao H-S, Xu Y-F, Kuang D-B, Su C-Y (2017f) Large-grained perovskite films via $FA_xMA_{1-x}Pb(I_xBr_{1-x})_3$ single crystal precursor for efficient solar cells. Nano Energy 34:264–270

Chen B-A, Lin J-T, Suen N-T, Tsao C-W, Chu T-C, Hsu Y-Y, Chan T-S, Chan Y-T, Yang J-S, Chiu C-W, Chen HM (2017g) In situ identification of photo-and moisture-dependent phase evolution of perovskite solar cells. ACS Energy Lett 2(2):342–348

Chen H, Ye F, Tang WT, He JJ, Yin MS, Wang YB, Xie FX, Bi EB, Yang XD, Gratzel M, Han LY (2017h) A solvent- and vacuum-free route to large-area perovskite films for efficient solar modules. Nature 550(7674):92

Chen R, Wang W, Bu TL, Ku ZL, Zhong J, Peng Y, Xiao S, You W, Huang F, Cheng Y, Fu Z (2019) Materials and structures for the electron transport layer of efficient and stable perovskite solar cells. Acta Phys-Chim Sin 35:401–407

Cheng Y, Yang Q-D, Xiao J, Xue Q, Li H-W, Guan Z, Yip H-L, Tsang S-W (2015) Decomposition of organometal halide perovskite films on zinc oxide nanoparticles. ACS Appl Mater Interfaces 7:19986–19993

Cheng N, Liu P, Qi F, Xiao YQ, Yu WJ, Yu ZH, Liu W, Guo SS, Zhao XZ (2016) Multi-walled carbon nanotubes act as charge transport channel to boost the efficiency of hole transport material free perovskite solar cells. J Power Sources 332:24–29

Choi H, Jeong J, Kim H-B, Kim S, Walker B, Kim G-H, Kim JY (2014) Cesium-doped methylammonium lead iodide perovskite light absorber for hybrid solar cells. Nano Energy 7:80–85

Choi J, Song S, Horantner MT, Snaith HJ, Park T (2016) Well-defined nanostructured, single-crystalline TiO_2 electron transport layer for efficient planar perovskite solar cells. ACS Nano 10:6029–6036

Christians JA, Fung RCM, Kamat PV (2014) An inorganic hole conductor for organo-lead halide perovskite solar cells. Improved hole conductivity with copper iodide. J Am Chem Soc 136:758–764

Ciro J, Ramirez D, Mejia Escobar MA, Montoya JF, Mesa S, Betancur R, Jaramillo F (2017) Self-functionalization behind a solution-processed NiO_x film used as hole transporting layer for efficient perovskite solar cells. ACS Appl Mater Interfaces 9:12348–12354

Coll M, Gomez A, Mas-Marza E, Almora O, Garcia-Belmonte G, Campoy-Quiles M, Bisquert J (2015) Polarization switching and light-enhanced piezoelectricity in lead halide perovskites. J Phys Chem Lett 6(8):1408–1413

Correa-Baena J-P, Abate A, Saliba M, Tress W, Jacobssonm TJ, Grätzelm M, Hagfeldtm A (2017) The rapid evolution of highly efficient perovskite solar cells. Energy Environ Sci 10:710–727

D'Innocenzo V, Grancini G, Alcocer MJP, Kandada ARS, Stranks SD, Lee MM, Lanzani G, Snaith HJ, Petrozza A (2014) Excitons versus free charges in organo-lead tri-halide perovskites. Nat Comm 5(1):3586

Dang Y, Liu Y, Sun Y, Yuan D, Liu X, Lu W, Liu G, Xia H, Tao X (2015) Bulk crystal growth of hybrid perovskite material $CH_3NH_3PbI_3$. CrystEngComm 17(3):665–670

Dasgupta U, Chatterjee S, Pal AJ (2017) Thin-film formation of 2D MoS_2 and its application as a hole-transport layer in planar perovskite solar cells. Sol Energy Mater Sol Cells 172:353–360

De Bastiani M, Dell'Erba G, Gandini M, D'Innocenzo V, Neutzner S, Kandada ARS, Grancini G, Binda M, Prato M, Ball JM, Caironi M, Petrozza A (2016) Ion migration and the role of preconditioning cycles in the stabilization of the J–V characteristics of inverted hybrid perovskite solar cells. Adv Energy Mater 6(2):1501453

Deng J, Mortazavi M, Medhekar NV, Liu JZ (2012) Band engineering of $Ni_{1-x}Mg_xO$ alloys for photocathodes of high efficiency dye-sensitized solar cells. J Appl Phys 112:123703
Deng Y, Dong Q, Bi C, Yuan Y, Huang J (2016) Air-stable, efficient mixed-cation perovskite solar cells with Cu electrode by scalable fabrication of active layer. Adv Energy Mater 6:1600372
deQuilettes DW, Vorpahl SM, Stranks SD, Nagaoka H, Eperon GE, Ziffer ME, Snaith HJ, Ginger DS (2015) Impact of microstructure on local carrier lifetime in perovskite solar cells. Science 348(6235):683–686
deQuilettes DW, Zhang W, Burlakov VM, Graham DJ, Leijtens T, Osherov A, Bulović V, Snaith HJ, Ginger DS, Stranks SD (2016) Photo-induced halide redistribution in organic–inorganic perovskite films. Nat Comm 7(1):11683
Di Giacomo F, Razza S, Matteocci F, D'Epifanio A, Licoccia S, Brown TM, Di Carlo A (2014) High efficiency $CH_3NH_3PbI_{(3-x)}Cl_x$ perovskite solar cells with poly(3-hexylthiophene) hole transport layer. J Power Sources 251:152–156
Do Sung S, Ojha DP, You JS, Lee J, Kim J, Lee WI (2015) 50 nm sized spherical TiO_2 nanocrystals for highly efficient mesoscopic perovskite solar cells. Nanoscale 7:8898–8906
Dong X, Hu H, Lin B, Ding J, Yuan N (2014) The effect of ALD-Zno layers on the formation of $CH_3NH_3PbI_3$ with different perovskite precursors and sintering temperatures. Chem Commun 50:14405–14408
Dong Q, Fang Y, Shao Y, Mulligan P, Qiu J, Cao L, Huang J (2015a) Electron-hole diffusion lengths >175 μm in solution-grown $CH_3NH_3PbI_3$ single crystals. Science 347:967–970
Dong Q, Shi Y, Wang K, Li Y, Wang S, Zhang H, Xing Y, Du Y, Bai X, Ma T (2015b) Insight into perovskite solar cells based on SnO_2 compact electron-selective layer. J Phys Chem C 119:10212–10217
Dong Q, Liu F, Wong MK, Djurišić AB, Ren Z, Shen Q, Ng A, Surya C, Chan WK (2016): Indium oxide-based perovskite solar cells. In: FH Teherani, DC Look, DJ Rogers (eds), Proceedings SPIE 9749, oxide-based materials and devices VII, vol 9749, p 97491S
Duong T, Mulmudi HK, Shen H, Wu Y, Barugkin C, Mayon YO, Nguyen HT, Macdonald D, Peng J, Lockrey M (2016) Structural engineering using rubidium iodide as a dopant under excess lead iodide conditions for high efficiency and stable perovskites. Nano Energy 30:330–340
Duong T, Wu Y, Shen H, Peng J, Fu X, Jacobs D, Wang E-C, Kho TC, Fong KC, Stocks M (2017) Rubidium multication perovskite with optimized bandgap for perovskite-silicon tandem with over 26% efficiency. Adv Energy Mater 7:1700228
Fang Y, Dong Q, Shao Y, Yuan Y, Huang J (2015) Highly narrowband perovskite single-crystal photodetectors enabled by surface-charge recombination. Nat Photonics 9:679–686
Frost JM, Butler KT, Brivio F, Hendon CH, van Schilfgaarde M, Walsh A (2014) Atomistic origins of high-performance in hybrid halide perovskite solar cells. Nano Lett 14(5):2584–2590
Gangishetty MK, Scott RWJ, Kelly TL (2016) Effect of relative humidity on crystal growth, device performance and hysteresis in planar heterojunction perovskite solar cells. Nanoscale 8(12):6300–6307
Gharibzadeh S, Nejand BA, Moshaii A, Mohammadian N, Alizadeh AH, Mohammadpour R, Ahmadi V, Alizadeh A (2016) Two-step physical deposition of a compact CuI hole-transport layer and the formation of an interfacial species in perovskite solar cells. Chemsuschem 9:1929–1937
Gheno A, Thu Pham TT, Di Bin C, Bouclé J, Ratier B, Vedraine S (2017) Printable WO_3 electron transporting layer for perovskite solar cells: influence on device performance and stability. Sol Energy Mater Sol Cells 161:347–354
Gholipour S, Correa-Baena JP, Domanski K, Matsui T, Steier L, Giordano F, Tajabadi F, Tress W, Saliba M, Abate A, Ali AM, Taghavinia N, Grätzel M, Hagfeldt A (2016) Highly efficient and stable perovskite solar cells based on a low-cost carbon cloth. Adv Energy Mater 6
Giesbrecht N, Schlipf J, Oesinghaus L, Binek A, Bein T, Müller-Buschbaum P, Docampo P (2016) Synthesis of perfectly oriented and micrometer-sized MAPbBr 3 perovskite crystals for thin-film photovoltaic applications. ACS Energy Lett 1(1):150–154

Grancini G, Roldan-Carmona C, Zimmermann I, Mosconi E, Lee X, Martineau D, Narbey S, Oswald F, De Angelis F, Gräetzel M, Nazeeruddin MK (2017) One-year stable perovskite solar cells by 2D/3D interface engineering. Nat Commun 8:15684

Grätzel M (2014) The light and shade of perovskite solar cells. Nat Mater 13(9):838–842

Green MA, Ho-Baillie A, Snaith HJ (2014) The emergence of perovskite solar cells. Nat Photonics 8(7):506–514

Habisreutinger SN, Leijtens T, Eperon GE, Stranks SD, Nicholas RJ, Snaith HJ (2014a) Carbon nanotube/polymer composites as a highly stable hole collection layer in perovskite solar cells. Nano Lett 14:5561–5568

Habisreutinger SN, Leijtens T, Eperon GE, Stranks SD, Nicholas RJ, Snaith HJ (2014b) Enhanced hole extraction in perovskite solar cells through carbon nanotubes. J Phys Chem Lett 5:4207–4212

Hadouchi W, Rousset J, Tondelier D, Geffroy B, Bonnassieux Y (2016) Zinc oxide as a hole blocking layer for perovskite solar cells deposited in atmospheric conditions. RSC Adv 6:67715–67723

Han GS, Chung HS, Kim DH, Kim BJ, Lee J-W, Park N-G, Cho IS, Lee J-K, Lee S, Jung HS (2015) Epitaxial 1D electron transport layers for high-performance perovskite solar cells. Nanoscale 7:15284–15290

Hao F, Stoumpos CC, Cao DH, Chang RPH, Kanatzidis MG (2014a) Lead-free solid-state organic–inorganic halide perovskite solar cells. Nat Photon 8:489–494

Hao F, Stoumpos CC, Chang RPH, Kanatzidis MG (2014b) Anomalous band gap behavior in mixed Sn and Pb perovskites enables broadening of absorption spectrum in solar cells. J Am Chem Soc 136:8094–8099

Hao F, Stoumpos CC, Liu Z, Chang RPH, Kanatzidis MG (2014c) Controllable perovskite crystallization at a gas-solid interface for hole conductor-free solar cells with steady power conversion efficiency over 10%. J Am Chem Soc 136(46):16411–16419

Hashmi SG, Martineau D, Dar MI, Myllymaki TTT, Sarikka T, Ulla V, Zakeeruddin SM, Grätzel M (2017) High performance carbon-based printed perovskite solar cells with humidity assisted thermal treatment. J Mater Chem A 5:12060–12067

Hawash Z, Ono LK, Raga SR, Lee MV, Qi Y (2015) Air exposure induced dopant redistribution and energy level shifts in spin-coated spiro-MeOTAD films. Chem Mater 27:562–569

Hawash Z, Ono LK, Qi Y (2018) Recent advances in spiro-MeOTAD hole transport material and its applications in organic-inorganic halide perovskite solar cells. Adv Mater Interfaces 5:1700623

He Q, Yao K, Wang X, Xia X, Leng S, Li F (2017) Room-temperature and solution-processable Cu-doped nickel oxide nanoparticles for efficient hole-transport layers of flexible large-area perovskite solar cells. ACS Appl Mater Interfaces 9:41887–41897

Heo JH, Lee MH, Han HJ, Patil BR, Yu JS, Im SH (2016) Highly efficient low temperature solution processable planar type $CH_3NH_3PbI_3$ perovskite flexible solar cells. J Mater Chem A 4:1572–1578

Hirst LC, Ekins-Daukes NJ (2011) Fundamental losses in solar cells. Prog Photovolt 19(3):286–293

Hodes G (2013) Perovskite-based solar cells. Science 342:317–318

Hoke ET, Slotcavage DJ, Dohner ER, Bowring AR, Karunadasa HI, McGehee MD (2015) Reversible photo-induced trap formation in mixed-halide hybrid perovskites for photovoltaics. Chem Sci 6(1):613–617

Hong S, Han A, Lee EC, Ko K, Park J, Song H, Han M-H, Han C (2015) A facile and low-cost fabrication of TiO_2 compact layer for efficient perovskite solar cells. Curr Appl Phys 15:574–579

Hou Y, Quiroz COR, Scheiner S, Chen W, Stubhan T, Hirsch A, Halik M, Brabec CJ (2015) Low-temperature and hysteresis-free electron-transporting layers for efficient, regular, and planar structure perovskite solar cells. Adv Energy Mater 5:1501056

Hou XM, Hu Y, Liu HW, Mei AY, Li X, Duan M, Zhang GA, Rong YG, Han HW (2017) Effect of guanidinium on mesoscopic perovskite solar cells. J Mater Chem A 5:73–78

https://www.iea.org/reports/world-energy-outlook-2019

Hu X, Zhang XD, Liang L, Bao J, Li S, Yang WL, Xie Y (2014a) High-performance flexible broadband photodetector based on organolead halide perovskite. Adv Funct Mater 24(46):7373–7380

Hu L, Peng J, Wang WW, Xia Z, Yuan JY, Lu JL, Huang XD, Ma WL, Song HB, Chen W, Cheng YB, Tang J (2014b) Sequential deposition of $CH_3NH_3PbI_3$ on planar NiO film for efficient planar perovskite solar cells. ACS Photonics 1:547–553
Hu L, Wang W, Liu H, Peng J, Cao H, Shao G, Xia Z, Ma W, Tang J (2015) PbS colloidal quantum dots as an effective hole transporter for planar heterojunction perovskite solar cells. J Mater Chem A 3:515–518
Hu C, Bai Y, Xiao S, Zhang T, Meng XY, Ng WK, Yang Y, Wong KS, Chen H, Yang S (2017a) Tuning the A-site cation composition of FA perovskites for efficient and stable NiO-based *p-i-n* perovskite solar cells. J Mater Chem A 5:21858–21865
Hu W, Liu T, Yin X, Liu H, Zhao X, Luo S, Guo Y, Yao Z, Wang J, Wang N, Lin H, Guo Z (2017b) Hematite electron-transporting layers for environmentally stable planar perovskite solar cells with enhanced energy conversion and lower hysteresis. J Mater Chem A 5:1434–1441
Hua Y, Zhang J, Xu B, Liu P, Cheng M, Kloo L, Johansson EMJ, Sveinbjornsson K, Aitola K, Boschloo G, Sun L (2016) Facile synthesis of fluorene-based hole transport materials for highly efficient perovskite solar cells and solid-state dye-sensitized solar cells. Nano Energy 26:108–113
Huang Z, Zeng X, Wang H, Zhang W, Li Y, Wang M, Cheng Y-B, Chen W (2014) Enhanced performance of p-type dye sensitized solar cells based on mesoporous $Ni_{1-x}Mg_xO$ ternary oxide films. RSC Adv 4:60670–60674
Huang L, Hu Z, Xu J, Zhang K, Zhang J, Zhang J, Zhu Y (2016a) Efficient and stable planar perovskite solar cells with a non hygroscopic small molecule oxidant doped hole transport layer. Electrochim Acta 196:328–336
Huang Y, Zhu J, Ding Y, Chen S, Zhang C, Dai S (2016b) TiO_2 sub-microsphere film as scaffold layer for efficient perovskite solar cells. ACS Appl Mater Interfaces 8:8162–8167
Huang J, Xu P, Liu J, You X-Z (2017) Sequential introduction of cations deriving large-grain $Cs_xFA_{1-x}PbI_3$ thin film for planar hybrid solar cells, insight into phase-segregation and thermal-healing behavior. Small 13:1603225
Huangfu M, Shen Y, Zhu G, Xu K, Cao M, Gu F, Wang L (2015) Copper iodide as inorganic hole conductor for perovskite solar cells with different thickness of mesoporous layer and hole transport layer. Appl Surf Sci 357:2234–2240
Hwang SH, Roh J, Lee J, Ryu J, Yun J, Jang J (2014) Size-controlled SiO_2 nanoparticles as scaffold layers in thin-film perovskite solar cells. J Mater ChemA 2:16429–16433
Im J-H, Lee C-R, Lee J-W, Park S-W, Park N-G (2013) 6.5% efficient perovskite quantum-dot-sensitized solar cell. Nanoscale 3(10):4088–4093
Im J, Stoumpos CC, Jin H, Freeman AJ, Kanatzidis MG (2015) Antagonism between spin-orbit coupling and steric effects causes anomalous band gap evolution in the perovskite photovoltaic materials $CH3NH3Sn_{1-x}Pb_xI3$. J Phys Chem Lett 6:3503–3509
International Energy Agency (2019) World Energy Outlook 2017. Technical report, 2017
Ip AH, Thon SM, Hoogland S, Voznyy O, Zhitomirsky D, Debnath R, Levina L, Rollny LR, Carey GH, Fischer A, Kemp KW, Kramer IJ, Ning Z, Labelle AJ, Chou KW, Amassian A, Sargent EH (2012) Hybrid passivated colloidal quantum dot solids. Nat Nanotech 7:577–582
Isikgor FH, Li B, Zhu H, Xu Q, Ouyang J (2016) High performance planar perovskite solar cells with a perovskite of mixed organic cations and mixed halides, $MA_{1-x}FA_xPbI_{3-y}Cl_y$. J Mater Chem A 4:12543–12553
Ito S, Tanaka S, Vahlman H, Nishino H, Manabe K, Lund P (2014a) Carbon-double-bond-free printed solar cells from $TiO_2/CH_3NH_3PbI_3$/CuSCN/Au: structural control and photoaging effects. ChemPhysChem 15:1194–1200
Ito S, Tanaka S, Manabe K, Nishino H (2014b) Effects of surface blocking layer of Sb_2S_3 on nanocrystalline TiO_2 for $CH_3NH_3PbI_3$ perovskite solar cells. J Phys Chem C 118:16995–17000
Ito S, Tanaka S, Nishino H (2015) Lead-halide perovskite solar cells by CH_3NH_3I dripping on PbI_2-CH_3NH_3I-DMSO precursor layer for planar and porous structures using CuSCN hole-transporting material. J Phys Chem Lett 6:881–886

Jacobsson TJ, Correa-Baena J-P, Pazoki M, Saliba M, Schenk K, Gratzel M, Hagfeldt A (2016) Exploration of the compositional space for mixed lead halogen perovskites for high efficiency solar cells. Energy Environ Sci 9:1706–1724
Jena AK, Kulkarni A, Miyasaka T (2019) Halide perovskite photovoltaics: background, status, and future prospects. Chem Rev 119(5):3036–3103
Jeng JY, Chiang YF, Lee MH, Peng SR, Guo TF, Chen P, Wen TC (2013) CH3NH3PbI3 perovskite/fullerene planar-heterojunction hybrid solar cells. Adv Mater 25:3727–3732
Jeng JY, Chen KC, Chiang TY, Lin PY, Tsai TD, Chang YC, Guo TF, Chen P, Wen TC, Hsu YJ (2014) Nickel oxide electrode interlayer in $CH_3NH_3PbI_3$ perovskite/PCBM planar-heterojunction hybrid solar cells. Adv Mater 26:4107–4113
Jeon NJ, Noh JH, Kim YC, Yang WS, Ryu S, Seok SI (2014) Solvent engineering for high-performance inorganic–organic hybrid perovskite solar cells. Nat Mater 13(9):897–903
Jeon I, Chiba T, Delacou C, Guo Y, Kaskela A, Reynaud O, Kauppinen EI, Maruyama S, Matsuo Y (2015a) Single-walled carbon nanotube film as electrode in indium-free planar heterojunction perovskite solar cells: investigation of electron-blocking layers and dopants. Nano Lett 15:6665–6671
Jeon NJ, Noh JH, Yang WS, Kim YC, Ryu S, Seo J, Seok SI (2015b) Compositional engineering of perovskite materials for high-performance solar cells. Nature 517(7535):476
Jeon I, Seo S, Sato Y, Delacou C, Anisimov A, Suenaga K, Kauppinen EI, Maruyama S, Matsuo Y (2017) Perovskite solar cells using carbon nanotubes both as cathode and as anode. J Phys Chem C 121:25743–25749
Jiang H, Kloc C (2013) Single-crystal growth of organic semiconductors. MRS Bull 38(1):28–33
Jiang Q, Zhang L, Wang H, Yang X, Meng J, Liu H, Yin Z, Wu J, Zhang X, You J (2016) Erratum: corrigendum: enhanced electron extraction using SnO_2 for high-efficiency planar-structure $HC(NH_2)2PbI_3$-based perovskite solar cells. Nat Energy 2:16177
Jiang X, Yu Z, Zhang Y, Lai J, Li J, Gurzadyan GG, Yang X, Sun L (2017) High-performance regular perovskite solar cells employing low-cost poly(ethylenedioxythiophene) as a hole-transporting material. Sci Rep 7:42564
Johnston MB, Herz LM (2016) Hybrid perovskites for photovoltaics: charge-carrier recombination, diffusion, and radiative efficiencies. Acc Chem Res 49(1):146–154
Jung JW, Chueh CC, Jen AKY (2015a) High-performance semi transparent perovskite solar cells with 10% power conversion efficiency and 25% average visible transmittance based on transparent CuSCN as the hole-transporting material. Adv Energy Mater 5:1500486
Jung JW, Chueh CC, Jen AK (2015b) A low-temperature, solution-processable, Cu-doped nickel oxide hole-transporting layer via the combustion method for high-performance thin-film perovskite solar cells. Adv Mater 27:7874–7880
Ke W, Fang G, Wang J, Qin P, Tao H, Lei H, Liu Q, Dai X, Zhao X (2014) Perovskite solar cell with an efficient TiO_2 compact film. ACS Appl Mater Interfaces 6:15959–15965
Ke W, Fang G, Liu Q, Xiong L, Qin P, Tao H, Wang J, Lei H, Li B, Wan J, Yang G, Yan Y (2015a) Low-temperature solution-processed tin oxide as an alternative electron transporting layer for efficient perovskite solar cells. J Am Chem Soc 137:6730–6733
Ke W, Zhao D, Cimaroli AJ, Grice CR, Qin P, Liu Q, Xiong L, Yan Y, Fang G (2015b) Effects of annealing temperature of tin oxide electron selective layers on the performance of perovskite solar cells. J Mater Chem A 3:24163–24168
Ke W, Zhao D, Grice CR, Cimaroli AJ, Ge J, Tao H, Lei H, Fang G, Yan Y (2015c) Efficient planar perovskite solar cells using room-temperature vacuum-processed C_{60} electron selective layers. J Mater Chem A 3:17971–17976
Ke W, Zhao D, Xiao C, Wang C, Cimaroli AJ, Grice CR, Yang M, Li Z, Jiang C-S, Al-Jassim M, Zhu K, Kanatzidis MG, Fang G, Yan Y (2016) Cooperative tin oxide fullerene electron selective layers for high-performance planar perovskite solar cells. J Mater Chem A 4:14276–14283
Ke WJ, Stoumpos CC, Kanatzidis MG (2019) "Unleaded" perovskites: Status Quo and future prospects of tin-based perovskite solar cells. Adv Mater 31(47):1803230

Kegelmann L, Wolff CM, Awino C, Lang F, Unger EL, Korte L, Dittrich T, Neher D, Rech B, Albrecht S (2017) It takes two to Tango—double-layer selective contacts in perovskite solar cells for improved device performance and reduced hysteresis. ACS Appl Mater Interfaces 9:17245–17255

Kim H-S, Lee C-R, Im J-H, Lee K-B, Moehl T, Marchioro A, Moon S-J, Humphry-Baker R, Yum J-H, Moser JE, Grätzel M, Park N-G (2012) Lead iodide perovskite sensitized all-solid-state submicron thin film mesoscopic solar cell with efficiency exceeding 9%. Sci Rep 2(1):591

Kim H-B, Choi H, Jeong J, Kim S, Walker B, Song S, Kim JY (2014a) Mixed solvents for the optimization of morphology in solution-processed, inverted-type perovskite/fullerene hybrid solar cells. Nanoscale 6(12):6679–6683

Kim JH, Liang PW, Williams ST, Cho N, Chueh CC, Glaz MS, Ginger DS, Jen AKY (2014b) High-performance and environmentally stable planar heterojunction perovskite solar cells based on a solution-processed copper-doped nickel oxide hole-transporting layer. Adv Mater 27:695–701

Kim J, Kim G, Kim TK, Kwon S, Back H, Lee J, Lee SH, Kang H, Lee K (2014c) Efficient planar-heterojunction perovskite solar cells achieved via interfacial modification of a sol–gel ZnO electron collection layer. J Mater Chem A 2:17291–17296

Kim JH, Liang PW, Williams ST, Cho N, Chueh CC, Glaz MS, Ginger DS, Jen AK (2015) High-performance and environmentally stable planar heterojunction perovskite solar cells based on a solution-processed copper-doped nickel oxide hole-transporting layer. Adv Mater 27:695–701

Kim J, Kim HP, Teridi MAM, Yusoff ARbM, Jang J (2016) Bandgap tuning of mixed organic cation utilizing chemical vapor deposition process. Sci Rep 6:37378

Kim H-S, Seo J-Y, Xie H, Lira-Cantu M, Zakeeruddin SM, Grätzel M, Hagfeldt A (2017) Effect of Cs-incorporated NiO_x on the performance of perovskite solar cells. ACS Omega 2:9074–9079

Kogo A, Numata Y, Ikegami M, Miyasaka T (2015) Nb_2O_5 blocking layer for high open-circuit voltage perovskite solar cells. Chem Lett 44:829–830

Kohnehpoushi S, Nazari P, Nejand BA, Eskandari M (2018) MoS_2: a two-dimensional hole-transporting material for high-efficiency, low-cost perovskite solar cells. Nanotechnology 29:205201

Kojima A, Teshima K, Shirai Y, Miyasaka T (2009) Organometal halide perovskites as visible-light sensitizers for photovoltaic cells. J Am Chem Soc 131(17):6050–6051

Kolny-Olesaik J, Weller H (2013) Synthesis and application of colloidal $CuInS_2$ semiconductor nanocrystals. ACS Appl Mater Interfaces 5:12221–12237

Ku Z, Rong Y, Xu M, Liu T, Han H (2013) Full printable processed mesoscopic $CH_3NH_3PbI_3/TiO_2$ heterojunction solar cells with carbon counter electrode. Sci Rep 3:3132

Kumar MH, Yantara N, Dharani S, Gräetzel M, Mhaisalkar S, Boix PP, Mathews N (2013) Flexible, low-temperature, solution processed ZnO-based perovskite solid state solar cells. Chem Commun 49:11089–11091

Kwon YS, Lim J, Yun H-J, Kim Y-H, Park T (2014) A diketopyrrolopyrrole-containing hole transporting conjugated polymer for use in efficient stable organic–inorganic hybrid solar cells based on a perovskite. Energy Environ Sci 7(4):1454–1460

Kwon U, Kim BG, Nguyen DC, Park JH, Ha NY, Kim SJ, Ko SH, Lee S, Lee D, Park HJ (2016) Solution-processible crystalline NiO nanoparticles for high-performance planar perovskite photovoltaic cells. Sci Rep 6:30759

Lang F, Gluba MA, Albrecht S, Rappich J, Korte L, Rech B, Nickel NH (2015) Perovskite solar cells with large-area CVD-graphene for tandem solar cells. J Phys Chem Lett 6:2745–2750

Lee MM, Teuscher J, Miyasaka T, Murakami TN, Snaith HJ (2012) Efficient hybrid solar cells based on meso-superstructured organometal halide perovskites. Science 338:643–647

Lee JW, Lee TY, Yoo PJ, Grätzel M, Mhaisalkar S, Park NG (2014) Rutile TiO_2-based perovskite solar cells. J Mater Chem A 2:9251–9259

Lee JW, Kim DH, Kim HS, Seo SW, Cho SM, Park NG (2015a) Formamidinium and cesium hybridization for photo- and moisture-stable perovskite solar cell. Adv Energy Mater 5(20):1501310

Lee J, Menamparambath MM, Hwang JY, Baik S (2015b) Hierarchically structured hole transport layers of spiro-OMeTAD and multiwalled carbon nanotubes for perovskite solar cells. Chem Sus Chem 8:2358–2362

Lee JW, Kim DH, Kim HS, Seo SW, Cho SM, Park NG (2015c) Formamidinium and cesium hybridization for photo- and moisture-stable perovskite solar cell. Adv Energy Mater 5:1501310

Lee K, Yoon C, Noh J, Jang J (2016) Morphology-controlled mesoporous SiO_2 nanorods for efficient scaffolds in organo-metal halide perovskite solar cells. Chem Commun 52:4231–4234

Leijtens T, Eperon GE, Pathak S, Abate A, Lee MM, Snaith HJ (2013) Overcoming ultraviolet light instability of sensitized TiO_2 with meso-superstructured organometal tri-halide perovskite solar cells. Nat Commun 4:2885

Li Z (2015) Stable perovskite solar cells based on WO_3 nanocrystals as hole transport layer. Chem Lett 44:1140–1141

Li WZ, Dong HP, Guo XD, Li N, Li JW, Niu GD, Wang LD (2014a) Graphene oxide as dual functional interface modifier for improving wettability and retarding recombination in hybrid perovskite solar cells. J Mater Chem A 2:20105–20111

Li Z, Kulkarni SA, Boix PP, Shi E, Cao A, Fu K, Batabyal SK, Zhang J, Xiong Q, Wong LH, Mathews N, Mhaisalkar SG (2014b) Laminated carbon nanotube networks for metal electrode-free efficient perovskite solar cells. ACS Nano 8:6797–6804

Li Y, Zhu J, Huang Y, Wei J, Liu F, Shao Z, Hu L, Chen S, Yang S, Tang J, Yao J, Dai S (2015a) Efficient inorganic solid solar cells composed of perovskite and PbS quantum dots. Nanoscale 7:9902–9907

Li Y, Zhu J, Huang Y, Liu F, Lv M, Chen SH, Hu LH, Tang JW, Yao JX, Dai SY (2015b) Mesoporous SnO_2 nanoparticle films as electron-transporting material in perovskite solar cells. RSC Adv 5:28424–28429

Li M, Wang Z-K, Yang Y-G, Hu Y, Feng S-L, Wang J-M, Gao X-Y, Liao L-S (2016a) Copper salts doped spiro-OMeTAD for high-performance perovskite solar cells. Adv Energy Mater 6:1601156

Li H, Cao K, Cui J, Liu S, Qiao X, Shen Y, Wang M (2016b) 14.7% efficient mesoscopic perovskite solar cells using single walled carbon nanotubes/carbon composite counter electrodes. Nanoscale 8:6379–6385

Li Z, Yang M, Park J-S, Wei S-H, Berry JJ, Zhu K (2016c) Stabilizing perovskite structures by tuning tolerance factor: formation of formamidinium and cesium lead iodide solid-state alloys. Chem Mater 28:284–292

Li G, Zhang T, Guo N, Xu F, Qian X, Zhao Y (2016d) Ion-exchange-induced 2D–3D conversion of $HMA_{1-x}FA_xPbI_3Cl$ perovskite into a high-quality $MA_{1-x}FA_xPbI_3$ perovskite. Angew Chem, Int Ed 55:13460–13464

Li Z, Tinkham J, Schulz P, Yang M, Kim DH, Berry J, Sellinger A, Zhu K (2017a) Acid additives enhancing the conductivity of spiro-OMeTAD toward high-efficiency and hys-teresis-less planar perovskite solar cells. Adv Energy Mater 7:1601451

Li J, Yao JX, Liao XY, Yu RL, Xia HR, Sun WT, Peng LM (2017b) A contact study in hole conductor free perovskite solar cells with low temperature processed carbon electrodes. RSC Adv 7:20732–20737

Liang L, Huang Z, Cai L, Chen W, Wang B, Chen K, Bai H, Tian Q, Fan B (2014) Magnetron sputtered zinc oxide nanorods as thickness-insensitive cathode interlayer for perovskite planar-heterojunction solar cells. ACS Appl Mater Interfaces 6:20585–20589

Liang PW, Chueh CC, Williams ST, Jen AKY (2015) Roles of fullerene-based interlayers in enhancing the performance of organometal perovskite thin-film solar cells. Adv Energy Mater 5:1402321

Liao W-Q, Zhang Y, Hu C-L, Mao J-G, Ye H-Y, Li P-F, Huang SD, Xiong R-G (2015) A lead-halide perovskite molecular ferroelectric semiconductor. Nat Commun 6(1):7338

Liao W, Zhao D, Yu Y, Shrestha N, Ghimire K, Grice CR, Wang C, Xiao Y, Cimaroli AJ, Ellingson RJ, Podraza NJ, Zhu K, Xiong R-G, Yan Y (2016) Fabrication of efficient low-bandgap perovskite solar cells by combining formamidinium tin iodide with methylammonium lead iodide. J Am Chem Soc 138:12360–12363

Ling X, Yuan J, Liu D, Wang Y, Zhang Y, Chen S, Wu H, Jin F, Wu F, Shi G, Tang X, Zheng J, Liu S (Frank), Liu Z, Ma W (2015) Room-temperature processed Nb_2O_5 as the electron-transporting layer for efficient planar perovskite solar cells. ACS Appl Mater Interfaces 9:23181–23188
Liu D, Kelly TL (2014) Perovskite solar cells with a planar heterojunction structure prepared using room-temperature solution processing techniques. Nat Photonics 8:133–138
Liu M, Johnston MB, Snaith HJ (2013) Efficient planar heterojunction perovskite solar cells by vapour deposition. Nature 501(7467):395–398
Liu Y, Yang Z, Cui D, Ren X, Sun J, Liu X, Zhang J, Wei Q, Fan H, Yu F, Zhang X, Zhao C, Liu S (2015a) Two-inch-sized perovskite CH3NH3PbX3 (X = Cl, Br, I) crystals: growth and characterization. Adv Mater 27(35):5176–5183
Liu ZH, Zhang M, Xu XB, Cai FS, Yuan HL, Bu LL, Li WH, Zhu AL, Zhao ZX, Wang MK, Cheng YB, He HS (2015b) NiO nanosheets as efficient top hole transporters for carbon counter electrode based perovskite solar cells. J Mater Chem A 3:24121–24127
Liu Z, Zhang M, Xu X, Bu L, Zhang W, Li W, Zhao Z, Wang M, Cheng YB, He H (2015c) p-Type mesoscopic NiO as an active interfacial layer for carbon counter electrode based perovskite solar cells. Dalton Trans 44:3967–3973
Liu T, Liu L, Hu M, Yang Y, Zhang L, Mei A, Han H (2015d) Critical parameters inTiO_2/ZrO_2/Carbon-based mesoscopic perovskite solar cell. J Power Sources 293:533–538
Liu T, Kim D, Han H, Yusoff AR, Jang J (2015e) Fine-tuning optical and electronic properties of graphene oxide for highly efficient perovskite solar cells. Nanoscale 7:10708–10718
Liu J, Gao C, Luo L, Ye Q, He X, Ouyang L, Guo X, Zhuang D, Liao C, Mei J, Lau W (2015f) Low-temperature, solution processed metal sulfide as an electron transport layer for efficient planar perovskite solar cells. J Mater Chem A 3:11750–11755
Liu J, Shirai Y, Yang X, Yue Y, Chen W, Wu Y, Islam A, Han L (2015g) High-quality mixed-organic-cation perovskites from a phase-pure non-stoichiometric intermediate $(FAI)_{1-x}$-PbI_2 for solar cells. Adv Mater 27:4918–4923
Liu JW, Pathak SK, Sakai N, Sheng R, Bai S, Wang ZP, Snaith HJ (2016a) Identification and mitigation of a critical interfacial instability in perovskite solar cells employing copper thiocyanate hole-transporter. Adv Mater Interfaces 3:1600571
Liu ZY, Sun B, Shi TL, Tang ZR, Liao GL (2016b) Enhanced photovoltaic performance and stability of carbon counter electrode based perovskite solar cells encapsulated by PDMS. J Mater Chem A 4:10700–10709
Liu Z, Shi T, Tang Z, Sun B, Liao G (2016c) Using a low-temperature carbon electrode for preparing hole-conductor-free perovskite heterojunction solar cells under high relative humidity. Nanoscale 8:7017–7023
Liu T, Yu L, Liu H, Hou Q, Wang C, He H, Li J, Wang N, Wang J, Guo Z (2017a) Ni nanobelts induced enhancement of hole transport and collection for high efficiency and ambient stable mesoscopic perovskite solar cells. J Mater Chem A 5:4292–4299
Liu S, Cao K, Li H, Song J, Han J, Shen Y, Wang M (2017b) Full printable perovskite solar cells based on mesoscopic TiO_2/Al_2O_3/NiO (carbon nanotubes) architecture. Sol Energy 144:158–165
Liu T, Zong Y, Zhou Y, Yang M, Li Z, Game OS, Zhu K, Zhu R, Gong Q, Padture NP (2017c) High-performance formamidinium-based perovskite solar cells via microstructure- mediated δ-to-α phase transformation. Chem Mater 29:3246–3250
Lu H, Ma Y, Gu B, Tian W, Li L (2015) Identifying the optimum thickness of electron transport layers for highly efficient perovskite planar solar cells. J Mater Chem A 3:16445–16452
Luo Q, Zhang Y, Liu CY, Li JB, Wang N, Lin H (2015) Iodide-reduced graphene oxide with dopant-free spiro-OMeTAD for ambient stable and high-efficiency perovskite solar cells. J Mater Chem A 3:15996–16004
Luo Q, Ma H, Zhang Y, Yin XW, Yao ZB, Wang N, Li JB, Fan SS, Jiang KL, Lin H (2016) Cross-stacked super aligned carbon nanotube electrodes for efficient hole conductor-free perovskite solar cells. J Mater Chem A 4:5569–5577

Luo H, Lin X, Hou X, Pan L, Huang S, Chen X (2017a) Efficient and air-stable planar perovskite solar cells formed on graphene-oxide-modified PEDOT:PSS hole transport layer. Nano- Micro Lett 9:39

Luo Q, Chen H, Lin Y, Du H, Hou Q, Hao F, Wang N, Guo Z, Huang J (2017b) Discrete Iron(III) oxide nanoislands for efficient and photostable perovskite solar cells. Adv Funct Mater 1702090:1–9

Lv M, Zhu J, Huang Y, Li Y, Shao Z, Xu Y, Dai S (2015) Colloidal $CuInS_2$ quantum dots as inorganic hole-transporting material in perovskite solar cells. ACS Appl Mater Interfaces 7:17482–17488

Ma J, Zheng X, Lei H, Ke W, Chen C, Chen Z, Yang G, Fang G (2017) Highly efficient and stable planar perovskite solar cells with large-scale manufacture of E-beam evaporated SnO_2 toward commercialization. Sol RRL 1:1700118

Madhavan VE, Zimmermann I, Roldan-Carmona C, Grancini G, Buffiere M, Belaidi A, Nazeeruddin MK (2016) Copper thiocyanate inorganic hole-transporting material for high-efficiency perovskite solar cells. ACS Energy Lett 1:1112–1117

Mahmood K, Swain BS, Amassian A (2014) Double-layered ZnO nanostructures for efficient perovskite solar cells. Nanoscale 6:14674–14678

Mahmood K, Swain BS, Kirmani AR, Amassian A (2015a) Highly efficient perovskite solar cells based on a nanostructured WO_3–TiO_2 core–shell electron transporting material. J Mater Chem A 3:9051–9057

Mahmood K, Swain BS, Amassian A (2015b) 16.1% Efficient hysteresis-free mesostructured perovskite solar cells based on synergistically improved ZnO nanorod arrays. Adv Energy Mater 5:1500568

Mahmud MA, Elumalai NK, Upama MB, Wang D, Wright M, Chan KH, Xu C, Haque F, Uddin A (2016) Single versus mixed organic cation for low temperature processed perovskite solar cells. Electrochim Acta 222:1510–1521

Mali SS, Shim CS, Park HK, Heo J, Patil PS, Hong CK (2015a) Ultrathin atomic layer deposited TiO_2 for surface passivation of hydrothermally grown 1D TiO_2 nanorod arrays for efficient solid-state perovskite solar cells. Chem Mater 27:1541–1551

Mali SS, Su Shim C, Kook Hong C (2015b) Highly porous Zinc Stannate (Zn_2SnO_4) nanofibers scaffold photoelectrodes for efficient methyl ammonium halide perovskite solar cells. Sci Rep 5:11424

Mamun AA, Ava TT, Zhang K, Baumgart H, Namkoong G (2017) New PCBM/carbon based electron transport layer for perovskite solar cells. Phys Chem Chem Phys 19:17960–17966

Matsui T, Seo J-Y, Saliba M, Zakeeruddin SM, Grätzel M (2017) Room-temperature formation of highly crystalline multication perovskites for efficient, low-cost solar cells. Adv Mater 29:1606258

McMeekin DP, Sadoughi G, Rehman W, Eperon GE, Saliba M, Hörantner MT, Haghighirad A, Sakai N, Korte L, Rech B (2016) A mixed-cation lead mixed-halide perovskite absorber for tandem solar cells. Science 351:151–155

Meng L, You J, Guo T-F, Yang Y (2016) Recent advances in the inverted planar structure of perovskite solar cells. Acc Chem Res 49(1):155–165

Mielczarek K, AA (2014) Perovskite based hybrid solar cells with transparent carbon nanotube electrodes, MRS Proc, 1667, Mrss14-1667-b09-82

Morton O (2006) Solar energy: a new day dawning? Silicon Valley sunrise. Nature 443(7107):19–22

Motti SG, Gandini M, Barker AJ, Ball JM, Srimath Kandada AR, Petrozza A (2016) Photoinduced emissive trap states in lead halide perovskite semiconductors. ACS Energy Lett 1(4):726–730

Murugadoss G, Kanda H, Tanaka S, Nishino H, Ito S, Imahoric H, Umeyama T (2016) An efficient electron transport material of tin oxide for planar structure perovskite solar cells. J Power Sources 307:891–897

Najafi L, Taheri B, Martin-Garcia B, Bellani S, Di Girolamo D, Agresti A, Oropesa-Nunez R, Pescetelli S, Vesce L, Calabro E, Prato M, Del Rio Castillo AE, Di Carlo A, Bonaccorso F (2018) MoS_2 quantum dot/graphene hybrids for advanced interface engineering of a $CH_3NH_3PbI_3$ perovskite solar cell with an efficiency of over 20. ACS Nano 12(11):10736–10754

Nazari FAP, Abdollahi Nejand B, Ahmadi V, Payandeh M, Salavati-Niasar M (2017) Physicochemical interface engineering of CuI/Cu as advanced potential hole-transporting materials/metal contact couples in hysteresis-free ultralow-cost and large-area perovskite solar cells. J Phys Chem C 121:21935–21944

Nejand BA, Ahmadi V, Gharibzadeh S, Shahverdi HR (2016) Cuprous oxide as a potential low-cost hole-transport material for stable perovskite solar cells. Chemsuschem 9:302–313

Nie W, Tsai H, Blancon JC, Liu F, Stoumpos CC, Traore B, Kepenekian M, Durand O, Katan C, Tretiak S, Crochet J, Ajayan PM, Kanatzidis M, Even J, Mohite AD (2018) Critical role of interface and crystallinity on the performance and photostability of perovskite solar cell on nickel oxide. Adv Mater 30:1703879

Niu G, Yu H, Li J, Wang D, Wang L (2016) Controlled orientation of perovskite films through mixed cations toward high performance perovskite solar cells. Nano Energy 27:87–94

Niu G, Li W, Li J, Liang X, Wang L (2017) Enhancement of Thermal Stability for Perovskite Solar Cells through Cesium Doping. RSC Adv 7:17473–17479

Noel NK, Abate A, Stranks SD, Parrott ES, Burlakov VM, Goriely A, Snaith HJ (2014) Enhanced photoluminescence and solar cell performance via Lewis base passivation of organic-inorganic lead halide perovskites. ACS Nano 8(10):9815–9821

Nouri E, Mohammadi MR, Lianos P (2017) Inverted perovskite solar cells basedon lithium-functionalized graphene oxide as an electron-transporting layer. Chem Commun 53:1630–1633

NREL solar energy chart: https://www.nrel.gov/pv/assets/pdfs/pv-efficiency-chart.20190103.pdf

O'Regan B, Schwartz DT, Zakeeruddin SM, Grätzel M (2000) Electrodeposited nanocomposite n–p heterojunctions for solid-state dye-sensitized photovoltaics. Adv Mater 12:1263–1267

Ogomi Y, Morita A, Tsukamoto S, Saitho T, Fujikawa N, Shen Q, Toyoda T, Yoshino K, Pandey SS, Ma T, Hayase S (2014) $CH_3NH_3Sn_xPb_{(1-x)}I_3$ Perovskite Solar Cells Covering up to 1060 nm. J Phys Chem Lett 5:1004–1011

Oh LS, Kim DH, Lee JA, Shin SS, Lee J, Park IJ, Ko MJ, Park N, Pyo SG, Hong KS, Kim JY (2015) Zn_2SnO_4-based photoelectrodes for organolead halide perovskite solar cells. J Phys Chem C 118:22991–22994

Ono LK, Qi Y (2016) Surface and interface aspects of organometal halide perovskite materials and solar cells. J Phys Chem Lett 7(22):4764–4794

Ono LK, Raga SR, Remeika M, Winchester AJ, Gabe A, Qi Y (2015) Pinhole-free hole transport layers significantly improve the stability of MAPbI3-based perovskite solar cells under operating conditions. J Mater Chem A 3(30):15451–15456

Ou QD, Li C, Wang QK, Li YQ, Tang JX (2017) Recent advances in energetics of metal halide perovskite interfaces. Adv Mater Interfaces 4:1600694

Pae SR, Byun S, Kim J, Kim M, Gereige I, Shin B (2018) Improving uniformity and reproducibility of hybrid perovskite solar cells via a low-temperature vacuum deposition process for NiO_x hole transport layers. ACS Appl Mater Interfaces 10:534–540

Pan ZX, Mora-Sero I, Shen Q, Zhang H, Li Y, Zhao K, Wang J, Zhong XH, Bisquert J (2014) High-efficiency "green" quantum dot solar cells. J Am Chem Soc 136:9203–9210

Park NG (2015) Perovskite solar cells: an emerging photovoltaic technology. Mater Today 18:65–72

Park BW, Seok SI (2019) Intrinsic instability of inorganic-organic hybrid halide perovskite materials. Adv Mater 31(20):1805337

Park IJ, Park MA, Kim DH, Park GD, Kim BJ, Son HJ, Ko MJ, Lee DK, Park T, Shin H, Park N-G, Jung HS, Young J (2015a) New hybrid hole extraction layer of perovskite solar cells with a planar *p–i–n* geometry. J Phys Chem 119:27285–27290

Park JH, Seo J, Park S, Shin SS, Kim YC, Jeon NJ, Shin HW, Ahn TK, Noh JH, Yoon SC, Hwang CS, Seok SI (2015b) Efficient $CH_3NH_3PbI_3$ perovskite solar cells employing nanostructured p-type NiO electrode formed by a pulsed laser deposition. Adv Mater 27:4013–4019

Park IJ, Park MA, Kim DH, Park GD, Kim BJ, Son HJ, Ko MJ, Lee D-K, Park T, Shin H, Park N-G, Jung HS, Kim JY (2015c) New hybrid hole extraction layer of perovskite solar cells with a planar *p-i-n* geometry. J Phys Chem C 119:27285–27290

Park IJ, Kang G, Park MA, Kim JS, Seo SW, Kim DH, Zhu K, Park T, Kim JY (2017a) Highly efficient and uniform 1 cm^2 perovskite solar cells with an electrochemically deposited NiO_x hole-extraction layer. Chemsuschem 10:2660–2667

Park YH, Jeong I, Bae S, Son HJ, Lee P, Lee J, Lee C-H, Ko MJ (2017b) Inorganic rubidium cation as an enhancer for photovoltaic performance and moisture stability of $HC(NH_2)_2PbI_3$ perovskite solar cells. Adv Funct Mater 27:1605988

Pathak SK, Abate A, Ruckdeschel P, Roose B, Gödel KC, Vaynzof Y, Santhala A, Watanabe SI, Hollman DJ, Noel N, Sepe A, Wiesner U, Friend R, Snaith HJ, Steiner U (2014) Performance and stability enhancement of dye-sensitized and perovskite solar cells by Al doping of TiO_2. Adv Funct Mater 24:6046–6055

Pattanasattayavong P, Gross NY, Zhao K, Ndjawa GON, Li J, Yan F, O'Regan BC, Amassian A, Anthopoulo TD (2013a) Hole-transporting transistors and circuits based on the transparent inorganic semiconductor copper(I) thiocyanate (CuSCN) processed from solution at room temperature. Adv Mater 25:1504–1509

Pattanasattayavong P, Ndjawa GON, Zhao K, Chou KW, Gross NY, O'Regan BC, Amassian A, Anthopoulos TD (2013b) Electric field-induced hole transport in copper(i) thiocyanate (CuSCN) thin-films processed from solution at room temperature. Chem Commun 49:4154–4156

Paulo S, Stoicaa G, Cambaraua W, Martinez-Ferrerob E, Palomares E (2016) Carbon quantum dots as new hole transport material for perovskite solar cells. Synth Met 222:17–22

Pellet N, Gao P, Gregori G, Yang T-Y, Nazeeruddin MK, Maier J, Grätzel M (2014) Mixed-organic-cation perovskite photovoltaics for enhanced solar-light harvesting. Angew Chem Int Ed 53:3151–3157

Petrus ML, Schlipf J, Li C, Gujar TP, Giesbrecht N, Muller-Buschbaum P, Thelakkat M, Bein T, Huttner S, Docampo P (2017) Capturing the sun: a review of the challenges and perspectives of perovskite solar cells. Adv. Energy Mater 7(16):1700264

Qin P, Tanaka S, Ito S, Tetreault N, Manabe K, Nishino H, Nazeeruddin MK, Grätzel M (2014a) Inorganic hole conductor-based lead halide perovskite solar cells with 12.4% conversion efficiency. Nat Commun 5:3834

Qin P, Domanski AL, Chandiran AK, Berger R, Butt HJ, Darm MI, Moehl T, Tetreault N, Gao P, Ahmad S, Nazeeruddin MK, Grätzel M (2014b) Yttrium-substituted nanocrystalline TiO_2 photoanodes for perovskite based heterojunction solar cells. Nanoscale 6:1508–1514

Qin PL, Lei HW, Zheng XL, Liu Q, Tao H, Yang G, Ke WJ, Xiong LB, Qin MC, Zhao XZ, Fang GJ (2016) Copper-doped chromium oxide hole-transporting layer for perovskite solar cells: interface engineering and performance improvement. Adv Mater Interfaces 3:1500799

Qiu L, Deng J, Lu X, Yang Z, Peng H (2014) Integrating perovskite solar cells into a flexible fiber. Angew Chem Int Ed Engl 53:10425–10428

Quan LN, Yuan M, Comin R, Voznyy O, Beauregard EM, Hoogland S, Buin A, Kirmani AR, Zhao K, Amassian A, Kim DH, Sargent EH (2016) Ligand-stabilized reduced-dimensionality perovskites. J Am Chem Soc 138:2649–2655

Rao H-S, Chen B-X, Li W-G, Xu Y-F, Chen H-Y, Kuang D-B, Su C-Y (2015) Improving the extraction of photogenerated electrons with SnO_2 nanocolloids for efficient planar perovskite solar cells. Adv Funct Mater 25:7200–7207

Rao H, Sun W, Ye S, Yan W, Li Y, Peng H, Liu Z, Bian Z, Huang C (2016a) Solution-Processed CuS NPs as an inorganic hole-selective contact material for inverted planar perovskite solar cells. ACS Appl Mater Interfaces 8:7800–7805

Rao H, Ye S, Sun W, Yan W, Li Y, Peng HT, Liu Z, Bian Z, Li Y, Huang C (2016b) A 19.0% efficiency achieved in CuO_x-based inverted $CH_3NH_3PbI_{3-x}Cl_x$ solar cells by an effective Cl doping method. Nanoenergy 27:51–57

Rehman W, McMeekin DP, Patel JB, Milot RL, Johnston MB, Snaith HJ, Herz LM (2017) Photovoltaic mixed-cation lead mixed-halide perovskites: links between crystallinity, photo-stability and electronic properties. Energy Environ Sci 10(1):361–369

Reyna Y, Salado M, Kazim S, Pérez-Tomas A, Ahmad S, Lira-Cantu M (2016) Performance and stability of mixed $FAPbI_{3(0.85)}MAPbBr_{3(0.15)}$ halide perovskite solar cells under outdoor conditions and the effect of low light irradiation. Nano Energy 30:570–579

Richardson G, O'Kane SEJ, Niemann RG, Peltola TA, Foster JM, Cameron PJ, Walker AB (2016) Can slow-moving ions explain hysteresis in the current–voltage curves of perovskite solar cells? Energy Environ Sci 9(4):1476–1485

Rong Y, Ku Z, Mei A, Liu T, Xu M, Ko S, Li X, Han H (2014) Hole-conductor-free mesoscopic $TiO_2/CH_3NH_3PbI_3$ heterojunction solar cells based on anatase nanosheets and carbon counter electrodes. J Phys Chem Lett 5:2160–2164

Rong Y, Hou X, Hu Y, Mei A, Liu L, Wang P, Han H (2017) Synergy of ammonium chloride and moisture on perovskite crystallization for efficient printable mesoscopic solar cells. Nat Commun 8:14555

Ryu S, Seo J, Shin SS, Kim YC, Jeon NJ, Noh JH, Il Seok S (2015) Fabrication of metal-oxide-free $CH_3NH_3PbI_3$ perovskite solar cells processed at low temperature. J Mater Chem A 3:3271–3275

Ryu J, Lee K, Yun J, Yu H, Lee J, Jang J (2017) Paintable carbon-based perovskite solar cells with engineered perovskite/carbon interface using carbon nanotubes dripping method. Small 13:1701225

Sahli F, Werner J, Kamino BA, Brauninger M, Monnard R, Paviet-Salomon B, Barraud L, Ding L, Leon JJD, Sacchetto D, Cattaneo G, Despeisse M, Boccard M, Nicolay S, Jeangros Q, Niesen B, Ballif C (2018) Fully textured monolithic perovskite/silicon tandem solar cells with 25.2% power conversion efficiency. Nat Mater 17(9):820

Saidaminov MI, Abdelhady AL, Murali B, Alarousu E, Burlakov VM, Peng W, Dursun I, Wang L, He Y, Maculan G, Goriely A, Wu T, Mohammed OF, Bakr OM (2015) High-quality bulk hybrid perovskite single crystals within minutes by inverse temperature crystallization. Nat Commun 6(1):7586

Saidaminov MI, Kim J, Jain A, Quintero-Bermudez R, Tan H, Long G, Tan F, Johnston A, Zhao Y, Voznyy O, Sargent EH (2018) Suppression of atomic vacancies via incorporation of isovalent small ions to increase the stability of halide perovskite solar cells in ambient air. Nat Energy 3(8):648–654

Saliba M, Matsui T, Seo J-Y, Domanski K, Correa-Baena J-P, Nazeeruddin MK, Zakeeruddin SM, Tress W, Abate A, Hagfeldt A (2016a) Cesium-containing triple cation perovskite solar cells: improved stability, reproducibility and high efficiency. Energy Environ Sci 9:1989–1997

Saliba M, Matsui T, Domanski K, Seo J-Y, Ummadisingu A, Zakeeruddin SM, Correa-Baena J-P, Tress WR, Abate A, Hagfeldt A, Grätzel M (2016b) Incorporation of rubidium cations into perovskite solar cells improves photovoltaic performance. Science 354:206–209

Saliba M, Matsui T, Domanski K, Seo J-Y, Ummadisingu A, Zakeeruddin SM, Correa-Baena J-P, Tress WR, Abate A, Hagfeldt A, Grätzel M (2016c) Incorporation of rubidium cations into perovskite solar cells improves photovoltaic performance. Science 354(6309):206–209

Sanehira EM, Marshall AR, Christians JA, Harvey SP, Ciesielski PN, Wheeler LM, Schulz P, Lin LY, Beard MC, Luther JM (2017) Enhanced mobility $CsPbI_3$ quantum dot arrays for record-efficiency, high-voltage photovoltaic cells. Sci Adv 3:eaao4204

Santra PK, Nair PV, Thomas KG, Kamat PV (2013) $CuInS_2$-sensitized quantum dot solar cell. Electrophoretic deposition, excited-state dynamics, and photovoltaic performance. J Phys Chem Lett 4:722–729

Sarkar A, Jeon NJ, Noh JH, Seok SI (2014) Well-organized mesoporous TiO_2 photoelectrodes by block copolymer-induced Sol-Gel assembly for inorganic-organic hybrid perovskite solar cells. J Phys Chem C 118:16688–16693

Sepalage GA, Meyer S, Pascoe A, Scully AD, Huang F, Bach U, Cheng Y-B, Spiccia L (2015) Copper (I) iodide as hole-conductor in planar perovskite solar cells: probing the origin of J-V hysteresis. Adv Funct Mater 25:5650–5661

Shao S, Liu F, Xie Z, Wang L (2010) High-efficiency hybrid polymer solar cells with inorganic P- and N-type semiconductor nanocrystals to collect photogenerated charges. J Phys Chem C 114:9161

Sherkar TS, Jan Anton Koster L (2016) Can ferroelectric polarization explain the high performance of hybrid halide perovskite solar cells? Phys Chem Chem Phys 18(1):331–338

Shi D, Adinolfi V, Comin R, Yuan M, Alarousu E, Buin A, Chen Y, Hoogland S, Rothenberger A, Katsiev K, Losovyj Y, Zhang X, Dowben PA, Mohammed OF, Sargent EH, Bakr OM (2015) Low trap-state density and long carrier diffusion in organolead trihalide perovskite single crystals. Science 347(6221):519–522

Shockley W, Queisser HJ (1961) Detailed balance limit of efficiency of p-n junction solar cells. J Appl Phys 32(3):510–519

Smecca E, Numata Y, Deretzis I, Pellegrino G, Boninelli S, Miyasaka T, La Magna A, Alberti A (2016) Stability of solution-processed $MAPbI_3$ and $FAPbI_3$ layers. Phys Chem Chem Phys 18(19):13413–13422

Snaith HJ (2013) Perovskites: the emergence of a new era for low-cost, high-efficiency solar cells. J Phys Chem Lett 4:3623–3630

Snaith HJ, Abate A, Ball JM, Eperon GE, Leijtens T, Noel NK, Stranks SD, Wang JT-W, Wojciechowski K, Zhang W (2014) Anomalous hysteresis in perovskite solar cells. J Phys Chem Lett 5(9):1511–1515

Son DY, Im JH, Kim HS, Park NG (2014) 11% efficient perovskite solar cell based on ZnO nanorods: an effective charge collection system. J Phys Chem C 118:16567–16573

Song J, Zheng E, Bian J, Wang XF, Tian W, Sanehira Y, Miyasaka T (2015a) Low-temperature SnO_2-based electron selective contact for efficient and stable perovskite solar cells. J Mater Chem A 3:10837–10844

Song J, Bian J, Zheng E, Wang X-F, Tian W, Miyasaka T (2015b) Efficient and environmentally stable perovskite solar cells based on ZnO electron collection layer. Chem Lett 44:610–612

Stranks SD, Eperon GE, Grancini G, Menelaou C, Alcocer MJP, Leijtens T, Herz LM, Petrozza A, Snaith HJ (2013) Electron-hole diffusion lengths exceeding 1 micrometer in an organometal trihalide perovskite absorber. Science 342(6156):341–344

Stylianakis MM, Maksudov T, Panagiotopoulos A, Kakavelakis G, Petridis K (2019) Inorganic and hybrid perovskite based laser devices: a review. Materials 12(6):859

Su T-S, Hsieh T-Y, Hong C-Y, Wei T-C (2015) Electrodeposited ultrathin TiO_2 blocking layers for efficient perovskite solar cells. Sci Rep 5:16098

Subbiah AS, Halder A, Ghosh S, Mahuli N, Hodes G, Sarkar SK (2014) Inorganic hole conducting layers for perovskite-based solar cells. J Phys Chem Lett 5:1748–1753

Sum TC, Mathews N (2014a) Advancements in perovskite solar cells: photophysics behind the photovoltaics. Energy Environ Sci 7(8):2518–2534

Sum TC, Mathews N (2014b) Advancements in perovskite solar cells: photophysics behind the photovoltaics. Energy Environ Sci 7:2518–2534

Sun W, Li Y, Ye S, Rao H, Yan W, Peng HT, Li Y, Liu Z, Wang S, Chen Z, Xiao L, Bian Z, Huang C (2016a) High-performance inverted planar heterojunction perovskite solar cells based on a solution-processed CuO_x hole transport layer. Nanoscale 8:10806–10813

Sun W, Ye S, Rao H, Li Y, Liu Z, Xiao L, Chen Z, Bian Z, Huang C (2016b) Room-temperature and solution-processed copper iodide as the hole transport layer for inverted planar perovskite solar cells. Nanoscale 8:15954–15960

Sveinbjornsson K, Aitola K, Zhang J, Johansson MB, Zhang X, Correa-Baena J-P, Hagfeldt A, Boschloo G, Johansson EMJ (2016) Ambient air-processed mixed-ion perovskites for high-efficiency solar cell. J Mater Chem A 4:16536–16545

Tainter GD, Hörantner MT, Pazos-Outón LM, Lamboll RD, Āboliņš H, Leijtens T, Mahesh S, Friend RH, Snaith HJ, Joyce HJ, Deschler F (2019) Long-range charge extraction in back-contact perovskite architectures via suppressed recombination. Joule 3(5):1301–1313

Tan ZK, Moghaddam RS, Lai ML, Docampo P, Higler R, Deschler F, Price M, Sadhanala A, Pazos LM, Credgington D, Hanusch F, Bein T, Snaith HJ, Friend RH (2014) Bright light-emitting diodes based on organometal halide perovskite. Nat Nanotechnol 9(9):687–692

Tang J, Kemp KW, Hoogland S, Jeong KS, Liu H, Levina L, Furukawa M, Wang X, Debnath R, Cha D, Chou KW, Fischer A, Amassian A, Asbury JB, Sargent EH (2011) Colloidal-quantum-dot photovoltaics using atomic-ligand passivation. Nat Mater 10:765

Tang J-F, Tseng Z-L, Chen L-C, Chu S-Y (2016) ZnO nanowalls grown at low-temperature for electron collection in high-efficiency perovskite solar cells. Sol Energy Mater Sol Cells 154:18–22

Tao C, Neutzner S, Colella L, Marras S, Srimath Kandada AR, Gandini M, De Bastiani M, Pace G, Manna L, Caironi M, Bertarelli C, Petrozza A (2015) 17.6% stabilized efficiency in low-temperature processed planar perovskite solar cells. Energy Environ Sci 8:2365–2370

Tavakoli MM, Tavakoli R, Hasanzadeh S, Mirfasih MH (2016) Interface engineering of perovskite solar cell using a reduced-graphene scaffold. J Phys Chem C 120:19531–19536

Tian H, Xu B, Chen H, Johansson EM, Boschloo G (2014) Solid-state perovskite-sensitized p-type mesoporous nickel oxide solar cells. Chemsuschem 7:2150–2153

Tsai HH, Nie WY, Blancon JC, Toumpos CCS, Asadpour R, Harutyunyan B, Neukirch AJ, Verduzco R, Crochet JJ, Tretiak S, Pedesseau L, Even J, Alam MA, Gupta G, Lou J, Ajayan PM, Bedzyk MJ, Kanatzidis MG, Mohite AD (2016a) High-efficiency two-dimensional Ruddlesden-Popper perovskite solar cells. Nature 536(7616):312–316

Tsai CM, Wu HP, Chang ST, Huang CF, Wang CH, Narra S, Yang YW, Wang CL, Hung CH, Diau EWG (2016b) Role of tin chloride in tin-rich mixed-halide perovskites applied as mesoscopic solar cells with a carbon counter electrode. ACS Energy Lett 1:1086–1093

Tsai CM, Wu GW, Narra S, Chang HM, Mohanta N, Wu HP, Wang CL, Diau EWG (2017) Control of preferred orientation with slow crystallization for carbon-based mesoscopic perovskite solar cells attaining efficiency 15%. J Mater Chem A 5:739–747

Tsujimoto K, Nguyen DC, Ito S, Nishino H, Matsuyoshi H, Konno A, Asoka Kumara GR, Tennakone K (2012) TiO_2 surface treatment effects by Mg^{2+}, Ba^{2+}, and Al^{3+} on Sb_2S_3 extremely thin absorber solar cells. J Phys Chem C 116:13465–13471 (2012)

Upama MB, Elumalai NK, Mahmud MA, Wang D, Haque F, Gonçales VR, Gooding JJ, Wright M, Xu C, Uddin A (2017) Role of fullerene electron transport layer on the morphology and optoelectronic properties of perovskite solar cells. Org Electron 50:279–289

Vanalakar SA, Agawane GL, Shin SW, Suryawanshi MP, Gurav KV, Jeon KS, Patil PS, Jeong CW, Kim JY, Kim JH (2015) A review on pulsed laser deposited CZTS thin films for solar cell applications. J Alloys Compd 619:109–121

Walsh A, Chen SY, Wei SH, Gong XG (2012) Kesterite thin-film solar cells: advances in materials modelling of Cu2ZnSnS4. Adv Energy Mater 2:400–409

Wang H, Kim DH (2017) Perovskite-based photodetectors: materials and devices. Chem Soc Rev 46(17):5204–5236

Wang KC, Jeng JY, Shen PS, Chang YC, Diau EW, Tsai CH, Chao TY, Hsu HC, Lin PY, Chen P, Guo TF, Wen TC (2014a) p-type mesoscopic nickel oxide/organometallic perovskite heterojunction solar cells. Sci Rep 4:4756

Wang KC, Shen PS, Li MH, Chen S, Lin MW, Chen P, Guo TF (2014b) Low-temperature sputtered nickel oxide compact thin film as effective electron blocking layer for mesoscopic NiO/$CH_3NH_3PbI_3$ perovskite heterojunction solar cells. ACS Appl Mater Interfaces 6:11851–11858

Wang H, Zeng X, Huang Z, Zhang W, Qiao X, Hu B, Zou X, Wang M, Cheng YB, Chen W (2014c) Boosting the photocurrent density of p-type solar cells based on organometal halide perovskite-sensitized mesoporous NiO photocathodes. ACS Appl Mater Interfaces 6:12609–12617

Wang L, Fu W, Gu Z, Fan C, Yang X, Li H, Chen H (2014d) Low temperature solution processed planar heterojunction perovskite solar cells with a CdSe nanocrystal as an electron transport/extraction layer. J Mater Chem C 2:9087–9090

Wang JTW, Ball JM, Barea EM, Abate A, Alexander-Webber JA, Huang J, Saliba M, Mora-Sero I, Bisquert J, Snaith HJ, Nicholas RJ (2014e) Low-temperature processed electron collection layers of graphene/TiO_2 nanocomposites in thin film perovskite solar cells. Nano Lett 14:724–730

Wang K, Shi Y, Dong Q, Li Y, Wang S, Yu X, Wu M, Ma T (2015a) Low-temperature and solution-processed amorphous WO_X as electron-selective layer for perovskite solar cells. J Phys Chem Lett 6:755–759
Wang H, Hu XY, Chen HX (2015b) The effect of carbon black in carbon counter electrode for $CH_3NH_3PbI_3/TiO_2$ heterojunction solar cells. RSC Adv 5:30192–30196
Wang XY, Li Z, Xu WJ, Kulkarni SA, Batabyal SK, Zhang S, Cao AY, Wong LH (2015c) TiO_2 nanotube arrays based flexible perovskite solar cells with transparent carbon nanotube electrode. Nano Energy 11:728–735
Wang J, Qin M, Tao H, Ke W, Chen Z, Wan J, Qin P, Xiong L, Lei H, Yu H, Fang G (2015d) Performance enhancement of perovskite solar cells with Mg-doped TiO_2 compact film as the hole-blocking layer. Appl Phys Lett 106:121104
Wang H, Sayeed BA, Wang T (2015e) Perovskite solar cells based on nanocrystalline SnO_2 material with extremely small particle sizes. Aust J Chem 68:1783–1788
Wang BB, Zhang ZG, Ye SY, Gao L, Yan TH, Bian ZQ, Huang CH, Li YF (2016a) Solution-processable cathode buffer layer for high-performance ITO/CuSCN-based planar heterojunction perovskite solar cell. Electrochim Acta 218:263–270
Wang BX, Liu TF, Zhou YB, Chen X, Yuan XB, Yang YY, WP Liu, Wang JM, Han HW, Tang YW (2016b) Hole-conductor-free perovskite solar cells with carbon counter electrodes based on ZnO nanorod arrays. Phys Chem Chem Phys 18:27078–27082
Wang F, Endo M, Mouri S, Miyauchi Y, Ohno Y, Wakamiya A, Murata Y, Matsuda K (2016c) Highly stable perovskite solar cells with an all-carbon hole transport layer. Nanoscale 8:11882–11888
Wang K, Shi Y, Li B, Zhao L, Wang W, Wang X, Bai X, Wang S, Hao C, Ma T (2016d) Amorphous Inorganic Electron-selective layers for efficient perovskite solar cells: feasible strategy towards room-temperature fabrication. Adv Mater 28:1891–1897
Wang C, Zhao D, Grice CR, Liao W, Yu Y, Cimaroli A, Shrestha N, Roland PJ, Chen J, Yu Z, Liu P, Cheng N, Ellingson R, Zhao X, Yan Y (2016e) Low-temperature plasma-enhanced atomic layer deposition of tin oxide electron selective layers for highly efficient planar perovskite solar cells. J Mater Chem A 4:12080–12087
Wang F, Ma J, Xie F, Li L, Chen J, Fan J, Zhao N (2016f) Organic cation-dependent degradation mechanism of organotin halide perovskites. Adv Funct Mater 26:3417–3423
Wang S, Jiang Y, Juarez-Perez EJ, Ono LK, Qi Y (2016g) Accelerated degradation of methylammonium lead iodide perovskites induced by exposure to iodine vapour. Nat Energy 2(1):16195
Wang HX, Yu Z, Jiang X, Li JJ, Cai B, Yang XC, Sun LC (2017a) Efficient and stable inverted planar perovskite solar cells employing CuI as hole-transporting layer prepared by solid–gas transformation. Energy Technol 5:1836–1843
Wang X, Deng L-L, Wang L-Y, Dai S-M, Xing Z, Zhan X-X, Lu X-Z, Xie S-Y, Huang R-B, Zheng L-S (2017b) Cerium oxide standing out as an electron transport layer for efficient and stable perovskite solar cells processed at low temperature. J Mater Chem A 5:1706–1712
Wang Z, McMeekin DP, Sakai N, van Reenen S, Wojciechowski K, Patel JB, Johnston MB, Snaith HJ (2017c) Efficient and air-stable mixed-cation lead mixed-halide perovskite solar cells with n-doped organic electron extraction layers. Adv Mater 29:1604186
Wang R, Mujahid M, Duan Y, Wang ZK, Xue JJ, Yang Y (2019) A review of perovskites solar cell stability. Adv Funct Mater 29(47):1808843
Wei Z, Chen H, Yan K, Yang S (2014) Inkjet printing and instant chemical transformation of a $CH_3NH_3PbI_3$/nanocarbon electrode and interface for planar perovskite solar cells. Angew Chem Int Ed Engl 53:13239–13243
Wei HY, Xiao JY, Yang YY, Lv ST, Shi JJ, Xu X, Dong J, Luo YH, Li DM, Meng QB (2015a) Free-standing flexible carbon electrode for highly efficient hole-conductor-free perovskite solar cells. Carbon 93:861–868
Wei ZH, Chen HN, Yan KY, Zheng XL, Yang SH (2015b) Hysteresis-free multi-walled carbon nanotube-based perovskite solar cells with a high fill factor. J Mater Chem A 3:24226–24231

Wei W, Hu BY, Jin FM, Jing ZZ, Li YX, Blanco AAG, Stacchiola DJ, Hu YH (2017) Potassium-chemical synthesis of 3D graphene from CO_2 and its excellent performance in HTM-free perovskite solar cells. J Mater Chem A 5:7749–7752

Wei-Chih L, Kun-Wei L, Tzung-Fang G, Jung L (2015) Perovskite-based solar cells with nickel-oxidized nickel oxide hole transfer layer. IEEE Trans Electron Devices 62:1590–1595

Wijeyasinghe N, Regoutz A, Eisner F, Du T, Tsetseris L, Lin YH, Faber H, Pattanasattayavong P, Li JH, Yan F, McLachlan MA, Payne DJ, Heeney M, Anthopoulos TD (2017) Copper (I) thiocyanate (CuSCN) hole-transport layers processed from aqueous precursor solutions and their application in thin-film transistors and highly efficient organic and organometal halide perovskite solar cells. Adv Funct Mater 27:1701818

Wojciechowski K, Leijtens T, Siprova S, Schlueter C, Hoärantner MT, Wang JTW, Li CZ, Jen AKY, Lee TL, Snaith HJ (2015) C_{60} as an efficient n-type compact layer in perovskite solar cells. J Phys Chem Lett 6:2399–2405

Wu Z, Bai S, Xiang J, Yuan Z, Yang Y, Cui W, Gao X, Liu Z, Jin Y, Sun B (2014a) Efficient planar heterojunction perovskite solar cells employing graphene oxide as hole conductor. Nanoscale 6:10505–10510

Wu Y, Yang X, Chen H, Zhang K, Qin C, Liu J, Peng W, Islam A, Bi E, Ye F, Yin M, Zhang P, Han L (2014b) Highly compact TiO_2 layer for efficient hole-blocking in perovskite solar cells. Appl Phys Express 7:52301

Wu Q, Xue C, Li Y, Zhou P, Liu W, Zhu J, Dai S, Zhu C, Yang S (2015a) Kesterite Cu_2ZnSnS_4 as a low-cost inorganic hole-transporting material for high-efficiency perovskite solar cells. ACS Appl Mater Interfaces 7:28466–28473

Wu WQ, Huang F, Chen D, Cheng YB, Caruso RA (2015b) Thin films of dendritic anatase titania nanowires enable effective hole-blocking and efficient light-harvesting for high-performance mesoscopic perovskite solar cells. Adv Funct Mater 25:3264–3272

Wu R, Yang B, Xiong J, Cao C, Huang Y, Wu F, Sun J, Zhou C, Huang H, Yang J (2015c) Dependence of device performance on the thickness of compact TiO_2 layer in perovskite/TiO_2 planar heterojunction solar cells. J Renew Sustain Energy 7:043105

Wu X, Trinh MT, Niesner D, Zhu H, Norman Z, Owen JS, Yaffe O, Kudisch BJ, Zhu XY (2015d) Trap states in lead iodide perovskites. J Am Chem Soc 137(5):2089–2096

Wu Q, Zhou W, Liu Q, Zhou P, Chen T, Lu Y, Qiao Q, Yang S (2016) Solution-processable ionic liquid as an independent or modifying electron transport layer for high-efficiency perovskite solar cells. ACS Appl Mater Interfaces 8:34464–34473

Wu Y, Xie F, Chen H, Yang X, Su H, Cai M, Zhou Z, Noda T, Han L (2017) Thermally stable $MAPbI_3$ perovskite solar cells with efficiency of 19.19% and area over 1 cm^2 achieved by additive engineering. Adv Mater 29:1701073

Xiao MD, Gao M, Huang FZ, Pascoe AR, Qin TS, Cheng YB, Bach U, Spiccia L (2016) Efficient perovskite solar cells employing inorganic interlayers. Chemnanomat 2:182–188

Xiao YQ, Cheng N, Kondamareddy KK, Wang CL, Liu P, Guo SS, Zhao XZ (2017) W-doped TiO_2 mesoporous electron transport layer for efficient hole transport material free perovskite solar cells employing carbon counter electrodes. J Power Sources 342:489–494

Xie FX, Chen CC, Wu YZ, Li X, Cai ML, Liu X, Yang XD, Han LY (2017) Vertical recrystallization for highly efficient and stable formamidinium-based inverted-structure perovskite solar cells. Energy Environ Sci 10:1942–1949

Xing G, Mathews N, Sun S, Lim SS, Lam YM, Grätzel M, Mhaisalkar S, Sum TC (2013) Long-range balanced electron- and hole-transport lengths in organic-inorganic $CH_3NH_3PbI_3$. Science 342(6156):344–347

Xing Y, Sun C, Yip HL, Bazan GC, Huang F, Cao Y (2016) New fullerene design enables efficient passivation of surface traps in high performance *p-i-n* heterojunction perovskite solar cells. Nano Energy 26:7–15

Xu X, Zhang H, Cao K, Cui J, Lu J, Zeng X, Shen Y, Wang M (2014) Lead methylammonium triiodide perovskite-based solar cells: an interfacial charge-transfer investigation. Chem Sus Chem 7:3088–3094

Xu X, Liu Z, Zuo Z, Zhang M, Zhao Z, Shen Y, Zhou H, Chen Q, Yang Y, Wang M (2015a) Hole selective NiO contact for efficient perovskite solar cells with carbon electrode. Nano Lett 15:2402–2408

Xu X, Zhang H, Shi J, Dong J, Luo Y, Li D, Meng Q (2015b) Highly efficient planar perovskite solar cells with a TiO_2/ZnO electron transport bilayer. J Mater Chem A 3:19288–19293

Xu L, Wan F, Rong Y, Chen H, He S, Xu X, Liu G, Han H, Yuan Y, Yang J, Gao Y, Yang B, Zhou C (2017) Stable monolithic hole-conductor-free perovskite solar cells using TiO_2 nanoparticle binding carbon films. Org Electron 45:131–138

Yan K, Wei Z, Li J, Chen H, Yi Y, Zheng X, Long X, Wang Z, Wang J, Xu J, Yang S (2015) High-performance graphene-based hole conductor-free perovskite solar cells: Schottky junction enhanced hole extraction and electron blocking. Small 11:2269–2274

Yang YY, Xiao JY, Wei HY, Zhu LF, Li DM, Luo YH, Wu HJ, Meng QB (2014) An all-carbon counter electrode for highly efficient hole-conductor-free organo-metal perovskite solar cells. RSC Adv 4:52825–52830

Yang WS, Noh JH, Jeon NJ, Kim YC, Ryu S, Seo J, Seok SI (2015a) High-performance photovoltaic perovskite layers fabricated through intramolecular exchange. Science 348(6240):1234–1237

Yang Y, Ri K, Mei AY, Liu LF, Hu M, Liu TF, Li X, Han HW (2015b) The size effect of TiO_2 nanoparticles on a printable mesoscopic perovskite solar cell. J Mater Chem A 3:9103–9107

Yang IS, You JS, Do Sung S, Chung CW, Kim J, Lee WI (2016a) Novel spherical TiO_2 aggregates with diameter of 100 nm for efficient mesoscopic perovskite solar cells. Nano Energy 20:272–282

Yang D, Yang R, Ren X, Zhu X, Yang Z, Li C, Liu SF (2016b) Hysteresis-suppressed high-efficiency flexible perovskite solar cells using solid-state ionic-liquids for effective electron transport. Adv Mater 28:5206–5213

Yang D, Zhou X, Yang R, Yang Z, Yu W, Wang X, Li C, Liu SF, Chang RPH (2016c) Surface optimization to eliminate hysteresis for record efficiency planar perovskite solar cells. Energy Environ Sci 9:3071–3078

Yang Z, Rajagopal A, Chueh C-C, Jo SB, Liu B, Zhao T, Jen AKY (2016d) Stable low-bandgap Pb–Sn binary perovskites for tandem solar cells. Adv Mater 28:8990–8997

Yang D, Zhou X, Yang R, Yang Z, Yu W, Wang X, Li C, Liu S, Chang RPH (2016e) Surface optimization to eliminate hysteresis for record efficiency planar perovskite solar cells. Energy Environ Sci 9(10):3071–3078

Yang YL, Chen HN, Zheng XL, Meng XY, Zhang T, Hu C, Bai Y, Xiao S, Yang SH (2017) Ultrasound-spray deposition of multi-walled carbon nanotubes on NiO nanoparticles-embedded perovskite layers for high-performance carbon-based perovskite solar cells. Nano Energy 42:322–333

Yang Y, Wu J, Wang X, Guo Q, Liu X, Sun W, Wei Y, Huang Y, Lan Z, Huang M, Lin J, Chen H, Wei Z (2020) Suppressing vacancy defects and grain boundaries via Ostwald Ripening for high-performance and stable perovskite solar cells. Adv Mater 32(7):1904347

Ye S, Sun W, Li Y, Yan W, Peng H, Bian Z, Liu Z, Huang C (2015) CuSCN-based inverted planar perovskite solar cell with an average PCE of 15.6%, Nano Lett 15:3723–3728

Yeo JS, Kang R, Lee S, Jeon YJ, Myoung N, Leek CL, Kim DY, Yun JM, Seo YH, Kim SS, Na SI (2015) Highly efficient and stable planar perovskite solar cells with reduced graphene oxide nanosheets as electrode interlayer. Nano Energy 12:96–104

Yi C, Luo J, Meloni S, Boziki A, Ashari-Astani N, Grätzel C, Zakeeruddin SM, Röthlisberger U, Grätzel M (2016) Entropic stabilization of mixed A-cation ABX_3 metal halide perovskites for high performance perovskite solar cells. Energy Environ Sci 9:656–662

Yin XT, Que MD, Xing YL, Que WX (2015) High efficiency hysteresis-less inverted planar heterojunction perovskite solar cells with a solution-derived NiO_x hole contact layer. J Mater Chem A 3:24495–24503

Yin X, Chen P, Que M, Xing Y, Que W, Niu C, Shao J (2016) Highly efficient flexible perovskite solar cells using solution-derived NiOx hole contacts. ACS Nano 10:3630–3636

Yoshikawa K, Yoshida W, Irie T, Kawasaki H, Konishi K, Ishibashi H, Asatani T, Adachi D, Kanematsu M, Uzu H, Yamamoto K (2017) Exceeding conversion efficiency of 26% by heterojunction interdigitated back contact solar cell with thin film Si technology. Sol Energy Mater Sol 173:37–42

You J, Hong Z, Yang Y, Chen Q, Cai M, Song T-B, Chen C-C, Lu S, Liu Y, Zhou H, Yang Y (2014) Low-temperature solution-processed perovskite solar cells with high efficiency and flexibility. ACS Nano 8:1674–1680

You P, Liu Z, Tai Q, Liu S, Yan F (2015) Efficient semi transparent perovskite solar cells with graphene electrodes. Adv Mater 27:3632–3638

You J, Meng L, Song TB, Guo TF, Yang YM, Chang WH, Hong Z, Chen H, Zhou H, Chen Q, Liu Y, De Marco N, Yang Y (2016) Improved air stability of perovskite solar cells via solution-processed metal oxide transport layers. Nat Nanotechnol 11:75–81

Yu W, Li F, Wang H, Alarousu E, Chen Y, Lin B, Wang L, Hedhili MN, Li Y, Wu K, Wang X, Mohammed OF, Wu T (2016a) Ultrathin Cu_2O as an efficient inorganic hole transporting material for perovskite solar cells. Nanoscale 8:6173–6179

Yu X, Chen S, Yan K, Cai X, Hu H, Peng M, Chen B, Dong B, Gao X, Zou D (2016b) Enhanced photovoltaic performance of perovskite solar cells with mesoporous SiO_2 scaffolds. J Power Sources 325:534–540

Yu Y, Wang C, Grice CR, Shrestha N, Chen J, Zhao D, Liao W. Cimaroli AJ, Roland PJ, Ellingson RJ (2016) Improving the performance of formamidinium and cesium lead triiodide perovskite solar cells using lead thiocyanate additives. ChemSusChem 9:3288–3297

Yuan Y, Huang J (2016) Ion migration in organometal trihalide perovskite and its impact on photovoltaic efficiency and stability. Acc Chem Res 49(2):286–293

Yue GQ, Chen D, Wang P, Zhang J, Hu ZY, Zhu YJ (2016) Low-temperature prepared carbon electrodes for hole-conductor-free mesoscopic perovskite solar cells. Electrochim Acta 218:84–90

Yue SZ, Liu K, Xu R, Li MC, Azam M, Ren K, Liu J, Sun Y, Wang ZJ, Cao DW, Yan XH, Qu SC, Lei Y, Wang ZG (2017) Efficacious engineering on charge extraction for realizing highly efficient perovskite solar cells. Energy Environ Sci 10:2570–2578

Zhang F, Yang X, Wang H, Cheng M, Zhao J, Sun L (2014) Structure engineering of hole-conductor free perovskite-based solar cells with low-temperature-processed commercial carbon paste as cathode. ACS Appl Mater Interfaces 6:16140–16146

Zhang LJ, Liu TF, Liu LF, Hu M, Yang Y, Mei AY, Han HW (2015a) The effect of carbon counter electrodes on fully printable mesoscopic perovskite solar cells. J Mater Chem A 3:9165–9170

Zhang F, Yang X, Cheng M, Li J, Wang W, Wang H, Sun L (2015b) Engineering of hole-selective contact for low temperature-processed carbon counter electrode-based perovskite solar cells. J Mater Chem A 3:24272–24280

Zhang J, Juárez-Pérez EJ, Mora-Seró I, Viana B, Pauporté T (2015c) Fast and low temperature growth of electron transport layers for efficient perovskite solar cells. J Mater Chem A 3:4909–4915

Zhang J, Shi C, Chen J, Ying C, Wu N, Wang M (2016a) Pyrolysis preparation of WO_3 thin films using ammonium metatungstate DMF/water solution for efficient compact layers in planar perovskite solar cells. J Semicond 37:033002

Zhang NN, Guo YJ, Yin X, He M, Zou XP (2016b) Spongy carbon film deposited on a separated substrate as counter electrode for perovskite-based solar cell. Mater Lett 182:248–252

Zhang FG, Yang XC, Cheng M, Wang WH, Sun LC (2016c) Boosting the efficiency and the stability of low-cost perovskite solar cells by using CuPc nanorods as hole transport material and carbon as counter electrode. Nano Energy 20:108–116

Zhang CX, Luo YD, Chen XH, Chen YW, Sun Z, Huang SM (2016d) Effective improvement of the photovoltaic performance of carbon-based perovskite solar cells by additional solvents. Nano-Micro Lett 8:347–357

Zhang LQ, Yang XL, Jiang Q, Wang PY, Yin ZG, Zhang XW, Tan HR, Yang Y, Wei MY, Sutherland BR, Sargent EH, You JB (2017a) Ultra-bright and highly efficient inorganic based perovskite light-emitting diodes. Nat Commun 8:15640

Zhang J, Hultqvist A, Zhang T, Jiang L, Ruan C, Yang L, Cheng Y, Edoff M, Johansson EMJ (2017b) Al_2O_3 underlayer prepared by atomic layer deposition for efficient perovskite solar cells. Chemsuschem 10:3810–3817

Zhang M, Yun JS, Ma Q, Zheng J, Lau CFJ, Deng X, Kim J, Kim D, Seidel J, Green MA (2017c) High-efficiency rubidium-incorporated perovskite solar cells by gas quenching. ACS Energy Lett 2:438–444

Zhao K, Munir R, Yan B, Yang Y, Kim T, Amassian A (2015) Solution-processed inorganic copper (I) thiocyanate (CuSCN) hole transporting layers for efficient *p-i-n* perovskite solar cells. J Mater Chem A 3:20554–20559

Zhao DW, Yu Y, Wang CL, Liao WQ, Shrestha N, Grice CR, Cimaroli AJ, Guan L, Ellingson RJ, Zhu K, Zhao XZ, Xiong RG, Yan YF (2017) Low-bandgap mixed tin-lead iodide perovskite absorbers with long carrier lifetimes for all-perovskite tandem solar cells. Nat. Energy 2(4):17018

Zheng X, Wu C, Jha SK, Li Z, Zhu K, Priya S (2016) Improved phase stability of formamidinium lead triiodide perovskite by strain relaxation. ACS Energy Lett 1:1014–1020

Zheng X, Chen H, Li Q, Yang Y, Wei Z, Bai Y, Qiu Y, Zhou D, Wong KS, Yang S (2017) Boron doping of multiwalled carbon nanotubes significantly enhances hole extraction in carbon-based perovskite solar cells. Nano Lett 17:2496–2505

Zhou HP, Hsu WC, Duan HS, Bob B, Yang WB, Song TB, Hsu CJ, Yang Y (2013) CZTS nanocrystals: a promising approach for next generation thin film photovoltaics. Energy Environ Sci 6:2822–2838

Zhou H, Shi Y, Dong Q, Zhang H, Xing Y, Wang K, Du Y, Ma T (2014a) Hole-conductor-free, metal-electrode-free $TiO_2/CH_3NH_3PbI_3$ heterojunction solar cells based on a low-temperature carbon electrode. J Phys Chem Lett 5:3241–3246

Zhou H, Chen Q, Li G, Luo S, Song TB, Duan HS, Hong Z, You J, Liu Y, Yang Y (2014b) Interface engineering of highly efficient perovskite solar cells. Science 345:542–546

Zhou H, Shi Y, Wang K, Dong Q, Bai X, Xing Y, Du Y, Ma T (2015) Low-temperature processed and carbon-based $ZnO/CH_3NH_3PbI_3/C$ planar heterojunction perovskite solar cells. J Phys Chem C 119:4600–4605

Zhu Z, Bai Y, Zhang T, Liu Z, Long X, Wei Z, Wang Z, Zhang L, Wang J, Yan F, Yang S (2014) High-performance hole-extraction layer of sol–gel-processed NiO nanocrystals for inverted planar perovskite solar cells. Angew Chem Int Ed 126:12571–12575

Zhu Z, Zheng X, Bai Y, Zhang T, Wang Z, Xiao S, Yang S (2015) Mesoporous SnO_2 single crystals as an effective electron collector for perovskite solar cells. Phys Chem Chem Phys 17:18265–18268

Zhu Z, Zhao D, Chueh C-C, Shi X, Li Z, Jen AKY (2018) Highly efficient and stable perovskite solar cells enabled by all-crosslinked charge-transporting layers. Joule 2:168–183

Zuo C, Ding L (2015) Solution-processed Cu_2O and CuO as hole transport materials for efficient perovskite solar cells. Small 11:5528–5532

Zuo CT, Bolink HJ, Han HW, Huang JS, Cahen D, Ding LM (2016) Advances in perovskite solar cells. Adv Sci 3(7):1500324

Textile-Based Dye-Sensitized Solar Cells: Fabrication, Characterization, and Challenges

P. Salinas, D. Ganta, J. Figueroa, and M. Cabrera

1 Introduction

Dye-synthesized solar cells are solar cells that are made from plant dyes (Ananth et al. 2014; Calogero et al. 2014, 2010,2012; Chang and Lo 2010; Chien and Hsu 2014; Ganta et al. 2019, 2017; Hernandez-Martinez et al. 2011; Kumara et al. 2004, 2013a, b; Lai et al. 2008; Mathew et al. 2014; Noor et al. 2014; Shanmugam et al. 2013; Teoli et al. 2016; Wang et al. 2006; Yusoff et al. 2014; Zhou et al. 2011). A DSSC consists of three main components, the plant-based dye photoanode, an electrolyte with a redox couple, and a counterelectrode. Figure 1 shows the basic structure of a typical DSSC, consisting of a cathode, an anode, and a coating of TiO_2. Scientists and researchers alike are attempting to discover new ways to increase the conductivity and longevity of the solar cell. DSSCs first came to popularity in the late 2010s, with many studies focusing on DSSCs composed of two pieces of FTO glass slides. There are advantages of plant-based DSSCs over conventional DSSCs. Plant-based DSSCs are fabricated from natural plant-based dyes, unlike the toxic chemical dyes used in the case of chemical-based DSSCs. Plant-based DSSCs are inexpensive, simple to fabricate and easy to handle unlike the chemical-based DSSCs. DSSCs do boast excellent conversion efficiency and are somewhat portable but not flexible. Thus, this causes their usefulness to drop, especially as wearable energy devices. With a fully textile-based DSSC, it can be flexible, portable and can be placed in any number of configurations and is more practical for real-world usage.

Recently, there have been several papers published regarding FTO-fabric hybrid DSSCs (HDSSC) (Lam et al. 2017; Liu et al. 2019; Sahito et al. 2015, 2016; Xu et al. 2014). The textile was being integrated into a DSSC, and the research faces many

P. Salinas (✉) · D. Ganta · J. Figueroa · M. Cabrera
School of Engineering, Texas A&M International University, Laredo, TX 78041, USA
e-mail: petersalinas@dusty.tamiu.edu

H. Tyagi et al. (eds.), *New Research Directions in Solar Energy Technologies*, Energy, Environment, and Sustainability,
https://doi.org/10.1007/978-981-16-0594-9_5

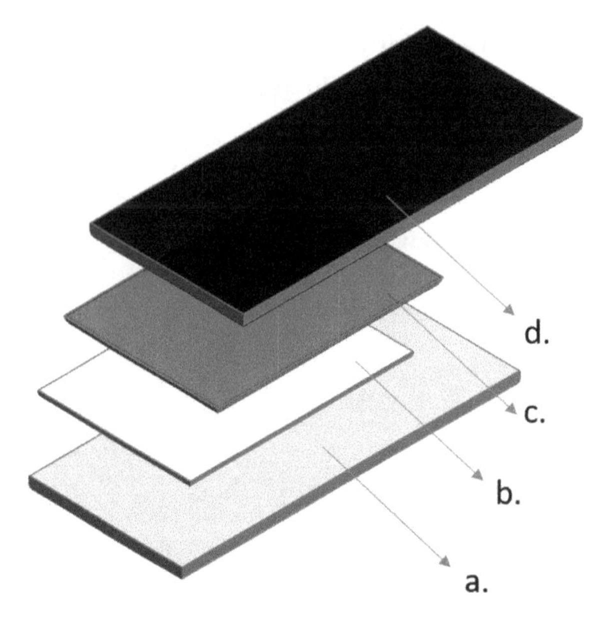

Fig. 1 Simple schematic of the components in a DSSC: **a** FTO glass anode, **b** FTO with TiO_2 nanomaterial paste, **c** addition of the plant dye to the FTO glass with TiO_2 nanomaterial paste, and **d** is the FTO cathode with a coating of a thin carbon film

challenges, the book chapter reviews some of the challenges faced in the fabrication of textile-based DSSC (TDSSC).

With large corporations such as Apple and Samsung—among others—popularizing products such as smartwatches, Bluetooth headphones, and virtual reality devices, the interest in and demand for wearable technology are at an all-time high. The research question posed to take DSSCs to the next level is to find a way to integrate a fully textile-based DSSC into a piece of clothing, enabling the clothing to store the charge for applications in a cell phone or similar small technologies. TDSSC is a DSSC composed solely of a textile fabric such as cotton. There is research attempting to fabricate a TDSSC that boasts good conductivity and a good enough efficiency to hold a charge. The goal for these is to implement them onto a piece of clothing such as a jacket or shirt to, for example, be able to charge a cellular or wearable device if a person is ever in need of a charge. In the medical field, a doctor requires that wearable medical devices have a continuous supply of energy to power them, helping save human life. A TDSSC is an environmentally friendly source of green energy, avoiding toxic waste recycled every year from the battery devices. TDSSCs can also see future applications in aerospace, military, outdoor equipment, and other flexible energy sources, so their applications are not just limited to clothing (Fu et al. 2018; Grancarić et al. 2018; Liang et al. 2018; Peng et al. 2018; Susrutha et al. 2015; Tsuboi et al. 2015). Figure 2 shows how a TDSSC would be composed, while Fig. 3 shows the composition of a conventional DSSC to serve as a comparison. The operation of a TDSSC includes the transport of electrons created by sunlight absorption via the

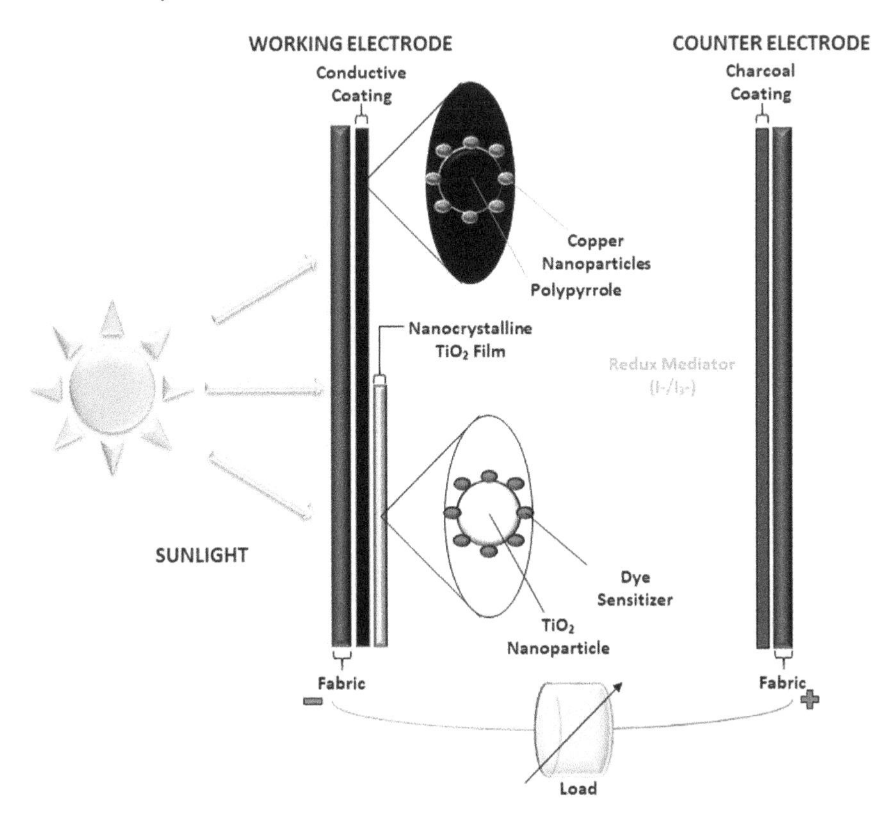

Fig. 2 Simple schematic of the operation of a TDSSC

plant dye. Electrons are generated because of the chemical reaction between the dye and sunlight.

The challenge with implementing a TDSSC into clothing is to find a proper method to make the material resistivity low enough to allow a charge to be held. In addition, a robust coating method that will evenly stay onto the chosen fabric has yet to be developed, and there is much to be discovered in terms of a TDSSC. The standard coating method used in most DSSCs is to add a TiO_2 paste onto the cathode of the cell, cure it using high heat, and then to add a dye that can absorb sunlight enough to create a sizable amount of voltage within the cell. Since this practice was introduced, scientists have implemented and tested new methods of coating a DSSC so that it can have a higher efficiency rating.

Additionally, the dye used in all DSSCs is an essential component since its efficiency, along with that of the TiO_2 paste, will be the deciding factors on whether the produced cell will be functional or unusable. To thoroughly test a dye's potential, and analysis using a spectrophotometer may be done. The spectrophotometer measures the amount of light absorbed by the dye as it passes through a selected range in nanometers. If too much light goes through the solution, then that dye

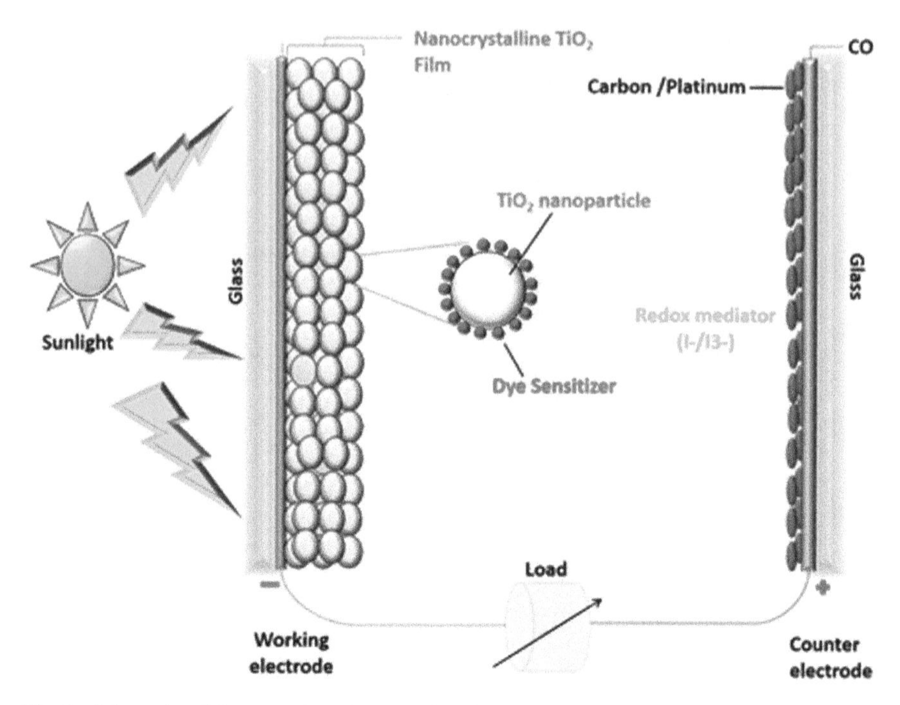

Fig. 3 Schematic of the operation of a typical DSSC. The image is reproduced with permission from Ganta et al. (2019) © 2012, Springer (Ganta et al. , 2019)

may not be suitable for use in the fabrication of DSSCs. Figure 4 demonstrates an example of a blackberry dye's UV–Vis Absorption Spectra using a UV 2450 UV–Vis Spectrophotometer.

As illustrated in Fig. 4, the optical absorption characteristics of the blackberry dye were measured on a range of 400–650 nm. The absorbance peaked at an approximate wavelength of 526 nm. This value matches the data values reported in the literature (DeSilva et al. 2017; Olea et al. 1999).

Once either type of cell is assembled, it is placed under natural or simulated sunlight and connected to a source measure unit with access to KickStart software in order to measure the relationship between the current and voltage. This relationship is characterized as an I-V curve shown through a current vs. voltage graph. The higher the current and voltage being produced by the cell, the better the potential of the cell will be. We reviewed some of the results published in the literature. As illustrated in Fig. 5, is the comparison of four differently prepared DSSCs, with a pair of a single-walled carbon nanotube (SWCNT) and Ag-PEDOT (poly(3,4-ethylenedioxythiophene)-poly(styrenesulfonate)) cells left in light, and a pair of SWCNT and AG-PEDOT left in the dark (Kye et al. 2018; Zhang et al. 2017). The cells, when left in light, will absorb more light as can be expected and exhibit the right voltage and current readings.

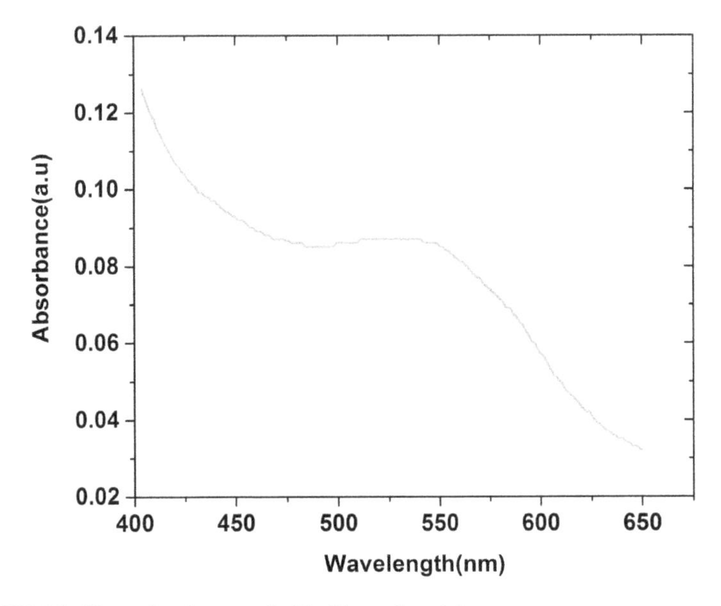

Fig. 4 UV–Vis Absorption Spectra of a blackberry-based dye

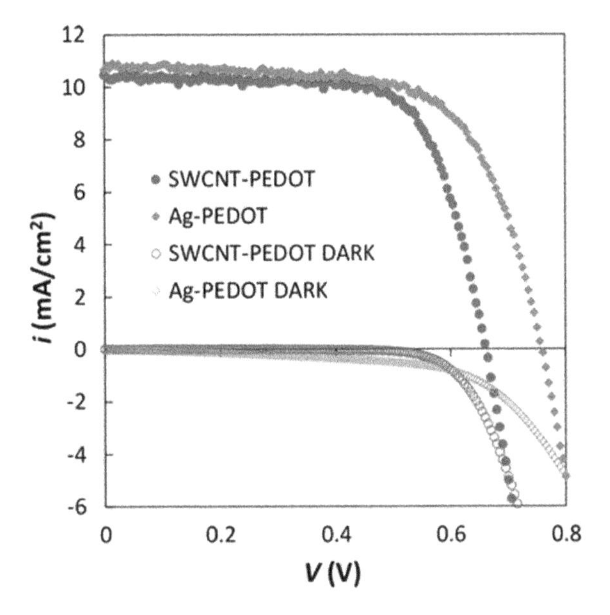

Fig. 5 I-V curves of the DSSC with different material counterelectrodes and their dark current curves. The image is reproduced with permission from Aitola et al. (2015) © 2012, Springer (Aitola et al. 2015)

2 Methods

2.1 Integration of Plant Dyes in Solar Cells

Natural plant dyes have been used to fabricate dye solar cells. A dye-sensitized solar cell's main component is the dye that is used. Differing dyes allow for different types of properties and different reactions to occur within the cell; therefore, the electrical output of the cell will be different depending on the type of dye being used on the cell. Issues when working with dye-sensitized solar cells primarily involve the foldability and stretchability of the fabric that the cell is being integrated on, as well as the ability for the coatings to stay with the material as the material gets washed. A paper cites Lam et al. (2017) in which they create a perovskite solar cell using PEDOT: PSS coating as an anode for the cell while using another PEDOT: PSS coating (low-conductive) on top of the anode layer and using a PCBM (phenyl-C61-butyric acid methyl ester) as the electron layer for the solar cell (Lam et al. 2017). This solar cell gained efficiencies of 5.72%.

Grassmann et al. (2019) used organic plant-based dyes in the creation of TDSSCs (Grassmann et al. 2019). The article primarily discusses different types of solar cells, including plant-based, dye-sensitized based, and organic-based. Each cell has its type of coating that makes it differ from the other cells due to each of the different coatings having its different type of property that affects the way that each cell acts. Two different types of DSSCs were assembled for electrical testing, ones that implement the use of counter electrodes, where the molten electrolyte was dripped on the dye and put together with conductively coated fabric, and ones that implemented the use of working electrodes, where the gelatin electrolyte was poured onto the graphite coated glass counter electrode (Grassmann et al. 2019). Before the cell was gelled, both electrodes were added together. Initial testing of the two types of DSSCs was done using two multimeters that measured short-circuit currents and open-circuit voltages that were under an 800 W halogen lamp.

Regarding differing types of DSSCs, the ones that contain a TiO_2 layer are more effective because of the high conversion efficiency potential, the chemical stability, and intense charge-transfer absorption in the entire visible light spectrum (Al-Alwani et al. 2017). Natural dyes are primarily used to their low cost, the efficiency of the dyes themselves, and the different types that can be studied. Different types of plant-based dyes show differing properties and energy-to-electric conversion efficiencies. This is all dependent on the type of dye that is being used for the solar cell, as the reason for each dye having a different efficiency is dependent on the properties of the dye, such as light absorption. Various researchers, including Al-Alwani et al. (2017) and Calogero et al. (2010, 2012), have shown that chlorophyll can absorb red, blue, and violet wavelengths and reflect green light, giving it its pigment (Calogero et al. 2010, 2012, 2014). *Pandannus Amaryllifolius (P. Amaryllifolius),* was ground up and used as a natural dye due to the high source of natural green extracts, theoretically allowing for a high amount of light absorption and green light reflection. TiO_2 was mixed with 3.0 ml of polyethylene glycol (PEG, MW 20,000). When mixed with

PEG, cracking on the surface of the cell was minimized, which usually occurs during high-temperature testing with solar cells. Triton X-100 was added to TiO_2 particles to adhere better. Results showed that the chlorophyll was suitable for being used as a photosensitizer in the visible-light region (Al-Alwani et al. 2017).

Researchers Gu et al. (2018) used different types of vegetables, natural plants, for the plant-based dyes; the vegetables extracted include purple cabbage, purple grape skin, mulberry, carrot, and potato (Gu et al. 2018). The amount of absorption and the overall efficiency of the dye is dependent on the pigment of the dye itself; purple cabbage showed absorption peaks of 336 and 531 nm. This is based on the different overall compositions of the vegetable itself, as shown with the carrot dye in which the much different composition led to 453 and 484 nm of absorption. Carrot, mulberry, and purple cabbage have a higher absorption rate when compared to potato and grape due to their better overall absorption of light; absorption density is proportional to the dye's concentration. Purple cabbage, however, showed the overall better optimal performance with the DSSC and the power conversion efficiency.

A disadvantage to natural dye sensitizers is that it typically contributes to poor cell stability. The natural plant-based dyes used in this conducted research primarily come from Indian plants due to unspecified limitations. Warmer colors, the color spectrum range of red, yellow, and orange, absorb a range between 470 and 550 nm that is not usually absorbed by chlorophyll, these are known as carotenoids. Carotenoids consist of eight different isoprenoid units. Carotenoids are widely produced in nature; 108 tons/year. Carotenoids typically have a higher absorption coefficient, but they generally give a reduced efficiency in a DSSC due to their poor dye regeneration. Chlorophyll is what gives plants the green color pigmentation, and it is usually unstable regarding being used with an acid or base, one of the disadvantages of it being used in a solar cell. For whatever dye is being used, it must follow the criteria that it should have an excellent anchoring to it, as well as good stability regarding being used with an acid or a base and of course, excellent electrical efficiency when converting the light into energy.

While organic dyes are used in DSSC fabrication, in the case of Mathew et al. (2014), he and his fellow researchers looked at how the implementation of a molecularly engineered porphyrin dye, SM315, performs as part of a DSSC. Previous DSSCs had only reached a peak PCE efficiency of 12.3% (Mathew et al. (2014). A previously molecularly engineered green dye used by the researchers, SM371, yielded an efficiency of 12%. Porphyrin-based dyes absorb sunlight very well in the Soret and Q bands but much else. The new engineered SM315 improved upon the previously used SM371 and made the already excellent absorption in the Soret and Q bands greater. The DSSCs prepared by the authors are made with TiO_2 paste, and no counterelectrode is mentioned. What the researchers were able to find through testing was that the SM315 posted a new efficiency high of 13%, 0.8 larger than the previously recorded high and 1% larger than SM371. The V_{oc} readings taken for SM315 were 0.96 V and the J_{SC} was 15.9 mA/cm^2, which is decent when compared to past results. The use of these two dyes is a steppingstone in improving the efficiency of dyes to possibly 1% or even 20% efficiency.

2.2 Fabrication and Coating

The necessity of having a good coating method when producing a solar cell is a primary indicator if the cell will be good or not. Many researchers solely study the effects that different coating methods can have on a solar cell's performance. In an article written by Azizi et al. (2016) looks at two different ways to prepare a DSSC; electrophoretic deposition process (EPD) and doctor blade technique (DB) (Azizi et al. 2016). For the EPD technique, 21 nm TiO_2 nanoparticles and phase composition of 80% anatase and 20% rutile were used. This technique is electrochemical based, in that they apply an electric field to two electrodes and charged particles, and then the particles accumulate to create a film. This film was used on one of the DSSCs with aluminum as the CE, and 60 V of power was applied to the FTO glass slide. For the doctor blade technique, 5 g of nanocrystalline titanium dioxide was mixed with 3 ml of diluted acetic acid and then ground for half an hour until it was in paste form. Both techniques used anthocyanin obtained from Karkade flowers as the dye. In total, six DSSCs were tested. What the authors looked at to compare the two techniques were the cracks that formed in the films. These cracks decreased the conversion efficiency in the DSSCs, so the numbers are not what they should be.

Like TDSSCs, fiber-shaped dye-sensitized solar cells (FDSSCs) with a Tribo-electric Nanogenerator (TENG) integrated are the focus of an article written by Pu et al. (2016). FDSSCs are an attractive DSSC application due to their low cost and decent energy production, and the capability to be sewn into fabrics adds to the attractiveness. TENG was chosen due to their ability to use human motion, such as the arms swinging or legs moving to create mechanical energy. This combined with an FDDSC that absorbs sunlight to use as energy, can be sewn into an article of clothing such as a shirt that opens doors for creating a textile that can power a small device like a cell phone. The TENG used is polyester cotton that had intricate patterns etched into the fabric by way of laser-scribing. It was then coated with Ni using a technique called electroless deposition (ELD). The FDSSCs used were fabricated by having two Ti wires coated in TiO_2, which are then sintered at 450 °C for 30 min. They are then left to soak in ethanol for 24 h. The counterelectrode for the cells is a Pt wire that is put into a transparent and flexible plastic tube filled with liquid electrolyte. The tube was filled with an electrolyte, acetonitrile, dimethyl imidazolium iodide, I_2, $LiClO_4$, 4-tert-butylpyridine, and guanidine thiocyanate solution that is then sealed with glue. Overall, the efficiency for the TDSSC-TENG combo was 6% with short circuit density, J_{sc}, of 10.6 mA/cm^2 and an average open-circuit voltage V_{oc}, of 0.6 V.

Research has been conducted and tested regarding perovskite solar cells (PVSCs) and the various ways in which they could be improved while using differing structure types, known as n-i-p and p-i-n. Different materials were tested and compared with one another to observe which material would work best with the different environmental effects that would be applied to it when being used in a perovskite solar cell along with different chemicals. Chemicals included would be with a formula of ABX3 (Lam et al. 2017). These chemicals include CH_3, NH_3, and PBI_3. The data recorded and examined primarily came from various charts according to measurement as well as multiple J-V curves. The PVSCs were set in a series connection to test if it can light up a diode on the same textile as the PVSC. This would allow for different ways of testing how much power and energy are being transferred within the cell. The cell was tested within the water by immersing the device in water with no variables affecting it. After a non-specified period, the textile was taken out of the water, showing no conflicts or issues after coming from the water. With this, it is shown that the cell that the researchers have created is able to work after being washed. Although not specified, considering the way that solar cells act over differing actions over an interval of time, there could be a chance that after repeated washings of the textile, there could be a chance that the effectiveness of the cell could significantly decrease before washing.

Another alternate coating method that researchers investigated developing is a spray that would act as the dye for the solar cell. Other types of coating methods for fabrics primarily include screen printing, inkjet printing, and dispenser printing (Li et al. 2019). The advantages of using the spray coating as opposed to a standard dye coating include the spray coating being incredibly lightweight and minimizes the impact on the feel of the fabric (Li et al. 2019). The primary goal of using a spray coating for a textile solar cell would be to keep the cell as durable as possible while being able to maintain a relatively high-power output. Optimization is tested through various types of cells, including PV2000 and Pt-4, with Pt-4 having better optimization results due to better thickness in the cell. After multiple cycles of the device, the efficiency continued to drop with each cycle. After a certain number of cycles, the researchers recorded efficiency of 0.05% and after more cycles afterward, it eventually decreased to 0.03%, showing that the device usually deteriorates after multiple washing cycles.

Due to rapid growth in the application of photovoltaic electric power generation, there has been an increasing demand for lightweight and flexible solar cells. Most of the research done in past years revolved around flexible plastics. However, the transparent conductive oxide (TCO) films used are brittle and can be easily fractured. Therefore, attention has been placed upon textiles due to their lightweight, flexibility, and mechanical properties (Yun et al. 2014). Wu et al. (2017) researched creating more lightweight solar textiles onto fabrics to advance better the quality of textile-based solar cells. A primary issue with creating one is that it is difficult and complicated in order to intricately weave a fiber-based solar cells into different textiles and fabrics. The solar cell constructed during this research primarily includes two electrodes, and two different layers: a light-harvesting active layer and a blocking or transport layer (Wu et al. 2017). After a certain amount of time or number of folds/actions taken, the efficiency and effectiveness of the solar cell will significantly decrease. When creating an ultra-lightweight wearable solar cell, it is essential to take the washability, stretchability, and foldability factors into consideration. The demand for lightweight power sources has increased within recent years, conducting research based on these types of solar cells much desired.

Since DSSCs have a low production cost and relatively high efficiency, they are of high interest. An advantage DSSCs have is their ability to separate their photoanode and counter electrode via a liquid electrolyte insertion. By sewing textile-structured electrodes onto fabrics like cotton, a highly flexible and efficient DSSC can be developed (Yun et al. 2014). The textile-based cell is made by weaving each electrode by the loom and depositing TiO_2, heat treating, sewing to form the core-integrated DSSC device, and dye loading. Both, the photoanode and counter electrode, consist of a 3:1 woven structure of stainless-steel ribbon and Ti wire and a woven structure of glass fiber yarns and more Ti wire. The counterelectrodes were prepared and deposited onto the designated textile and then by the deposition of a paste before being heat treated. The textile-structured electrodes were then attached using a sewing machine. Overall, even though it had a high current density, the efficiency was approximately 4.3%, of which 5.3% corresponded to the maximum efficiency. At bending conditions, the photovoltaic performance was maintained at 80% of its original value with a curvature radius of 10 cm and 30% with a curvature radius of 4 mm. After 1000 bending cycles over the curvature of 1 cm, the performance remained at 70% of its original value. It was determined that the pattern of the weave and the material of the textile could influence the performance of the DSSC.

In a study by Yang et al. (2014), a new and general method of production for a stretchable, wearable photovoltaic textile based on elastic, electrically conducting fibers were developed (Yang et al. 2014). To prepare the fiber electrodes, first, an aligned multi-walled carbon nanotube (MWCNT) sheet was wound onto rubber fibers. Then, a Ti wire was twisted onto the MWCNT fiber and was coated in a photoactive material to make a wire-shaped DSSC. These DSCCs were then woven onto a textile. The DSSC had a conversion efficiency rate of 7.13% and was well maintained under tension. The conducting fibers were found to be highly stretchable and flexible, retained their structure, and their electrical resistance was nearly unchanged when stretched. The resistance of the fibers was further tested at various helical angles. Due to a deformation rate of 50%, the resistance values did not recover to their original value (Yang et al. 2014). These fibers can be used to create stretchable electronics. This design allows for the flexibility of the DSSC. These can be woven into various textiles and maintain a high rate of elasticity and an efficient photovoltaic performance. The MWCNT arrays are made from a chemical vapor deposition of Fe/Al_2O_3 on a silicon substrate at 740 °C (Yang et al. 2014). The carbon source was ethylene, and the thickness of the arrays used was of 250 μm.

Zhang et al. (2017) reported DSSC textiles fabricated using polybutylene terephthalate (PBT) polymer yarns woven into different fabrics (Zhang et al. 2017). This achieved a power conversion efficiency (PCE) of 1.3% per fiber, the highest reported at the time. Unfortunately, due to the bending limitations, the fibers tend to break and loose performance as they degrade. Furthermore, when woven into textiles, they tend to lose PCE due to partial shading. Additionally, since they are liquid-based DSSCs, they suffer from leakage, corrosion, or stability issues. This, however, could be avoided by using a solid electrolyte with a PCE of 2.7% for Ag and 2.8% for AgNW. By using the architecture of a solid-state (ss) DSSC on FTO glass and spin

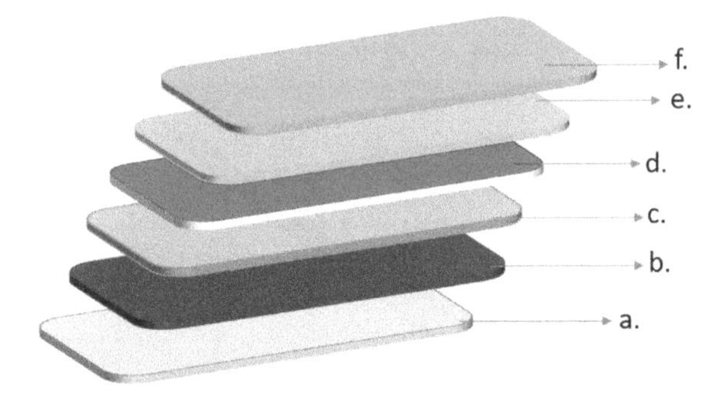

Fig. 6 Composition of the textile-based ssDSSC anode. **a** Textile/FTO, **b** polyimide, **c** silver, **d** TiO_2 with dye, **e** PEDOT: PSS, and **f** AgNWs

coating with a solid electrolyte, the PCE was 3.7% (Liu et al. 2019). An investigation was held regarding the implementation of a DSSC on Kapton, which yielded an efficiency of 7.03%. However, FTO glass was used as the electrolyte which made it inflexible. Additionally, it was determined that using a liquid electrolyte may not be suitable in fabricating TDSSCs because the fabric is porous, which causes the electrolyte to soak through and evaporate. TiO_2 was similarly investigated leading to an efficiency of 7.41%. However, it was found that the TiO_2 layer requires a minimum temperature of 450 °C to sinter the film and achieve the recorded efficiency. Consequently, the fabrication of a 2-D ssDSSC woven into high-temperature glass fiber textile substrates is presented in this article Liu et al. (2019). The fabrication consists of five steps: first, the roughness of the fabric must be reduced via a flexible polyimide layer. The use of liquid polyimide differs from the use of polyurethane. Then, the silver bottom electrode must be screen printed onto the dried polyimide. Then, there is a decomposition of the TiO_2 (Liu et al. 2019). Then, a TiO_2 film is applied during spraying. Finally, it is spray-coated with PEDOT: PSS and then AgNW over it. This structure can be seen as illustrated in Fig. 6.

The ssDSSC on FTO glass had an efficiency of 2.8% with a V_{OC} of 0.44 V and a current density, J_{SC}, of 18.5 mA/cm^2. The surface roughness was then further reduced, allowing for a smoother surface, effectively reducing the resistance. This led to a maximum PCE of 0.4% with a V_{OC} of 0.31 V and a J_{SC} of 5.2 mA/cm^2 (Liu et al. 2019). The J/V curve for the two different ssDSSCs fabricated by the researchers can be found in Fig. 7. The fabric ssDSSC indicates there is resistance to charge movement between the functional layers due to the excessive and uneven film thicknesses (Liu et al. 2019). This set of *J/V* curves is good because both ssDSSCs reach 1 V of power generated which for a non-conventional solar cell is difficult to reach at the current state of affairs.

A DSSC is incomplete without the addition of a CE, due to it causing the electrical reaction that produces the voltage required to power devices. In an article written by Xin et al. (2011), they implemented a copper–zinc tin sulfide counterelectrode. The research implements the copper–zinc tin CE by way of a solution-based synthesis approach (Xin et al. 2011). They first started by dissolving the CZSC in oleylamine

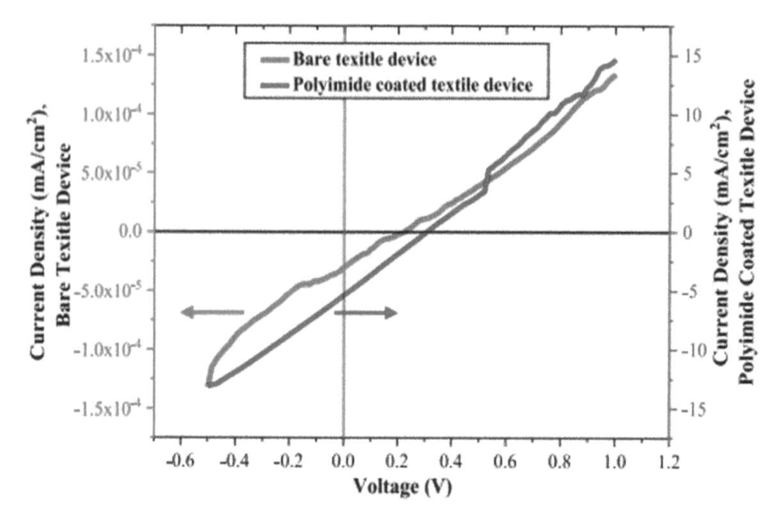

Fig. 7 The J/V curves of ssDSSCs fabricated on bare glass textiles and polyimide-coated glass (Liu et al. 2019)

and purifying them at 130 °C. They then injected a sulfur solution and stirred it at 225 °C for an hour. The solution was then centrifuged to yield nanoparticles of CZSC. At this point, the solution is ready to be placed onto DSSC glass slides as a CE. The DSSC created by the authors that yielded the best results was the spin-coated post-treated DSSC, with a V_{OC} of 0.80 V, J_{SC} of 17.7 mA/cm^2, and a PCE of 7.37%. The testing of various DSSCs helped the authors determine that the optimum thickness of a CZSC layer is 1–2 μm. The authors concluded that CZSC is much more cost-efficient, easy to produce than other alternatives such as platinum, and the potential for eliminating expensive CEs for DSSCs is on the horizon.

A paper by Sahito et al. (2015) details using graphene and carbon nanotubes to coat the counterelectrode used in a DSSC (Sahito et al. 2015). Graphene oxide was created from graphene that was synthesized by way of the Hummers method (Sahito et al. 2015, 2016; Song et al. 2011; Berendjchi et al. 2016). This was then combined with sulfuric acid in an ice bath and stirred for 30 min. A reaction occurred after 4 h of continuous stirring, which was then quelled by dropping the temperature to 10 °C. Afterward, hydrogen peroxide was stirred for 30 min. Afterward, water was added, and the solution was sonicated for 30 min. The fabric used, cotton, was then soaked into a slightly diluted and sonicated graphene oxide solution. After drying for 20 min at 80 °C, the fabrics were treated with hydroiodic acid fumes for 20 min. The fabric was then rinsed with DI water until a neutral pH was achieved and dried at 100 °C for 30 min. The results showed that the CE created has a resistance of 55 ohms/sq., which is like that of a standard FTO glass slide. Though the cell created showed an efficiency reading of 2.52% with a J_{SC} of 9.08 mA/cm^2, it is still commendable and can be improved upon by implementing other methods. Figure 8 demonstrates the graphite-coated fabric (GCF) CE when compared to more expensive platinum, Pt, coated CE. The platinum-coated is far superior, but the cost-effectiveness and

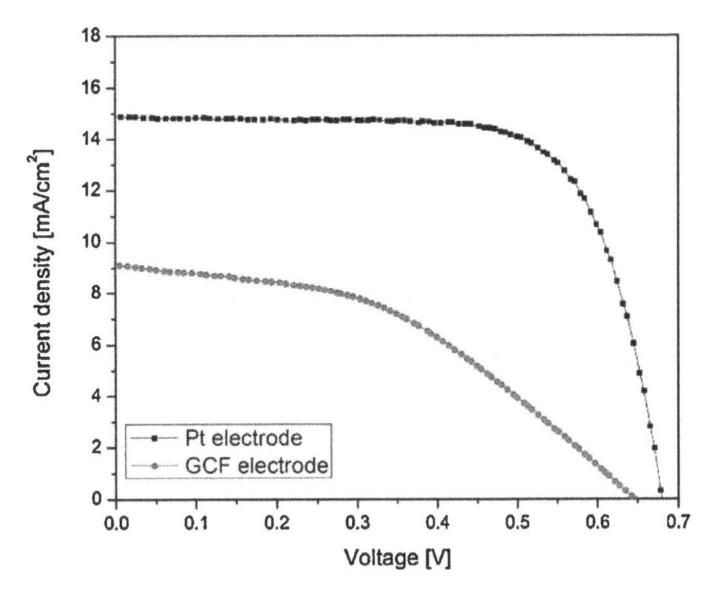

Fig. 8 Graph of voltage versus current density with two different types of electrodes used on a DSSC. The image is reproduced with permission from Sahito et al. (2015) © 2015, Elsevier

improvability of the graphene-coated CE make it more attractive to use in further experimentation.

Polypyrrole is used as the chemical in the creation of a DSSC in an article by Xu et al. (2016). A Ni, nickel plating, on the chosen fabric, cotton, used. The pyrrole was synthesized using polymerization (Xu et al. 2014 , 2016; Zhang et al. 2017; Calvo et al. 2002; Bu et al. 2013; Nagai and Segawa 2004). Cotton, platinum (Pt) foil, and KCl saturated calomel electrode was used as the working, counter, and reference electrodes, respectively. The pyrrole was polymerized for 1, 1.5, and 2.5 h. The surface resistance readings for the polypyrrole coated cotton read 7.01 ohms/cm^2, 7.16 ohms/cm^2, and 7.42 ohms/cm^2, respectively. The efficiency of each polymerized pyrrole showed readings of 3.41f%, 3.83%, 3.73% and the J_{SC} of each read 7.39 mA/cm^2, 7.85 mA/cm^2, and 7.90 mA/cm^2. The V_{OC} values read 720 mV, 740 mV, and 720 mV, respectively. It can be concluded that the 1.5-h polymerized polypyrrole is the most viable coating to be used in the creation of DSSCs.

Table 1 summarizes the produced open-circuit voltages (V_{OC}), photovoltaic energy conversion efficiencies (η), and current densities (J_{SC}) of several hybrid DSSCs—composed of an FTO anode and a textile cathode—and TDSSCs reported from the literature. They were compared to their corresponding conventional DSSC counterparts.

Table 1 Comparison of the performance of TDSSCs, HDSSCs, and DSSCs

Type of solar cell	Substrate material for anode/cathode	V_{oc} (V)	η (%)	J_{sc} (mA/cm^2)	Ref
DSSC	FTO	0.56	1.50	10.6	37
HDSSC	FTO/cotton	0.52	1.00	8.06	37
TDSSC	Cotton	0.50	0.40	3.44	37
DSSC	FTO	0.44	2.83	18.6	22
TDSSC	Glass fiber	0.31	0.40	5.20	22
DSSC	FTO	0.66	7.20	14.9	23
HDSSC	FTO/cotton	0.64	2.52	9.08	23
DSSC	FTO	0.67	8.44	15.9	24
HDSSC	FTO/cotton	0.66	6.93	14.8	24
DSSC	FTO	0.70	6.16	15.1	25
HDSSC	FTO/cotton	0.65	3.30	9.60	25

3 Challenges and Perspectives

In recent years, multiple advancements have been made regarding TDSSC research. Li et al. (2019) and his group were able to optimize a coating method in the fabrication of a TDSSC that utilizes a spray to coat the textiles. This coating possesses a power conversion of 0.4% and allows for improved durability and life expectancy (Li et al. 2019). Furthermore, Grassmann et al. (2019) put forth the idea of using a gelled electrolyte and developed one using gelatin that produced some impressive results (Grassmann et al. 2019). Most interestingly, Liu and his group investigated a novel fabrication method that screen prints a solid-state DSSC directly onto a woven glass fiber textile, allowing for a prolonged avoidance of oxidation and obtaining an efficiency of 0.4% (Liu et al. (2019).

We conducted preliminary experiments for the fabrication of TDSSCs and the challenges faced involved: increasing the conductivity of the textile, sufficient adherence of solution to fabric, a proper curing process, proper dilution and application of chemicals, limited plant-dye shelf life, the hindrance with the flexibility of cells, and liquid electrolyte application. A significant issue was encountered in the very first step of producing a TDSSC: making the resistance of the desired fabric low enough so that a sufficient charge could be produced. The resistance of an FTO glass slide is, on average, between 30 and 60 ohms. Various methods exist to lower the resistivity of the fabric in order to replicate such resistivity measurements. One such method is to coat both sides of the cell. There are two ways to go about this method. The first is to create a conductive polypyrrole paste and coat one side of the fabric. The difficulty in this is making sure the fabric can adequately absorb the paste. After each coating, it is advised to rinse the fabric with DI water then apply another coating. The second is to create an aqueous solution of polypyrrole and leave the fabric soaking in it for any specified amount of time, between 1 and 24 h. The issue with both methods is that

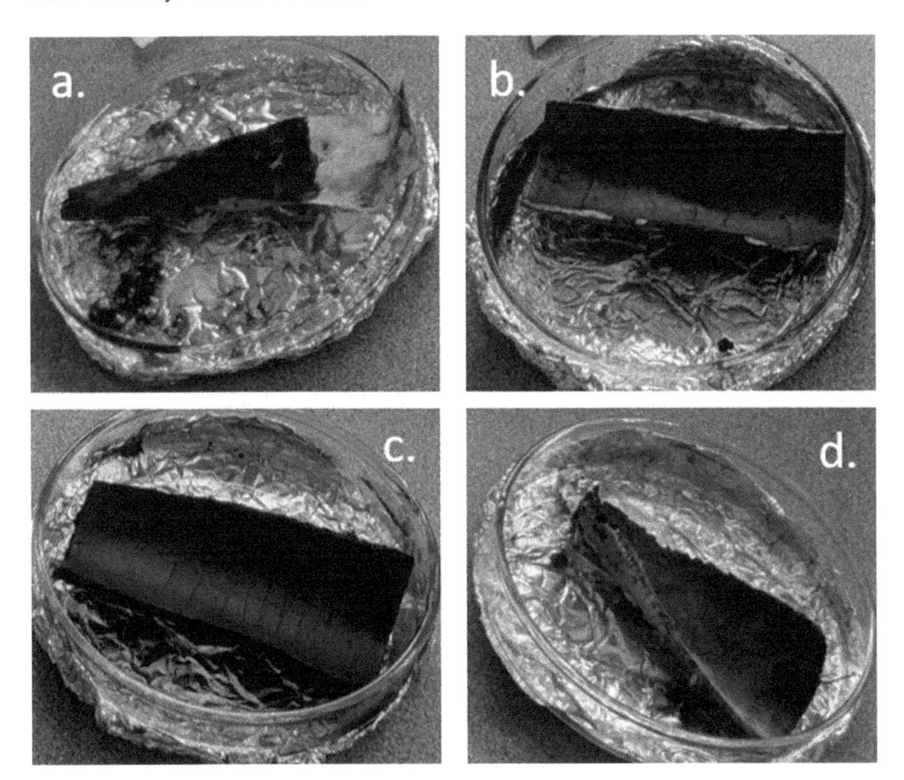

Fig. 9 Different fabrics coated with a polypyrrole paste: **a** nylon, **b** chamois, **c** leather, and **d** canvas

after a certain amount of coatings, the resistance in the fabric does not go down any further. Therefore, choosing a proper coating method can be quite limiting. Shown in Fig. 9 are various fabrics coated with polypyrrole paste previously discussed. In the image, the coating is smooth, but after a day or so, the coating starts to crack. Unfortunately, if this occurs, the fabric must be cleaned to apply a new coating, rendering previous results, essentially, useless. The fabrics shown in the image are nylon, chamois, leather, and canvas.

When coating material such as nylon, it is difficult to use the aqueous solution due to the fabric's hydrophobic qualities. If left in the solution, it will absorb it but will not give favorable resistance readings. Figure 10 shows how the coating will look after a day of drying.

Curing the polypyrrole paste onto the fabrics on a hot plate at 50 °C for anywhere between five to ten minutes—depending on the specific fabric and amount of coatings—has proven to help reduce the cracking of the coating. However, by introducing high heat to the fabrics, the risk of burning the material is significantly increased. Consequently, this added risk also negatively affects the paste, rendering the entire anode useless if such a case were to occur. This consequence is shown in Fig. 11.

Fig. 10 Coating of polypyrrole on the textile over time: **a** prior, **b** after a day has passed, showing mild to severe cracking on the surface

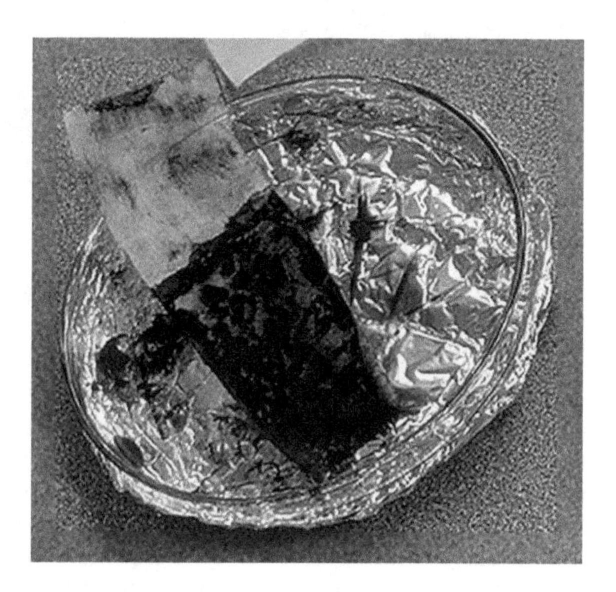

Fig. 11 Result of a nylon anode due to the over-curing

Therefore, it can be inferred that by using lower heat at an extended period in the curing process, the burning issue could be mitigated.

Similarly, when applying the necessary layer of TiO_2 paste onto the anode, curing must take place in order for it to adhere to the fabric. The same issues that plague the curing process of the polypyrrole substance onto the fabric are the leading challenges here as well. If not careful, the fabric will burn, consequently rending the conductive paste and the TiO_2 on it, useless. This is shown in Fig. 12.

A most recent issue encountered is the improper dilution of nitric acid when preparing the TiO_2 paste. This caused an intense chemical reaction that created smoke when applied onto the anode components of the cell. Shortly after, said component began to melt through the conductive coating, then proceeded to burn and contort or

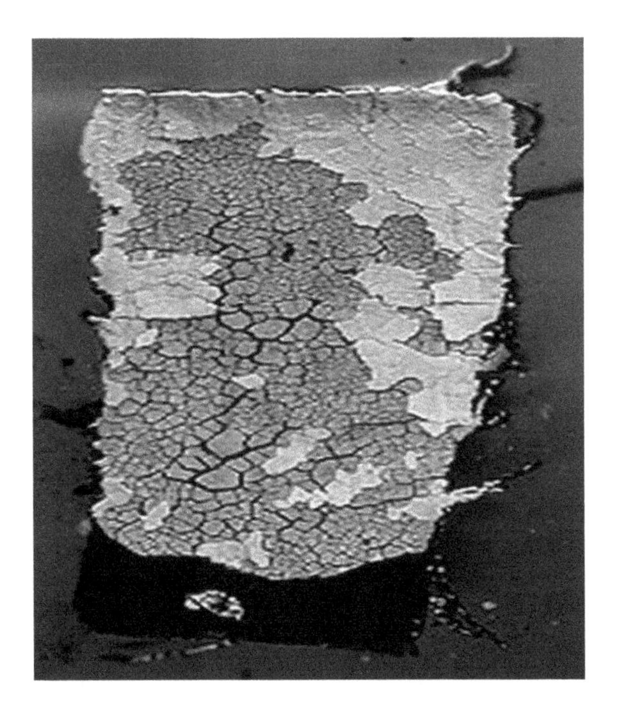

Fig. 12 Fragmented, cured TiO_2 paste peeling off from the coated canvas fabric

melt away its fabric, depending on the material. The result of the chemical reactions can be seen in Figs. 13 and 14.

Additionally, the plant-based dyes also have a short shelf life and must be used within that lifespan to maintain its efficiency. Due to the dyes being made of organic plant leaves or fruit, it begins to spoil around the month mark of its conception if refrigerated or approximately a week if left at room temperature. This process is aided by the dye's contact with oxygen and sunlight, as well as its separation from the ethanol within its vial due to the differing densities. Shown in Fig. 15 is an example of a decomposing dye.

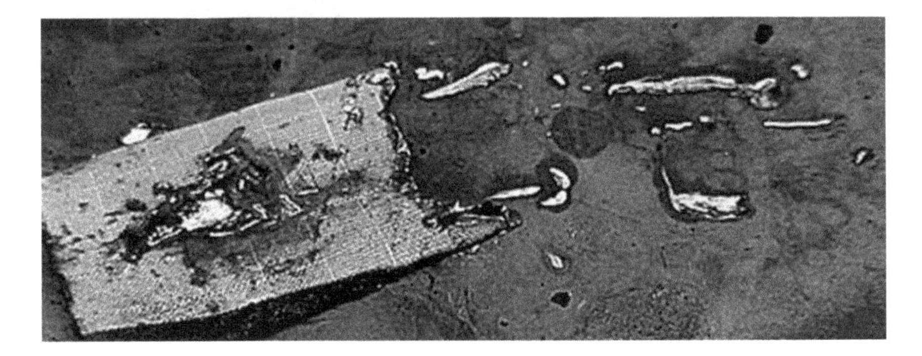

Fig. 13 Nylon anode melted away after improper TiO_2 paste coating

Fig. 14 Leather anode burnt and contorted after improper TiO_2 paste coating

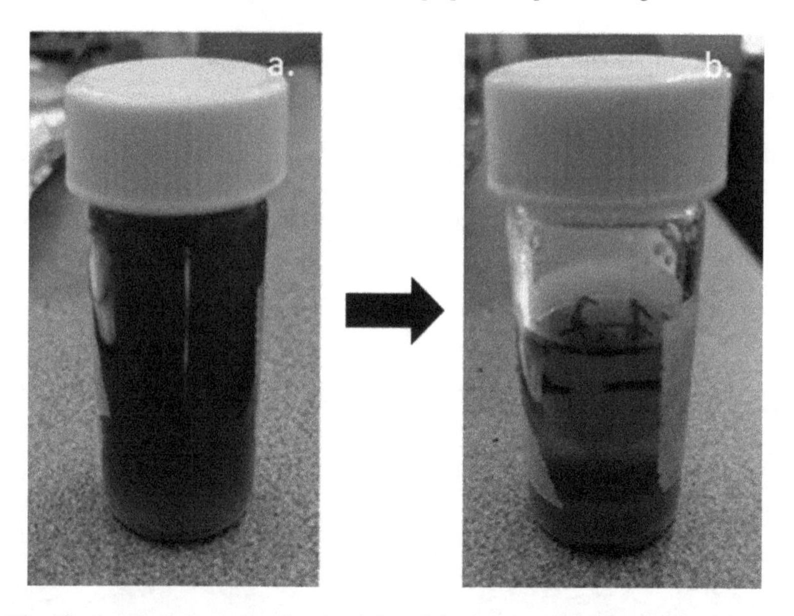

Fig. 15 **a** Fresh spinach-based dye, **b** spinach-based dye in state of decomposition

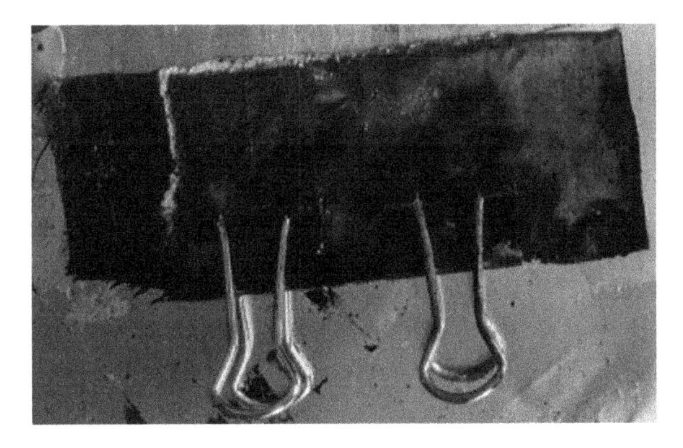

Fig. 16 Assembled TDSSC cell held together by binder clips before the application of the electrolyte

Due to the fabrics' flexibility, it is challenging to keep the anode and cathode in tight contact, as would be a conventional FTO glass DSSC. To mitigate this problem, 2–3 binder clips were used per cell, as seen in Fig. 16, in preparation for the addition of the electrolyte. However, this did not yield the best results since the cells were not entirely held together, but rather only partially.

Once the cells are assembled and prepared for testing, the analysis must be done as soon as possible after the liquid electrolyte is incorporated for the best possible results. This is due to the fast evaporation rate of the electrolyte and drying of the conductive paste. Unlike an FTO glass-based DSSC, since there is a lack of pressure to keep the electrolyte from escaping the internal workings of the cell, the TDSSC will decay and cease to function within several hours. Similarly, the moisture within the conductive paste of the cell will begin to escape, causing the paste to dry and crack, leading it to fall or peel off the fabric, rendering what was once a functional TDSSC into a non-functional material.

4 Conclusions

In summary, various TDSSC fabrication and coating methods which use natural plant-based or laboratory-manufactured dyes as sensitizers were reviewed. DSSCs and, more specifically, TDSSCs are a fundamental step towards making green energy extensively available by way of portable energy conversion. The environment will benefit significantly from making plant-based solar energy more popular in the industry and something as compact as a textile solar cell can go a long way in helping maintain a healthier environment.

The future of DSSCs and their implementation onto fabrics seem to be on the right path, and, soon, the proper scientific development methods will be further advanced.

The current conversion efficiencies for textile-based DSSCs are still lower than traditional DSSCs, but the challenges are being addressed in future work. Additionally, implementation into fabric will become much more streamlined once researchers have figured out the proper way to address the fabrication problems. Humans will be able to use this technology to improve both the quality of life and well-being. The majority of the people including doctors will have access to this inexpensive and sustainable green energy technology.

While different coating techniques were reported to coat each component of the TDSSC, the method that yields the best efficiency, and that is most cost-effective has yet to be identified. The copious amounts of plant-based and laboratory-manufactured dyes available for use in the TDSSC fabrication make it challenging to pinpoint the exact dye that should be used in any configuration. Therefore, it is unlikely that there will ever be one specific dye that will be universally used. The discovery of new materials is still in the works in order to improve the energy transport across the materials in the TDSSCs.

The challenges currently holding back advancement in the successful creation of TDSSCs, especially in the coating and fabrication methods, were numerous, but are not impossible to hurdle over. It is only a matter of time until researchers find a way past these hurdles and begin to manufacture textiles that can turn sunlight into electrical charge. These textiles can never truly replace commercial electricity, but having the ability to produce energy to power an electronic device on the go, such as a cell phone, will improve the quality of life of everyday activities.

References

Aitola K, Zhang J, Vlachopoulos N, Halme J, Kaskela A, Nasibulin AG, Kauppinen EI, Boschloo G, Hagfeldt A (2015) Carbon nanotube film replacing silver in high-efficiency solid-state dye solar cells employing polymer hole conductor. J Solid State Electrochem 19:3139–3144

Al-Alwani MA, Mohamad AB, Kadhum AAH, Ludin NA, Safie N, Razali M, Ismail M, Sopian K (2017) Natural dye extracted from Pandannus amaryllifolius leaves as sensitizer in fabrication of dye-sensitized solar cells. Int J Electrochem Sci 12:747–761

Ananth S, Vivek P, Arumanayagam T, Murugakoothan P (2014) Natural dye extract of lawsonia inermis seed as photo sensitizer for titanium dioxide based dye sensitized solar cells. Spectrochim Acta Part a Mol Biomol Spectrosc 128:420–426

Azizi T, Touihri A, Karoui MB, Gharbi R (2016) Comparative study between dye-synthesized solar cells prepared by electrophoretic and doctor blade techniques. Optik 127:4400–4404

Berendjchi A, Khajavi R, Yousefi AA, Yazdanshenas ME (2016) Improved continuity of reduced graphene oxide on polyester fabric by use of polypyrrole to achieve a highly electro-conductive and flexible substrate. Appl Surf Sci 363:264–272

Bu C, Tai Q, Liu Y, Guo S, Zhao X (2013) A transparent and stable polypyrrole counter electrode for dye-sensitized solar cell. J Power Sources 221:78–83

Calogero G, Di Marco G, Cazzanti S, Caramori S, Argazzi R, Di Carlo A, Bignozzi CA (2010) Efficient dye-sensitized solar cells using red turnip and purple wild sicilian prickly pear fruits. Int J Mol Sci 11:254–267

Calogero G, Yum J-H, Sinopoli A, Di Marco G, Grätzel M, Nazeeruddin MK (2012) Anthocyanins and betalains as light-harvesting pigments for dye-sensitized solar cells. Sol Energy 86:1563–1575

Calogero G, Citro I, Di Marco G, Minicante SA, Morabito M, Genovese G (2014) Brown seaweed pigment as a dye source for photoelectrochemical solar cells. Spectrochim Acta Part a Mol Biomol Spectrosc 117:702–706

Calvo P, Rodrıguez J, Grande H, Mecerreyes D, Pomposo J (2002) Chemical oxidative polymerization of pyrrole in the presence of m-hydroxybenzoic acid-and m-hydroxycinnamic acid-related compounds. Synth Met 126:111–116

Chang H, Lo Y-J (2010) Pomegranate leaves and mulberry fruit as natural sensitizers for dye-sensitized solar cells. Sol Energy 84:1833–1837

Chien C-Y, Hsu B-D (2014) Performance enhancement of dye-sensitized solar cells based on anthocyanin by carbohydrates. Sol Energy 108:403–411

DeSilva LA, Pitigala P, Gaquere-Parker A, Landry R, Hasbun J, Martin V, Bandara T, Perera A (2017) Broad absorption natural dye (Mondo-Grass berry) for dye sensitized solar cell. J Mater Sci: Mater Electron 28:7724–7729

Fu X, Xu L, Li J, Sun X, Peng H (2018) Flexible solar cells based on carbon nanomaterials. Carbon 139:1063–1073

Ganta D, Jara J, Villanueva R (2017) Dye-sensitized solar cells using Aloe Vera and Cladode of Cactus extracts as natural sensitizers. Chem Phys Lett 679:97–101

Ganta D, Combrink K, Villanueva R (2019) Natural dye-sensitized solar cells: fabrication, characterization, and challenges. In: Tyagi H, Agarwal AK, Chakraborty PR, Powar S (eds) Advances in solar energy research. Springer Singapore, Singapore, pp 129–155

Grancarić AM, Jerković I, Koncar V, Cochrane C, Kelly FM, Soulat D, Legrand X (2018) Conductive polymers for smart textile applications. J Ind Text 48:612–642

Grassmann C, Grethe T, Krause A, Großerhode C, Storck JL, Ehrmann A, Van Langenhove L, Schwarz-Pfeiffer A (2019) Textile based dye-sensitized solar cells with natural dyes. In: AUTEX2019, Autex

Gu P, Yang D, Zhu X, Sun H, Li J (2018) Fabrication and characterization of dye-sensitized solar cells based on natural plants. Chem Phys Lett 693:16–22

Hernandez-Martinez AR, Estevez M, Vargas S, Quintanilla F, Rodriguez R (2011) New dye-sensitized solar cells obtained from extracted bracts of Bougainvillea glabra and spectabilis betalain pigments by different purification processes. Int J Mol Sci 12:5565–5576

Kumara N, Ekanayake P, Lim A, Iskandar M, Ming LC (2013a) Study of the enhancement of cell performance of dye sensitized solar cells sensitized with Nephelium lappaceum (F: Sapindaceae). J SolEnergy Eng 135:031014

Kumara G, Okuya M, Murakami K, Kaneko S, Jayaweera V, Tennakone K (2004) Dye-sensitized solid-state solar cells made from magnesiumoxide-coated nanocrystalline titanium dioxide films: enhancement of the efficiency. J Photochem Photobiol A: Chem 164:183–185

Kumara N, Ekanayake P, Lim A, Liew LYC, Iskandar M, Ming LC, Senadeera G (2013b) Layered co-sensitization for enhancement of conversion efficiency of natural dye sensitized solar cells. J Alloy Compd 581:186–191

Kye MJ, Cho J, Yu JC, Chang Y-W, Han J, Lee E, Lim HS, Lim JA (2018) "Drop-on-textile" patternable aqueous PEDOT composite ink providing highly stretchable and wash-resistant electrodes for electronic textiles. Dyes Pigm 155:150–158

Lai WH, Su YH, Teoh LG, Hon MH (2008) Commercial and natural dyes as photosensitizers for a water-based dye-sensitized solar cell loaded with gold nanoparticles. J Photochem Photobiol A: Chem 195:307–313

Lam J-Y, Chen J-Y, Tsai P-C, Hsieh Y-T, Chueh C-C, Tung S-H, Chen W-C (2017) A stable, efficient textile-based flexible perovskite solar cell with improved washable and deployable capabilities for wearable device applications. RSC Adv 7:54361–54368

Liang X, Long G, Fu C, Pang M, Xi Y, Li J, Han W, Wei G, Ji Y (2018) High performance all-solid-state flexible supercapacitor for wearable storage device application. Chem Eng J 345:186–195

Liu J, Li Y, Yong S, Arumugam S, Beeby S (2019) Flexible printed monolithic-structured solid-state dye sensitized solar cells on woven glass fibre textile for wearable energy harvesting applications. Sci Rep 9 (1362)

Li Y, Arumugam S, Krishnan C, Charlton MD, Beeby SP (2019) Encapsulated textile organic solar cells fabricated by spray coating. ChemistrySelect 4:407–412

Mathew S, Yella A, Gao P, Humphry-Baker R, Curchod BF, Ashari-Astani N, Tavernelli I, Rothlisberger U, Nazeeruddin MK, Grätzel M (2014) Dye-sensitized solar cells with 13% efficiency achieved through the molecular engineering of porphyrin sensitizers. Nat Chem 6:242

Nagai H, Segawa H (2004) Energy-storable dye-sensitized solar cell with a polypyrrole electrode, Chemical communications 974–975

Noor M, Buraidah M, Careem M, Majid S, Arof A (2014) An optimized poly (vinylidene fluoride-hexafluoropropylene)–NaI gel polymer electrolyte and its application in natural dye sensitized solar cells. Electrochim Acta 121:159–167

Olea A, Ponce G, Sebastian P (1999) Electron transfer via organic dyes for solar conversion. Sol Energy Mater Sol Cells 59:137–143

Peng M, Dong B, Zou D (2018) Three dimensional photovoltaic fibers for wearable energy harvesting and conversion. J Energy Chem 27:611–621

Pu X, Song W, Liu M, Sun C, Du C, Jiang C, Huang X, Zou D, Hu W, Wang ZL (2016) Wearable power-textiles by integrating fabric triboelectric nanogenerators and fiber-shaped dye-sensitized solar cells. Adv Energy Mater 6:1601048

Sahito IA, Sun KC, Arbab AA, Qadir MB, Jeong SH (2015) Graphene coated cotton fabric as textile structured counter electrode for DSSC. Electrochim Acta 173:164–171

Sahito IA, Sun KC, Arbab AA, Qadir MB, Choi YS, Jeong SH (2016) Flexible and conductive cotton fabric counter electrode coated with graphene nanosheets for high efficiency dye sensitized solar cell. J Power Sources 319:90–98

Shanmugam V, Manoharan S, Anandan S, Murugan R (2013) Performance of dye-sensitized solar cells fabricated with extracts from fruits of ivy gourd and flowers of red frangipani as sensitizers. Spectrochim Acta Part a Mol Biomol Spectrosc 104:35–40

Song J, Yin Z, Yang Z, Amaladass P, Wu S, Ye J, Zhao Y, Deng WQ, Zhang H, Liu XW (2011) Enhancement of photogenerated electron transport in dye-sensitized solar cells with introduction of a reduced graphene oxide–TiO2 junction. Chem–A Eur J 17:10832–10837

Susrutha B, Giribabu L, Singh SP (2015) Recent advances in flexible perovskite solar cells. Chem Commun 51:14696–14707

Teoli F, Lucioli S, Nota P, Frattarelli A, Matteocci F, Di Carlo A, Caboni E, Forni C (2016) Role of pH and pigment concentration for natural dye-sensitized solar cells treated with anthocyanin extracts of common fruits. J Photochem Photobiol A: Chem 316:24–30

Tsuboi K, Matsumoto H, Fukawa T, Tanioka A, Sugino K, Ikeda Y, Yonezawa S, Gennaka S, Kimura M (2015) Simulation study on optical absorption property of fiber-and fabric-shaped organic thin-film solar cells with resin sealing layer. Sen-I Gakkaishi 71:121–126

Wang X-F, Matsuda A, Koyama Y, Nagae H, Sasaki S-I, Tamiaki H, Wada Y (2006) Effects of plant carotenoid spacers on the performance of a dye-sensitized solar cell using a chlorophyll derivative: enhancement of photocurrent determined by one electron-oxidation potential of each carotenoid. Chem Phys Lett 423:470–475

Wu C, Kim TW, Guo T, Li F (2017) Wearable ultra-lightweight solar textiles based on transparent electronic fabrics. Nano Energy 32:367–373

Xin X, He M, Han W, Jung J, Lin Z (2011) Low-cost copper zinc tin sulfide counter electrodes for high-efficiency dye-sensitized solar cells. Angew Chem Int Ed 50:11739–11742

Xu J, Li M, Wu L, Sun Y, Zhu L, Gu S, Liu L, Bai Z, Fang D, Xu W (2014) A flexible polypyrrole-coated fabric counter electrode for dye-sensitized solar cells. J Power Sources 257:230–236

Xu Q, Li M, Yan P, Wei C, Fang L, Wei W, Bao H, Xu J, Xu W (2016) Polypyrrole-coated cotton fabrics prepared by electrochemical polymerization as textile counter electrode for dye-sensitized solar cells. Org Electron 29:107–113

Yang Z, Deng J, Sun X, Li H, Peng H (2014) Stretchable, wearable dye-sensitized solar cells. Adv Mater 26:2643–2647

Yun MJ, Cha SI, Seo SH, Lee DY (2014) Highly flexible dye-sensitized solar cells produced by sewing textile electrodes on cloth. Scient Rep 4:5322

Yusoff A, Kumara N, Lim A, Ekanayake P, Tennakoon KU (2014) Impacts of temperature on the stability of tropical plant pigments as sensitizers for dye sensitized solar cells. J Biophys 2014 (2014)

Zhang X, Zhang B, Ouyang X, Chen L, Wu H (2017) Polymer solar cells employing water-soluble polypyrrole nanoparticles as dopants of PEDOT: PSS with enhanced efficiency and stability. J Phys Chem C 121:18378–18384

Zhou H, Wu L, Gao Y, Ma T (2011) Dye-sensitized solar cells using 20 natural dyes as sensitizers. J Photochem Photobiol A: Chem 219:188–194

Building Energy Harvesting Powered by Solar Thermal Energy

Sampad Ghosh, Sivasankaran Harish, and Bidyut Baran Saha

1 Introduction

The rapid development of the building and industrial structures in the urban areas causes to deteriorate the urban environment. Due to the changes in the heat balance, the air temperature in densely constructed urban areas is higher than the temperature of the surrounding suburban and rural areas. This effect is referred to as the urban heat island (UHI) effect (Yamamoto 2006). Luke Howard first identified the UHI effect in London and later on in other European countries, as well as in the USA and Asia. Indeed, the UHI effect exists in almost every big city in the world (Priyadarsini 2012). UHI has adverse effects not only on climate changes, atmospheric pollution (Lan and Zhan 2017), earth's resources (Brazel et al. 2007), but also it has a negative impact on human well-being.

S. Ghosh · B. B. Saha (✉)
Interdisciplinary Graduate School of Engineering Sciences, Kyushu University, 6-1 Kasuga-koen, Kasuga-shi, Fukuoka 816-8580, Japan
e-mail: saha.baran.bidyut.213@m.kyushu-u.ac.jp

S. Ghosh · S. Harish · B. B. Saha
International Institute for Carbon-Neutral Energy Research (WPI-I^{2}CNER), Kyushu University, 744 Motooka, Nishi-ku, Fukuoka 819-0395, Japan

S. Ghosh
Department of Electrical and Electronic Engineering, Chittagong University of Engineering and Technology, Chittagong-4349, Bangladesh

B. B. Saha
Department of Mechanical Engineering, Kyushu University, 744 Motooka, Nishi-ku, Fukuoka 819-0395, Japan

H. Tyagi et al. (eds.), *New Research Directions in Solar Energy Technologies*, Energy, Environment, and Sustainability,
https://doi.org/10.1007/978-981-16-0594-9_6

It is clear that higher energy consumption is expected as the temperature rises in urban areas, in particular for cooling purposes in summer. It is reported that for an increase of 1 °C in urban temperature, power consumption would increase by around 0.5% in Tokyo, Japan, whereas it may increase to 3.5% in some big cities of the USA (Santamouris 2001). On the contrary, global warming could lead to a significant shift from power consumption.

The incident of solar radiation on urban surfaces is absorbed and then converted into sensible heat. Urban surfaces are mostly covered by pavements and buildings (including vertical walls, roofs), and these two contributed 35% and 20% areas, respectively (Levinson et al. 2010). Recently, energy consumption in buildings has appealed a great deal of attention since buildings (including commercial, educational, residential) are one of the fastest growing electricity consumption sectors globally. For instance, commercial buildings in the USA consume about 18% of total electricity consumption (Hidalgo-Leon et al. 2018). Currently, it is seen that conventional structures are moving toward smart buildings, and people are conscious about the installment of the well-established renewable energy system (e.g., solar plants, wind farms). However, still, researchers are finding alternative devices that can utilize a small amount of environmental energy to convert electrical energy for low-power applications such as sensors and MEMS. These devices can use a different form of energy (e.g., kinetic energy, thermal energy, and so on) as an input. Those device's technology is called energy harvesting, also known as energy scavenging. This chapter provides a brief description of energy harvesting from solar thermal energy, and it is crucial to consider this technology in buildings instead of batteries.

1.1 Limitations of Batteries in Buildings

Wireless sensor network (WSN) system, which is considered as the future of modern society, is powered by batteries (Ryu et al. 2019). However, there are several drawbacks to using batteries in building autonomous systems.

(a) It is expected that the electronic device has a much higher lifetime than the battery. This means there is a requirement for battery replacement and during that, it may create a disruption of service. On the other hand, usage of battery may arise environment and health hazard issues when disposed directly to the ambient and not handled properly, respectively.
(b) Sometimes there is a possibility to forget replacing batteries. If this may happen in any critical system, such as a smoke/gas detector in a building, the resulting consequences will be massive.
(c) Since there is always a chemical reaction inside of a battery, batteries self-discharge may limit the operational lifetime of very low-power electronic systems.
(d) In many electronic systems, the battery size often surpasses the device size (Matiko et al. 2014).

Energy harvesting performance from thermal energy is basically linked to the amount and nature of the energy source that exists in the environment. When designing a thermal energy harvesting system, basic knowledge through phenomena to measurement has to be known in advance.

2 Energy Harvesting in Buildings: State of the Art

2.1 *Thermoelectric Energy Harvesting*

Enormous free thermal energy is available in the environment, which is coming either from waste heat sources (e.g., engines, furnaces, boilers, etc.) or sunlight via natural heating. Thermoelectric device or generator (TEG) is an effective way that allows converting thermal energy into electrical energy. Figure 1 shows a simple thermoelectric generator. TEG is made of several thermoelectric elements connected electrically in series but thermally in parallel. When there is a temperature gradient between two sides, the majority carrier flows from the hot side to the cold side, resulting in a generation of voltage. TEG has different advantages, like low maintenance costs, low operating costs, high reliability, longer life, less noise, compact in size, no movable components, and zero greenhouse gas emissions (Hidalgo-Leon et al. 2018). These devices are based on the Seebeck effect, Peltier effect, and Thomson effect.

Seebeck Effect. The thermoelectric effects underlying the conversion of thermoelectric energy can be easily discussed by Fig. 2. It can be considered as a circuit formed by two dissimilar materials (*a* and *b*), which are connected electrically in series and

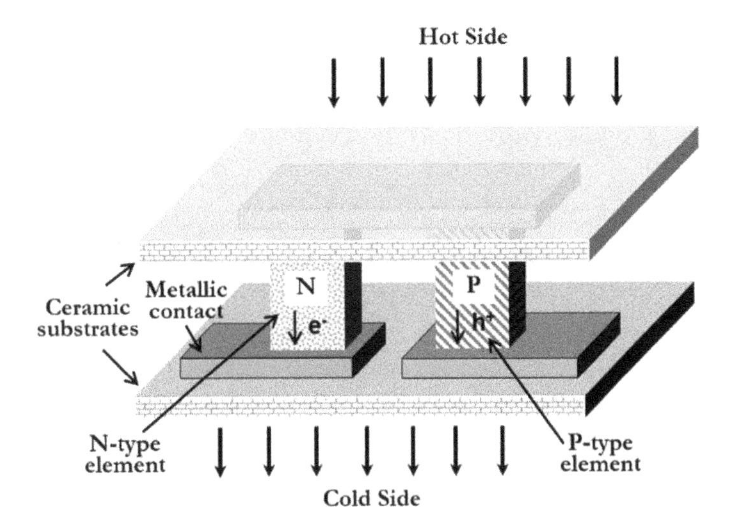

Fig. 1 Schematic representation of a thermoelectric generator

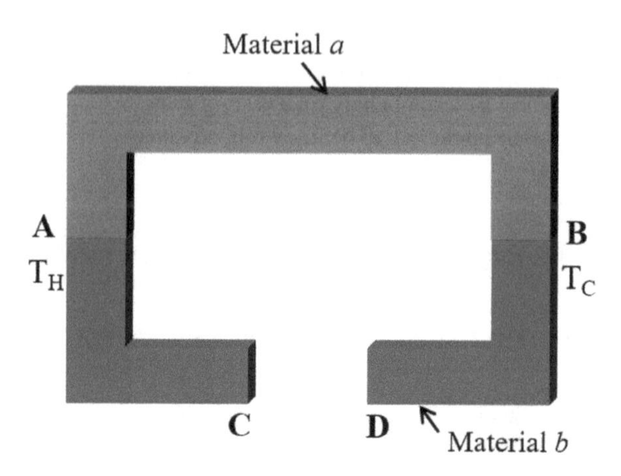

Fig. 2 Schematic of a fundamental thermocouple

thermally in parallel. If the junctions at A and B are maintained at different temperatures T_H and T_C (where, $T_H > T_C$), an open-circuit voltage (V) is developed by $V = \alpha(T_H - T_C)$ or $\alpha = V/\Delta T$ between C and D, which defines the differential Seebeck coefficient α_{ab} between the elements *a* and *b*. The relationship is linear for small ΔT. Sometimes the Seebeck coefficient is denoted by *S* and measured in VK^{-1} or more often μVK^{-1}. Seebeck coefficient is an intrinsic property of materials that are related to their electronic properties and not dependent on geometry (Ponnamma et al. 2017). For p-type semiconductors, it is positive, and for n-type ones, it is negative.

Peltier Effect. If the reverse situation is considered in Fig. 2, i.e., an external voltage is applied across C and D, a current *I* flows around the circuit, which causes a rate of heating *q* at one junction and a rate of cooling $-q$ at other junction. The ratio of *I* to *q* defines the Peltier coefficient given as $\Pi = I/q$ with unit in volts.

Thomson Effect. It relates the generated heat *q* with current flowing *I* where there is a temperature difference ΔT and is defined by $q = \beta I \Delta T$ where β is the Thomson coefficient with a unit in VK^{-1}.

2.2 *Thermoelectric Generation and the Figure of Merit*

It is well defined that thermoelectric materials should have crystal-like electronic properties and glass-like thermal properties, but it is challenging to realize in practice. The most frequently used materials are semiconductors where p- and n-type are electrically connected in series but thermally in parallel. Figure 3 shows a structure of a thermoelectric module assembled from an array of thermoelements. The thermoelements are connected by a metallic contact, and it is essential to have that low resistance.

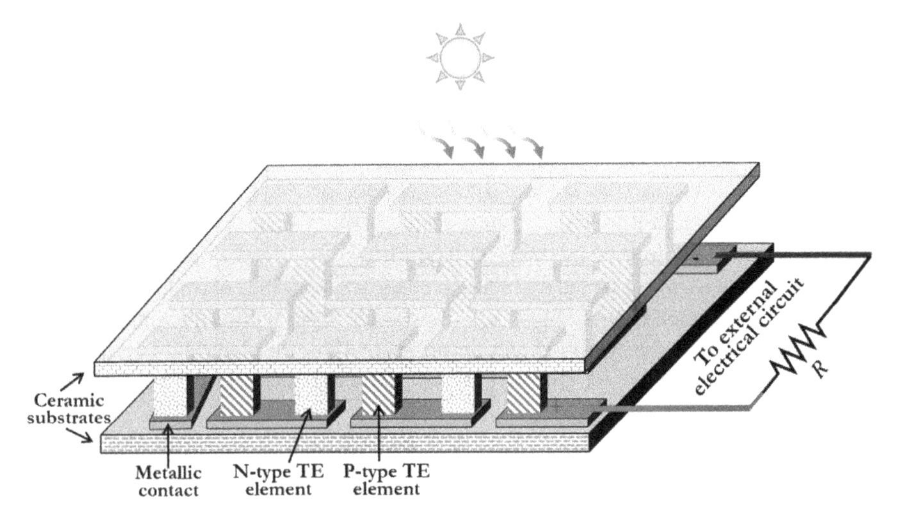

Fig. 3 A thermoelectric module connected with external load

The performance of the thermoelectric device is dependent on the combination of p- and n-type materials used in device fabrication. The effectiveness of thermoelectric material is described by a thermoelectric figure of merit (ZT) without dimensions, defined as

$$\mathrm{ZT} = \frac{S^2\sigma T}{\kappa} = \frac{S^2\sigma T}{\kappa_e + \kappa_l} = \frac{\mathrm{PF}}{\kappa_e + \kappa_l}T \tag{1}$$

where S represents the Seebeck coefficient, σ is the electrical conductivity, κ_e is the electronic contribution of thermal conductivity, κ_l is the lattice contribution of thermal conductivity, T is the absolute temperature, and PF is the power factor. To achieve a high value of ZT, the material should have a high value of Seebeck coefficient, high electrical conductivity and low thermal conductivity. However, these three parameters are interlinked together like a magic triangle since the improvement of one parameter degrades another. So, in practice, a trade-off between these parameters is to be considered.

Usually, the figure of merit is used to calculate the efficiency. The efficiency is defined as

$$\eta = \eta_{\mathrm{carnot}}\frac{\sqrt{1+\mathrm{ZT}}-1}{\sqrt{1+\mathrm{ZT}}+(T_{\mathrm{C}}/T_{\mathrm{H}})} \tag{2}$$

where $\eta_{\mathrm{carnot}} = \frac{T_{\mathrm{H}}-T_{\mathrm{C}}}{T_{\mathrm{H}}}$, ZT, T_{H} and T_{C} are Carnot efficiency, figure of merit, hot side, and cold side temperature, respectively. Thus, the performance of a device can be improved by increasing the thermoelectric figure of merit.

3 Thermoelectric Measurement of Bulk Samples

3.1 Electrical Conductivity Measurement

The thermoelectric materials are mostly semiconductor, and oxide layers can easily form on the surface. It may create a non-ohmic contact issue, i.e., it can ruin good electrical contact. A four-probe method (see Fig. 4) is a good practice to measure electrical conductivity where current is injected through one pair of probe, and voltage is measured via other probes. Materials dimension (e.g., width, thickness, length) should be measured preciously to get an accurate value of electrical conductivity. The electrical conductivity is obtained by the following equation.

$$\sigma = \left(\frac{I}{V}\right) \times \left(\frac{L_0}{A}\right) \tag{3}$$

where L_0 is the distance between the measuring leads, and A is the cross-sectional area. During measurement, it is recommended to have sufficient distance between voltage and current leads. The recommended distance is considered as (Rowe and Rowe 2006)

$$L - L_0 \geq 2t \tag{4}$$

where L is the total length of the sample, and t is the thickness of the sample.

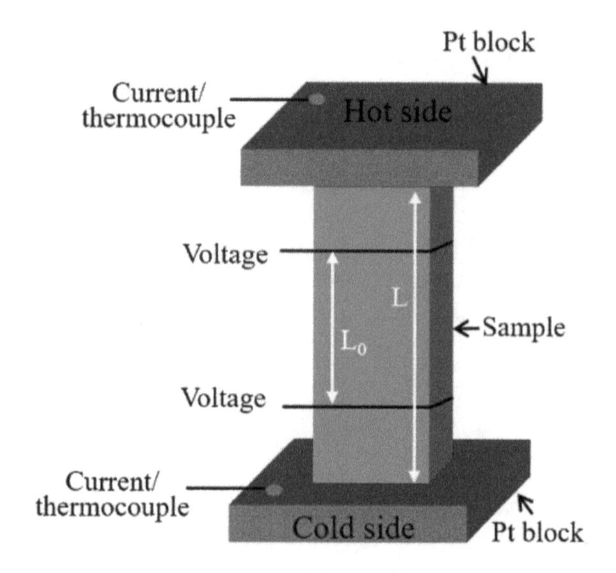

Fig. 4 A four-probe setup for electrical conductivity and Seebeck coefficient measurement

3.2 Seebeck Coefficient Measurement

The Seebeck coefficient, like electrical conductivity, is an intrinsic property of the material, and its accurate measurement is necessary. Moreover, it gives the information of the charge carrier. A positive value of the Seebeck coefficient implies that the majority carriers are hole, while negative value corresponds to the electron as the charge carriers. The Seebeck coefficient is independent, corresponds to the sample geometry and is given by the voltage and temperature gradient ratio along with the sample.

$$S = \frac{V}{\Delta T} = \frac{V}{T_{\mathrm{H}} - T_{\mathrm{C}}} \tag{5}$$

where T_{H} and T_{C} are the hot and cold side temperature, respectively.

3.3 Thermal Conductivity Measurement—Laser Flash Method

Sometimes, it is difficult to measure the high accuracy thermal conductivity of a sample. There are many methods and techniques such as steady-state technique, comparative technique, three-omega method and laser flash method have been used for thermal conductivity measurement. Among them, the laser flash method is widely used. In this method, a short duration of the laser pulse is irradiated in on the side of the sample, and the temperature rise is observed at another side of the sample by an IR camera. The thermal diffusivity (D) is determined by the following equation.

$$D = 0.1388\left(\frac{d^2}{t_{1/2}}\right) \tag{6}$$

where d is the sample thickness, and $t_{1/2}$ is the time for reaching half the maximum temperature rise.

The specific heat capacity (C_p) is measured by separate equipment called differential scanning calorimeter (DSC). It gives the C_p value of a sample after comparing it with standard alpha-alumina (α-Al_2O_3). Details about the C_p measurement are described later in this chapter. Then by measuring the sample density (ρ), the thermal conductivity (κ) can be determined from the given equation.

$$\kappa = \rho D C_p \tag{7}$$

Of course, some laser flash apparatus has the facility to measure both thermal diffusivity and specific heat capacity to give thermal conductivity directly. Otherwise, the thermal conductivity can be calculated using Eq. (7).

3.4 Hall Coefficient, Carrier Concentration and Mobility

It is necessary to measure the carrier concentration by measuring the Hall effect since it defines the optimum value of the power factor for the thermoelectric materials. The Hall voltage (V_H) which is generated due to the Lorentz force is given by

$$V_H = \frac{BIb}{neA} = \frac{R_H BI}{t} \tag{8}$$

where $R_H = 1/ne$ is the Hall coefficient, B is the magnetic field, I is current, b is the sample width, A is the cross-sectional area, and t is the sample thickness. Also, the electrical conductivity is related to the Hall coefficient and mobility as

$$\mu = R_H \sigma \tag{9}$$

In general, the van der Pauw configuration (Hemenger 1973) is widely used for measuring the Hall effect where four contacts close to the edge are created on the sample.

4 Cement Composites as Thermoelectric Material

Conventional thermoelectric materials are mostly semiconductors and the drawbacks usually associated with them are the high cost of their manufacturing, scarcity, and increased weight. Cement, a common structural material, has both the structural and non-structural function. Non-structural functions included sensing, heating, energy storage, and generation (Chung 2019). Energy generation from cement composites is attractive, which is based on the thermoelectric (TE) phenomenon. Cement is considered for thermoelectric materials because of its excellent stability in the environment, non-toxicity, low synthesis cost, flexibility to be integrated into different forms, and inherent lower thermal conductivity. Also, energy harvesting features zero energy consumption and environmentally friendly. However, compared with traditional thermoelectric materials, cement as a TE material has received less attention due to low thermoelectric efficiency and figure of merit resulting from low electrical conductivity and low Seebeck coefficient. The main task is, therefore, to develop TE cement composites that have superb thermoelectric properties.

Sun et al. first reported the Seebeck coefficient of $+5.5\ \mu V°C^{-1}$ for carbon fiber reinforced concrete (Sun et al. 1998). Since then, many researchers have been tried to increase the Seebeck value for cement composites and reported extensively. Bromine (Br) intercalated carbon fibers ($-17\ \mu V°C^{-1}$) (Wen and Chung 2000), expanded graphite ($-54.5\ \mu V°C^{-1}$) (Wei et al. 2018a), stainless steel fibers ($+3\ \mu V°C^{-1}$) (Wen and Chung 2004), steel slag carbon fibers ($+14.4\ \mu V°C^{-1}$) (Zuo et al. 2013), carbon nanotubes ($+57.98\ \mu V°C^{-1}$) (Wei et al. 2018b), steel fiber silica fume ($+68\ \mu V°C^{-1}$) (Chung 2001), carbon fiber with mineral admixtures ($+127.4\ \mu V°C^{-1}$)

(Bahar and Salih 2008), pyrolytic carbon fibers (Wei et al. 2016), bismuth telluride (Bi_2Te_3) (Yao and Xia 2014), $Ca_3Co_4O_9$ (+58.6 $\mu V°C^{-1}$) (Wei et al. 2013), Bi_2O_3 (+100.28 $\mu V°C^{-1}$) (Wei et al. 2014), Fe_2O_3 (+2500 $\mu V°C^{-1}$) (Ji et al. 2016), ZnO (+3300 $\mu V°C^{-1}$) (Ji et al. 2016), MnO_2 (−5490 $\mu V°C^{-1}$) (Ji et al. 2018) were introduced into the cement matrix. Although there has been huge progress in the Seebeck coefficient of cement composites, still, the figure of merit values does not be improved greatly due to the low electrical conductivity.

Chen et al. and Jian et al. measured the electrical conductivity of carbon fiber reinforced cement composite and found 0.005 Scm^{-1} (Chen et al. 2004) and 2×10^{-5} Scm^{-1} (Wei et al. 2018a), respectively. Later on, electrical conductivity value is improved to 0.818 Scm^{-1} and 1.0 Scm^{-1} by adding carbon nanotubes (Wei et al. 2018b) and expanded graphite synthesized from intercalated graphite (Brandt et al. 2018), respectively. Further enhancement is done by adding graphene oxide and ferrofluid (10.40 Scm^{-1}) (Singh et al. 2011), expanded graphite (24.8 Scm^{-1}) (Wei et al. 2018a) and exfoliated graphite (76.9 Scm^{-1}) (Muthusamy et al. 2010). From the above, it is observed that higher Seebeck coefficient and electrical conductivity could be achieved individually. However, it is challenging to improve them simultaneously in the same sample since thermoelectric properties (Seebeck coefficient, electrical and thermal conductivity) formed a magic triangle, i.e., improvement in one causes to degrade others.

As mentioned earlier, cement has inherently low thermal conductivity, which is beneficial to have a high figure of merit. In spite, over the years, many researchers have studied to reduce the thermal conductivity of cement composites further. By adding graphite, black carbon, and black iron oxide into cement, the thermal conductivity values are found 0.24, 0.25, and 0.20 $Wm^{-1} K^{-1}$, respectively (Wen et al. 2012). The introduction of rice husk provided 0.64 $Wm^{-1} K^{-1}$ (Doko et al. 2013), whereas 0.74 $Wm^{-1} K^{-1}$ (Jittabut et al. 2016) is obtained by changing the water to cement ratio. However, when carbon nanotubes and expanded graphite are included in cement, the thermal conductivity values increased to 0.841 $Wm^{-1} K^{-1}$ (Wei et al. 2018a) and 3.213 $Wm^{-1} K^{-1}$ (Wei et al. 2018a), respectively. So, the important thing is to keep the thermal conductivity to a relatively lower value to have a higher figure of merit (ZT).

After getting the above-discussed thermoelectric properties, i.e., Seebeck coefficient, electrical conductivity, and thermal conductivity, one can easily calculate the figure of merit and efficiency by using Eqs. (1) and (2), respectively. From the literature review, it is found that very few works have presented the numerical value of ZT for cement composites. It is reported that adding carbon nanotubes (Wei et al. 2018b), carbon fiber (Wei et al. 2014), and expanded graphite (Wei et al. 2018a) has provided the maximum figure of merit as 9.33×10^{-5}, 3.11×10^{-3}, and 6.82×10^{-4}, respectively.

In our work, we have introduced graphene nanoplatelets (GnP) into cement for the first time to improve the thermoelectric properties of cement composites. Graphene is added because it possesses many exciting features such as lightweight, thermoelectric effect, high electrical conductivity, exceptional mechanical strength, relatively straightforward manufacturing, and cheaper than carbon nanofibers and nanotubes

(Cataldi et al. 2018). The following section includes the fabrication of graphene-based cement composites as well as their measurement of thermoelectric properties followed by the figure of merit calculation.

4.1 *Specimen Fabrication*

Cement (portland type) and graphene nanoplatelets (GnP) are purchased from Toyo Matelan Co. Ltd., Japan and XG Sciences, USA, respectively. The average diameter of as-received GnP (H-grade) is 25 μm. The other attributes of the purchased GnP are the average thickness of about 15 nm, the surface area of 50 to 80 m^2g^{-1}, and a specific density of 2.2 gcm^{-3}. To have the optimum concentration of graphene and investigate the effect of graphene content, different concentration such as 5, 10, 15, and 20 wt% by mass of cement is introduced into the composites.

A Planetary Ball Mills (Fritsch Japan Co. Ltd) blends the appropriate amount of GnP and cement. To facilitate the blending, 30 g Zirconia (Zr) balls are added in the blending cups. The blending recipe is followed: rotation speed—500 rpm, cycle time—60 min, pause between cycles—5 min, total no. of the cycle—12, reversible rotation—ON. This blending is vital since it is a dry dispersive process and allows for having a potentially uniform distribution of graphene in the composites. After mixing, a vibrating sieve machine (Nitto, ANF-30) separates Zr balls from the mixture. The sieving time of approximately 20 min is spent for each batch, and an aperture size of 106 μm (140 mesh) is used. This machine can not only help to have an even particle size in the composite, but also help to reproduce precise data. The ratio of water to cement is fixed at 0.1 irrespective of the GnP concentration (Ghosh et al. 2020).

Appropriate quantities of blended materials suitably mixed with water are transferred in cylindrical steel die (φ20 $\times$ 30 mm^2) and compressed by a pressing system (NT-200H, NPa system) afterward. Approximately, 40 MPa pressure is applied for several minutes on the loaded samples to make it bulk (φ20 $\times$ 4 mm^2) (Ghosh et al. 2019). The disc-shaped sample is cured at room temperature and then dried at 200 °C for several hours to remove moisture and other contaminants from the samples. The above-described sample preparation is schematically represented in Fig. 5.

4.2 *Thermoelectric Properties*

Electrical Properties. The cement composite's Seebeck coefficient and electrical conductivity are measured at the same time by a four-probe Seebeck measuring system (RZ2001i) from Ozawa science, Japan. The disc-shaped samples are cut by a micro-cutter machine (Maruto Co. Ltd., Japan) to get the desired rectangle shape (4 $\times$ 4 $\times$ 10 mm^3), as shown in Fig. 5. This cutter is equipped with a diamond saw, run by a DC motor, rotated at 20 rpm for this work. After that, Pt wire is mounted on the sample near to the two edges, as shown in Fig. 6. For accurate measurement, Pt wire

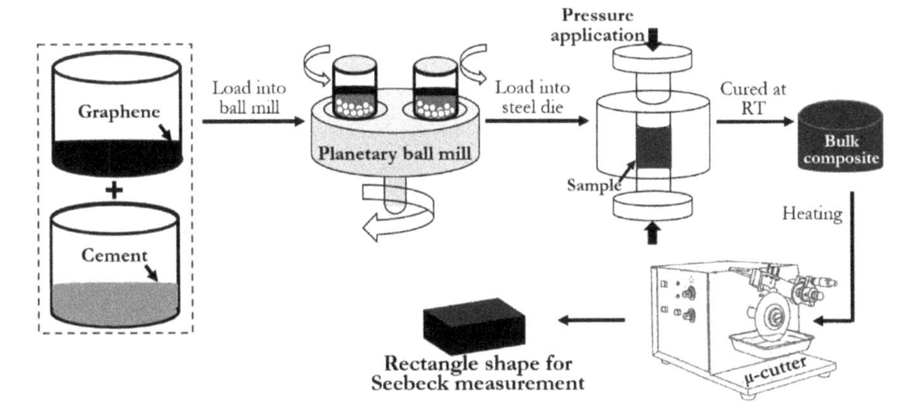

Fig. 5 Preparation of cement composite

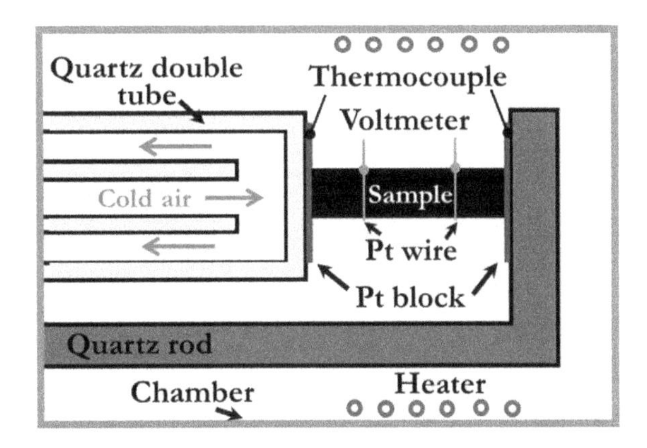

Fig. 6 Schematic representation of RZ2001i with the sample inside

was carefully surrounded by Pt paste. The samples are clamped between the Pt block (see Fig. 6), and heat is applied from one side of the samples while the other side kept at room temperature. The hot side is heated automatically from ambient temperature to 75 °C with a heating rate of 0.01 °Cs^{-1} to maintain the system's thermal balance. The schematic representation of a part of the RZ2001i with a sample loaded inside is demonstrated in Fig. 6.

It was not possible to measure thermoelectric properties such as electrical conductivity and Seebeck coefficient of cement without the GnP content. Therefore, they are considered as nil. The temperature dependence of electrical conductivity, Seebeck coefficient, and power factor for cement composites are presented in Fig. 7. The electrical conductivity of the composites increases with increasing GnP contents, which are expected, and the average values are around 3.0, 8.5, 11.5, and 16.0 Scm^{-1} with 5, 10, 15, and 20 wt% addition of graphene, respectively. These values are higher than the electrical conductivity of cement (10^{-7} Scm^{-1}) and within the range of semiconductor's electrical conductivity (10^{-9}–10^{4} Scm^{-1}) (Singh et al. 2011).

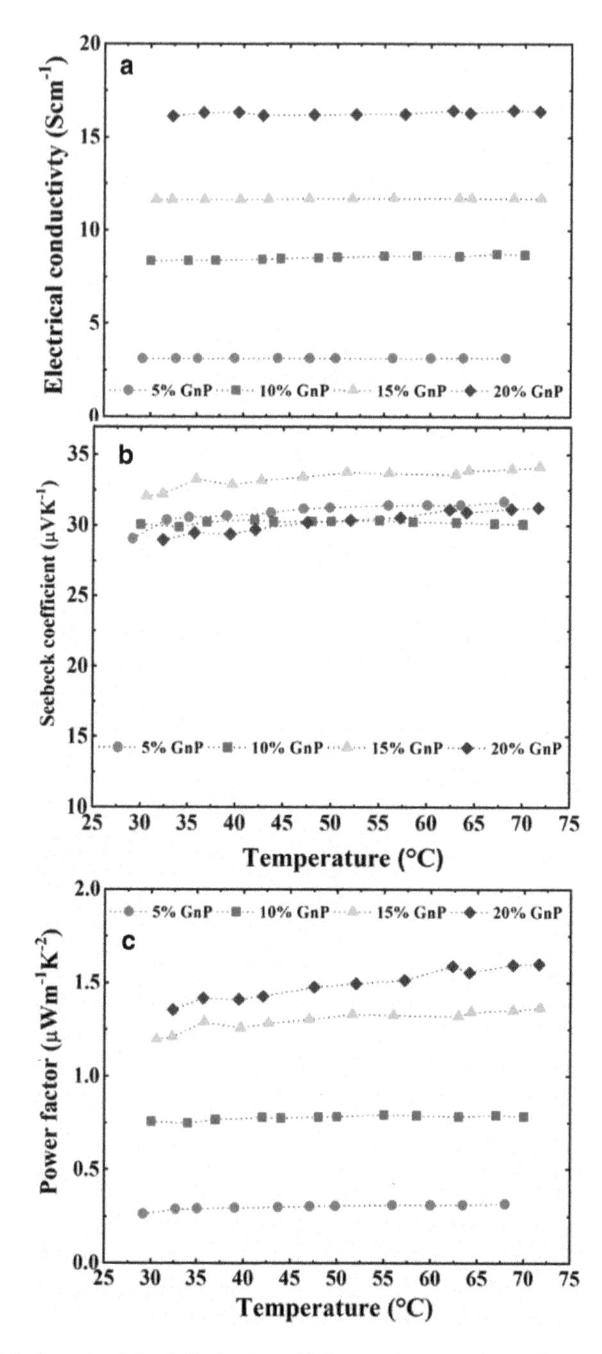

Fig. 7 **a** Electrical conductivity, **b** Seebeck coefficient, and **c** power factor for cement composites with temperature. Images reproduced with permission from reference (Ghosh et al. 2019). Copyright 2019, Elsevier

Also, conductivity marginally increases with temperature suggesting typical semiconductor behavior, and of course, the charge carriers are mainly contributed by the graphene. The Seebeck coefficient is positive for all composites, meaning that the prepared cement composites are p-type semiconductors with the hole being the majority carriers. Seebeck values increase with 5, 10, and 15 wt% GnP inclusions but decrease with 20 wt%. The most significant Seebeck coefficient of about 35 μVK^{-1} is obtained at 70 °C with 15 wt% GnP inclusions. However, although the composites showed enormous electrical conductivity at maximum GnP content (20 wt%), their Seebeck coefficient is lower than the composites with 15 wt% GnP. It can be attributed to a relatively higher concentration of carrier in the composites because there is a coupling effect between the electrical conductivity and Seebeck coefficient for thermoelectric materials (Wen and Chung 1999). Cement composites with a high value of graphene contents exhibited a higher power factor value. The power factor displays almost the linear temperature relation.

Thermal Properties. Thermal diffusivity of graphene-based cement composites with the size of $10 \times 10 \times 2$ mm^3 is carried out by Netzsch LFA 457 MicroFlash, Germany. The main components of LFA 457 are laser source, heating furnace, infrared (IR) detector, and data acquisition system (Min et al. 2007). The measurement is carried out in an ambient state with a room temperature range up to 75 °C. A short duration high-intensity laser pulse is applied to one side of the sample. The absorbed heat travels through the sample's total height and causes an increase in the other side's temperature. This is detected by an IR detector. Finally, the built-in software processes all the gathered data by solving a set of equations (Pal et al. 2019) and generated the required information.

The differential scanning calorimeter (DSC) produced by Shimadzu Corporation Ltd., Japan, is used to measure specific heat capacities at temperatures between 25 and 75 °C. This machine can achieve a peak power with high sensitivity and low noise (<1 μW). It has a thick flat base to grip the sample and reference pans. Associated signals obtained by thermocouples and others are processed by a thermal analyzer. This analyzer then sent all the data to a computer where necessary software is loaded for further processing. Aluminum (Al) pans and lids with an identical shape and weight are used to house the samples. First, the empty pan and lid are weighted using a microbalance (Model AND BM-22, USA). Standard alpha-alumina (α-Al_2O_3) powder purchased from Sigma-Aldrich is used for calibration of the heating rate and specific heat capacity. After putting alumina into the pan, it is sealed with lid by using the DSC sample sealing tool (TA instruments) and subsequently measured the weight. Same procedures like sealing, the weighting is also carried out for the cement composite. The actual weight of the sample is calculated by subtracting the empty pan's weight from the weight, including the pan and composite. During the heat capacity experiment, a flow rate of 50 ml of nitrogen gas per minute is provided for the maintenance of inert conditions. Finally, thermal conductivity (κ) is determined from the specific heat capacity, thermal diffusivity, and density according to Eq. (7). The sample density is calculated from the sample geometry ($10 \times 10 \times 2$ mm^3) and mass.

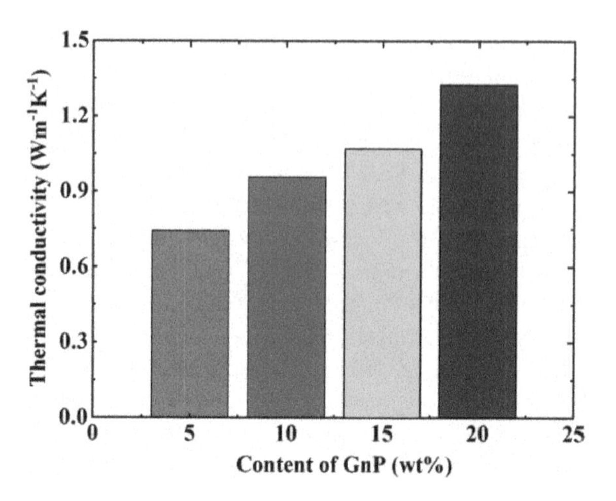

Fig. 8 The thermal conductivity for cement composites with 5, 10, 15, and 20 wt% graphene

In Fig. 8, the average thermal conductivity (κ) of samples with 5, 10, 15, and 20 wt% graphene is 0.75, 0.95, 1.1, and 1.3 Wm^{-1} K^{-1}, respectively. All these values are somewhat higher than those of thermal conductivity of cement (0.7 Wm^{-1} K^{-1}) and increase with the increase of the graphene content. It is worthy of mentioning that the thermal conductivity of the cement composite with graphene varies little at room temperature to 75 °C. This is mainly due to the reciprocal temperature dependence of specific heat capacity and thermal diffusivity. The obtained thermal conductivity consists of both propagating (lattice vibrations or phonon scattering) and diffusive vibrational modes (electrons) (Wei et al. 2018a). For a degenerate semiconductor, thermal conductivity represented by Wiedemann–Franz (WF) law (Kittel 1971) is linked to electrical conductivity as follows.

$$\kappa_e = \sigma L_0 T \tag{10}$$

where κ_e, σ, L_0, T is the electronic contribution of thermal conductivity, electrical conductivity, Lorenz number (2.44 × 10^{-8} WS^{-1} K^{-2} (Kittel 1971)), and temperature, respectively. According to the above equation, the electronic contribution of thermal conductivity is calculated for all composites at room temperature. The values of κ_e for x-cement composite ($x = 5, 10, 15, 20$ wt% of GnP) are 2.2 × 10^{-3}, 6.2 × 10^{-3}, 8.4 × 10^{-3} and 11.6 × 10^{-3} Wm^{-1} K^{-1}, respectively, which are much less than the average thermal conductivity (κ). The contribution of κ_e to κ is suggested to be much smaller than lattice contribution of thermal conductivity (κ_l), indicating κ_l is dominant in graphene-based cement composites. Due to the high phonon scattering between graphene and cement, cement composite has less thermal conductivity than traditional thermoelectric oxides and metals, generally 2.0 Wm^{-1} K^{-1} at least (Wei et al. 2018a).

4.3 Hall Effect

Measurement of the Hall data is often used to determine material types (n-type or p-type), majority carrier density, and mobility. The van der Pauw method (Hemenger 1973) is employed for measuring the Hall coefficient of cement composite. In this technique, four ohmic contacts are made on each sample and then connect carefully with four probes of the Hall system (Nanometrics, USA). These probes are used for current injection and voltage measurement under a strong magnetic field, which is applied in both positive and negative z-direction. The analysis is done in the atmospheric condition. The sample size is 10×10 mm^2, with a thickness of about 1 mm. Lower thickness (<1 mm) for accurate results is highly preferred.

Hall data, like carrier mobility and density for the cement composites with different graphene contents, is shown in Fig. 9. It is found that the carrier mobility and density both are inversely related with increasing of graphene contents. This is expected theoretically, as mentioned in Eq. (9). The carrier density of the composites is +0.5, +2.5, +6.5, and +7.6 cm^{-3} with 5, 10, 15, and 20 wt% of graphene, respectively. Hall coefficient (R_H) is calculated from $R_H = 1/ne$ where n is the carrier density and e is the charge of the carrier. Carrier mobility (μ) is obtained using Eq. (9) by putting electrical conductivity and Hall coefficient. The mobility values are 3.7, 2.1, 1.0, and 1.3 cm^{2}V^{-1} s^{-1} for x-composites ($x = 5$, 10, 15, 20 wt% of GnP). It is also found that the Hall coefficient is all positive values indicating all the composites are p-type and conducting carrier are holes. These results are in good agreement with Seebeck measurement, where it was found that the majority carriers are holes. Hence, the measured Hall data corroborated the Seebeck measurement values.

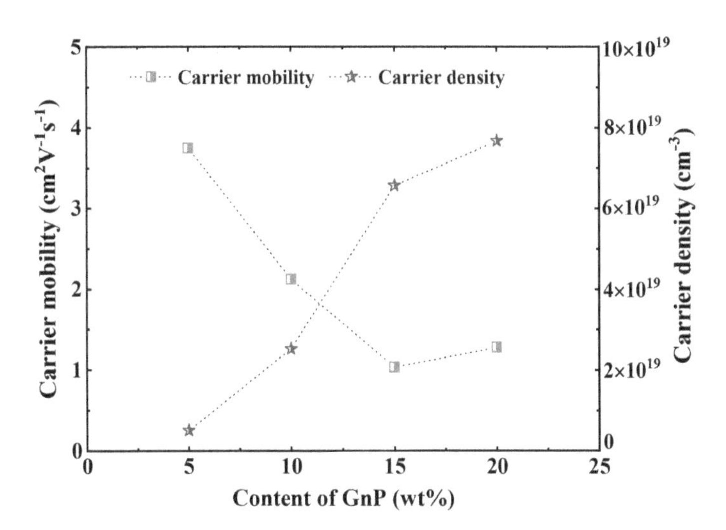

Fig. 9 Carrier mobility and density with 5, 10, 15, and 20 wt% graphene

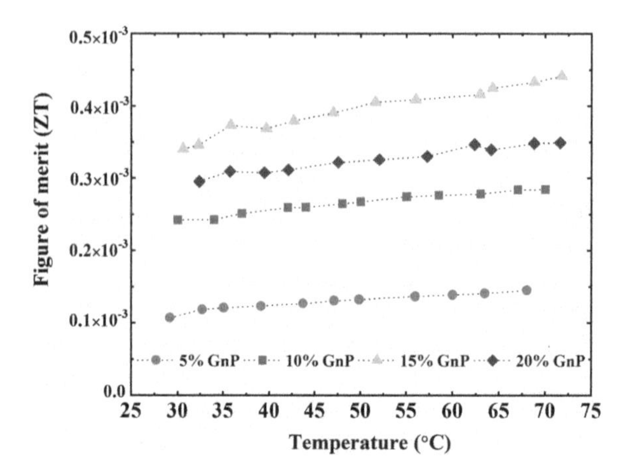

Fig. 10 Temperature dependence thermoelectric figure of merit. Image reproduced with permission from reference (Ghosh et al. 2019). Copyright 2019, Elsevier

4.4 *Figure of Merit*

Figure 10 illustrates the figure of merit (ZT) for the cement composites with a temperature change. It is noteworthy to mention that the value of ZT increases with temperature. ZT values are calculated using Eq. (1). It is found that there is remarkable progress in terms of maximum ZT value for 15 wt% composite compared to the others. It is anticipated since they exhibited the most significant Seebeck coefficient, and the Seebeck coefficient has a squared contribution in ZT, as shown in Eq. (1). The maximum ZT value as 0.44×10^{-3} is achieved with 15 wt% graphene contents whereas 0.15×10^{-3}, 0.28×10^{-3}, and 0.35×10^{-3} are obtained with 5, 10, and 20 wt% GnP contents. Although the obtained ZT value for the graphene-based cement composite is small because of the lower electrical conductivity and Seebeck coefficient, it is promising and substantial compared with other cement composites reported (Wei et al. 2018a, b, 2014). The obtained ZT with graphene is in the same order of magnitude as of the highest reported ZT till present (Wei et al. 2014). Moreover, the obtained ZT is comparable with some oxide thermoelectric materials like $Ca_3Co_4O_9$ (Demirel et al. 2012), $Bi_2Sr_2Co_2O_y$ (Hsu et al. 2012), and metal chalcogenide like Bi_2O_2Se (Zhang et al. 2013).

5 Conclusions

Due to the urban heat island (UHI), people living in the urban area continuously experience miserable weather and significantly increase energy consumption. This chapter discusses the building energy harvesting materials based on the temperature gradient to deal with UHI alleviation as well as energy generation. Graphene-based cement composites have low thermal conductivity and considerably high Seebeck coefficient, resulting in a high value of the figure of merit and consequently, applicable in thermoelectric devices. The properties such as electrical conductivity, Seebeck coefficient, thermal conductivity, carrier density, and mobility for graphene enhanced cement composites are addressed in this chapter. Since there is a tremendous amount of heat energy trapped by many building surfaces in the world, which is literally free, effective use of these places can reveal an opportunity for energy harvesting. In short, this brief study of eco-friendly building material may open new opportunities for harvesting electrical energy to fed small powered electronics in the near future.

References

Bahar D, Salih Y (2008) Thermoelectric behavior of carbon fiber reinforced lightweight concrete with mineral admixtures. New Carbon Mater 23:21–24

Brandt AM, Olek J, Glinicki MA, Leung CKY (2018) Brittle matrix composites 10, 1st edn. Woodhead Publishing, Cambridge

Brazel A et al (2007) Determinants of changes in the regional urban heat island in metropolitan Phoenix (Arizona, USA) between 1990 and 2004. Clim Res 33:171–182

Cataldi P, Athanassiou A, Bayer I (2018) Graphene nanoplatelets-based advanced materials and recent progress in sustainable applications. Appl Sci 8:1438

Chen B, Wu K, Yao W (2004) Conductivity of carbon fiber reinforced cement-based composites. Cem Concr Compos 26:291–297

Chung DDL (2001) Cement-matrix composites for thermal engineering. Appl Therm Eng 21:1607–1619

Chung DDL (2019) A review of multifunctional polymer-matrix structural composites. Compos Part B Eng 160:644–660

Demirel S, Aksan MA, Altin S (2012) Low temperature electrical and thermal transport properties of the $Ca_{3-x}Sb_xCo_4O_9$ system. J Mater Sci Mater Electron 23:2251–2256

Doko VK, Hougan AC, Adjovi EC, Ayite DY, Bedja K (2013) Study of thermal conductivity of light concrete based on rice husks. Int J Sci Eng Res 4:1735–1739

Ghosh S, Harish S, Saha BB (2019) Thermoelectric properties of graphene and carbon nanotube. Proc Int Exch Innov Conf Eng Sci 5:30–31

Ghosh S, Harish S, Rocky KA, Ohtaki M, Saha BB (2019) Graphene enhanced thermoelectric properties of cement based composites for building energy harvesting. Energy Build 202:109419

Ghosh S, Harish S, Ohtaki M, Saha BB (2020) Enhanced figure of merit of cement composites with graphene and ZnO nanoinclusions for efficient energy harvesting in buildings. Energy 198:117396

Hemenger PM (1973) Measurement of high resistivity semiconductors using the van der Pauw method. Rev Sci Instrum 44:698–700

Hidalgo-Leon R et al (2018) Energy harvesting technologies: analysis of their potential for supplying power to sensors in buildings. In: 2018 IEEE 3rd Ecuador technical chapters Meeting (ETCM 2018), pp 1–6

Hsu HC et al (2012) Enhanced thermoelectric figure-of-merit ZT for hole-doped $Bi_2Sr_2Co_2O_y$ through Pb substitution. J Appl Phys 111 (2012)

Ji T, Zhang X, Li W (2016) Enhanced thermoelectric effect of cement composite by addition of metallic oxide nanopowders for energy harvesting in buildings. Constr Build Mater 115:576–581

Ji T, Zhang X, Zhang X, Zhang Y, Li W (2018) Effect of manganese dioxide nanorods on the thermoelectric properties of cement composites. J Mater Civ Eng 30:04018224-1–8

Jittabut P, Pinitsoontorn S, Thongbai P, Amornkitbamrung V, Chindaprasirt P (2016) Effect of nano-silica addition on the mechanical properties and thermal conductivity of cement composites. Chiang Mai J Sci 43 (2016)
Kittel C (1971) Introduction to solid state physics, 8th edn. Wiley, NW
Lan Y, Zhan Q (2017) How do urban buildings impact summer air temperature? The effects of building configurations in space and time. Build Environ 125:88–98
Levinson R et al (2010) A novel technique for the production of cool colored concrete tile and asphalt shingle roofing products. Sol Energy Mater Sol Cells 94:946–954
Matiko JW, Grabham NJ, Beeby SP, Tudor MJ (2014) Review of the application of energy harvesting in buildings. Meas Sci Technol 25
Min S, Blumm J, Lindemann A (2007) A new laser flash system for measurement of the thermophysical properties. Thermochim Acta 455:46–49
Muthusamy S, Wang S, Chung DDL (2010) Unprecedented vibration damping with high values of loss modulus and loss tangent, exhibited by cement-matrix graphite network composite. Carbon N Y 48:1457–1464
Pal A, Uddin K, Thu K, Saha BB (2019) Activated carbon and graphene nanoplatelets based novel composite for performance enhancement of adsorption cooling cycle. Energy Convers Manag 180:134–148
Ponnamma D, Ogunleye GJ, Sharma P, AlMaadeed MA (2017) Piezo- and thermoelectric materials from biopolymer composites. In: Biopolymer composites in electronics. Elsevier Inc.
Priyadarsini R (2012) Urban heat island and its impact on building energy consumption. Adv Build Energy Res 1:267–276
Rowe DM, Rowe EDM (2006) Thermoelectrics handbook: macro to nano, 1st edn. Taylor & Francis, NW
Ryu H, Yoon HJ, Kim SW (2019) Hybrid energy harvesters: toward sustainable energy harvesting. Adv. Mater. 31:1–19
Santamouris M et al (2001) On the impact of urban climate on the energy consumption of buildings. Sol Energy 70:201–216
Singh AP, Mishra M, Chandra A, Dhawan SK (2011) Graphene oxide/ferrofluid/cement composites for electromagnetic interference shielding application. Nanotechnology 22:465701
Sun M, Li Z, Mao Q, Shen D (1998) Thermoelectric percolation phenomena in carbon fiber-reinforced concrete. Cem Concr Res 28:1707–1712
Wei J, Hao L, He GP, Yang CL (2013) Thermoelectric power of carbon fiber reinforced cement composites enhanced by $Ca_3Co_4O_9$. Appl Mech Mater. 320:354–357
Wei J, Hao L, He G, Yang C (2014) Enhanced thermoelectric effect of carbon fiber reinforced cement composites by metallic oxide/cement interface. Ceram Int 40:8261–8263
Wei J, Zhang Q, Zhao L, Hao L, Yang C (2016) Enhanced thermoelectric properties of carbon fiber reinforced cement composites. Ceram Int 42:11568–11573
Wei J, Zhao L, Zhang Q, Nie Z, Hao L (2018) Enhanced thermoelectric properties of cement-based composites with expanded graphite for climate adaptation and large-scale energy harvesting. Energy Build 159:66–74
Wei J et al (2018) Thermoelectric properties of carbon nanotube reinforced cement-based composites fabricated by compression shear. Ceram Int 44:5829–5833
Wen S, Chung DDL (1999) Seebeck effect in carbon fiber-reinforced cement. Cem Concr Res 29:1989–1993
Wen S, Chung DDL (2000) Enhancing the seebeck effect in carbon fiber-reinforced cement by using intercalated carbon fibers. Cem Concr Res 30:1295–1298
Wen S, Chung DDL (2004) Effect of fiber content on the thermoelectric behavior of cement. J Mater Sci 39:4103–4106
Wen LM, Wang Q, Song P, Pan LG (2012) Effect of graphite/carbon black/iron oxide black on thermal properties of cement-based composites. Sol. Energy 21:37–40
Yamamoto Y (2006) Measures to mitigate urban heat islands. Sci Technol Trends Q Rev 18:65–83

Yao W, Xia Q (2014) Preparation and thermoelectric properties of bismuth telluride-carbon fiber reinforced cement composites. J Funct Mater 45:15134–15137

Zhang K et al (2013) Synthesis and thermoelectric properties of Bi_2O_2Se nanosheets. Mater Res Bull 48:3968–3972

Zuo JQ, Yao W, Qin JJ (2013) Enhancing the thermoelectric properties in carbon fiber/cement composites by using steel slag. Key Eng Mater 539:103–107

A Brief Review on Self-cleaning Coatings for Photovoltaic Systems

Sutha Senthil and K. R. Ravi

1 Introduction

Solar energy is one of the copious renewable sources among other renewable resources. India is receiving ample amount of solar energy with an average intensity of about 200 MW/km^2. Having geographical area of approximately 3 million km square, it is estimated that if at least 10% of this area are effectively utilized, the approximate energy production would be 8 million MW, which is equivalent to 5909 Mtoe per year (https://www.indiaenergyportal.org/subthemes.php?text=solar). Thus, for harvesting incident solar energy radiation, solar panels have been installed in the semi-arid regions, arid regions, in the wastelands, and on the rooftops.

When the solar panel is installed in outdoor environment, dust particles in the air and in the environment accumulate on the surface, which seems to reduce the conversion efficiency by 10–40%. In India, the dust deposition reduced the solar to electrical energy conversion by $0.37 \pm 0.09\%$ day^{-1} and by $5.12 \pm 0.55\%$ per particulate matter mass per meter square. It is also estimated that dust deposition reduced the annual solar power generation by minimum 8%, which led to an annual loss of approximately 8.4 crores INR (Valerino et al. 2020). In addition, the presence of rain, wind, and humidity has a significant influence on PV module's surface, as they promote deposition, accumulation, and adhesion of dust particles. The presence of moisture on the PV surface absorbs the incident solar radiation and reduces its

S. Senthil
Department of Electronics and Communication Engineering, Sri Venkateswara College of Engineering, Sriperumbudur, Chennai, India

K. R. Ravi (✉)
Department of Metallurgical and Materials Engineering, Indian Institute of Technology Jodhpur, Jodhpur, India
e-mail: ravikr@iitj.ac.in

H. Tyagi et al. (eds.), *New Research Directions in Solar Energy Technologies*, Energy, Environment, and Sustainability,
https://doi.org/10.1007/978-981-16-0594-9_7

intensity in reaching the active part of the solar cell. The light drizzling of rainfall enhances the deposition of dust, whereas the heavy rainfall has the capacity to clean the PV surface effectively. The forces responsible for adhesion due to humidity include gravitational forces, capillary forces, and electrostatic forces, which lead to power reduction up to 80%. Capillary forces and Van der Waal's forces are responsible for 98% adhesion in the case of high humidity conditions and dry conditions, respectively (Mekhilef et al. 2012; Ghosh 2020).

In India, it was reported that the annual average reduction of the solar panel power output was about 12% in Delhi and over 9% in Kolkata. In addition, different types of solar cells such as GaAs, CdTe, and perovskite have different power output responses, which were worse compared to the standard silicon panels. The high-performance perovskite-based solar panels were responded with worst average power reduction of over 17% in Delhi (https://www.weforum.org/agenda/2018/09/air-pollution-can-put-a-dent-in-solar-power/). It was described that rooftop panels in urban regions of Delhi generated 30% less power compared to the rooftop panels installed 40 km away. It was also found that the soiling in Delhi lead to power losses of 0.3 to 0.6% per day and pollution related blockage lead to production losses of about 6% annually (Nobre et al. 2016). The western parts of India, Gujarat and Rajasthan, are more prone to soiling issues, owing to the presence of Thar Desert. The fine sand particles deposit on the PV panels and reduce the incident solar radiation to reach the active part up to 50%. Due to frequent dust storms in these regions, the dust particles are transported to neighboring states such as Punjab and Haryana (Boddupalli et al. 2017; Yadav et al. 2014). Hence, it is particularly stressed that the cleaning of solar panels cover glass is mandatory.

Several methods have been proposed to maintain the efficiency of the solar panel from dust accumulation, comprising of labor-intensive cleaning, natural source of cleaning, and by motorized mode of cleaning. According to Ghosh (2020) and Ghazi et al. (2014), the solar panels in the colder region such as Himalayan region can be cleaned in six months cycle using wet type cleaning technique. The solar panels in the warm and humid conditions of southern part of India need to be cleaned in monthly cycle using wipers, and in humid equator regions of central part of India, the solar panels must be cleaned in three months cycle using anti-reflective self-cleaning coatings. The solar panels in the hot and dry regions of western part of India should be cleaned in weekly cycle using dry robotic type of device.

For cleaning the solar panels, approximately 35,000 INR would be charged for operation and maintenance (O&M) to keep the panels operating at peak performance (https://www.homeadvisor.com/). In the last decade, self-cleaning coatings have been explored for cleaning the solar panel surfaces, thereby reducing O&M costs. This chapter discusses the role of self-cleaning coatings on solar panel surfaces based on the results published in the years 2018 and 2019. Self-cleaning coatings are subdivided into two main categories: (1) Superhydrophilicity and (2) Superhydrophobicity. Superhydrophilicity is a property inspired from the Pitcher plant, where the surfaces have water contact angle of roughly 0° and the droplets slides off from the surface easily. Superhydrophobic surfaces are created by bio-mimicking the surface of lotus leaves. The combination of micro and nanostructured roughness features covered with low surface energy hydrophobic material, enable the surface to remain clean in the dirty environment. Such surfaces exhibit the characteristics such as water contact angle (WCA) larger than 150°, roll-off angle/sliding angle (SA) less than 5°, and contact angle hysteresis (CAH) less than 10°. The self-cleaning coatings offer numerous advantages over conventional cleaning methods. It includes small quantity of pure water is needed for cleaning the panels in water-scarce areas, low maintenance cost and does not go along with heavy equipment for cleaning. These unique properties of self-cleaning coatings bring in possible solutions for maintaining the solar panel efficiency over its lifespan (https://www.masterbuilder.co.in/self-cleaning-coating-solar-panel-applications/).

Despite of having advantages, there are several practical problems associated with it. First, the coating must be either superhydrophobic or superhydrophilic to minimize the need for cleaning. Second, the coating must have high optical transmission for solar radiation over the range of wavelengths from 400 to 1100 nm (https://www.sciencedirect.com/topics/engineering/solar-cell). To meet these constraints, the coating thickness must be in the range of nano or sub-micron, and the particle size must be in the nanoscale range. Third, the coating must withstand the conditions of outdoor environment; hence the coating must possess environmental and mechanical durability. These factors limit the selection of materials for the fabrication of self-cleaning coatings on solar panel surfaces. Hence, this chapter tries to answer the following questions based on the results published in the literature in the year 2018 and 2019: How do self-cleaning surfaces are created on solar panel cover glass? What are the required properties to be met? How the durability of the surfaces characterized? How did the coatings have impact on photovoltaic performance in clean and in dusty environment?

2 Wettability Properties

It is one of the fundamental properties of the solid surface, and it is determined by the equilibrium of the three phases (solid (s), liquid (l) and vapor (v)) at triple point contact area. The degree of this property is determined by the forces (adhesive and cohesive) exist between the solid and the liquid at equilibrium. For example, a drop of water on a solid substrate can bead up, while hexane on the same substrate can more likely to spread.

Contact angles of liquid with the solid substrate are a measure of wettability. In 1805, Young (1805) described the contact angles of liquid on a smooth, homogeneous surfaces, as by Eq. (1),

$$\gamma_{sv} = \gamma_{sl} + \gamma_{lv} \cos\theta_Y \quad (1)$$

where $\theta_{Y \text{ and }} \gamma$ are Young's contact angle and interfacial tension among the three phases, respectively. Young's equation was based on topographically smooth and chemically homogenous surfaces, demonstrating one equilibrium contact angle as

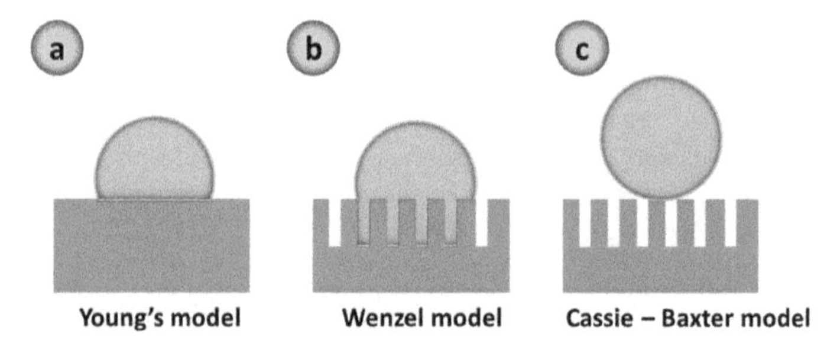

Fig. 1 Theoretical models on wettability

in Fig. 1a. However, real surfaces are topographically rough and exhibit a range of contact angles as in Fig. 1b, c.

The correlation between surface roughness and contact angle was defined by Wenzel and Cassie-Baxter equations. Wenzel equation (1936) stated that enhancing the surface roughness would enhance the wettability. For instance, a hydrophobic (hydrophilic) surface becomes highly hydrophobic (highly hydrophilic) with enhancement in topographical surface roughness. Wenzel model could be explained by Eq. (2),

$$\cos\theta_W = r\cos\theta_Y \tag{2}$$

where θ_W and θ_Y are Wenzel and Young's contact angle, respectively. r is the surface roughness ratio and can be termed as the ratio between the real and projected solid surface area. The r value is 1 and greater than 1 for smooth and rough surfaces, respectively. In Wenzel model, the liquid follows the contours of rough surfaces as in Fig. 1b.

In Cassie-Baxter model (1944), the liquid does not penetrate the hollow grooves as in Fig. 1c, and it can be explained with the following Eq. (3):

$$\cos\theta_{CB} = f_1\cos\theta_{Y1} + f_2\cos\theta_{Y2} \tag{3}$$

where f_1 and f_2 are fractional area of solid surface and air, respectively. As the contact angle of liquid against air is 180° and $f_1 + f_2 = 1$, Eq. (3) became:

$$\cos\theta_{CB} = f_1(\cos\theta_Y + 1) - 1 \tag{4}$$

It is emphasized that to avoid the diffusion of liquid through the contours of the rough surface, the geometrical characteristics of the surfaces such as size, shape, and spacing must be designed accordingly.

In the case of water as test liquid, the surface is hydrophobic when the contact angle is above 90° and hydrophilic when it is below the same value. A surface is superhydrophobic when the contact angle is above 150° and superhydrophilic when it is below 10°. The schematic representation of different water contact angles (WCA)

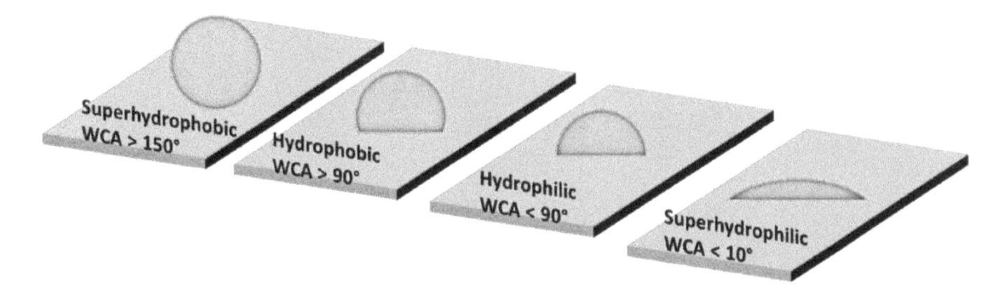

Fig. 2 Schematic representation of different water contact angles

is given in the Fig. 2. Table 1 summarizes the self-cleaning properties of different types of wettable surfaces reported in the years 2018 and 2019 for the solar panel applications (Joshi et al. 2019; Arabatzis et al. 2018; Adak et al. 2018; Zhang et al. 2018, 2019a, b; Moffitt et al. 2019; Kim et al. 2019; Adachi et al. 2018; Rahman and Padavettan 2012; Pan et al. 2019; Agustín-Sáenz et al. 2019; Li et al. 2018, 2019; Zhi and Zhang 2018; Gurav et al. 2010; Hassan et al. 2019, 2020; Nayshevsky et al. 2019; Zhan et al. 2018; De et al. 2018; Dong et al. 2018; Zhao et al. 2018; Satapathy et al. 2018; Polizos et al. 2018; Alam et al. 2018a; b; Mahmud 2009; Rombaut et al. 2019; Abu-Dheir et al. 2019; Zhu et al. 2018; Yousaf et al. 2018; Tombesi et al. 2019; Wang et al. 2019; Majhy et al. 2018; Wu et al. 2018; Bake et al. 2018; Ahmad and Eshaghi 2018; Han et al. 2016; Roslizar et al. 2019; Luo et al. 2019; Raut et al. 2011; Karunakaran et al. 2011; Milionis et al. 2016).

From Table 1, it can be seen that most of the superhydrophilic coatings were based on titania and other materials include SiO_2, SiO_2–MgF_2, SiO_2–TiO_2 and titanium oxynitride (Joshi et al. 2019; Arabatzis et al. 2018; Adak et al. 2018; Zhang et al. 2018; Moffitt et al. 2019). These include commercial coatings such as Surfashield G (Arabatzis et al. 2018) and Wattglass (Moffitt et al. 2019). In surfashield G, which composed of nanocrystalline titania and silica, the WCA value was measured as 6°. The hydrogen bonding of water molecules with the TiO_2 surface through van der Waals forces was explained for low WCA and anti-contamination of glass surfaces (Arabatzis et al. 2018). In another commercial product, Wattglass surfaces, due to the superhydrophilic nature of the silica coatings, WCA images were unable to be measured (Moffitt et al. 2019). Adak et al. (2018) fabricated porous TiO_2 films using dip coating and enhanced its transparent superhydrophilic self-cleaning properties via plasma treatment (ultra-high purified N_2 gas).

The wettability behavior of atmospheric pressure chemical vapor deposition (APCVD) deposited titanium oxy nitride films was studied with respect to different ratio of N/O, which was obtained by controlling thermal heat treatment conditions (https://www.sciencedirect.com/topics/engineering/solar-cell). There was a decrease in N/Ti ratio and increase in O/Ti ratio when increasing the heat treatment temperature. The as-deposited TiN film was hydrophobic and after heat treatment under visible irradiation, the WCA was decreased below 20°, owing to the oxidation of coatings. This behavior was suggested due to creation of surface oxygen vacancies during oxidation and the transition of Ti^{4+} to Ti^{3+} sites when the wavelength of the

Table 1 Summary of the wettability properties of self-cleaning coatings

Materials	Methods of fabrication	Wettability	Values (WCA/SA/CAH)	Ref
Wattglass–SiO_2 nps (commercial)	–	Superhydrophilic	–	Moffitt et al. (2019)
SiO_2–MgF_2	Dip coating	Superhydrophilic	<6	Joshi et al. (2019)
SurfaShield G (commercial)	Air sprayer	Superhydrophilic	6	Arabatzis et al. (2018)
TiO_2/Triblock copolymer Pluronic®	Dip coating, plasma treatment	Superhydrophilic	<5	Adak et al. (2018)
SiO_2 nps, SiH_4 gas	Hypersonic impact deposition system – plasma chamber + spray chamber	Superhydrophilic	<5	Alam et al. (2018a, b)
Titanium oxynitride (TiN_xO_y)	Atmospheric Pressure Chemical Vapor Deposition	Superhydrophilic	102 to 4	Zhang et al. (2018)
Polyimide (PI)/PDMS (porogen)	Sputtering, thermal wetting, spin coating, etching	Hydrophobic	115	Rombaut et al. (2019)
SiO_2 nps /tetraethyl orthosilicate (TEOS), isobutyltrimethoxysilane (OTES)	Solvent casting	Hydrophobic	136/–/38	Hassan et al. (2019)
fluorinated ethylene propylene (FEP)	Hot embossing	Hydrophobic	97/–/46	Nayshevsky et al. (2019)
PPFC, NSN	Roll-to-roll sputtering	Hydrophobic	105/–/–	Kim et al. (2019)
SiO_2 nps/hexadecyltrimethoxysilane (HDTMS)	Spin coating	Hydrophobic	147/–/<10	Zhang et al. (2019c)
SiO_2 nps/hexamethyldisilazane (HMDS)	Dip coating	Hydrophobic	<96	Agustín-Sáenz et al. (2019)

(continued)

Table 1 (continued)

Materials	Methods of fabrication	Wettability	Values (WCA/SA/CAH)	Ref
Hollow silica nanospheres/methyltriethoxysilane (MTES)	Dip coating	Hydrophobic	134/–/–	Zhang et al. (2019c)
Colloidal silica PL-7, silicon oil impregnation/MTMS	Dip coating	Hydrophobic	<129/–/32	Abu-Dheir et al. (2019)
silica, PEG4000, PDMS	Dip coating, CVD	Hydrophobic	145/5/–	Zhu et al. (2018)
SiO_2 nps (220 nm, 75 nm, 30 nm), n-Octadecyltrichlorosilane	Dip coating	Hydrophobic	142/–/–	Yousaf et al. (2018)
TiO_2/SiO_2 coatings/trimethoxy (1H, 1H, 2H, 2H-nonafluorohexyl) silane	Dip coating	Hydrophobic	33/–/–	Adachi et al. (2018)
SiO_2 nps, 1H, 1H, 2H, 2H Perfluorooctyltriethoxysilane	Aerosol-assisted chemical vapor deposition	Superhydrophobic	165/<1/–	Tombesi et al. (2019)
SiO_2 nps/3-aminopropyltrimethoxysilane, (AMPTS), isobuthytrimethoxysilane (OTES)	Dip coating	Superhydrophobic	158/–/2	Hassan et al. (2020)
SiO_2 nps/Heptadecafluoro-1, 1, 2, 2-tetradecyl) triethoxysilane (ACFAS)	Bar coater	Superhydrophobic	151/–/–	Wang et al. (2019)
SiO_2 nps/Hexamethyl disilazane (HMDS)	Drop coating	Superhydrophobic	151/–/–	Pan et al. (2019)
SiO_2 nps/organic silicon resin	–	Superhydrophobic	153/9/–	Zhang et al. (2019c)
SiO_2 nps/Hexamethyldisilazane (HMDS)	Dip coating	Superhydrophobic	162/6/19	Li et al. (2018)
PTFE	Electron beam evaporation	Superhydrophobic	168/1/–	Zhan et al. (2018)
PDMS/candle soot	Spin coating	Superhydrophobic	161/–/–	Majhy et al. (2018)
Hydrophobic and fluorinated SiO_2 nps/parachloro benzotrifluoride (PCBTF)	Spray coating	Superhydrophobic	166/–/–	Polizos et al. (2018)

(continued)

Table 1 (continued)

Materials	Methods of fabrication	Wettability	Values (WCA/SA/CAH)	Ref
Anti-fingerprint solution (commercial)	Heat treatment for phase separation, etching	Superhydrophobic	162/3/–	Zhan et al. (2018)
Glycidoxypropyltrimethoxysilane (KH-560), Hexamethyl disilazane (HMDS), octa decylamine (ODA), Poly (ethylene glycol) (PEG-2000), Tetraethoxysilane (TEOS)	Dip coating	Superhydrophobic	158/–/1	Zhi and Zhang (2018)
Silica nanocrystal and poly (furfuryl alcohol)	Spray coating	Superhydrophobic	164/–/–	Zhao et al. (2018)
PDMS, silica, PFOTS	Spin coating	Superhydrophobic	165/–/–	Wu et al. (2018)
Polysiloxane/silica	Spray coating	Superhydrophobic	167/1/–	Dong et al. (2018)
SiO_2 nps /1H,1H,2H,2HPerfluorooctyl-trichlorosilane (PFOTS)	Spray coating	Superhydrophobic	169/1/6.5	Bake et al. (2018)
Trimethoxymethylsilane/Cholorotrimethylsilane	Dip coating	Superhydrophobic	153/6.5/–	Ahmad and Eshaghi (2018)
SiO_2 nps/polyethylene	Dip coating	Superhydrophobic	170/3.8/–	Satapathy et al. (2018)
SketchNanoGard Sol-R (commercial)	–	Superhydrophilic and anti-static	–	https://www.sketchnano.com/sketchnanogard-self-cleaning/
Perma-Clean (commercial)	–	Superhydrophilic and anti-static	–	https://www.solarnovus.com/perma-clean-self-cleaning-anti-static-coating-for-solar-panels_N9725.html
EXCEL StayClean (commercial)	Wiping	Superhydrophilic and anti-static	–	https://www.xlcoatings.com/solar-panel-cleaner-stay-clean

irradiation was greater than 387 nm. Similarly, in the case of MgF_2 porous coatings, the decrease in WCA was observed with increase in the curing temperature (Ghazi et al. 2014). Thus, temperature has a major role in the wettability properties of the coatings.

In contrast to the superhydrophilic coatings, hydrophobic coatings were also evaluated for the cleaning of solar panel cover glass applications. Roll-to-roll sputtering technique is one of the versatile approaches for the fabrication of large-area rigid as well as flexible coatings. Kim et al. (2019) employed this technique for fabricating hydrophobic multilayer films composed of plasma-polymerized fluorocarbon (PPFC), $Nb_2O_5/SiO_2/Nb_2O_5$ (NSN). Adachi et al. (2018) dip coated the hydrophobic TiO_2/SiO_2 coatings on polycarbonate substrates. The morphology of the coatings was found to be dense to porous structure depending on the specific concentration of SiO_2 (0–40%). The self-cleaning competence of the fabricated coatings was confirmed by the decomposition of the organic pollutant (oleic acid) in the presence of UV light radiation. Once exposed to UV light, such surface recovered to become superhydrophilic, thus demonstrating the strong self-cleaning ability.

In superhydrophobic surfaces, most of the coatings were based on silica, since the material can be synthesized in sizes ranging from 5 to 2000 nm, and were optically transparent, inert to pH, and environmentally benign (Rahman and Padavettan 2012). The chemical modification of silica surfaces was done via grafting of silane coupling agents such as 1H,1H,2H,2H Perfluorooctyl-triethoxysilane, aminopropyltrimethoxysilane (AMPTS), Heptadecafluoro-1,1,2,2-tetradecyl) triethoxysilane (ACFAS), organic silicon resin, parachlorobenzotrifluoride (PCBTF), 1H,1H,2H,2H Perfluorooctyl-trichlorosilane, trimethoxymethylsilane/Cholorotrimethylsilane. Thus, WCA of such surfaces was greater than 150°, some of them were exceeded 165°. Pan et al. (2019) fabricated hierarchical micro nanostructures via the drop coating method and proposed that the surface geometrical structure had greater influence on surface wettability rather than the surface energy. It should be mentioned here that the fabrication protocol plays a significant role in determining the wettability characteristics of the coatings. It was found that the silica nanoparticles along with hexamethyldisilazane (HMDS) were hydrophobic/superhydrophobic behavior according to the acid/base catalyzed fabrication method (Agustín-Sáenz et al. 2019; Li et al. 2018; Zhi and Zhang 2018). This can be explained as follows: The structure of the silica sol is formed from the consecutive hydrolysis and condensation reactions. The hydrolysis reaction is catalyzed by the addition of acid/base catalyst. At low pH levels (highly acidic, HCl (Agustín-Sáenz et al. 2019)), the silica particle tends to form a smooth, linear chain with low cross-link density. As the pH value increases (highly basic, ammonia and amine (Zhang et al. 2018; Moffitt et al. 2019)), the silica particle tends to form rough, porous, more branched, and higher cross-link density (Gurav et al. 2010). The resultant silica surface modified with HMDS prepared via acid catalyst were smooth and hydrophobic (Agustín-Sáenz et al. 2019), whereas the surface modified with base catalyst (Li et al. 2018; Zhi and Zhang 2018) were rough and superhydrophobic in nature.

Hasan et al. (2019, 2020) investigated the wettability property of dusty hydrophobic surface at inclination, by collecting the real dust particles in the following locations every three months: (i) King Fahd University of Petroleum and Minerals in the Kingdom of Saudi Arabia and (ii) Dammam area in Saudi Arabia. The research work investigated and explored the dust removal mechanism based on transition and cloaking time. The transition time represented the droplet dwelling time during spreading and retracting on the hydrophobic surface. The cloaking time represented the droplet spreading time on dust particle surface under interfacial tension. It was proposed that when the transition time of a water droplet on a hydrophobic/superhydrophobic surface was larger than the cloaking time of the dust particles by the water droplet, the dust particles were easily collected from the surface. Nayshevsky et al. (2019) attempted to fabricate hydrophilic patterns 2-mm diameter on hydrophobic fluorinated ethylene propylene (FEP) coated solar cover glass to increase dew collection efficiency and subsequently tested the surfaces in simulated dew environment. It was observed that the nucleation of water droplets was greater on a hydrophilic surface than on hydrophobic surface. It was observed that the water collection rate increased by a factor of 2, owing to the hybrid hydrophilic (nucleation zones) and hydrophobic (high mobility zones) surfaces.

Zhan et al. (2018) employed a commercially available anti-fingerprint solution, which comprises of perfluoropolyether (PFPE) and a silane coupling agent for the fabrication of superhydrophobic films. De et al. (2018) and Dong et al. (2018) utilized a well-established commercial material such as Polytetrafluoroethylene (PTFE) and polysiloxane, respectively, for their superhydrophobic materials research. Zhao et al. (2018) employed furfuryl alcohol, which can be polymerized into poly furfuryl alcohol for environment-friendly production of superhydrophobic surfaces. Satapathy et al. (2018) fabricated porous and non-porous superhydrophobic linear low-density polyethylene (LLDPE) polymer coatings with SiO_2 as a nanofiller and attained WCA around 170°. Apart from silane, Polizos et al. (2018) utilized parachlorobenzotrifluoride for the chemical surface modification, and WCA of 166° was achieved. Wu et al. (2018) presented two kinds of re-entrant geometrical model structures, namely hexagonally triangular protrusions and rectangular micropillars, depending upon the dimensions of the silica templates. The apparent contact angles based on the geometrical model were in close relation with the observed experimental contact angles. In addition to the flat surfaces, the research group fabricated superhydrophobic polymer structures on curved concave and convex surfaces.

In summary, there is a need to achieve partially wetting state (metastable state) for superhydrophobic surfaces in addition to surface roughness; hence, WCA will reach 150°. However, there is no need for metastable state in superhydrophilic surfaces which in turn depends only upon surface roughness; thus, WCA can easily reach 0°. Such superhydrophilic surfaces are effective in cleaning hydrophobic dust particles and biological contaminants; however, they rely on the usage of large quantities of water, which is cumbersome in the desert and water-scarce areas. In India, PV panels are cleaned in two cycles in all seasons except the monsoon season where one cycle of cleaning is carried out. Pure water of around 2.5 L per panel per cycle is the average requirement to keep the panel at peak performance. Considering a 10 MWp capacity plant has around 33,000 panels, the water requirement per month will be 165,000 L for two cycles cleaning. This enormous quantity of water cannot be used to clean the panels in arid, medium to high water-scarce areas (https://solarpost.in). Such constraints are not applicable for superhydrophobic surfaces, as they can have applications in desert and water-scarce areas.

Superhydrophilic surfaces are high surface energy surfaces, thus can be easily coated on different types of substrates. The water molecules effectively spread on superhydrophilic surfaces because the surface energy of such surfaces is equivalent to water. Superhydrophilic self-cleaning surfaces in combination with photocatalyst TiO_2 remain clean from the materials whose surface energy greater than that of water. In contrast, superhydrophobic surfaces are low surface energy surfaces, thus there is a need to prepare rough surfaces, irrespective of the substrate type and shape. In recent years, superhydrophilic/superhydrophobic surfaces are fabricated with an additional property namely anti-static property. The anti-static property of the coatings is characterized by the high electrical conductivity, thereby disabling the attraction of fine dust particles in the air and prevents the dirt build-up on PV surfaces. Especially in industrial areas and desert regions, the combination of superhydrophilic/superhydrophobic surfaces and anti-static property will protect the PV panels from fly ash and sand particles. Commercial coatings such as SketchNanoGard Sol-R (https://www.sketchnano.com/sketchnanogard-self-cleaning/), Perma-Clean Solar from DryWired (https://www.solarnovus.com/perma-clean-self-cleaning-anti-static-coating-for-solar-panels_N9725.html), Excel StayClean (https://www.xlcoatings.com/solar-panel-cleaner-stay-clean) were designed to increase the efficiency of solar panels through its anti-static properties and superhydrophilic properties by preventing dust and dirt build up.

3 Optical Properties

The optical transmittance of the self-cleaning coatings is another crucial parameter for cover glass applications since the cover glass is the outermost physical structure of the solar panel. When the percentage of light transmitted through the cover glass is high, the power conversion efficiency (PCE) is high. To increase the transmission and reduce reflection, antireflection (AR) coatings serve this purpose. Therefore, self-cleaning coatings accompanied by AR property in the wavelength of 320–1100 nm are essential for cover glass applications (Pan et al. 2019).

The phenomenon of reflection occurs when the transmission of light changes its direction while traveling from one medium to another, and this change of direction is determined by a parameter—refractive index (RI) of a material/coating and the antireflection property can be achieved if the difference of RI is small. In the case of single-layer AR coatings, RI should be between air and substrate. In addition, the optical thickness (t) of the coatings should be $\lambda/4n_c$, where λ is the wavelength of incident light radiation and n_c is the RI. These kinds of coatings are challenging to fabricate; hence recent studies focused on double-layer and multilayer coatings. In the double-layer AR coating, the coating comprises a high RI layer (n_1) and a low RI layer (n_2). This kind of coating structure provides AR property only in the narrow band of wavelengths, and hence, it is not suitable for solar panel applications. Hence, multilayer structure has been efficiently employed for broadband AR property.

Another method of achieving AR property is the gradient refractive index profile structure, which can be explained as a single non-uniform layer with a gradual change of RI from the coating upper surface to the substrate lower surface. This film has broadband and omnidirectional AR property, having great potential for solar panel applications. RI of the coating can be still lowered by inducing porosity into the coatings. Here, RI is dependent on the volume fraction of porosity, and it can be lowered by lowering the solid fraction, thereby increasing the proportion of air (Raut et al. 2011). However, the RI of non-homogeneous surface rely on surface topological features, because light gets scattered by surface roughness profile. Thus, optical transmittance property and superhydrophobic self-cleaning property are considered as complementary properties in terms of surface roughness, thus achieving their coexistence remains challenging. Therefore, it is important to create suitable nanostructures smaller than 100 nm that have both competitive properties for solar panel cover glass applications (Karunakaran et al. 2011).

The detailed summary of the optical parameters of the self-cleaning coatings from the literatures is presented in Table 2. Li et al. (2019) fabricated superhydrophobic porous silica nanostructures using simple spin coating technique. The coating cycle had a great influence on the microstructure, which in turn on the properties of surfaces. The coarse and void structures of the microstructures and their optimal balance were found to be the reason for low RI (1.15) and high transmittance (99%). Similarly, Pan et el. (2019) fabricated hierarchical micro nanostructures via drop coating, whose transmittance value exceeds 95% over the wavelength region 300–800 nm. Zhang et al. (2019a) used yet another simple dip-coating technique to fabricate highly transparent (optical transmittance exceeds 95%) near superhydrophobic coatings with hollow silica nanospheres.

Joshi et al. (2019) fabricated porous SiO_2–MgF_2-based superhydrophilic coatings and observed that fine-tuning of porosity at proper thickness is a key to reduce the RI and reflectance of the glass surface. The in-situ formation of HF during sol preparation promoted etching of the film. Beyond that, the inherent low RI property of MgF_2 aided in further reducing the RI of the resultant composite film, paving the way for transparent self-cleaning coatings. Kim et al. (2019) simulated the anti-reflective properties of multilayer coatings through optical simulation using essential Macleod program. The thickness of the multilayer thin film was optimized to minimize the total reflectance at each interface. From the simulation and calculated results, it was observed that increasing the thickness of fluorocarbon layer from 30 to 70 nm, the reflectance of the multilayer decreased from 6.71% to 1.80%, owing to optical index matching of fluorocarbon. Rombaut et al. (2019) developed a mathematical model to optimize the optical properties of polyimide coatings using COMSOL multiphysics. The topographical characteristics of nanohole structures such as average spacing of 200 nm, diameter of 135 nm, and an angle of 30° were optimized from the simulation results and had a good agreement with the experimental results.

Table 2 Summary of the optical properties of self-cleaning coatings

Materials/methods	Morphology	WCA°	*T*%	RI	Thickness, Porosity, R	Ref
SiO_2 nps, 1H, 1H, 2H, 2H Perfluoro octyl triethoxysilane, aerosol assisted chemical vapor deposition (AACVD)	Multilayer films; lower layer—square shape nanoparticles—20 nm to 100 nm; middle layer—spherical shape nanoparticles—300 nm to 1 μm	165	*T*%—90 300–700 nm		–	Wang et al. (2019)
SiO_2 nps	Porous coating, spherical shape nanoparticles—23 nm	–	*T*%—94; 300–700 nm	1.3	100–200 nm, 40%, –	Moffitt et al. (2019)
Polyimide (PI)/PDMS (porogen), dewetting	Nanohole—spacing of 200 nm and diameter of 135 nm	115	*T*%—>85; 380–800 nm		60 nm, –, < 10%	Rombaut et al. (2019)
SiO_2 nps/silane (3-amino propyl tri methoxy silane, iso buthy tri methoxy silane, dip coating	Silica nanoparticles—30 nm	158	*T*%—80; 300–800 nm			Gurav et al. (2010)
SiO_2 nps /Hepta decafluoro-1, 1, 2, 2-tetradecyl) triethoxy silane (ACFAS), bar coater	Silica nanoparticles	151	*T*%-92.1 @ 500 nm			Wang et al. (2019)
SiO_2 nps/hexadecyl trimethoxysilane (HDTMS), spin coating	Nano porous structured film	147	T_{max}%—99 @ 580 nm. T_{avg} % > 95%; 432–900 nm	1.15	146 nm,–, 1%	Li et al. (2019)
SiO_2 nps/HMDS, drop coating	Micro-nano surface structures	151	*T*%—90.1 to 97%; 350–800 nm		–	Pan et al. (2019)

(continued)

Table 2 (continued)

Materials/methods	Morphology	WCA°	*T*%	RI	Thickness, Porosity, R	Ref
SiO_2 nps/(HMDS), dip coating	Monolayer porous film/bilayer with inner denser and outer porous film	<96		1.43	143 nm, 46%, –	Agustín-Sáenz et al. (2019)
SiO_2 nps/organic silicon resin	Micro-nano surface structures	153	*T*%—91.4%; 400–800 nm		–	Zhang et al. (2019a)
Hollow silica nano spheres/MTES, dip coating	Hollow silica spheres	134	Max—99.24, avg—95.61	2	–	Zhang et al. (2019b)
SiO_2 nps/MTMS hydrolysate, cast coating	Silica nps	<110	89.3–90.3; 300–1700 nm		3 μm, –, –	Mahmud (2009)
Hot embossing	Random microtextures	156		1.34	–	Roslizar et al. (2019)
Colloidal silica PL-7, silicon oil impregnation/MTMS, dip coating	Porous like connected structures of size 50 nm	< 130	90%, roughness—0.65 μm	1.21	0.9–1.4 μm, –, –	Abu-Dheir et al. (2019)
SiO_2-MgF_2, dip coating	Porous structures	<6	*T*%—96%; 4.7% and 4.9% imp; 300–1500 nm	1.28–1.39	69–134 nm, 10–41%, <5%	Joshi et al. (2019)
SiO_2 nps/HMDS, dip coating	Silica nanoparticles	162	*T*% > 99.7; AR		21 nm, –, –	Li et al. (2018)
PTFE, electron beam evaporation	Two layers—inner denser layer and outer porous layer	168	AR T_{max}—98% and T_{avg}—97%	1.13 and 1.18		De et al. (2018)
PDMS, candle soot, spin coating	100 nm size rough grooved structure	161	*T*%—80			Majhy et al. (2018)
silica, PEG4000, PDMS, dip coating, CVD	Groovy linear chain network	145	T_{max} %—96.44; 350—600 nm			Han et al. (2016)

(continued)

Table 2 (continued)

Materials/methods	Morphology	WCA°	*T*%	RI	Thickness, Porosity, R	Ref
SurfaShield G, air sprayer	Nanoparticles	6	*T*% increase compared to bare glass		150 nm, –, –	Arabatzis et al. (2018)
TiO_2/Triblock copolymer Pluronic® plasma treatment	Porous film	< 5	Roughness—1.08 to 1.30 nm	1.27	63–82 nm, 73–85%, –	Adak et al. (2018)
Hydrophobic and fluorinated silica nanoparticles/PCBTF, spray coating	Nanoparticles	166	*T*% ~ to uncoated one; 190–1200 nm			Polizos et al. (2018)
Anti-fingerprint solution, phase separation	Granular with a pore size of 10–30 nm	162	*T*% > 93.2		500–1000 nm, –, < 10%	Zhan et al. (2018)
Glycidoxypropyltrimethoxysilane (KH-560), HMDS, ODA, PEG-2000), TEOS, dip coating	Crosslinked network of nanopores of size 150 nm	158	*T*%–93, roughness—3.98 nm;	1.34	244 nm, –, –	Zhi and Zhang (2018)
Silica nps, silane gas, aerosol spray deposition system	Porous	< 5	*T*%—98, rms roughness—28.7 nm; 300–1400 nm	1.15, 1.3	133 nm and 115 nm, 60% and 30%, –	Alam et al. (2018a)
SiO_2 nps, silane glass, hypersonic impact deposition system - plasma chamber + spray chamber	Porous	< 5	Single (96.5) double (98); 300—1200 nm	1.35, 1.2	115 and 130 nm, –, –	Alam et al. (2018b)
Silica nanocrystal and poly (furfuryl alcohol), spray coating	Spherical shaped particles of size 800 nm	164	*T*% > 80			Zhao et al. (2018)
PDMS, silica, PFOTS, spin coating	Hexagonal porous structures with triangular protrusions or hexagonal rectangular micropillars	165	*T*% > 80%; 400–800 nm			Wu et al. (2018)

(continued)

Table 2 (continued)

Materials/methods	Morphology	WCA°	*T*%	RI	Thickness, Porosity, R	Ref
Polysiloxane/silica, spray coating	Cross linked nanoparticles	167			10–15 μm, –, –	Dong et al. (2018)
SiO_2 nps /1H,1H,2H,2HPerfluorooctyl-trichlorosilane (PFOTS), spray coating	Micro-nano surface structures	169	*T*%–70 at 400 nm, 85% at 800 nm; roughness—120 nm			Bake et al. (2018)
SiO_2 nps (220 nm, 75 nm, 30 nm), n-Octa decyltrichlorosilane, dip coating	Assembly of silica nanoparticles into mono and multi layers with voids in between	142	Roughness—109.6 nm, *T*% ~ bare glass	1.404	716.10 nm, –, –	Yousaf et al. (2018)
Titanium oxynitride (TiN_xO_y), Atmospheric Pressure CVD (APCVD)	Needle-like or flocculent-like surface morphology	102 to 4	*T*%—27		500 nm, –, 68%	Zhang et al. (2018)
TiO_2/SiO_2 coatings/trimethoxy (1H,1H,2H,2H-nonafluorohexyl) silane, dip coating	Dense to porous structure	33	*T*% > 87; 400–800 nm			Rahman and Padavettan (2012)
Trimethoxymethylsilane/Choloro trimethylsilane, dip coating	nps 40 to 80 nm	153	T%–95% bare glass–91%; 380–760 nm	1.23	690 nm, –, 8.84%	Ahmad and Eshaghi (2018)
Dewetting	Average nps ~31.4 to ~78.7 nm;	–			3 to 8 nm, –, <3%	Luo et al. (2019)
Plasma-polymerized fluorocarbon (PPFC), $Nb_2O_5/SiO_2/Nb_2O_5$ (NSN), roll-to-roll sputtering system	Uniform dense structure	105			PPFC (70 nm), Nb_2O_5(47 nm), SiO_2 (20 nm), –, 5.5%	Kim et al. (2019)

In the context of vacuum-based sophisticated techniques, the process parameters can be varied to achieve the final desirable properties of the coatings. Alam et al. (2018a, b) fabricated porous silica nanoparticulate superhydrophilic coatings using an aerosol hypersonic spray deposition system and achieved transmittance greater than 95%. De et al. (2018) fabricated superhydrophobic antireflection coatings based on PTFE on glass substrates via electron beam evaporation. The porous gradient coatings remained anti-reflective over the range of incidence angles 0° to 60° with the average transmittance of 97%. It was explained that porous nature of the film structure deposited at high deposition rate and low RI were responsible for the AR property of the coatings. Huh et al. (2019) fabricated an anti-reflective film with moth-eye pattern on PV surfaces using nanoimprint lithography. The fabricated surface was placed in outdoor atmosphere in 7 months and observed that the film exhibited 5% increase in transmittance and power conversion efficiency of 2.85%, when compared to control surface.

From Table 2, it is understood that the optical properties of the coatings depend on the morphology of the structure used, type of fabrication employed, thickness, refractive index, porosity of the coatings. It is also evident that this rare combination of properties can be achieved irrespective of the sophisticated, non-sophisticated, vacuum, and non-vacuum-based fabrication techniques. In current years, sol gel technology is combined with dip coating for the fabrication of anti-reflective coating for large-area surfaces. The most used materials are silica, titania, zirconia, magnesium fluoride, zinc oxide, and silicon nitride with refractive indices of 1.46, 2.43, 2.15, 1.32–1.38, 2.00 and 2.05, respectively (Sarkın et al. 2020). These materials with inherent properties and optimal structural variations lead to efficient anti-reflective films suppressing reflectance. Alternatively, sputtering techniques and CVD, lithography, screen printing, and roll-to-roll methods have been used in AR film depositions (Shanmugam et al. 2020). Most of the commercial PV panels are coated with MgF_2-based anti-reflective coating deposited by vacuum-based technology. This material is chosen for its low refractive index of 1.34–1.39, insoluble when deposited on hot substrates, good adhesion to glass substrates, remained transparent (for layers up to 1 μm thickness) from UV (1200 Å) to IR (~8 μm) region (https://materion.com). The MgF_2 coatings reduced the reflectance of the surface to 1.3%, whereas the uncoated surface exhibited the reflectance of 4% (www.photonics.com). ZnS/MgF_2-based AR coatings on PV surface fabricated by glancing angle deposition improved the power conversion efficiency from 9.91 to 13.3% (Oh et al. 2020).

4 Durability Properties

Natural self-cleaning surfaces can retain their functional properties throughout their lifetime by self-healing. One of the major drawbacks of artificial self-cleaning coatings is that they lose their properties when exposed to harsh environmental conditions. In most cases, the self-cleaning coatings lost its functionality due to the peeling of coatings from the substrate, scratching of coatings, application of load on the coated substrate causing indentation, the impact of raindrops, damage of coatings due to erosion caused by sand particles, dew, humidity, temperature, and UV radiation. It is apparent that if the coatings are not durable enough, they are no longer qualified for long-term practical applications. This section and Table 3 summarizes the diverse characterization techniques employed in the recently published articles to evaluate the durability performance of self-cleaning coatings.

4.1 Abrasion Test

In the abrasion test, the coatings can be easily abraded if the adhesion between the underlying substrate and the fabricated coating layers are poor. Zhang et al. (2019a) evaluated the abrasion resistance of the coatings according to ISO 11998. Both sides of the coatings exhibited negligible degradation in properties without any macroscopic scratch, even after 1000 cycles. The reason was ascribed to its closely packed hybrid silica nanostructures and the strong bonding of the coatings with the substrate. Tombesi et al. (2019) investigated the friction resistance of the three-layer silica coating structure fabricated through AACVD using a sandpaper abrasion test. It was observed that even after five cycles which were loaded with 100 g of weight, the silica stacking surface retained superhydrophobic property. However, beyond five cycles, the coating gradually lost its superhydrophobic nature, owing to the damage of the micro/nanostructure. It was also suggested that this problem can be mitigated by (i) introducing an adhesive layer between the substrate and the coatings (ii) incorporating high mechanically durable precursors in the coatings.

To determine the friction resistance of fabricated coatings, Zhan et al. (2018) used an eraser (area 1 cm^2 loaded with a weight of 500 g) as a frictional surface, which was rubbed at 10 mm/s. The nanostructures were damaged by the eraser, causing WCA less than 112° and SA as 42° after 30 cycles. However, a significant improvement in the abrasion resistance was observed when the heat treatment temperature and its duration were increased to 700 °C for 2 h. The surface nanostructures were reorganized with heat treatment; thus, even after 50 abrasion cycles, the coatings remained sticky superhydrophobic and transparent, even though AR property was lost.

Another abrasion test, sandblasting was carried out by Arabatzis et al. (2018) according to IEC 60668-2-68. The research team utilized four combinations of vital parameters such as types of dust particles, its dimensions, speed of wind, and its time duration, to test the mechanical robustness of the coatings. The coating remained stable when sand of diameter 100 μm used and less than 5 m/s wind speed and less than 8 h of testing duration. Beyond that, the stability of the coating was poor, ascribed due to the change in the coating structure. Polizos et al. (2018) assessed the durability of the spray-coated silica nanoparticles by performing falling sand abrasion testing (sand particles 55 μm diameter) according to ISO 12103-1 standard. The optical transmittance of the fabricated samples was reduced after abrasion, however, the value remained higher compared to the uncoated surface. Further, the curing time of the coatings was optimized between 30 and 60 min for better durable resistance, based on the abrasion model. Yet another falling sand abrasion test and sandpaper abrasion test was performed by Wu et al. (2018) to analyze the scratch resistance of the hexagonally ordered porous SU-8 photoresist-based polymer film surface. For the sandpaper scratch test, 1500-mesh sandpaper loaded with 500 g weight was dragged in one direction at 1 cms^{-1}. The coating exhibited no obvious change after falling sand and sandpaper scratch test. The inherent flexibility and mechanical robustness of SU-8 photoresist-based polymer film retained its superhydrophobic property even after seven bending cycles. The results showed that the inherent property of the materials should be considered while selecting the materials for the fabrication of durable solar application coatings.

The durability of the superhydrophilic TiO_2/SiO_2 coatings was investigated using Taber abrasion test, using force of 4.9 N at 60 RPM. From the haze measurements, it was observed that 43% and 35% of haze percentage were observed for uncoated and coated samples, respectively. This increase in haze percentage in the uncoated samples was due to the severe scratches and uneven surface structures, which in turn promoted light scattering (Adachi et al. 2018). Similarly, Bake et al. (2018) conducted abrasion test using 240 GRIT size sandpaper on spray-coated silica nanostructures, and it was observed that the coating lost its self-cleaning properties after few cycles owing to the damage of the surface roughness. Even though the coating remained superhydrophobic after five cycles, there was 20.3% reduction in roll-off angle and 284.6% increase in CAH.

4.2 Crockmeter Test

Rombaut et al. (2019) evaluated the durability of the polyimide coatings via crockmeter test using microfiber cloth with a force of 9 N over 2 cm^2 for certain number of cycles. After more than 500 passes, the polyimide nanostructures remained unchanged, owing to the inherent mechanical strength of polyimide and the presence of an adhesion promoter (APTES) between the coating material and the substrate. Similarly, Satapathy et al. (2018) examined the mechanical strength of LLDPE/SiO_2 superhydrophobic coatings by rubbing microfiber cloth with a force of 0.5 N. The coating remained unchanged up to 42 number of cycles, due to the inherent material properties of LLDPE polymer matrix and its porous surface of LLDPE/SiO_2. However, the roll-off angle was increased from 8° to 90°, exhibiting sticky superhydrophobic nature.

4.3 Pencil Hardness Test

Pencil hardness test is performed to assess the mechanical strength of the thin films as per ASTM D3363. If the coating bonds with the substrate through van der Waals interactions, the film cannot withstand 1B pencil hardness. It has been reported that a good index for durable self-cleaning surface should withstand at least 2H (Milionis et al. 2016). Here, a specific hardness pencil at a 45° angle loaded with an applied force is pressed strongly and moved along with a constant speed. In the research work by Zhi and Zhang found that dip-coated nano porous silica surface withstood 4H pencil hardness test. However, after the 4H pencil test, the transmittance decreased below 95%, since the porous structures were covered by the nanoparticles (Zhi and Zhang 2018). The scratch resistance of the porous silica-based anti-reflective self-cleaning nanostructured coatings were also measured by other researchers, and the corresponding scratch hardness of scale value greater than 3H were obtained for outdoor applications (Joshi et al. 2019; Adak et al. 2018; Zhao et al. 2018; Li et al. 2019; Zhu et al. 2018).

4.4 Nanoindentation

In addition to the pencil hardness, the mechanical properties of the superhydrophobic coating were evaluated using nanoindentation technique. Mechanical durability of dip-coated TiO_2 surfaces subjected to plasma treatment by Adak et al. (2018) was measured using nanoindentation. For a given applied load, it was observed that the depth of indentation was greater for untreated films than N plasma treated films, indicating the beneficial effect of plasma treatment on the mechanical robustness of the coating. The formation of O–Ti–N bonds after plasma treatment significantly increased the mechanical properties of the coatings.

4.5 Tape Test

Joshi et al. (2019) conducted crosshatch tape test as per ASTM F2452 to determine the adhesion strength of the superhydrophilic porous SiO_2/MgF_2 coatings. It was observed that small fragments were removed off at intersections. Zhao et al. (2018) conducted cross-cut tape adhesion test according to ASTM D3359, equivalent to ISO 2409 on tempered superhydrophobic organic–inorganic silica films fabricated by sequentially reinforced additive coating (SRAC). No separated regions were stripped off from the superhydrophobic film, indicating better adhesion of the coating with the substrate. The high mechanical strength of the coating was suggested due to the inclusion of Na ion during the tempering process and enhancement of inter-particles bonding by annealing the film at 670 °C for 40 min. Satapathy et al. (2018) determined the adhesion strength of the porous superhydrophobic nanocomposite coatings using an adhesive tape peel-off test. It was observed that porous superhydrophobic LLDPE/SiO_2 coatings had better adhesion resistance than non-porous coatings. Further, LLDPE/SiO2 coatings were stronger than LLDPE coatings, attributed to the presence of SiO_2 nanoparticles. On the downside, the presence of SiO_2 nanoparticles also contributed to the easy abrasion of the coated surface. This limitation was overcome by the research team by the fabrication of porous LLDPE/SiO_2 coatings, thus incorporating both better adhesion and abrasion resistance in the coating.

4.6 *Water Impact Test*

Few researchers performed water impact test to understand the impact resistance of the coatings. The water droplets were impacted on dip-coated porous silica coatings, with a velocity $v = 2.7$ m/s. The water jet damaged the porous surface and decreased the transmittance to 94% (Zhi and Zhang 2018). Similarly, water droplets were impacted with a velocity of 4.25 m/s on spray-coated silica nanoparticles surface, and it was evident that the wettability and transmittance properties of the coatings did not show any significant changes. This is because the silica nanoparticles were not removed during the water jet test and did not degrade the surface functionalized CF_3 groups (Bake et al. 2018).

4.7 *Chemical Stability*

It is yet another important test that should be carried out in determining the long-term stability of self-cleaning coating when they are subjected to acidic, alkaline, neutral conditions. In this test, the coating is immersed in corrosive solutions with a wide range of pH levels, and the desirable properties are evaluated. Tombesi et al. (2019) performed corrosion testing of AACVD deposited three-layer silica coatings by immersing in acidic and basic solutions for 72 h. The WCA and SA of the surface remain unaffected until 48 h. Beyond 48 h, there was a slight decrement in WCA and increment in SA. Such changes were prominent in alkaline environment owing to the increase in the generation of the surface hydroxyl groups. Likewise, the films were immersed in solvents such as ethanol and toluene for 5 h. There was no obvious change in WCA and SA. In the same way, Zhu et al. (2018) also immersed the porous PDMS modified silica coatings in the acidic, neutral, and alkaline environments for 10 days separately. Alike (Tombesi et al. 2019), it was found that WCA did not change significantly when immersed in acidic and neutral environments. In the alkaline environment, there was a 20° decrease in WCA as the time duration increased to 10 days, owing to the change in surface structure of the coatings contributed by the reaction of SiO_2 nanoparticles in alkaline medium (Zhu et al. 2018).

Adak et al. (2018) checked the adhesion strength of the porous superhydrophilic TiO_2 coatings by immersed in hot 5% saline water for 60 min, according to the test for coating on solar glass covers (Milionis et al. 2016). The coating remained optically transparent and superhydrophilic before and after such experiments. The chemical stability of the porous LLDPE/SiO_2 superhydrophobic coatings fabricated by dip coating was studied by Satapathy et al. (2018). In strongly acidic conditions, the acid solution entered the pores and polymer matrix swelled leading to the loss of wettability properties within 15 min of immersion. However, the WCA remained unchanged in weak acidic as well as alkaline solution (pH = 6–8) and got damaged in moderate alkaline solution (pH = 9).

The hexagonal structures fabricated by Wu et al. (2018) remained superhydrophobic in acid and alkaline environments. The chemical stability of the PDMS-based transparent self-cleaning surface fabricated by Majhy et al. (2018) was tested by dipping the surface in various solvents such as isopropyl alcohol, acetone, ethanol, and phosphate buffer solution (PBS) for 10 min. The coating remained superhydrophobic in isopropyl alcohol, acetone, and ethanol solvents. In PBS, there was a 10° decrease in WCA and no prominent change in the WCA after dipping in benzoic acid and acetic acid, indicated that the surface can be used in different chemical environments.

4.8 Thermal, UV and Environment Stability

When coatings are subjected to outdoor environments for prolonged time, a wide variety of physicochemical processes called 'weathering' will take place. Ultraviolet (UV) radiation and moisture are the critical factors for coating degradation under weathering exposure. During UV absorption process, when the amount of energy absorbed by the coating material is greater than the bond energy of material, there will be a significant degradation in the optical and wettability properties (Mahmud 2009). Hence, UV resistant coatings are one of the highly desirable for solar applications. The AACVD deposited three-layer silica-based superhydrophobic coatings fabricated by Tombesi et al. (2019) were subjected to a temperature of 300 °C and then at 400 °C for 5 h. The coatings resisted prolonged thermal stress and remained in self-cleaning superhydrophobic nature. In addition, the samples exhibited excellent UV stability for 72 h. Environmental weathering stability was performed on hybrid SiO_2–MgF_2 coatings by Joshi et al. (2019) using a weatherometer following ASTM standard G155. The porous superhydrophilic films showed high durability with less than 0.5% loss in transmission.

The research group of Agustín-Sáenza et al. (2019) performed damp-heat (DH) accelerated aging test for porous silica coatings according to IEC 62108 (1000 h, 85 °C and 85% RH). The porous AR coatings sintered at 550 °C exhibited transmittance loss of 1.8%. The influence of sintering temperature over structural relaxation to form stable structure was explained from the reported results. However, the coating lost its AR property as RI increased after aging due to the meso-structural variation. Zhao et al. (2018) conducted an outdoor test by placing the fabricated films in the open air at 45° angle in Wuhan, China for 3 months. The weather slightly decreased the WCA to ~154° after 3 months, while maintaining superhydrophobicity. Zhang et al. (2019a) evaluated the humidity resistance of the hydrophobic, AR hollow silica nanosphere-based coating by a highly accelerated temperature and humidity stress test (HAST). This test was carried out at 121 °C and 97% RH, where 24 h HAST is equivalent to a conventional 1000 h damp-heat (85 °C/85% relative humidity) test. The absorption of moisture and corresponding shrinkage in the coatings exhibited blue shift, and a relative degradation of 0.56% in T_{avg}, owing to the change in optical properties of the coatings. However, the hybrid coatings remained superior over the control coatings, regarding the excellent humidity resistance after 120 h HAST.

Zhang et al. (2019b) conducted two durability tests, thermal shock/humidity series (THS) and high-low temperature cycle/electron beam (EB) radiation/vacuum ultraviolet (VUV) radiation series (HEVS) test for their dust repellent hydrophobic cast coated silica coatings. In the THS test, the samples were dipped in liquid nitrogen for 10 min and heat treated at 100 °C oven for 5 min. The above process was repeated up to six cycles and subjected at 60 °C and RH of 92% for 96 h. In HEVS test, the samples were placed in a rapid temperature change test chamber to undergo 200 times of high-low temperature cycle (high temperature: 85.0 °C, low temperature: −145 °C, temperature change rate: 11.5 °C/min). Further in EB test, the samples were exposed to 1 meV EB with dosage of 1×10^{15} e/cm^2. Thereafter, the samples were exposed to UV radiation, equivalent to 3 months of UV dose on Mars. The optical transmittance of 92% remained same after THS test, and WCA was slightly reduced by 3°. No change in coating structure was observed for HEVS test. Further, VUV radiation led the sample surface to be light yellow and superhydrophilic. Further, there was a decrease optical transmittance due to change in the decomposition of organic coatings.

The change in optical transmittance of surfaces in natural environments was studied by leaving for one week in an outdoor environment (Abu-Dheir et al. 2019). The optical transmittance of the surface with silicon oil was significantly reduced after the outdoor tests, due to the settlements of dust particles on the surface. It was summarized that the use of silicon oil impregnation improved the optical transmittance significantly; however, it reduced the WCA and had poor outdoor stability.

A variety of methods have been used to evaluate the durability of self-cleaning coatings for solar panel cover glass ranging from chemical stability, thermal stability, abrasion resistance, water jet test, pencil hardness, UV and weather resistance, nanoindentation, sandblasting, and tape test. However, the lack of standardization leads to the comparison of different reports impossible. From the reported articles, it can be said that abrasion resistance of the surfaces must be carried out, and the surface must be exposed to commercial standardized abradants of different materials with variable hardness and texture.

Another important factor in characterizing the durability is that considering the changes in dynamic contact angles such as roll-off angle or tilting angle and contact angle hysteresis. Reporting the static contact angle does not define the abrasion resistance of the coatings since the advancing angles always remain high. However, such abrading effect is reflected promptly in receding contact angles, which in turn have undesirable effect in contact angle hysteresis. The optical durability of the self-cleaning coatings can be assessed according to the military specification MIL-C-675C and MIL-C-48497A. The tests required to determine the durability of anti-reflective films corresponding to MIL-C-675C are salt solubility, humidity, salt spray

Table 3 Summary of durability properties of self-cleaning coatings

Materials/methods	Corrosion testing	Heat treatment/aging test/UV treatment	UV treatment	Mechanical durability	Ref
SiO2 nps, 1H, 1H, 2H, 2H Perfluoro octyl triethoxysilane, AACVD	No change in CA and SA	After 5 h the sample remained superhydrophobic. No changes in wettability after UV treatment		**Sandpaper abrasion test**: At 5 cycles, the surface remained superhydrophobic. After 10 and 15 cycles, WCA were 138° and 120°, respectively. After 15 cycles, the coatings became hydrophobic	Tombesi et al. (2019)
Polyimide (PI)/PDMS (porogen), dewetting				**Crockmeter test**: nanostructure remains practically unchanged	Rombaut et al. (2019)
SiO_2 nps/hexadecyl trimethoxysilane (HDTMS), spin coating				**Pencil hardness test**: After the 3H pencil hardness test, no significant scratches were observed	Li et al. (2019)
SiO_2 nps/(HMDS), dip coating		**Damp-heat (DH) accelerated aging test**: WCA—between 80° and 90°			Agustín-Sáenz et al. (2019)

(continued)

Table 3 (continued)

Materials/methods	Corrosion testing	Heat treatment/aging test/UV treatment	UV treatment	Mechanical durability	Ref
Hollow silica nano spheres/MTES, dip coating		**Highly accelerated temperature and humidity stress test (HAST)**: 0.6% decrease of T% after 120 h		**Abrasion test**: 0.01% loss in T_{avg} and the coatings structure remained stable after 1000 cycles	Zhang et al. (2019a)
SiO_2 nps/MTMS hydrolysate, cast coating		**Thermal shock/humidity series (THS) test**: The transparency did not change. Decrease in WCA. **High-low temperature cycle/electron beam radiation/vacuum ultraviolet radiation series (HEVS) test**: no change in appearance after 200 cycles. The samples were light yellow in color without cracks. Tavg slightly declined from 92.89% to 91.62%. WCA obviously reduced			Zhang et al. (2019b)

(continued)

Table 3 (continued)

Materials/methods	Corrosion testing	Heat treatment/aging test/UV treatment	UV treatment	Mechanical durability	Ref
SiO_2–MgF_2, dip coating			**UV and weather stabilities**: showed high durability and $T\% < 0.5\%$	**Pencil hardness (ASTM D3363) and crosshatch tape test (ASTM F2452)**: Pencil hardness—4H and adhesion rating—4	Joshi et al. (2019)
PDMS, candle soot, spin coating	The coating remained superhydrophobic in IPA, acetone, ethanol solvents. In PBS, 10° decrease in WCA. no change in the WCA in benzoic acid and acetic acid			**Abrasion test**: The WCA showed less than 10% change in the WCA. **Water jet test**: The coating remained superhydrophobic with WCA of 160° ± 1°	Majhy et al. (2018)
Silica, PEG4000, PDMS, dip coating, CVD	**Acid and alkali resistance test**: No variation in WCA in acidic and neutral but shifted down in alkaline environment			**Pencil Hardness test**: Hardness value decreased with the content of PEG from 1 to 6H. **Smudge test**: Effective removal of smudge	Joshi et al. (2019)

(continued)

Table 3 (continued)

Materials/methods	Corrosion testing	Heat treatment/aging test/UV treatment	UV treatment	Mechanical durability	Ref
SurfaShield G, air sprayer		**In door thermal cycling**: after 200 cycles, drop in Tavg from 94.94% to 93.35%, but higher than uncoated glass (92.58%). Morphology remained unaffected	**UV test**: T% decreased to 92.20%. The morphology of the coating was not affected	**Sand blasting**: Poorly stability as the sandblasting parameters became worse	Arabatzis et al. (2018)
TiO_2/Triblock copolymer Pluronic®, plasma treatment	**Adhesion strength**: There is no peeling observed	**Open air environment**: for over one month. Negligible decrease in optical transmittance	**UV test**: samples were UV stable	**Pencil Hardness**: 3H for single layer and 5H for double-layer coatings. **Nano indentation**: For an applied load > 40 μN, increase in the indentation depths for untreated films. Hardness and elastic modulus for untreated films were 0.86–1.17 GPa and 38.8–63.4 GPa, respectively. After plasma treatment, the values increased to 0.92–2.02 GPa and 30.4–47.7 GPa, respectively	Adak et al. (2018)

(continued)

Table 3 (continued)

Materials/methods	Corrosion testing	Heat treatment/aging test/UV treatment	UV treatment	Mechanical durability	Ref
Hydrophobic and fluorinated silica nps/PCBTF, spray coating				**Abrasion resistance**: T% of the abraded sample is slightly lower than that of the non-abraded sample but higher than the uncoated sample	Polizos et al. (2018)
Anti-fingerprint solution (commercial), phase separation				**Abrasion resistance**: WCA < 112° and SA increased to 42° after 30 cycles. After 50 abrasion cycles, WCA decreased to 151°. AR effect of the coating decreased after abrasion; $T_{avg} > 91\%$	Zhan et al. (2018)
Glycidoxypropyltrimethoxysilane (KH-560), HMDS, ODA, PEG-2000), TEOS, dip coating	Stable in aqueous HCl and aqueous NaOH solutions			**Pencil hardness test**: 4H. T% decreases apparently. However, the surface is still AR. **water drop impact test**: WCA ~ 150°; T% reduced to 94%	Hassan et al. (2020)
Silica, furfuryl alcohol, spray coating		**Outdoor test:** placed in open air for 3 months. WCA decreased to ~ 154°		**Pencil hardness**: Stable even after 4H. **Tape adhesion test**: stable	Zhao et al. (2018)

(continued)

Table 3 (continued)

Materials/methods	Corrosion testing	Heat treatment/aging test/UV treatment	UV treatment	Mechanical durability	Ref
PDMS, silica, PFOTS, spin coating	**Corrosion test**: excellent liquid-repellency after 96 h			**Sand Abrasion test**: no obvious change in WCA. **Flexibility test**: remained superhydrophobic after 7 bending cycles	Wu et al. (2018)
SiO_2 nps /1H,1H,2H,2HPerfluorooctyl-trichlorosilane (PFOTS), spray coating		**Thermal stability**: Annealing at temp greater than 300 °C reduced WCA to 137°. Annealing at 300 °C improved T% by 6%		**Water jet test**: No change in WCA, SA, CAH. **Abrasion test**: No change in WCA after 5 cycles, while 20.3% reduction in SA and 284.6% increase in CAH	Bake et al. (2018)
TiO_2/SiO_2 coatings/trimethoxy(1H, 1H, 2H, 2H-nonafluorohexyl) silane, dip coating				**Abrasion test**: For pristine haze % increased from 0.41% to 43%. For FAS-deposited coatings, haze % increased from 0.33% to 35%	Adachi et al. (2018)
Trimethoxymethylsilane/Choloro trimethylsilane, dip coating		**Thermal stability:** stable up to 450 °C and above this temperature, hydrophobic became superhydrophilic at 600 °C			Ahmad and Eshaghi (2018)

(continued)

Table 3 (continued)

Materials/methods	Corrosion testing	Heat treatment/aging test/UV treatment	UV treatment	Mechanical durability	Ref
SiO_2 nps embedded linear low-density polyethylene (LLDPE), dip coating	**Chemical stability**: Stable	**Thermal stability test: stable** up to 100 °C and at 120 °C, turned into hydrophobic (125.1 ± 2.1°)		**Abrasion resistance**: non-porous coating damaged after 10 cycles, while porous coating was stable up to 30 cycles. **Tape peeling test**: stable at least for 8 cycles. **Water jet impact test**: No damage for the flow rate of 2.6 m/s	Satapathy et al. (2018)

fog, severe abrasion according to MIL-E-12397, moderate abrasion, and adhesion. The durability requirements for single-layer and multilayer interference coatings are assessed according to the specification MIL-C-48497A. The necessary tests to be carried out are adhesion, humidity, moderate abrasion, temperature, solubility, and cleanability. The optional tests include severe abrasion, salt, and water solubility.

5 Photovoltaic Performance

Deposition of particulate matters on the cover glass, decline the overall power generation of PV panels on a periodical manner. The degree of power output reduction depends on the properties of the deposited dust such as specific mass and size and the geographical location and local environment of the solar power plants. The larger the mass and smaller the size of the dust particles, block incident solar radiation and significantly degrade the power conversion efficiency of the solar panel.

A field study was carried out with dusts that were collected from multiple solar panels located at Indian Institute of Technology (IIT), Gandhinagar in 2016 between January and March. Analysis of the dust samples revealed that 92% were from atmospheric dust particles and the remaining was composed of organic carbon, black carbon, and ions produced from combustion related human activities. It was also found that atmospheric dust had less influence in reducing solar energy power generation compared to combustion related dust, owing to their smaller size and sticky nature (Bergin 2017).

Interfacial interaction mechanism between these dust particles and the glass surface involves several types. Firstly, long-range attractive interactions, such as van der Waals, gravitational force, and magnetic attraction, which are responsible for settling the particles and forming contact area for adhesion on the glass surface. Gravitational force is responsible for settling large particles and van der Waals forces dominate for smaller particles. Secondly, short-range interactions, such as chemical bonds and hydrogen bonds, contribute to forming adhesion, after establishing the contact area for adhesion (Darwish et al. 2015). In this section, the photovoltaic performance of various self-cleaning coatings soiled with different types of dust particles was discussed.

Wang et al. (2019) analyzed the effect of dust accumulation on the superhydrophobic silica coatings on the cover glass of PV modules. Here, the natural dust deposition was simulated using laterite particles (25 μm diameter), and the different amounts of sodium chloride were dry-mixed with laterite particles. It was observed that irrespective of the presence of the coatings, dust deposition affected all the samples. However, maximum power reduction observed was 10.9% and 3.8% for the uncoated and coated samples, respectively. Irrespective of the salt content, the superhydrophobic film surpasses the power efficiency of the PV modules with the uncoated samples. The photovoltaic performance of silica-based AR coatings deposited by spin coating on dye sensitized solar cell was measured by Li et al. (2019) and observed that 6.78% and 10.12% increase of short circuit current and power conversion efficiency in coated samples, respectively. The research group used carbon powder as the dust particles for the photovoltaic performance evaluation. Kim et al. (2019) evaluated the role of PBFC70/NSN superhydrophobic coating deposited by roll-to-roll processing in the photovoltaic performance of the Perovskite solar cell. The short circuit current density and PCE were higher than those of bare perovskite solar cells.

Pan et al. (2019) carried out the photovoltaic evaluation of superhydrophobic samples with artificial dust whose chemical components were SiO_2, Al_2O_3, Fe_2O_3, and Na_2O. Following the dust deposition for 60 min, the efficiency reduction for the bare glass sample was about 15.88% while the PV efficiency reduction for the fabricated samples was about 5.7%, respectively. The reason was attributed to the micro-nano surface of superhydrophobic coatings. A similar observation was also made for superhydrophobic silica-coated glass by the research team of Pan et al. (2019). Here, the authors collected the real dust in South China University of Technology from Guangzhou, China, and compared the transmittance of coated glass substrate with respect to test dust and real dust. It was observed that the optical transmittance of coated and uncoated glass samples remained similar over the wavelength range of 350–800 nm. The coated substrate reduced the effects of dust deposition on solar PV performance, as for example, PV efficiency reduction for uncoated glass 2.8% and for superhydrophobic glass was 0.8% for tilt angle 30°.

Luo et al. (2019) studied the variation of J-V curves for the modules with moth-eye glass (dewetting and etched glass) and the flat glass at 0° to 60° incident angles of light radiation. As the incident angles increased from 0° to 60°, PCE increased dramatically from 4.6 to 9.9%. At the same time, the superhydrophobic coating performance was superior at higher tilt angle. For example, in the case of superhydrophobic silica-coated glass, efficiency reduction is 3.2%, 0.8%, 0.3% for $\theta = 30°, 45°, 60°$, respectively. On the other hand, efficiency reduction is 6.9%, 2.8%, 2.3% for $\theta = 30°, 45°, 60°$ for bare glass. In addition, the team evaluated the performance of superhydrophobic coating performance on a variety of solar cells. It was observed that efficiency reduction of the polycrystalline silicon PV cell (9% uncoated; 4.1% coated) was the higher than amorphous silicon PV cell (8.4% uncoated; 3.8% coated) and mono-crystalline silicon PV cell (6.9% uncoated; 4.1% coated).

Zhang et al. (2019a) investigated the effects of AR hybrid hollow silica nanosphere on crystalline Si modules. J_{sc} and PCE were increased by 2.79% and 4.40%, respectively, compared with those of uncoated modules. In addition, dust accumulation was lesser in coated modules, owing to closed porous structure of the AR coatings. The average degradations of 1.01% and 1.15% in J_{sc} and PCE, respectively, were observed in coated modules, which were much lower than uncoated modules.

The research group of Klampaftis studied the performance of the PV mini modules with textured FEP. A 0.6% improvement in J_{sc} was observed in the case of textured FEP, owing to the microtexture nature (Roslizar et al. 2019). The research team of Arabatzis et al. (2018) monitored the role of air sprayed TiO_2/SiO_2 superhydrophobic coatings role on PV characteristics in various seasons in Greece and in western China. The team found that 5% and 6% of power gain was observed in the entire test period in Greece and China, respectively, for coated panels.

The above results show that the presence of AR self-cleaning. In the dusty environment, the solar panels with self-cleaning cover glass recovered the PCE from the accumulation of dust particles, owing to its unique morphological features. The amount of dust accumulation depends on the inclination angle of the solar panels. When the inclination angle is large, the dust on the panels can be easily eradicated by gentle blowing wind in combination with the presence of self-cleaning coatings. An important aspect observed from the published literatures is that many research works use artificial dust particles, which are not acceptable for simulating the natural dust composition. The accumulation of dust, its types, and properties on the PV panels is strongly correlated with the local environmental and geographical conditions. Thus, it is suggested that researchers should carry out the performance evaluation of self-cleaning coatings in outdoor environment. At least, the research works should be carried out with the dust collected from the local environment, apart from using artificial dust particles.

6 Summary and Outlook

Although many research works have been carried out to fabricate self-cleaning surfaces, the fundamental aspects have to be clarified for creating the optimum coating surfaces, and several significant challenges are yet to be addressed toward solar energy utilization. The design and fabrication of surfaces that simultaneously show antireflection, self-cleaning, and durability properties are thus considered an immediate requirement to develop dust repellent and easy cleaning surfaces for solar panels using a self-cleaning approach. The relationship among the composition, morphological structure, and multifunctional properties of surfaces should be explained by combining the theoretical simulations with current analytical tools. So far, few fabricated surfaces favored the competitive properties via hierarchical micro-nano structures (Pan et al. 2019), gradient layered structures (De et al. 2018), sub-micron nanoparticles (Li et al. 2018; Ahmad and Eshaghi 2018), and nano porous (Alam et al. 2018a, b) surfaces.

Some of the complex fabrication processes of the coatings reported in the literature, for example, sputtering, electron beam evaporation, APCVD, AACVD, etc., are not appropriate to large-scale applications. Hence, novel and simplistic manufacturing methods should be formulated for functional applications. Dip and spray-coating, phase separation, dewetting, and roll-to-roll process are some of the ideal methods in the direction of cost-effective and large-scale manufacturing of self-cleaning coatings. The fabrication of multilayer AR coatings is relatively easy, while it also stances challenge to broadband AR property, if the coating structure comprises of alternate high and low RI layers. This problem can be effectively eradicated by preparing single-layer AR coating via hybrid TiO_2/SiO_2 (Adachi et al. 2018) and SiO_2/MgF_2 (Joshi et al. 2019) sols and its subsequent coating through simple dip coating. Another crucial factor, porosity can be induced by using porogen material like PDMS (Rombaut et al. 2019) and PEG (Zhi and Zhang 2018) in the coating solution and its subsequent heating to a certain temperature leads to porous surfaces. At the same time, Joshi et al. (2019) generated the porous films due to etching promoted by in-situ formation of HF, which reduced the RI and exhibiting the broadband AR property.

Most of the studies conducted on self-cleaning coating for solar panel applications are focused on increasing light transmission, reducing reflection, and tuning the wettability of the coatings. A limited number of studies have been conducted to investigate the durability aspects of coating, soon, the research should be directed toward the fabrication of durable coatings. The superhydrophilic-based self-cleaning coatings demonstrated better durability when compared to hydrophobic and superhydrophobic coatings. The possible reason could be the inadequate durability of silane-based low surface energy materials. Attempts to fabricate silane free self-cleaning surfaces should be welcomed since the thermal and mechanical stability of the silane is inferior. For instance, fabrication of hydrophobic, nanohole polyimide surfaces through the dewetting process was appreciable; however, further improvement in the process was needed to realize the effective self-cleaning behavior.

Fabrication of self-cleaning surfaces with thermal shock, UV, scratch, and abrasion resistance is needed for solar coatings to protect from the local environment. Thus, environmental aspects should be considered when preparing such surfaces, since solar panels are installed in industrial, desert, and residential areas. In addition, dusts from different environmental areas should be collected and analyzed. In cold countries, the interfacial interaction among snow, ice, and the cover glass of solar panels, remain a complex phenomenon since snow and ice exist in approximately 80 different variations (Jelle et al. 2016). Like the tropical conditions, the deposition of snow/ice depends on factors such as atmospheric and climatic conditions, local weather, temperature, pressure, incident solar radiation, and various pollutants. Icephobicity-based research works should be carried out to prevent and delay ice formation and nucleation on the PV cover glass.

In desert regions, superhydrophobicity-based self-cleaning coatings are preferable, as they employ small quantity of water as compared to superhydrophilic coatings. In India, PV panels are cleaned in two cycles in all seasons except the monsoon season where one cycle cleaning is carried out. Considerable quantity of pure water of around 2.5 L per panel per cycle is the average requirement to keep the panel at peak performance. Considering a 10 MWp capacity plant has around 33,000 panels, the water requirement per month will be 165,000 L for two cycles cleaning. This enormous quantity of water cannot be used to clean the panels in arid, medium to high water-scarce areas. In such cases, superhydrophobic self-cleaning coatings will provide a favorable solution to mitigate dust deposition. In material perspective, the researchers should bear in mind that the coating should have excellent abrasion resistance and hardness, as the arid and semi-arid regions are more susceptible for sandstorms.

In marine environment, the self-cleaning coatings should have high chemical stability, thus corrosion test including the wide range of pH must be carried out. Commercial marine coatings utilize fluoropolymer-based materials for protective coatings. Finally, multifunctional surfaces with either enhanced durability, self-healing features, and/or easily repairable are preferable for solar panel coatings.

References

Abu-Dheir N, Rifai A, Yilbas BS, Yousaf MR, Al-Sharafi A, Ali H, Khaled M, Al-Aqeeli N (2019) Sol-gel coating of colloidal particles deposited glass surface pertinent to self cleaning applications. Prog Org Coat 127:202–210. https://doi.org/10.1016/j.porgcoat.2018.11.022

Adachi T, Latthe SS, Gosavic SW, Roy N, Suzuki N, Ikari H, Kato K, Katsumata K, Nakata K, Furudate M, Inoue T, Kondo T, Yuasa M, Fujishima A, Terashima C (2018) Photocatalytic, superhydrophilic, self-cleaning TiO_2 coating on cheap, lightweight, flexible polycarbonate substrates. Appl Surf Sci 458:917–923. https://doi.org/10.1016/j.apsusc.2018.07.172

Adak D, Ghosh S, Chakraborty P, Srivatsa KMK, Mondal A, Saha H, Mukherjee R, Bhattacharyya R (2018) Non lithographic block copolymer directed self-assembled and plasma treated self-cleaning transparent coating for photovoltaic modules and other solar energy devices. Sol Energy Mater Sol Cells 188:127–139. https://doi.org/10.1016/j.solmat.2018.08.011

Agustín-Sáenz C, Machado M, Zubillaga O, Tercjak A (2019) Hydrophobic and spectrally broadband antireflective methyl-silylated silica coatings with high performance stability for concentrated solar applications. Sol Energy Mater Sol Cells 200:109962. https://doi.org/10.1016/j.solmat.2019

Ahmad MM, Eshaghi A (2018) Fabrication of antireflective superhydrophobic thin film based on the TMMS with self-cleaning and anti-icing properties. Prog Org Coat 122:199–206. https://doi.org/10.1016/j.porgcoat.2018.06.001

Alam K, Ali S, Saher S, Humayun M (2018a) Silica nano-particulate coating having self-cleaning and antireflective properties for PV modules. In: IEEE 21st International Multi-Topic Conference (INMIC), pp 1–5. https://doi.org/10.1109/INMIC.2018.8595701.

Alam K, Saher S, Ali S, Noman M, Qamar A, Hamayun M (2018b) Superhydrophilic, antifogging and antireflecting nanoparticulate coating for solar PV modules. In: International Conference on Power Generation Systems and Renewable Energy Technologies (PGSRET), pp 1–6. https://doi.org/10.1109/PGSRET.2018.8685951

Arabatzis I, Todorova N, Fasaki I, Tsesmeli C, Peppas A, Li WX, Zhao Z (2018) Photocatalytic, self-cleaning, antireflective coating for photovoltaic panels: characterization and monitoring in real conditions. Sol Energy 159:251–259. https://doi.org/10.1016/j.solener.2017.10.088

Bake A, Meraha N, Matin A, Gondal M, Qahtan T, Abu-Dheira N (2018) Preparation of transparent and robust superhydrophobic surfaces for self cleaning applications. Prog Org Coat 122:170–179. https://doi.org/10.1016/j.porgcoat.2018.05.018

Bergin MH, Ghoroi C, Dixit D, Schauer JJ, Shindell DT (2017) Large reductions in solar energy production due to dust and particulate air pollution. Environ Sci Technol Lett 4:339−344. https://doi.org/10.1021/acs.estlett.7b00197

Boddupalli N, Singh G, Chandra L, Bandyopadhyay B (2017) Reprint of: dealing with dust—some challenges and solutions for enabling solar energy in desert regions. Sol. Energy 154:134–143. https://doi.org/10.1016/j.solener.2017.04.073

Cassie ABD, Baxter S (1944) Wettability of porous surfaces. Trans Faraday Soc 40:546. https://doi.org/10.1039/TF9444000546

Darwish ZA, Kazem HA, Sopian K, Al-Goul MA, Alawadhi H (2015) Effect of dust pollutant type on photovoltaic performance. Renew Sustain Energy Rev 41:735–744. https://doi.org/10.1016/j.rser.2014.08.068

De R, Misal JS, Shinde DD, Polaki SR, Singh R, Som T, Sahoo NK, Rao KD (2018) A fast and facile fabrication of PTFE based superhydrophobic and ultra wideband angle insensitive anti-reflection coatings. Rapid Res Lett 12(6):1800041. https://doi.org/10.1002/pssr.201800041

Dong S, Li Y, Tian N, Li B, Yang Y, Li L, Zhang J (2018) Scalable preparation of superamphiphobic coatings with ultralow sliding angles and high liquid impact resistance. ACS Appl Mater Interfaces 10(49):41878–41882. https://doi.org/10.1021/acsami.8b17825

Ghazi S, Sayigh A, Ip K (2014) Dust effect on flat surfaces—a review paper. Renew Sustain Energy Rev 33:742–751. https://doi.org/10.1016/j.rser.2014.02.016

Ghosh A (2020) Soiling losses: a barrier for India's energy security dependency from photovoltaic power. Challenges 11(1):9. https://doi.org/10.3390/challe11010009

Gurav JL, Jung IK, Park HH, Kang ES, Nadargi DY (2010) Silica aerogel: synthesis and applications. J Nanomater. https://doi.org/10.1155/2010/409310

Han ZW, Wang Z, Feng XM, Li B, Mu ZZ, Zhang JQ, Niu SC, Ren LQ (2016) Antireflective surface inspired from biology: a review. Biosurf Biotribol 2(4):137–150. https://doi.org/10.1016/j.bsbt.2016.11.002

Hassan G, Yilbas BS, Al-Sharafi A, Al-Qahtani H (2019) Self-cleaning of a hydrophobic surface by a rolling water droplet. Sci Rep 9:5744. https://doi.org/10.1038/s41598-019-42318-3

Hassan G, Yilbas BS, Al-Sharafi A, Sahin AZ, Al-Qahtani H (2020) Solar energy harvesting and self-cleaning of surfaces by an impacting water droplet. Int J Energy Res 44(1):388–401. https://doi.org/10.1002/er.4935

https://www.solarnovus.com/perma-clean-self-cleaning-anti-static-coating-for-solar-panels_N9725.html

https://www.sketchnano.com/sketchnanogard-self-cleaning/

https://www.masterbuilder.co.in/self-cleaning-coating-solar-panel-applications/

https://www.xlcoatings.com/solar-panel-cleaner-stay-clean

https://materion.com

https://www.homeadvisor.com/

https://www.weforum.org/agenda/2018/09/air-pollution-can-put-a-dent-in-solar-power/

https://solarpost.in

https://www.indiaenergyportal.org/subthemes.php?text=solar

https://www.sciencedirect.com/topics/engineering/solar-cell

Huh D, Choi HJ, Byun M, Kim K, Lee H (2019) Long-term analysis of PV module with large-area patterned antireflective film. Renew Energy 135:525–528. https://doi.org/10.1016/j.renene.2018.12.055

Jelle BP, Gao T, Mofid SA, Kolås T, Stenst PM, Ng S (2016) Avoiding snow and ice formation on exterior solar cell surfaces—a review of research pathways and opportunities. Procedia Eng 145:699–706

Joshi DN, Atchuta SR, Reddy L, Kumar YN, Arkoti SS (2019) Super-hydrophilic broadband anti-reflective coating with high weather stability for solar and optical applications. Sol Energy Mater Sol Cells 200:110023. https://doi.org/10.1016/j.solmat.2019.110023

Karunakaran RG, Lu CH, Zhang Z, Yang S (2011) Highly transparent Superhydrophobic surfaces from the coassembly of nanoparticles (<100 nm). Langmuir 2(7):4594–4602. https://doi.org/10.1021/la104067c

Kim M, Kang T, Kim SH, Jung EH, Park HH, Seo J, Lee SJ (2019) Antireflective, self-cleaning and protective film by continuous sputtering of a plasma polymer on inorganic multilayer for perovskite solar cells application. Sol Energy Mater Sol Cells 191:55–61. https://doi.org/10.1016/j.solmat.2018.10.020

Li X, Wang Y, Wang R, Wang S, Zang D, Geng X (2018) A Dip-decoating process for producing transparent bi-superhydrophobic and wrinkled water surfaces. Adv Mater Interfaces 5(15):1800356. https://doi.org/10.1002/admi.201800356

Li W, Tan X, Zhu J, Xiang P, Xiao T, Tian L, Yang A, Wang M, Chene X (2019) Broadband antireflective and superhydrophobic coatings for solar cells. Mater Today Energy 12:348–355. https://doi.org/10.1016/j.mtener.2019.03.006

Luo X, Lu L, Yin M, Fang X, Chen X, Li D, Yang L, Lic G, Maa J (2019) Antireflective and self-cleaning glass with robust moth-eye surface nanostructures for photovoltaic utilization. Mater Res Bull 109:183–189. https://doi.org/10.1016/j.materresbull.2018.09.029

Mahmud GA (2009) Increasing the coating resistance against UV degradation and corrosion using nanocomposite coating. A Thesis.

Majhy B, Iqbal R, Sen AK (2018) Facile fabrication and mechanistic understanding of a transparent reversible superhydrophobic—superhydrophilic surface. Sci Rep 8:18018. https://doi.org/10.1038/s41598-018-37016-5

Mekhilef S, Saidur R, Kamalisarvestani M (2012) Effect of dust, humidity and air velocity on efficiency of photovoltaic cells. Renew Sustain Energy Rev 16:2920–2925

Milionis A, Loth E, Bayer IS (2016) Recent advances in the mechanical durability of superhydrophobic materials. Adv Colloid Interface Sci 229:57–79. https://doi.org/10.1016/j.cis.2015.12.007

Moffitt SL, Fleming RA, Thompson CS, Titus CJ, Kim E, Leu L, Toney MF, Schelhas LT (2019) Advanced X-ray scattering and spectroscopy characterization of an anti-soiling coating for solar module glass. ACS Appl Energy Mater 2(11):7870–7878. https://doi.org/10.1021/acsaem.9b01316

Nayshevsky I, Xu Q, Lyons AM (2019) Hydrophobic–hydrophilic surfaces exhibiting dropwise condensation for anti-soiling applications. IEEE J Photovolt 9(1):302–307. https://doi.org/10.1109/JPHOTOV.2018.2882636

Nobre AM, Dave D, Khor A, Malhotra R, Karthik S, Peters IM, Reindl T (2016) Advanced analyses of loss mechanisms for PV systems in Delhi, India. In: 32nd European Photovoltaic Solar Energy Conference and Exhibition, pp 1673–1677. https://doi.org/10.4229/EUPVSEC20162016-5CO.16.5

Oh G, Kim Y, Lee SJ, Kim EK (2020) Broadband antireflective coatings for high efficiency InGaP/GaAs/InGaAsP/InGaAs multi-junction solar cells. Sol Energy Mater Sol Cells 207:110359. https://doi.org/10.1016/j.solmat.2019.110359

Pan A, Lu H, Zhang L (2019) Experimental investigation of dust deposition reduction on solar cell covering glass by different self-cleaning coatings. Energy 181:645–653. https://doi.org/10.1016/j.energy.2019.05.223

Polizos G, Janga GG, Smith DB, List FA, Lassiter MG, Park J, Datskos PG (2018) Transparent superhydrophobic surfaces using a spray coating process. Sol Energy Mater Sol Cells 176:405–410. https://doi.org/10.1016/j.solmat.2017.10.029

Rahman IA, Padavettan V (2012) Synthesis of silica nanoparticles by sol-gel: size-dependent properties, surface modification, and applications in silica-polymer nanocomposites—a review. J Nanomater. https://doi.org/10.1155/2012/132424

Raut HK, Ganesh VA, Nair AS, Ramakrishna S (2011) Anti-reflective coatings: a critical, in-depth review. Energy Environ Sci 4:3779–3804. https://doi.org/10.1039/C1EE01297E

Rombaut J, Fernandez M, Mazumder P, Pruneri V (2019) Nanostructured hybrid-material transparent surface with antireflection properties and a facile fabrication process. ACS Omega 4(22):19840–19846. https://doi.org/10.1021/acsomega.9b02775

Roslizar A, Dottermusch S, Vüllers F, Kavalenka MN, Guttmann M, Schneider M, Paetzold UW, Hölscher H, Richards BS, Klampaftis E (2019) Self-cleaning performance of superhydrophobic hot-embossed fluoropolymer films for photovoltaic modules. Sol Energy Mater Sol Cells 189:188–196. https://doi.org/10.1016/j.solmat.2018.09.017

Sarkın AS, Ekren N, Sağlam S (2020) A review of anti-reflection and self-cleaning coatings on photovoltaic panels. Sol Energy 199:63–73. https://doi.org/10.1016/j.solener.2020.01.084

Satapathy M, Varshney P, Nanda D, Mohapatra SS, Behera A, Kumar A (2018) Fabrication of durable porous and non-porous superhydrophobic LLDPE/SiO_2 nanoparticles coatings with excellent self-cleaning property. Surf Coat Technol 341:31–39. https://doi.org/10.1016/j.surfcoat.2017.07.025

Shanmugam N, Pugazhendhi R, Elavarasan RM, Kasiviswanathan P, Das N (2020) Anti-reflective coating materials: a holistic review from PV perspective. Energies 13:2631. https://doi.org/10.3390/en13102631

Tombesi A, Li S, Sathasivam S, Page K, Heale FL, Pettinari C, Carmalt CJ, Parkin IP (2019) Aerosol-assisted chemical vapour deposition of transparent superhydrophobic film by using mixed functional alkoxysilanes. Sci Rep 9:7549. https://doi.org/10.1038/s41598-019-43386-1

Valerino M, Bergin M, Ghoroi C, Ratnaparkhi A, Smestad GP (2020) Low-cost solar PV soiling sensor validation and size resolved soiling impacts: a comprehensive field study in Western India. Sol Energy 204:307–315. https://doi.org/10.1016/j.solener.2020.03.118

Wang P, Kong M, Wang L, Ni L (2019) The effect of the superhydrophobic film on the generation efficiency of photovoltaic modules affected by salt-containing dust deposition. IEEE J Photovolt 9(6):1727–1732. https://doi.org/10.1109/JPHOTOV.2019.2930909

Wenzel RN (1936) Resistance of solid surfaces to wetting by water. Ind Eng Chem 28:988. https://doi.org/10.1021/ie50320a024

Wu Y, Zeng J, Si Y, Chen M, Wu L (2018) Large-area preparation of robust and transparent super-omniphobic polymer films. ACS Nano 12(10):10338–10346. https://doi.org/10.1021/acsnano.8b05600

www.photonics.com

Yadav NK, Pala D, Chandra L (2014) On the understanding and analyses of dust deposition on Heliostat. Energy, Procedia 57:3004–3013. https://doi.org/10.1016/j.egypro.2014.10.336

Young T (1805) An essay on the cohesion of fluids. Philos Trans R Soc Lond 95:65–87. https://doi.org/10.1098/rstl.1805.0005

Yousaf MR, Yilbas BS, Ali H (2018) Assessment of optical transmittance of oil impregnated and non-wetted surfaces in outdoor environment towards solar energy harvesting. Sol Energy 163:25–31. https://doi.org/10.1016/j.solener.2018.01.079

Zhan W, Wang W, Xiao Z, Yu X, Zhang Y (2018) Water-free dedusting on antireflective glass with durable superhydrophobicity. Surf Coat Technol 356:123–131. https://doi.org/10.1016/j.surfcoat.2018.09.064

Zhang X, Zhao G, Zhou J, Lin X, Zhang T, Duan G, Wang J, Han G (2018) Apcvd prepared TiNxOy films with energy-saving and self-cleaning functions. J Alloy Compd 746:445–452. https://doi.org/10.1016/j.jallcom.2018.02.202

Zhang J, Ai L, Lin S, Lan P, Lu Y, Dai N, Tan R, Fan B, Song W (2019a) Preparation of humidity, abrasion, and dust resistant antireflection coatings for photovoltaic modules via dual precursor modification and hybridization of hollow silica nanosphere. Sol Energy Mater Sol Cells 192:188–196. https://doi.org/10.1016/j.solmat.2018.12.032

Zhang J, Wang W, Zhou S, Yang H, Chen C (2019b) Transparent dust removal coatings for solar cell on mars and its anti-dust mechanism. Prog Org Coat 134:312–322. https://doi.org/10.1016/j.porgcoat.2019.05.028

Zhang L, Pan A, Cai R, Lu H (2019c) Indoor experiments of dust deposition reduction on solar cell covering glass by transparent super-hydrophobic coating with different tilt angles. Sol Energy 188:1146–1155. https://doi.org/10.1016/j.solener.2019.07.026

Zhao S, Zhao J, Wen M, Yao M, Wang F, Huang F, Zhang Q, Cheng YB, Zhong J (2018) Sequentially reinforced additive coating for transparent and durable superhydrophobic glass. Langmuir 34(38):11316–11324. https://doi.org/10.1021/acs.langmuir.8b01960

Zhi J, Zhang LZ (2018) Durable superhydrophobic surface with highly antireflective and self cleaning properties for the glass covers of solar cells. Sol Energy Mater Sol Cells 176:405–410. https://doi.org/10.1016/j.solmat.2017.10.029

Zhu Y, Chen L, Zhang C, Guan Z (2018) Preparation of hydrophobic antireflective SiO_2coating with deposition of PDMS from water-based SiO_2-PEG sol. Appl Surf Sci 457:522–528. https://doi.org/10.1016/j.apsusc.2018.06.177

Applications of Solar Energy

Hybrid Electrical-Solar Oven: A New Perspective

Sushant Pandey, Shruti Goswami, Prashant Saini, Satvasheel Powar, and Atul Dhar

1 Introduction

In the twentieth century, cooking appliances were mainly fueled by conventional resources. With the recent advancements in technology, these appliances are gradually shifting toward the usage of electricity. This shift can be observed in heating applications, food processing, and cooking applications. Since this change in food processing and cooking applications, several appliances like OTGs (Oven, Toaster, Grill), microwave ovens, induction cooktop, electric kettle, etc., have been introduced in the market. The forecast for non-renewable resource depletion led to the development of solar energy-based appliances. The developed solar cookers are available in the market with a variety of cooking options. In this chapter, we will be focusing on electric OTG ovens and solar cooking appliances.

In 1490, the first account of an oven entirely made of brick, and tile was found in Alsace, France. Over time, ovens underwent many changes from wood, iron, coal, gas, and even electricity (https://www.thoughtco.com/history-of-the-oven-from-cast-iron-to-electric-1992212). Each design had its motivation and purpose. The wood-burning stoves improved by adding fire chambers that helped in the release of contaminant smoke. The cast-iron stove came next, in the early 1700s. Researchers made several design iterations to its design. In the early nineteenth century, coal oven came into existence; however, its cylindrical shape and cast-iron body made it bulky. In another design/version, these ovens were accompanied by gas ovens.

The gas ovens became immensely popular among households, as gas was readily available by that time. Initial electric ovens invented in the late nineteenth century. But due to the lack of convenient electricity supply, these ovens did not become popular

S. Pandey · S. Goswami · P. Saini · S. Powar (✉) · A. Dhar
School of Engineering, Indian Institute of Technology Mandi, Mandi, Himachal Pradesh 175005, India
e-mail: Satvasheel@iitmandi.ac.in

H. Tyagi et al. (eds.), *New Research Directions in Solar Energy Technologies*, Energy, Environment, and Sustainability,
https://doi.org/10.1007/978-981-16-0594-9_8

at that time. It was only in the twentieth century when electricity was conveniently available, and these electric ovens were also available in a better design that led to their sales in the market. In 1946, the modern microwave oven was discovered by an American engineer Percy Spencer. These were high tech and available with a more user-friendly design. Even though it was a newer technology than electric oven using resistance heating, there are certain distinguishing features, which make both the ovens equally accessible in the market. Since electric oven heats the entire space, these electric ovens were quite like the conventional gas-stove cooking. This makes the ovens suitable for almost all kinds of cuisine, especially for the baking of food. On the other hand, a microwave does not bake but is ideal for reheating, roasting, and defrosting food. Hence, electric OTG ovens have a wide application among households, restaurants, and the baking industry, while microwave ovens are mostly confined to households. As per a survey reported by the Statista, as of March 2019 (https://www.statista.com/outlook/16010500/102/cookers-ovens/europe), India was on the second position in terms of the revenue generated by the sales of cookers and ovens with a market volume of US $11,451 million. Bizcommunity report forecasts (https://www.bizcommunity.com/Article/1/162/176273.html) that India will retain its current market growth trend, with an anticipated compound annual growth rate (CAGR) of +0.5% for the period from 2017 to 2025, which expected to lead the market volume to 135 million tons global bread and bakery consumption by 2025. This data not only shows the demand for cooking appliances, but also shows the scope of new technologies in the field of food processing and cooking.

In this chapter, a hybrid solution is introduced, which bridges the features and eliminates issues faced in electric ovens and solar cooking appliances individually. This hybrid oven is a combination of the best of the two appliances.

2 Solar Cooking Appliance

Solar cooking appliances were initially introduced in the market to provide an alternate method of cooking, using solar thermal energy. These appliances work on the principle of concentrating solar thermal energy and using this heat for cooking the food. These devices are reliant on sunlight to work; there is no other fuel required. On clear sunny days, it is possible to cook, for example, rice in a small panel solar cooker (Ronge et al. 2016).

There are majorly three necessary components in any solar cooking device, namely (Aadiwal et al. 2017),

(a) Concentrator
(b) Absorber
(c) Retainer.

The concentrator, as the name suggests, is responsible for concentrating the sunlight at a fixed point so that the concentrated sunlight can produce the required heat to cook the food. A concentrator is made of reflecting materials like silver, chromium, or aluminum, which has a shiny and smooth surface. The presence of a reflector speeds up the process of heat accumulation. Once the heat is concentrated to a point through the concentrator, this heat must be absorbed by the cookware. Thus, most solar cookers are made of thin surfaces and painted black on the inside. Dark colors assure excellent heat absorption while the thin surface design ensures that heat transfer is quick and even. Since the absorbed heat is retained in the cookware, it is essential to insulate the cookware properly. The retainer is either a lid or a cover to ensure that the heat is trapped inside the vessel to accumulate and maintain the required temperature to cook the food. Although a Swiss Physicist, Horace de Saussure, introduced the concept of the first solar oven in 1767, it was only in the 1980s when solar cooking appliances came into the market. After the 1980s, much cheaper, popular versions of solar cookers have come up (Ronge et al. 2016). There are mainly four types of solar cookers as follows:

(a) Box cooker
(b) Panel cooker
(c) Parabolic cooker
(d) Vacuum tube cooker.

(a) Box solar cooker

Box cookers are the most common and inexpensive type of solar cookers. This type of solar cooker has a simple construction design (Aadiwal et al. 2017). A conventional solar cooker has two boxes, one inside the other box with insulation in between them. The inner side of the inner box is painted with black color. The reflector has a mirror and a transparent cover on top of it. The primary function of the box-type solar cooker is to trap the heat inside that helps in the cooking of the food, which is kept inside the vessel. The heat-retaining process inside the box is based on the greenhouse effect. The short wavelength sunlight enters the box and is absorbed by the black color inside. And long-wavelength heat energy is radiated from the black surfaces. In the box cooker, heat loss occurs in all three forms, namely conduction, convection, and radiation. The hot air inside the cooker tries to rise due to its lower density. Any crack and lose surface casings result in the loss of heat, which lowers the overall temperature inside the cooker. Heat loss by radiation is mainly due to the heating of the cooker's body, which radiates heat into its surroundings as the surrounding is at a lower temperature. While the transparent lid traps the heat, the main reason for heat loss in this type of cooker is conduction from the surface of the box and vessels. Conduction is reduced by increasing the thermal resistance of the surface. The main advantage of this cooker is that it is a 100% emission-free device. However, the temperature is low, and it cannot store solar heat for later use. Diagram of a typical box-type solar cooker is shown in Fig. 1.

(b) Panel solar cooker

The panel solar cooker has a flat panel, which reflects and focuses sunlight for cooking and heating (Aadiwal et al. 2017). Since it mostly utilizes only the sunlight falling vertically on the cooker, the best cooking time for this type of cooker is mid of the

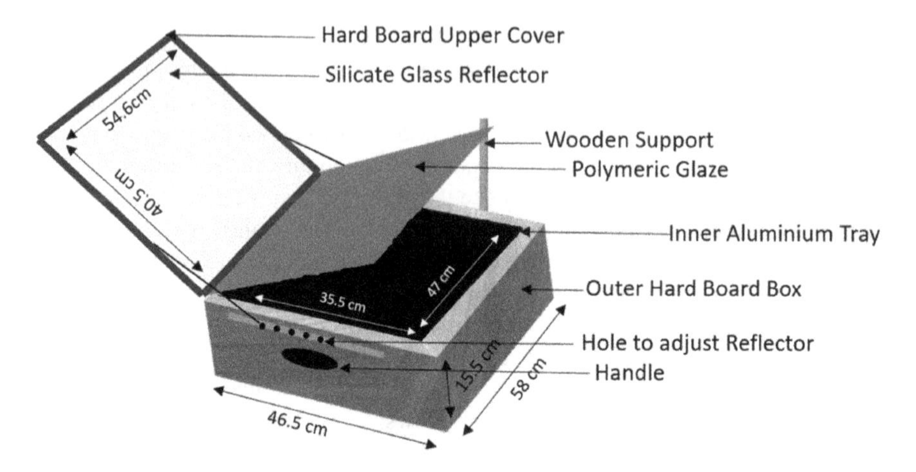

Fig. 1 Schematic diagram with typical dimensions of box-type solar cooker

day, when the sun is overhead. Such types of solar cookers typically rely on a large reflective panel, but these panels are unstable in high winds. Also, it cannot retain much heat when the weather is cloudy. It is the least popular type of solar cooker. Figure 2 shows a diagram of a panel solar cooker.

(c) Parabolic solar cooker

The parabolic solar cooker is a better option to attain higher temperatures (Aadiwal et al. 2017). It can achieve very high temperatures, which may lead to the burning of food. As shown in Fig. 3, these cookers are used with a solar collector dish, which is responsible for collecting solar radiation in the central focus point of the dish. A regular pressure cooker can be placed in the central tray of the dish. The bottom of the pressure cooker is painted black for better heat absorption.

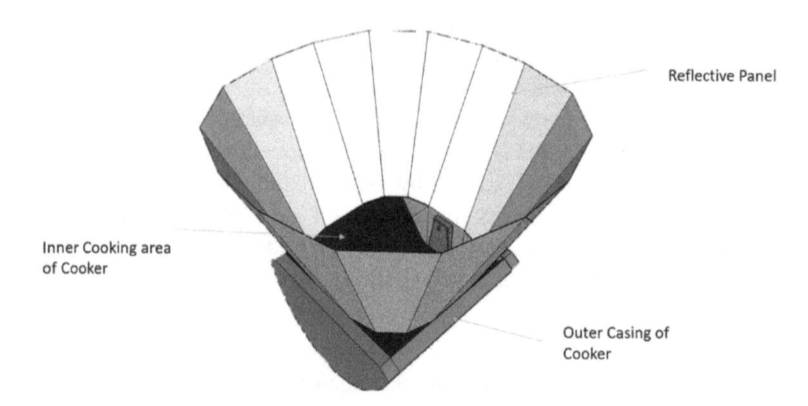

Fig. 2 Diagram of solar panel cooker

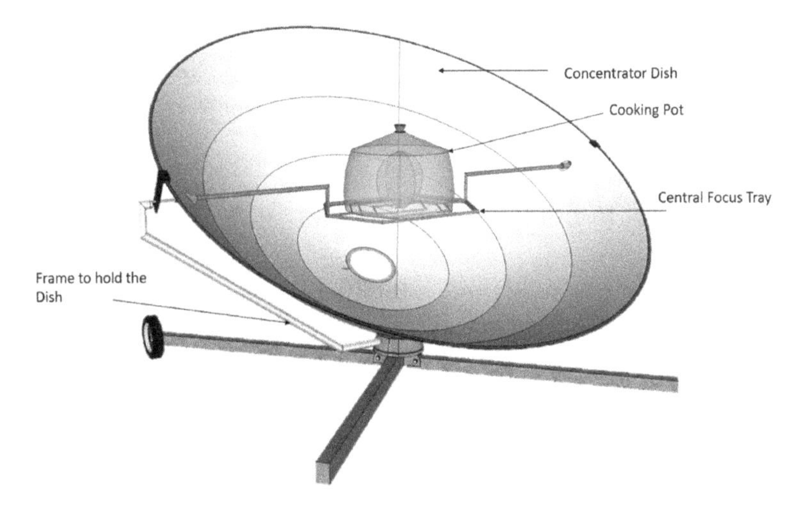

Fig. 3 Parabolic type solar cooker

(d) Vacuum tube solar cooker

This type of cooker is generally very user friendly as it requires one-dimensional solar tracking or no tracking at all. Their temperature attainability and efficiency are high. In Fig. 4 we can see, this type of cooker consists of a pair of tubes, one inside the other. The space between the outer tube and the inner tube is sealed to create a

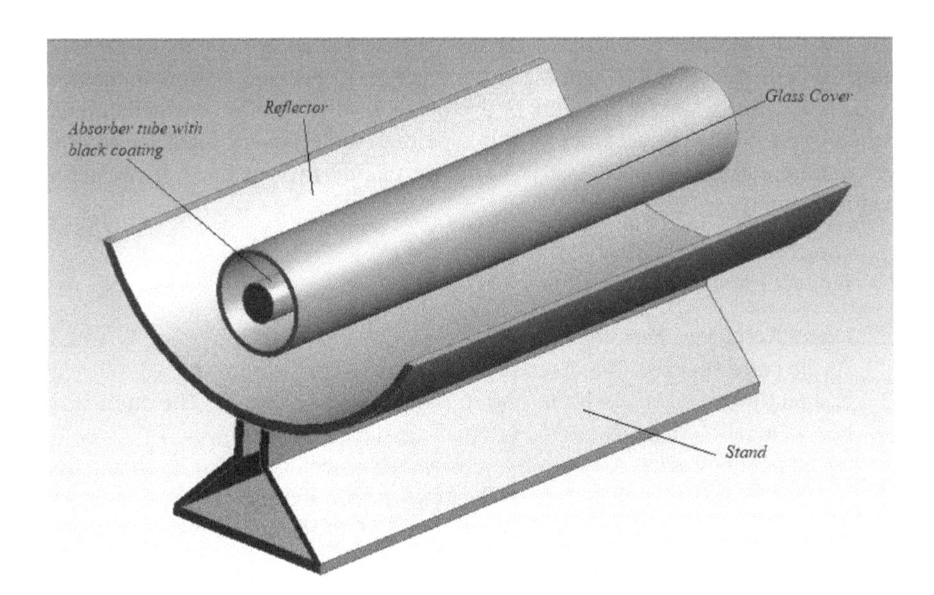

Fig. 4 Vacuum type solar cooker

vacuum. The food is kept in the inner tube, which is painted black for better heat absorption. The vacuum between the tubes provides heat insulation, which ensures that heat is trapped inside the tube for a more extended period.

3 The Concept Behind Electrical OTGs

OTG stands for oven, toaster, and griller. This type of oven has all the features of all three devices, i.e., oven, toaster, and griller. While providing enough cooking space, it also comes with easy user-controllable functions. From the design perspective, the oven comes in a double casing structure. The gap between the two casings ensures that the heat produced inside the oven while cooking does not reach the outer surface. This protection ensures user safety, preventing the user from having burns if he gets exposed to the outer surface of the oven. The outer casing and the oven door frame are made of galvanized iron or steel. The main reason behind using this material is its poor conductivity of heat. The inside frame consists of five sides of an aluminum alloy sheet. These sheets make the oven lightweight. The door frame consists of a transparent toughened glass door that has a high resistance to heat; while at the same time, it enables the user to see the meal being cooked inside the oven.

The primary underlying technology in any electric OTG is resistance heating. Resistance heating is a type of power frequency electrical heating. The driving principle behind this type of heating is the power that is dissipated in the form of heat, commonly known as the I^2R loss. Whenever current is passed through a resistive material, it produces power losses, and this power loss is dissipated in the form of heat into the atmosphere. It is one of the most efficient heating methods. There is no upper limit to the attainable temperature in this type of heating. It is recommended to use protection against overcurrent or overheating, which can be provided through switch-gears. Since the accuracy of this type of heating is very high, the desired temperature can be achieved easily. There are mainly two methods of resistance heating, namely

1. Direct resistance heating
2. Indirect resistance heating.

1. Direct Resistance Heating:
   As the name suggests, direct resistance heating is a process in which the current is directly passed through the object, which is to be heated. The amount of heat generation depends on the internal resistance of the object. This type of phenomenon restricts the option to choose the object as it must be conductive and highly resistive. This type of heating is used throughout the heating of long rods, billets of ferrous metals before forging, and continuous annealing of ferrous and non-ferrous wire.

2. Indirect Resistance Heating:
   In indirect resistance heating, a resistive element capable of producing power loss equivalent to the heat required is used as resistance, and heat is delivered to the material either by radiation or by convection. This type of heating is used in room heaters, immersion rods, ovens, etc.

Thereafter, comes the role of heat transfer (Rohsenow and Hartnett 1999). Heat transfer can occur in three ways, namely

1. Conduction: It is the transfer of heat from one part of a substance to another part of the same substance at different levels of temperature or from one substance to another via physical contact when there is a difference in temperature between the two.
   The rate of heat transfer if given by (Olugbade and Ojo 2019),

$$Q = \frac{kA\Delta T}{\Delta X} \tag{1}$$

   where

   - Q The rate of heat flow
   - K Thermal conductivity of the material
   - ΔT Temperature difference between the surfaces of metal
   - ΔX Thickness of the material
   - A Area of the surface.

2. Convection: It is the transfer of heat from a body to a fluid in motion. The motion of a fluid can be natural or forced by a pump or fan. The rate of heat transfer in this mode can be expressed as (Olugbade and Ojo 2019),

$$Q = hA[T_2 - T_1] \tag{2}$$

   where

   - h Coefficient of convective heat transfer
   - A Area of the surface not perpendicular to the direction of heat flow
   - T_2 Temperature of the body
   - T_1 Temperature of fluid.

3. Radiation: The electromagnetic radiation emitted by a substance in the form of heat by virtue of its temperature and at the expense of its internal energy is called radiation. Mathematically, the rate of heat transfer in this mode can be expressed in a simple form as (Olugbade and Ojo 2019),

$$Q = A\sigma T^4 \tag{3}$$

   where

Q Heat flux, energy per time
A Area of heat flux intensity
σ Stefan Boltzmann constant (5.67×10^{-8}) W/m^2(K^4)
T Absolute Temperature.

In a conventional electric oven, the required temperature in the oven is acquired by an indirect method of resistance heating. The heating rods present in the oven are responsible for supplying the heat. The heat is dissipated in the oven through convection and radiation. This heat is evenly circulated in the oven with the help of a fan commonly known as the convection fan. The temperature inside the oven is controlled by a thermostat-based control system. For roasting the meat, a rotisserie rod mechanism is also present in the OTGs. The user has the provision to set time and temperature for cooking. Some OTGs also provide an option to select modes like griller, toaster, or oven.

3.1 Literature Review of Existing Electric Ovens

Olugbade and Ojo (2019) provided detailed information for the fabrication, material used, and fabrication processes of an electric oven. In testing, his oven was found to be more efficient in terms of cooking time for baking and cooking food items. When the temperature setting is kept higher, the designed oven can bake food slightly quicker than the existing ovens of that time. The cooking time found in the experimental results is shown in Table 1.

Adeyinka et al. (2018) presented an improved design of the baking oven. It is portable, efficient, and cheaper than the existing ovens. A thermostat and interlock switch has been incorporated for better performance. It works on the principle of resistance heating. As per the experimental results, the heating element used can provide a minimum of 1,797.6 kJ of heat energy in the cooking chamber in 80 min. The improvement of this oven led to its faster baking time. It can be seen in Fig. 5 that the baking of fish was done for different sizes and compared with other existing ovens.

Table 1 Cooking time comparison between designed and already existing ovens (Olugbade and Ojo 2019)

Size of cake	Designed project	Already existing oven
Small size	20 min	25 min
Average size	50 min	55 min
Large size	1 h 15 min	1 h 30 min

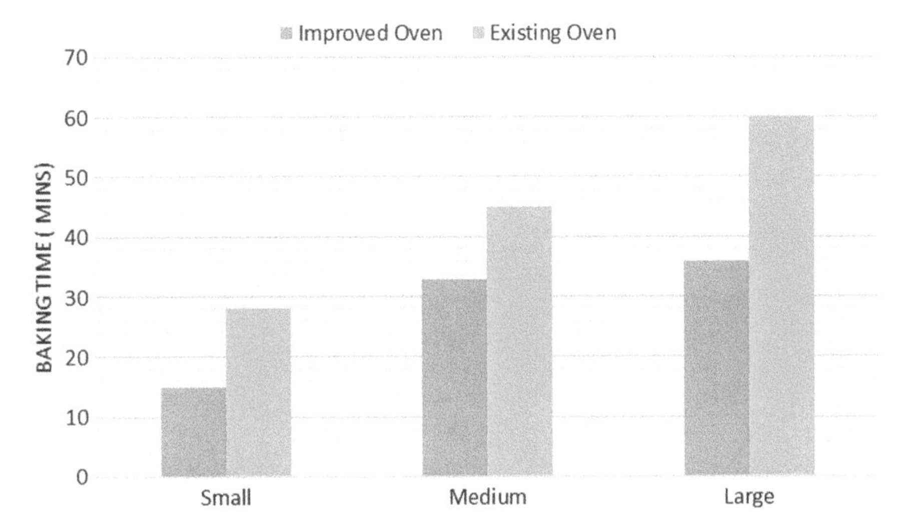

Fig. 5 Comparison between the cooking time of the existing oven and improved oven (Adeyinka et al. 2018)

4 Need for the Hybrid Oven

Electric ovens are an environment-friendly method of cooking. Since it is powered by electricity, it has zero emissions. Solar cooking appliances available in the market are not only environmentally friendly but also affordable, as they do not require any external fuel. Sunlight being an abundant source of energy makes solar cooking appliances free of running cost. To find out about its popularity among people, we carried out a primary survey using an online survey form. The study was intended to get an understanding of the interviewee's experience on the solar cooking appliance, their electric oven usage, and the interviewee's potential interest in hybrid oven appliances. The survey was carried out across various states of India. It was found that most households have an electrical OTG or microwave oven, but they hardly use it for cooking meals. The most common use was to reheat pre-cooked food. A handful responded to have used a solar cooking appliance at least once. Despite being environment friendly, these cooking options cannot convince users to shift from conventional gas stove-based cooking. For a better understanding of the underlying reasons, it is vital to know their differences and demerits that make them less popular.

Mainly two types of ovens are found among general households, namely OTGs and microwave ovens. OTGs being more popular in restaurants and bakery industries, fail to take their place in domestic applications. The main reason behind that is that OTGs are based on the concept of resistance heating, which consumes a significant amount of electric power. These ovens face a very tough competition with microwave ovens. Microwave ovens have a technological advantage as microwave ovens used microwave emitter-based technology. This leads to faster cooking as well as lesser electricity consumption. Though OTGs are better when it comes to cooking and they provide much more variety of cooking options, people tend to choose more time and cost-efficient options over better cooking. Microwave, on its own, has a few cooking advantages over OTGs. The process of defrosting, reheating cooked food, etc., and some user-friendly features like presets and programmable cooking options tend to give it an edge.

Despite requiring no running cost, solar cooking appliances are also not a compelling choice for households. Most of the solar cooking appliances available in the market have a bulky design, which makes them less portable. Portability is an essential requirement for a solar cooking appliance as they are to be carried in and out of the house for cooking. Also, their dependency on sunlight makes them a very unreliable option for cooking. Even on sunny days, cooking can be done only at noon. For the rest of the meals, some other cooking methods must be used. Whereas, other options such as an electric oven or gas stove can be used for cooking anytime. Hence, solar cooking appliances still have meager sales in the market.

In the last decade, the concept of a hybrid oven has been introduced. These ovens eliminate the main problem of other cooking appliances. Hybrid ovens are also a great way of making solar cooking efficient. Few manufacturers have already introduced this technology in the global market but there are hardly any such options available in India.

5 Literature Review of Existing Hybrid Ovens and Cookers

Joshi and Jani (2013) introduce the concept of photo-thermo solar cooker. It is made by fixing a DC heater inside a stainless-steel casserole. A 12 V, 40 Ah battery powers the heating mechanism of the cooker. This battery is charged by a 75-W solar panel, which is placed in direct sunlight. The purpose of this cooker is to be portable, robust, and at the same time, provide an option for 24-h cooking. The provision of electrically boosted heaters makes solar cooking possible even in days with less sunlight. The designed cooker was compared to various other solar cookers available in the market, and it was found that its hybrid design gives it an edge over the other cookers in terms of achieving set temperature and cooking time. And this hybrid cooker is also commercially viable.

Ibrahim and Jose (2016) designed a hybrid indirect solar cooker. The main components of this system were a solar cooker with an evacuated tube collector (ETC), latent heat storage, heat exchanger, and an alternate electric heating source. This system must be customized for the given location and application. The ETC line contains water as heat transfer fluid. The heat collected in ETC is supplied to the latent heat storage unit (PCM), from where it reaches the cooking vessel. This latent heat storage helps in utilizing solar thermal heat even at night. The PCM storage unit is also connected to an electrical heating plate for auxiliary heat supply. The whole system is controlled by a PIC micro-controller. The main aim of this system is to utilize the maximum solar radiation and make it usable even when it is not directly

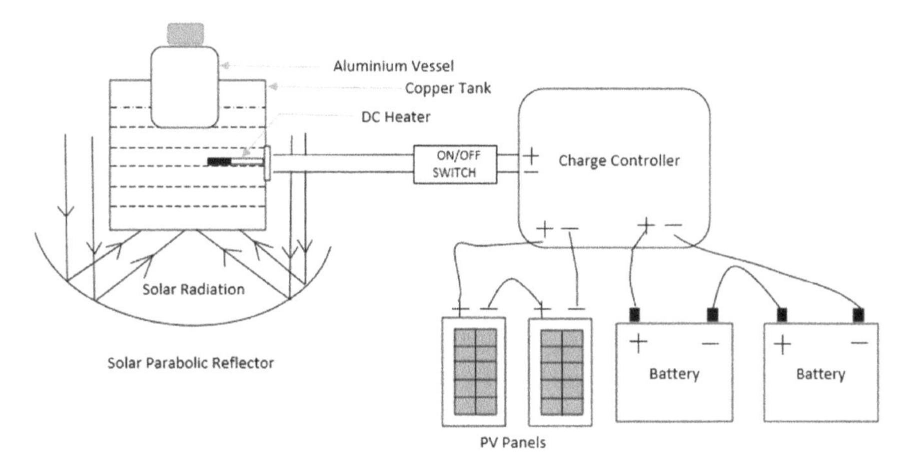

Fig. 6 Experimental setup of hybrid solar cooker (Gawali and Papade 2015)

available. The auxiliary electrical supply for heat compensation gives it an added benefit.

Gawali and Papade (2015) designed and fabricated a hybrid solar cooker. The cooking vessel is made of aluminum and is placed in a copper tank filled with oil. A solar parabolic reflector surrounds this tank. For an alternate source of heat, a DC heater was connected to the tank, which is powered by a battery. This battery is charged using solar panels. The experimental arrangement of this cooker can be seen in Fig. 6. The main purpose of this hybrid cooker is to eliminate the problem of solar cooking on cloudy days. Thermal energy storage (TES) system and auxiliary DC power-assisted heating of this setup make it different from the other existing hybrid solutions.

Shahzad (2013) present the design, development, and performance study of a circular type hybrid solar oven named electric cum solar oven (ECSO). In this, an electric plate has been attached at the bottom of a circular solar oven. The concept of this oven is like the box-type solar cooker except for the shape and missing reflector. Wooden frame and foam linings have been used for better insulation. The inner side of the oven was painted black for better heat absorption. In experiments, it was found that solar cooking works efficiently at noon with electrical assistance. The performance of this oven was compared to Azam et al. (2009) and found to be faster in terms of cooking time, as shown in Table 2.

Chukwuneke et al. (2018) present a unique concept of a hybrid oven in which the gas-stove cooking and electrical heating have been incorporated together. The oven has a gas burner in the bottom and heating element on the sides of the oven. The inner casing of the oven is made of aluminum sheet, and outer walls are of galvanized steel metal. It is connected to a gas cylinder for gas supply. The oven was made for domestic applications. By using both supplies, i.e., gas and electricity, high temperatures can be achieved easily. At a high temperature, cooking time reduces significantly.

Table 2 Cooking time of different food items (Shahzad 2013)

Cooking operation	Food Items	Quantity	Time (min)
Baking and frying	Fruit cake	½ kg	45
	Frying eggs	2	20
Boiling	Boiling of eggs	2	25
	Boiling of milk	500 ml	20
Other edibles	Pudding	½ kg	25
	Custard	175 g in 500 ml of milk	35

6 Design and Development of a New Hybrid Oven

The proposed hybrid oven is a fully functional OTG oven with provision for use as a solar cooking appliance. As can be seen in Fig. 7, the design of the oven is inspired by the conventional user-friendly domestic OTGs, which is to be also used with a solar collector dish as an accessory for cooking in sunlight and hybrid mode. This hybrid mode is a uniquely designed feature of the oven.

6.1 *Design Aspects of Oven*

On the basis of the design, the oven comprises of the following five components as shown in Fig. 7:

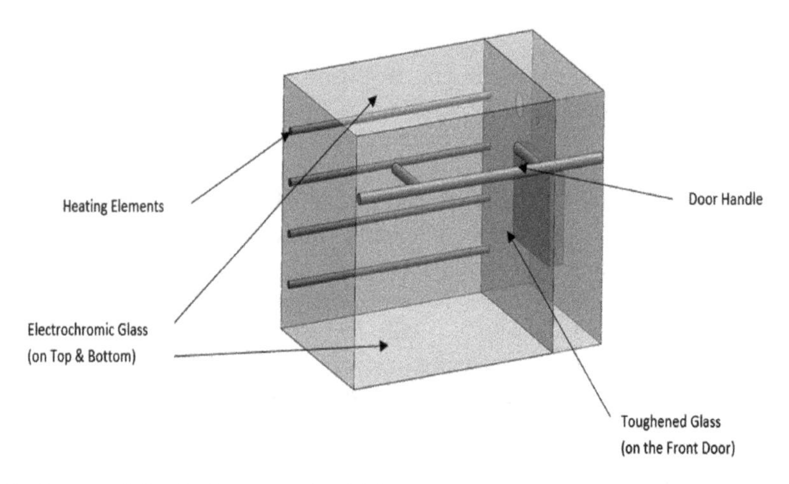

Fig. 7 Conceptual sketch of a proposed hybrid oven

1. **Outer casing**: A galvanized iron case is used, which covers three sides of the oven, i.e., top and two sides. It is used to minimize the heat leakage from the heating chamber and keep the outer surface of the oven cool.
2. **The heating chamber and internal body**: The internal heating chamber has three aluminum alloy-based walls, two walls made of electrochromic glass (i.e., bottom and rare wall), and the front door of the oven made of toughened glass, cover the remaining open side. The electrochromic glass is placed on the back and the bottom of the oven, which allows transferring the solar radiation to the heating chamber. The remaining body of the oven is also made of aluminum alloy.
3. **Control system**: Microcontroller-based control system has been used. It is mainly responsible for regulating the heat, functioning of different electrical and electronic components, and automatic switching between solar thermal and electrical supply when operating in hybrid mode.
4. **Heating elements**: Rod type, resistance heating-based elements have been used inside the heating chamber. These elements convert electrical energy into heat.
5. **Convection fan**: A low-power synchronous motor-based fan has been used for evenly distributing the heat by circulating the air in the heating chamber of the oven. Since it maximizes convection, hence called convection fan.
6. **Light bulb**: A normal low-powered yellow light bulb has been used inside the heating chamber of the oven. The main purpose of this bulb is to illuminate the cooking area while cooking indoor or in the dark area.
7. **Lithium-ion Battery**: The lithium-ion battery is used to power the control system when the oven operates in solar mode since there is no other electric power connected to the oven.
8. **Solar collector dish**: It is a hemispherical dish with a diameter of 1.4 m (Adil Ahmed et al. 2015). The inward curvature of the dish is made shiny and reflective to reflect the maximum solar radiation falling on it. The focus of the reflected solar radiation is the central area where a tray is present. This tray holds the object that is to be heated. In our case, the oven is placed in that tray. The primary reflective area of the dish is made of 36 fins, to give an easy assembly to the dish, in the shape of a hemisphere. The complete hemispherical structure rests on a mobile frame with wheels. There are two tracking brackets present on either side of the dish. These tracking brackets are used for positioning of the inclination angle of the dish. To use the dish, the oven must be placed in the central tray of the dish. Also, the dish should be in the outdoor area where the maximum sunlight is received. The dish is moved around the central axis so that the tracking brackets on the side are normal to the sun. If the outward pointed nuts do not cast a shadow on the tracking bracket, it means they are normal to the sun. During cooking, when the position of the sun changes throughout the day, the dish position must be adjusted approximately every hour so that the tracking brackets are always normal to the sun shown in Fig. 8. The position where tracking brackets are normal to the sun generates the maximum heat in the focus area of the dish. The reflective surface of the dish should be clean

Fig. 8 **a** When the tracking bracket is normal to the sun, **b** when the tracking bracket is not normal to the sun

before cooking. Accumulation of dust on the reflective surface also leads to a significant decrease in the temperature at the focus area.

6.2 Operational Dynamics

The basic principles of heat transfer, resistance heating, and light reflection govern the operation of this oven.

The oven provides the option of three modes of operation that can be used based on the availability of sunlight and an electrical power source. There is no need to manually select the mode, as the smart control system detects the mode itself depending on the type/types of power being supplied to the oven. Following are three modes of operation of the oven, namely

1. Electric mode: In this mode of operation, the oven works just like any other conventional electric OTG. The oven is connected to the electrical power supply. The food is cooked through the resistance heating provided by the heating rods present in the oven. The suitable temperature for preparing each item is controlled through a control system.
2. Solar mode: In the solar mode of operation, the oven does not require an electric power supply. The food is cooked just by concentrating the solar energy. Solar energy is concentrated on the oven with the help of a solar collector dish. To regulate the temperature in this mode, the role of electrochromic glass comes into play. The electrochromic glass allows solar radiation to enter the heating chamber until the desirable temperature attained. As the temperature exceeds the desired temperature, the electrochromic glass prevents solar radiation from

entering the heating chamber. To regulate the temperature when no power supply is connected to the oven, the control system is powered with a lithium-ion battery to keep the control system functional at all times irrespective of power supply given to the oven.

3. Hybrid mode: The hybrid mode of operation is the main unique feature of this oven. This feature is designed to improve the cooking time taken by solar cooking appliances and make a solar cooking feature of the oven functional during cloudy days. Also, it reduces the electrical power consumption as compared to electrical mode. The control system utilizes the electrical power supply only for heat compensation, i.e., if solar thermal heat is not enough to reach the set temperature, then the control system turns on the electrical heating. If the temperature inside the oven exceeds the set temperature, the control system first turns off the electrical input. If the temperature is still higher than the set limit, it prevents extra solar energy from entering the oven through the electrochromic glass.

The aim of developing this oven was not only to provide an option for clean energy cooking but also to provide users an option that is compelling enough for them to shift to electrical and solar cooking appliances. The most common problem identified in the literature and survey was that most people do not like solar cooking appliances because of the hassle involved in their use. Whereas, the most common uses for an electrical oven were found to be reheating of food and boiling of milk. The proposed oven provides the provision of nifty features like IoT-based control, food burn alarm, etc. These features enable the user to have a cooking experience that is not only modern and compelling, but also very easy and convenient for them. This will help in shifting the tendency of users to prefer appliances like an oven over the conventional gas-stove-based methods.

6.3 Comparison of the Proposed Hybrid Oven with the Existing Literatures

Joshi et al. [11]	The cooker had a small size and limited cooking provisions.
Hassan et al. [12]	The design involves a setup of bulky components making it less feasible for domestic use.
Gawali et al. [13]	Bulky design and limited cooking provisions.
Shazad et al. [14]	The design was able to capture and retain less solar thermal heat.
Chuckwuneke et al. [16]	Doesn't solve the problem of using non-renewable energy for cooking as LPG is used in the proposed design.

Proposed Hybrid Oven

It is an OTG oven which is similar to conventional OTG and Microwave Ovens in design, hence have enough space and the provision to cook anything that can cooked in a conventional OTG.

The oven has been designed using light weight materials that makes it easily portable.

The oven is designed to retain maximum heat and therefore, it has very limited heat loss.

It works on the combination of electricity and solar thermal radiation hence, it is clean energy device.

7 Experimental Results

Experiments were conducted to analyze the performance of the hybrid oven in all three modes. Various design aspects have been studied and tested for optimum performance.

Initially, the oven was tested for its capacity to attain maximum temperature in solar mode. The highest temperature found in April month on an average sunny day in a hilly area town was approximately 130 °C. This allowed the cooking of food items that are based on boiling like rice, idli, etc. Usually, the solar cookers available in the market have a limitation of the shape and size of the cooking area. As this oven itself acts as a solar cooker, it gives you the provision to cook anything that does not require very high temperatures. Also, the presence of a battery enables other features of oven-like food burn alarm, buzzer on the completion of the timer, etc., makes cooking in the sun very easy. The person cooking the food does not have to stand in the sun while cooking as the oven can be controlled using an IOT-based app. In electric mode, the cooking experience was like the conventional electrical OTGs available in the market. Several types of food like rice, boiling of milk, toast, pizza, cake, etc., were cooked in the oven to observe its performance in the electrical mode. The same food was simultaneously cooked in a Wonderchef 19L OTG oven bought from the market. The results in terms of cooking performance were found to be very similar. The focus of this oven is its performance in hybrid mode. In this mode, not only it gives faster cooking than the available solar cookers, but it also uses significantly less electricity as compared to conventional electrical OTGs. This is because it utilizes both solar energy as well as electricity in hybrid mode. For the experiment in this mode, the food items like rice, boiling of milk, pizza, cakes, toast, and a few more were cooked in the oven. The same food items were simultaneously cooked in Wonderchef 19L OTG oven for comparison. The results have shown an average energy saving of 52% in hybrid mode when compared to the Wonderchef oven. The cooking time was also found to be approximately half when compared to the cooking time of the same food items in the solar mode. These results make this mode not only energy-efficient, but also very convenient to use. It gives the proposed oven an edge over both electrical as well as solar cooking appliances. The oven being compact cuboid in shape gives it the advantage of portability over other commonly used box-type solar cookers. The materials used in the body of the oven have been chosen to make the oven lightweight.

8 Conclusions

According to a report by Statista, 25 million units of electric cookers and ovens are added every 5 years globally (https://www.statista.com/outlook/16010500/102/cookers-ovens/europe). If ovens like the proposed design can capture even 1% of the global market share, and assuming that such ovens are operated for let us say 100 h in hybrid mode in a year and they are rated at 2 kW each on an average, then it may save 25 million kWh of electricity globally every year only on the newly added units of the oven in the last five years. This, in turn, will also save approximately 12,375 tonnes of CO_2 emitted, producing the electricity that is being saved in hybrid mode (https://www.eia.gov/environment/emissions/carbon/). These estimations are calculated only considering the usual households. As we know, the global bakery industries rely on electrical ovens for their bakery products. We can only imagine how much energy would be saved if ovens in the bakery industry are replaced by hybrid ones. Such results show that products like these not only have a huge penitential in the market, but also they are the need of the hour especially when we are facing a global energy crisis. For developing nations like India, products like these which promote the use of clean energy and also help to mitigate environmental challenges can be a good way of contributing towards the United Nation's sustainable development goals, for example, #Goal 9 Industry, Innovation and Infrastructure and #Goal 13 Climate Action.

References

Aadiwal R, Hassani M, Kumar P (2017) An overview study of solar cookers. Spec Issue Int J Electron Commun Soft Comput Sci Eng 2011:2277–9477

Adeyinka A, Olusegun O, Taiye A, Mojeed L, Heritage O (2018) Development and performance evaluation of dual powered baking oven. Adv Res 17(3):1–15

Adil Ahmed S, Prasanna Rao NS, Murthy PLS, and Terani BP (2015) Detail study of parabolic solar cooker SK-14, pp 24–27

Author's personal copy: A review of vacuum tube based solar cookers with the experimental determination of energy and exergy efficiencies of a single vacuum tube based prototype

Azam M, Javed MY, Musadiq M, Zahira R (2009) Fabrication and performance study of slope type electric cum solar oven. Pak J Agric Sci 46(3):228–231

Chukwuneke JL, Nwuzor IC, Anisiji EO, Digitemie IE (2018) Design and fabrication of a dual powered baking oven. Adv Res 16(4):1–8

Cookers & Ovens—Europe | Statista Market Forecast. [Online]. Available: https://www.statista.com/outlook/16010500/102/cookers-ovens/europe. Accessed: 27 June 2020

Dryden IGC (1982) The efficient use of energy, Butterworth-Heinemann, Second Edition, UK, https://doi.org/10.1016/C2013-0-00885-7

Gawali SR, Papade CV (2015) Hybrid solar cooker. Int J Eng Res V4(06):763–766 (2015)

Global bread and bakery consumption continues to experience modest growth. [Online]. Available: https://www.bizcommunity.com/Article/1/162/176273.html. Accessed: 27 June 2020

History of the Oven from Cast Iron to Electric. [Online]. Available: https://www.thoughtco.com/history-of-the-oven-from-cast-iron-to-electric-1992212. Accessed: 27 June 2020

Ibrahim BHK, Jose V (2016) Hybrid indirect solar cooker with latent heat storage. Int J Eng Sci Res Technol, no. July

Joshi SB, Jani AR (2013) Photovoltaic and thermal hybridized solar cooker. ISRN Renew Energy 2013(June 2013):1–5

Olugbade T, Ojo O (2019) Development and performance evaluation of an improved electric baking oven, no. April 2019

Rohsenow WM, Hartnett JR (1999) Handbook of heat transfer, vol. 36, no. 06

Ronge H, Niture V, Ghodake S (2016) A review paper on utilization of solar energy for cooking. Imp Int J Eco-Friendly Technol 1(1):121–124

Shahzad (2013) Design and development of efficient domestic electric cum solar oven. J Basic Appl Sci, no. April

U.S. Energy Information Administration (EIA)—Ap. [Online]. Available: https://www.eia.gov/environment/emissions/carbon/. Accessed: 27 June 2020

Performance Analysis of Vacuum Insulation Panels Using Real Gas Equation for Mitigating Solar Heat Gain in Buildings

Divyanshu Sood, Pranaynil Saikia, Marmik Pancholi, and Dibakar Rakshit

1 Introduction

Vacuum insulation panels (VIPs) consist of dry core material embedded in a vacuum-tight cover called envelope (Liang et al. 2017), as shown in Fig. 1. The thermal conductivity of VIP varies with the core material. The suitable core materials for VIPs are fumed silica, polyurethane (PU) foam, glass fiber, and so on (Capozzoli et al. 2015). The fumed silica gives the best performance in terms of thermal conductivity when it is kept dry and evacuated. VIPs can provide 5–20 times lower thermal conductivity than that of conventional insulating materials (Brunner et al. 2014). Although the overall thermal conductivity of VIP depends considerably on the core material, it gets affected by the envelope material which is responsible for preventing the air and water permeation. In general, the nanostructure core material is desired for the core of VIP (Molleti et al. 2018; Liang et al. 2019). To maintain the desired thermal conductivity of the core, it is mandatory to maintain the vacuum inside the VIP. Getters and desiccants are used to maintain a suitable vacuum inside the VIP (Kalnæs and Jelle 2014).

The envelope maintains a vacuum inside the VIP by preventing the intake of water and air gases (Benson et al. 1994). Various envelope materials are used for this purpose. The envelope materials can either be metal film (MF) or metalized films (MFs) (Wegger et al. 2011). Metal films are laminated with a layer of polyethylene

D. Sood · P. Saikia · D. Rakshit (✉)
Centre for Energy Studies, Indian Institute of Technology Delhi, Hauz Khas, New Delhi 110016, India
e-mail: dibakar@iitd.ac.in

M. Pancholi
School of Engineering and Applied Sciences, Ahmedabad University, Ahmedabad, Gujrat 380009, India

H. Tyagi et al. (eds.), *New Research Directions in Solar Energy Technologies*,
Energy, Environment, and Sustainability,
https://doi.org/10.1007/978-981-16-0594-9_9

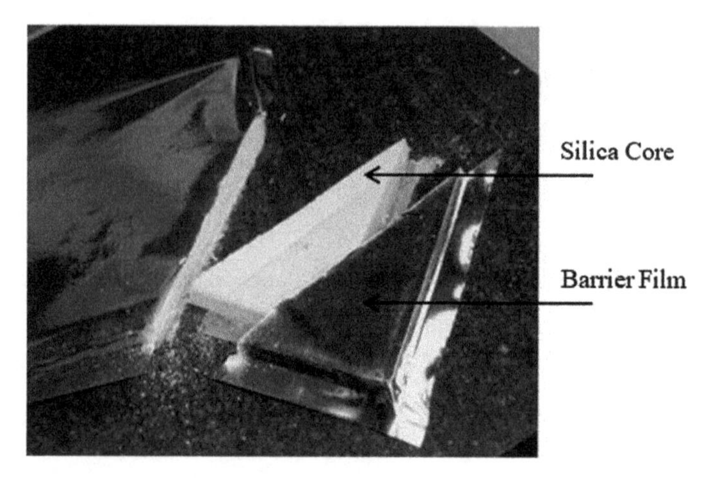

Fig. 1 Vacuum insulation panel (Simmler and Brunner 2005a)

terephthalate (PET) and metalized films have up to three layers of aluminum metalized polyethylene terephthalate (PET) or polypropylene (Tenpierik and Cauberg 2010). The envelope supports the vacuum for a longer period of time. The thermal conductivity of the core is a function of both air and water permeation. The pressure increase due to the intake of air and water across foil cover strongly depends on climatic conditions such as temperature, pressure, and relative humidity around VIP and also on the geometry of the panel (Alotaibi and Riffat 2014).

The total heat transfer through the VIP can take place via four different heat transfer processes: heat transfer via radiation, heat transfer via conduction in the core, heat transfer due to gas conduction, and heat transfer due to gas convection (Baetens et al. 2010). For convection to happen, bulk motion of molecules is required. In the present scenario, the concentration of molecules is very less inside the VIP, therefore the convection inside the VIP is not considered. So, the total heat flow is the summation of above-mentioned different heat transfer processes as given by Eq. (1),

$$Q_{\text{total}} = Q_r + Q_{cd} + Q_g \tag{1}$$

All these parameters should be minimum for insulation in buildings that results in an overall low thermal conductivity (Fricke et al. 2006). Lower the thermal conductivity of the insulation material, higher will be the thermal resistance which indicates less heat transfer. Thermal conductivity of core material depends on the amount of gas and water permeated into VIP. The permeation should be minimum and this can be achieved by selecting suitable envelope material with proper sealing (Kan et al. 2015). The increase in overall thermal conductivity of the VIP over time is called thermal conductivity aging. The thermal conductivity of the vacuum insulation panel depends upon gas permeation with time, water vapors formed inside the envelope, and water adsorbed on fumed silica core (Schwab et al. 2005a). To reduce the effect of parameters affecting the performance of VIP, getters and desiccants are used. The getters remove the gas molecules from the evacuated space either by chemically combining with them or by adsorbing them whereas desiccants bind the water molecules (Xu et al. 2018). Batard et al. conducted a study on the thermal performance of VIP installed in buildings in winter and autumn season (cold climate) (Batard et al. 2018a). In the present study, the climate considered is hot and dry which is a major distinction from the previous work of Batard et al.

The ideal gas law considers gas molecule as point particle that does not occupy any space and is not attracted or repelled by other gas molecules, i.e., intermolecular interactions are nearly zero. The ideal gas equation is represented as Eq. (2),

$$Pv = \frac{RT}{M} \tag{2}$$

To justify volume that a real gas molecule occupies, Van der Waals equation substitutes volume in ideal gas equation with $(V - b)$. The second modification in ideal gas law explains the molecular force of attraction. Van der Waals provided intermolecular attraction by adding a term a/V_m^2 to the observed pressure P in the equation. The constant 'a' provides a correction for the intermolecular forces and 'b' is a correction for finite molecular size. The Van der Waals equation can, therefore, be written as Eq. (3):

$$\left(P + \frac{n^2 a}{V^2}\right)(V - nb) = nRT \tag{3}$$

The present analysis emphasizes on using the Van der Waals equation for the performance analyses of VIP in terms of variation in internal pressure, thermal conductivity, service life. The previous studies used ideal gas equation for VIP analyses which does not provide a precise solution because, for air, to be considered as an ideal gas low pressure and high temperature is required. The key aspects of the present study were to investigate the performance of VIP considering air as a real gas to achieve more precise results and also to examine the effect of ambient temperature which is the function of solar radiation intensity falling on the surface.

2 Mathematical Model

The thermal performance of the VIPs is highly reliant on vacuum inside the envelope and gets disturbed by air and water permeation. The combined effect of air and water permeation inside the VIP increases the inside pressure and therefore thermal conductivity of core. This will further shorten the service life of a VIP. Under high temperature and low pressure, real gas behaves as an ideal gas. Here, the VIP is used for building application. The gas pressure is low inside the panel, which partially fulfills the ideal gas behavior requirements. However, the operating temperature of the VIP for the current application (20–50 °C) is much below the high-temperature requirement for gas to behave ideally. Furthermore, with time the concentration of gas molecules increases inside the VIP with a subsequent increase in pressure. Under these circumstances, considering Van der Waals equation for the analysis of VIPs leads to a more realistic approach. In this mathematical model, both molecular volume and intermolecular force of attraction factors are included for calculating the increase in inside pressure due to air permeation which affects the thermal conductivity and service life significantly. Also, the study has been carried out for hot and dry climate (relative humidity < 55%) (Nayak and Prajapati 2006), so the effect of water on the degradation of the fumed silica core can be neglected and only solar radiation is considered for the analysis.

2.1 Determination of Air Permeation

The VIP envelope consists of different layers, so it is difficult to calculate the definite value of air permeation through it. There will always be some uncertainty in it. It can be expressed in terms of air transmission rate (ATR) and is given by Eq. (4). The air permeates through the surface of the laminate cover per panel area (ATR_A) from both the front and the rear side and also along the circumference of the panel (ATR_L). The increase in pressure inside VIP affects the service life of the panel. The pressure is increased due to the intake of air through the envelope (Yrieix and Pons 2018).

$$\mathrm{ATR}_{\mathrm{total}} = \left(A \times Q_{g,A} + L \times Q_{g,L}\right) \times \Delta P_{g,0} = \mathrm{ATR}_A.A + \mathrm{ATR}_L.L \tag{4}$$

Real gas equation considers the volume occupied by molecules of air and also the pressure change due to intermolecular forces of attraction. Using the real gas equation to solve for $P_{a(\mathrm{int})}$ as represented in Eq. (5).

$$P_{a(\mathrm{int})} = \frac{RTn}{V_{\mathrm{effective}} - nb} - \frac{an^2}{(V_{\mathrm{effective}})^2} \tag{5}$$

As a result of gas permeation, the pressure inside the core of the VIP increases and can be found out by following relation (Yrieix and Pons 2018; Simmler and Brunner 2005b).

$$\frac{\mathrm{d}p_{\mathrm{gas}}}{\mathrm{d}t} = \frac{Q_{\mathrm{gas,total}}\Delta p_{\mathrm{gas}}}{V_{\mathrm{effective}}}\left(\frac{T_m p_0}{T_0}\right)t = \frac{\mathrm{ATR}_{\mathrm{total}}}{V_{\mathrm{effective}}}\left(\frac{T_m p_0}{T_0}\right)t \tag{6}$$

where

$$V_{\mathrm{effective}} = V_{\mathrm{total}} \times \mathrm{Porosity} \tag{7}$$

Analytically solving Eq. (6) and assuming initial pressure inside the VIP to be zero, the final equation for the internal pressure change with time is given by Eq. (8),

$$p(t) = p_{\text{applied}} - \left(p_{\text{applied}} - p_{\text{initial}}\right) \times \mathrm{e}^{\frac{T_m \times p_0 \times Q_{\text{gas,total}}}{T_0 \times V_{\text{effective}}} \times t} \tag{8}$$

2.2 Determination of Thermal Conductivity

The high thermal performance of VIP is achieved when the core remains in the dry and evacuated state. The relation between the gas pressure and thermal conductivity is a result of the Knudsen number (Kan et al. 2015) and is formulated as Eq. (9). Knudsen number is a dimensionless number, given by Eq. (10), which is defined as the ratio of the mean free path in the gas phase to mean pore diameter. It plays a significant role in insulation materials where gases are under low pressure. Since the porosity of silica core is very high (>90%) and is under very low pressure, the air at ambient pressure will play an important role in degrading the quality of silica. However, considering the narrow pore size in porous silica, the free air conduction $\lambda_{g,0}$ is reduced due to the Knudsen effect. β is a constant which depends upon the nature of gas and temperature and its value is close to 1.5 (Hans et al. 2005).

$$\lambda_g = \frac{\lambda_{g,0}}{1 + 2 \times \beta \times Kn} \tag{9}$$

where Kn is the Knudsen number.

$$Kn = \frac{l_{\text{mean}}}{\delta} \tag{10}$$

$$l_{\text{mean}} = \frac{k_B T}{\sqrt{2}\pi d_g^2 p_g} \tag{11}$$

Each layer in the construction can be treated as thermal resistance and the thermal conductivity with time can be calculated using Eq. (12) (Nayak and Prajapati 2006). Equivalent thermal conductivity of VIP is then the sum of thermal conductivity in dry and evacuated state and thermal conductivity due to the air intake.

$$\lambda_{\text{core}}(t) = \lambda_{\text{evac}} + \lambda_g(t) \tag{12}$$

Also, due to the thermal bridge effect at edges, the thermal conductivity of VIP changes. In this study, the thermal bridge effect at edges is only considered as the surface area of VIP is large. So, adding a thermal bridge effect into Eq. (12), the equivalent thermal conductivity of VIP can be represented by Eq. (13). The thermal

Table 1 Input parameters

Properties	Barrier envelope materials			Source
	AF	MF-2	MF-3	
ATR_A ($cm^3\ m^{-2}\ day^{-1}$)	–	–	0.0080	Hans et al. (2005)
ATR_L ($cm^3\ m^{-1}\ day^{-1}$)	0.0018	0.0039	0.0091	Hans et al. (2005)
Activation energy (E_a) ($kJ\ mol^{-1}$)	26	28	35	Schwab et al. (2005d)
$\psi(d)$ ($W\ m^{-1}\ K^{-1}$)	0.07	0.01	0.01	Nayak and Prajapati (2006)
Dry core density ($kg\ m^{-3}$)	200			Quenard and Sallee (2005)
Specific heat ($J\ kg^{-1}\ K^{-1}$)	800			Herek and Nsofor (2014)
$\lambda_{g,0}$ ($W\ m^{-1}\ K^{-1}$)	25.7×10^{-3}			Schwab et al. (2005b)
$P_{1/2,g}$ (Pa)	593			Schwab et al. (2005c)
Van der Waals constant '*a*'	134,762.25 × 10-6 Pa $m^6\ mol^{-2}$			Handout (n.d.)
Van der Waals constant '*b*'	36.6 × 10-6 $m^3\ mol^{-1}$			Handout (n.d.)

bridge at edge depends upon linear thermal transmittance, the thickness of the edge, the total length of an edge, and total area of the edges (Nayak and Prajapati 2006).

$$\lambda_{eq}(t) = \lambda_{evac} + \lambda_g(t) + \psi(d) \times \frac{d \times l}{A} \tag{13}$$

It is assumed that the pressure increase inside the VIP is caused by dry air infusion (Schwab et al. 2005c). Consider the thermal conductivity of the center of the panel in the evacuated and dry state equal to $4 \times 10^{-3}\ W\ m^{-1}\ K^{-1}$ (Batard et al. 2018b). Further the above-mentioned factors in Eq. (13) can be calculated using Eq. (14) (Tenpierik and Van Der Spoel 2007):

$$\lambda_g(t) = \frac{\lambda_{g,0}}{1 + \frac{p_{1/2,g}}{p_g(t)}} \tag{14}$$

where $p_g = p_{g,e}\left(1 - e^{\frac{-t}{\tau_g}}\right)$. τ_g is the time constant for air and can be estimated by:

$$\tau_g = \frac{\varepsilon V}{\text{ATR}(T, \emptyset)} \times \frac{T_0}{p_0 T} \tag{15}$$

According to the Arrhenius equation, the ambient temperature greatly affects the air permeance. The equation explains the logarithmic relation between the temperature and activation energy, represented by Eq. (16).

$$Q(T) = Q(T_0)e^{\frac{E_a}{R}\left(\frac{1}{T_0} - \frac{1}{T}\right)} \tag{16}$$

For solving various equations as discussed, input parameters are shown in Table 1. The values of ATR are at 23 °C, 50% RH, and 1 bar atmospheric pressure for different envelope materials. The size of the panel is $100 \times 100 \times 2\ \text{cm}^3$.

3 Heat Transfer Analysis

The transient one-dimensional conduction heat transfer in a concrete wall having VIP has been carried out in which temperature is a function of x (wall thickness) and t (time). The total thickness of the wall was chosen to be 260 mm including the thickness of VIP as 20 and 240 mm as the thickness of the concrete wall (Saikia et al. 2018a, b). The domain is divided into N elements of Δx each as shown in Fig. 2. The aim of study is to show the variation in heat gain inside the room when VIP analysis is done considering air as a real gas.

In central difference approximation of the second derivative with respect to x, the nodal equation will involve T_{i-1}, T_i and T_{i+1}. In explicit form, the second derivative with respect to x is evaluated with all the known temperatures at time t. In its nodal form, the equation may be written as Eq. (17) (Venkateshan 2009). The analysis was carried out for 21,600 s with a time step of 1 s. The smaller time step was chosen for better precision.

$$T_{i,j+1} = F_o\left[T_{i-1,j} + T_{i+1,j}\right] + [1 - 2\text{Fo}]T_{i,j} \tag{17}$$

where Fo is Fourier number and is formulated as $\text{Fo} = \frac{\alpha \Delta t}{\Delta x^2}$, $t = (j-1)\Delta t$ at $t = 0$, $j = 1$.

The two ends of the wall are open to the environment and convectively transfer heat to the surroundings at a temperature T_∞ via heat transfer coefficient h. The nodal equation representing the temperature at the wall ends is as follows

$$T_{1,j+1} = T_{1,j}[1 - 2\text{Fo}(1 + \text{Bi})] + 2F_o \times \text{Bi} \times T_\infty + 2F_o T_{2,j} \tag{18}$$

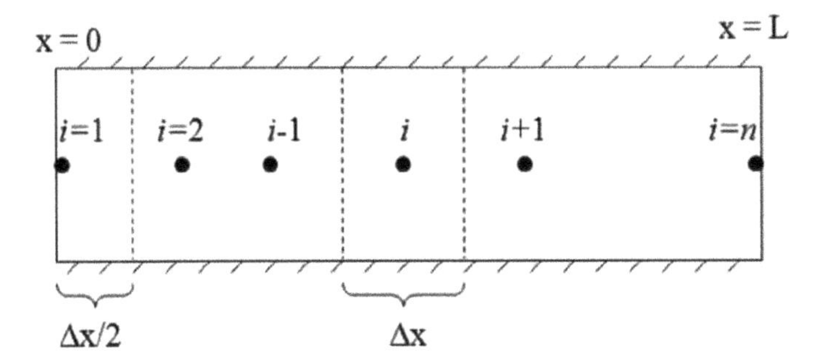

Fig. 2 1-D transient heat transfer along the wall

where Bi is the elemental Biot number and is given by $\mathrm{Bi} = \frac{h\Delta x}{k}$.

Sol–air temperature is computed by Eq. (19) to combine the effects of solar radiation and ambient temperature for heat transfer analysis (Sharma and Rakshit 2017; Krarti and Ihm 2009). The analysis is done for the Jodhpur city of Rajasthan (Nayak and Prajapati 2006). The sol–air temperature was calculated considering solar radiation for 6 h (11 AM to 5 PM) for the month of May. The average ambient temperature was around 312.55 K, and the corresponding sol–air temperature was calculated to be 329.6 K. The initial temperature was 303 K at the very first node. Either solar irradiance or temperature on the south wall is maximum for this period of time.

$$T_{\mathrm{sol}} = T_0 + \frac{\alpha_s \times I}{h} \tag{19}$$

The thermal performance of the VIP in a concrete embedded VIP wall is evaluated using transient heat transfer analysis. The deviation in thermal performance of VIP is evaluated in terms of reduced heat gain and variation in inside and outside temperature which changes due to the location of the VIP into the concrete wall. The optimum location of VIP is investigated by positioning the VIP at three different locations, i.e., the inner side of the wall, center of the wall, and on the outer side of the concrete VIP wall assembly as shown in Fig. 3. Heat ingress through a wall carrying VIP with different thermal conductivities obtained by considering air as an ideal gas and a real gas is compared for the given geographical conditions.

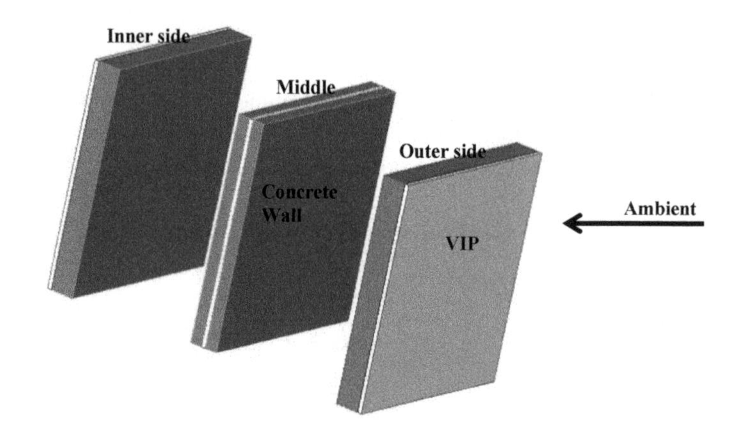

Fig. 3 Different location of VIP in the concrete wall

4 Ideal Gas Analogy

If volume occupied by air molecules and intermolecular forces of attraction, i.e., '*a*' and '*b*' is assumed to be negligible, then the air acts as an ideal gas. In the present model, by equating Van der Waals constants, i.e., a and b to zero in Eq. (5) and calculating further, a similar trend in results with relative error in the permissible range were attained as that using the ideal gas equation. Moreover, further calculations authenticate the present model more accurately as the thermal conductivity and internal pressure lies in the allowable deviation range as calculated using ideal gas, as illustrated by Figs. 4 and 5, respectively.

Also, the grid independence study was performed to optimize the number of nodes to carry the heat transfer analysis for VIP-concrete composite wall.

The numerical model developed for transient heat transfer study was tested for grid independence in order to institute the reliability of the model demonstrated in Fig. 6. For this, the number of nodes in the VIP wall is varied keeping the wall thickness the same. The VIP is placed on the inside of the wall and south wall irradiance is taken. The number of nodes in the VIP wall is varied from 26 to 234 nodes with an increment of 26 nodes in every step. Further, calculations are performed with 182 nodes after which the graph becomes asymptotic with dQ value of 4.31 kJ.

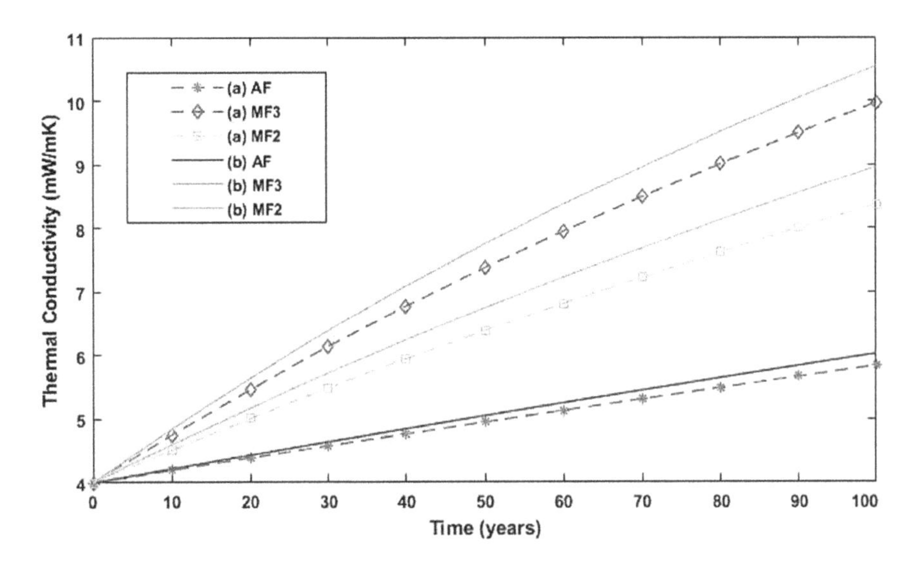

Fig. 4 **a** Thermal conductivity variation with time as per (Simmler and Brunner 2005a) (dotted lines) and **b** as per the present study (continuous lines)

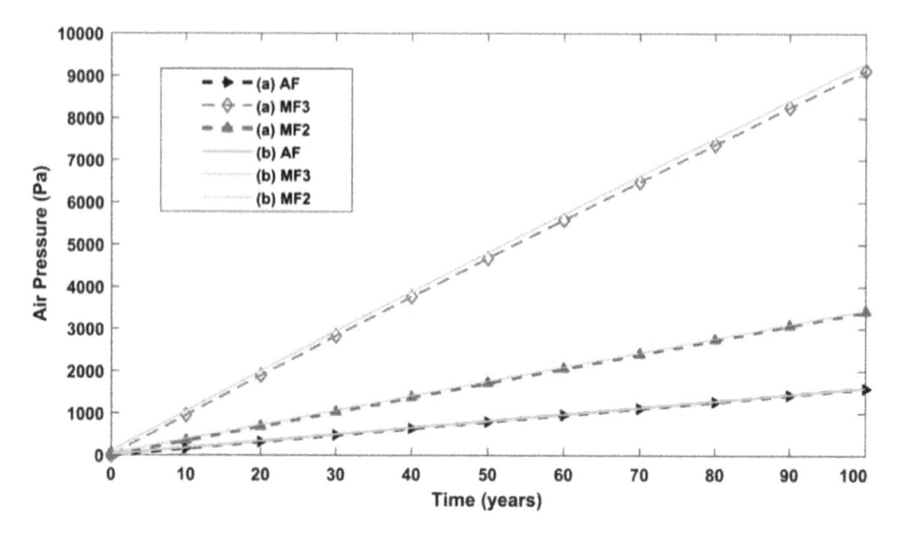

Fig. 5 **a** Variation in inside air pressure with time as (Simmler and Brunner 2005a) (dotted lines) and **b** as per the present study (continuous lines)

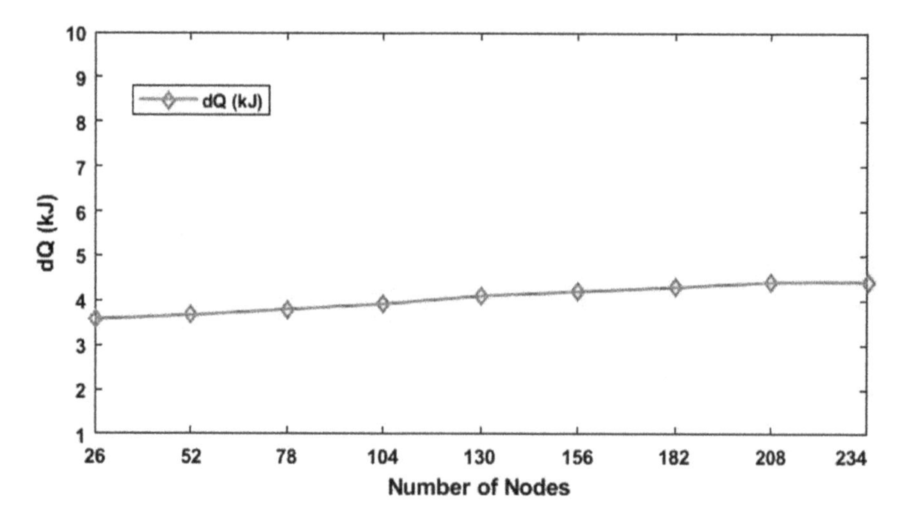

Fig. 6 Grid independence study

5 Results and Discussion

5.1 Comparison Analysis

The values of ATR for envelope material are used and calculations are performed using equations discussed previously. The ATR values are at 23 °C and 50% RH. By

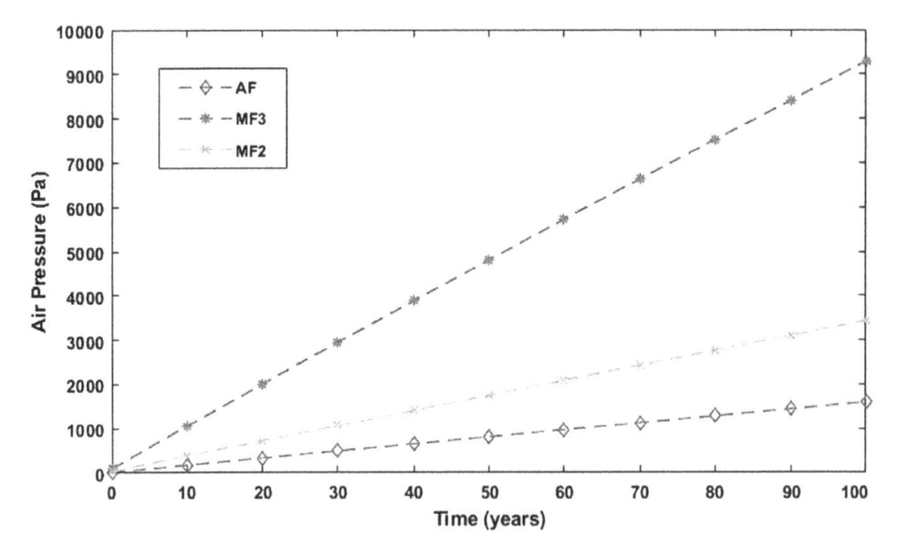

Fig. 7 Variation in inside air pressure with time for different envelope material

using Eq. (8), the air pressure with respect to time is calculated. Values are recorded assuming constant climatic conditions during the entire period. Similar variations in the parameters are observed when Van der Waals constants are considered. The study conducted by Wegger et al. (2011) neglected the thermal bridge effect (Wegger et al. 2011). To compare the present findings with the results of Wegger et al. (2011), the thermal bridge effect is neglected in Sect. 5.1.

Figure 7 depicts the variation in air pressure inside the VIP with time. Equation (8) is used to determine the variation of inside air pressure with time. The pressure increase is due to the air permeation inside the VIP through the envelope pores and sealing. It is because of the reason that different envelope materials have different resistance to air permeation because of the presence of multiple layers. The results got deviated from the prevailing results of the ideal gas equation when calculations were carried out using Van der Waals equation. Since Van der Waals equation considers the volume occupied by the air molecules that degrade the core of the VIP. Hence, the thermal conductivity increases.

The result comparison for the ideal gas equation and Van der Waals equation is mentioned in Table 2 consisting of different envelope materials. The curve for AF embedded VIP shows that internal air pressure is 1.62% more when analyzed using Van der Waals equation as compared to ideal gas for 50 years. With the increase in load on panel, i.e., atmospheric pressure, the rate of air ingression increases. Air permeation also depends upon the type of envelope, the thickness of envelope material, and geometrical specifications of the VIP. For a relatively shorter period of time (less than 30 years), the difference in pressure values calculated using the two different equations is not marginal. For buildings located in regions susceptible to frequent natural calamities (earthquake, tornado, etc.), the overall service life of the

Table 2 Variation in inside pressure (in Pa) due to air permeation

Envelope material		15 years	30 years	45 years	50 years
AF	Ideal	240.08	479.6	718.544	798.1
	Real	255.86	494.2	731.8935	811
	% increase (%)	6.5	3.04	1.8	1.62
MF-2	Ideal	519.4	1036.1	1550.1	1720.9
	Real	553.8	1068	1579.6	1749.5
	% increase (%)	6.62	3.08	1.9	1.66
MF-3	Ideal	1424.8	2829.6	4214.6	4671.9
	Real	1530.1	2945.5	4340.9	4801.6
	% increase (%)	7.39	4.10	2.99	2.78

building will be shorter and within this period the pressure calculation with the real gas equation is justified as illustrated by the difference in Table 2.

Equation (13) is used to calculate the thermal conductivity of VIP and results are shown in Fig. 8. The trend shows the direct relation between the time and thermal conductivity of the VIP. Internal pressure is a function of air ingression through envelope which further affects the thermal conductivity. The fumed silica core adsorbs the air molecules on its surface that are responsible for core degradation and consequently the performance of the VIP. The thermal conductivity values are highest for MF-3 envelope embedded VIP followed by MF-2 and are lowest for AF. Moreover, Table 3 compares the thermal conductivity result calculated by using the ideal gas

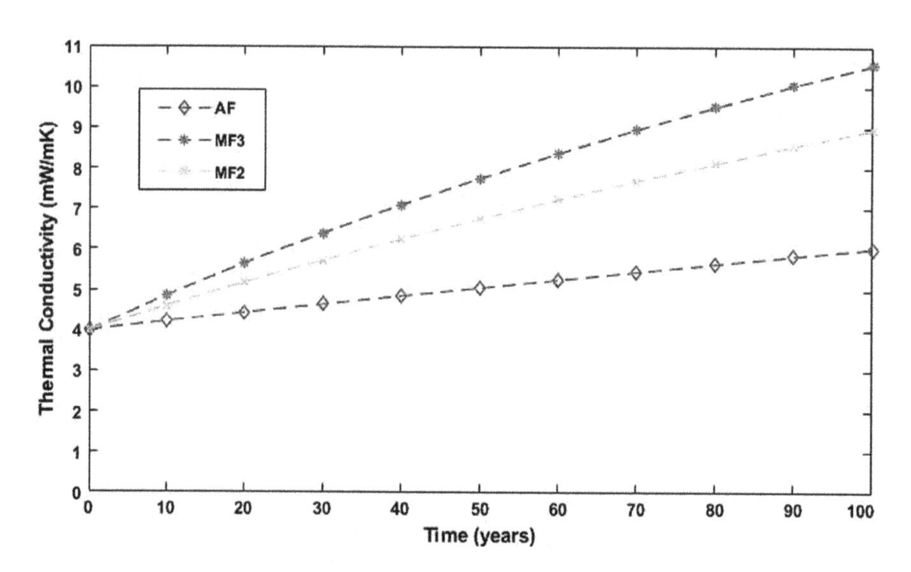

Fig. 8 Thermal conductivity variations with time for different envelope materials

Table 3 Variation in thermal conductivity (mW m^{-1} K^{-1})

Envelope material		15 years	30 years	45 years	50 years
AF	Ideal	4.2930	4.5794	4.8594	4.9514
	Real	4.3252	4.6422	4.9514	5.0527
	% increase	0.75	1.37	1.89	2.05
MF-2	Ideal	4.7668	5.489	6.1705	6.3891
	Real	4.8901	5.7205	6.4970	6.7447
	% increase	2.58	4.22	5.29	5.57
MF-3	Ideal	5.1152	6.1373	7.0776	7.3745
	Real	5.2530	6.3891	7.4241	7.7489
	% increase	2.69	4.10	4.89	5.08

Table 4 Service life of VIP

Envelope material	Service life (years)		% decrease in service life
	Ideal gas equation	Real gas equation	
AF	250	220	12
MF-2	92	77	16.3
MF-3	65	53	18.46

equation and the Van der Waals equation. It has been found that the variation is between 2 and 6% for the service life of 50 years.

Table 4 compares the results for the service life of VIP for different envelope materials. The increase of thermal conductivity over time reduces the service life of the VIP. Service life is defined as the time elapsed from the moment of manufacturing until the moment thermal conductivity exceeds limiting value as prescribed. Service life depends on the permeation inside VIP and external environmental conditions. Limiting value is set at 8×10^{-3} W m^{-1} K^{-1} (Schwab et al. 2005c). The results vary appreciably when calculated using the real gas equation. The maximum variation is observed in VIP with MF-3 envelope material, i.e., 18.46%. Using VIP with AF envelope material gives the highest service life, i.e., 220 years. Environmental conditions such as relative humidity, temperature, and solar irradiance affect the aging of the VIP. Getters and desiccants are provided inside the VIP which adsorb the air molecules so that the core remains in the dry and evacuated state for a longer period of time. This measure reduces thermal conductivity aging and increases service life.

The area of the wall is 4×3 m^2 (Ramos 1990). To evaluate the heat gain inside the room through the wall, the VIP was placed at three locations, viz. inner, center, and outer side. When the VIP was placed on the innermost side of the wall, the VIP concrete wall combination results in minimum heat gain inside the room through the wall and values are presented in Table 5. It is because of the reason that VIP is not directly exposed to solar radiation which reduces the direct environmental impact on it and also improves its thermal performance.

Table 5 Heat gain at different locations of VIP

Location of VIP	Heat gain (kJ)		
	AF	MF-2	MF-3
Inside	115.9091	93.2487	93.2487
Center	13,750.912	13,750.912	13,750.912
Outside	6472.786	14,388.838	14,546.298

Table 6 Heat gain comparison for MF-2 embedded VIP

Envelope material	Time (years)	Thermal conductivity (mW m^{-1} K^{-1})		dT (K)		Heat transfer, Q (kJ)		dQ (real-ideal) (kJ)
		Ideal	Real	Ideal	Real	Ideal	Real	
MF-2	10	4.516	4.600	0.039	0.039	89.153	89.647	0.494
	20	5.012	5.173	0.041	0.042	93.405	94.893	1.487
	30	5.489	5.720	0.043	0.044	97.951	100.427	2.475
	40	5.947	6.243	0.045	0.047	102.92	106.357	3.430
	50	6.389	6.744	0.048	0.049	108.15	112.475	4.317
	60	6.814	7.224	0.050	0.052	113.36	118.540	5.174
	70	7.224	7.684	0.052	0.055	118.54	124.630	6.089
	80	7.620	8.123	0.054	0.057	123.76	130.610	6.847
	90	8.002	8.550	0.057	0.060	128.89	136.520	7.626
	100	8.370	8.953	0.059	0.063	133.99	142.145	8.155

The heat gains through VIP–concrete composite wall when calculated using two different equations for MF-2 embedded VIP placed on the inner side of the wall are shown in Table 6. There was an 8.15 kJ of additional heat gain inside the room when the VIP–concrete composite wall was analyzed considering air as a real gas. Aforementioned thermal conductivity calculated using Van der Waals equation is 5–6% more when analyzed considering air as an ideal gas. Thermal conductivity directly affects the heat transfer rate as it permits more heat to pass through the wall.

5.2 *Heat Transfer Analysis Considering Thermal Bridge*

To calculate the heat gain inside the room through the concrete–VIP composite wall, the equivalent thermal conductivity of the VIP is calculated using Eq. (13), which also accounts for the thermal bridge effect. The thermal bridge has a significant effect on the thermal conductivity of AF embedded VIP as compared to other two envelope materials, i.e., MF-2 and MF-3 as shown in Tables 7, 8 and 9. Thermal bridging has an adverse effect on the overall performance of the VIP. The AF laminate reduces

Table 7 Heat transfer for AF embedded VIP

Time (years)	Thermal conductivity (mW m^{-1} K^{-1})	dT (K)			Heat transfer (kJ)
		T_{wall} (K)	T_i (K)	dT	
20	7.2317	297.0526	297	0.0526	118.666
40	7.6492	297.0551	297	0.0551	124.147
60	8.0530	297.0575	297	0.0575	129.633
80	8.4438	297.0599	297	0.0599	135.015
100	8.8224	297.0622	297	0.0622	140.286

Table 8 Heat transfer for MF-2 embedded VIP

Time (years)	Thermal conductivity (mW m^{-1} K^{-1})	dT (K)			Heat transfer (kJ)
		T_{wall} (K)	T_i (K)	dT	
20	5.5733	297.0439	297	0.0439	98.9727
40	6.6438	297.0493	297	0.0493	111.2145
60	7.6245	297.0549	297	0.0549	123.8394
80	8.5273	297.0604	297	0.0604	136.1448
100	9.3583	297.0656	297	0.0656	147.9302

Table 9 Heat transfer for MF-3 embedded VIP

Time (years)	Thermal conductivity (mW m^{-1} K^{-1})	dT (K)			Heat transfer (kJ)
		T_{wall} (K)	T_i (K)	dT	
20	6.0438	297.0461	297	0.0461	104.0097
40	7.4895	297.0541	297	0.0541	122.0332
60	8.7708	297.0619	297	0.0619	139.5855
80	9.9142	297.0692	297	0.0692	155.9366
100	10.9408	297.0758	297	0.0758	170.9716

the air permeation significantly and thereby increasing the service life of the VIP. However, because of the high thermal conductivity of aluminum (210 W m^{-1} K^{-1}) as compared to core thermal conductivity (0.004 W m^{-1} K^{-1}), the contribution of thermal bridging is higher. In contrast, the MF laminates show more increase in thermal conductivity but are less affected by thermal bridging due to polymer layers. Therefore, the combined effect of air permeation and thermal bridging could suggest that MF laminates are more suitable for building applications for the service life of 40–50 years.

6 Conclusions

1. A comprehensive study of VIP's aging characteristics has been conducted. The study illustrates the mathematical modeling of thermal performance of VIP considering air as real gas. The results vary when solved using Van der Waals equation instead of the ideal gas equation. The results for pressure increase inside the VIP and thermal conductivity deviate by 3–7% and 2–5% for MF-3 embedded VIP, respectively.
2. The service life of VIP is affected by 12% in AF embedded VIP and maximum decrease is observed for MF-3 embedded VIP, i.e., 18%. Therefore, using AF embedded VIP is the best option for longer service life as inferred from the findings of the present study.
3. It is observed that when VIP is placed on the inner side of the wall, it results in minimum heat gain inside the room through the wall as compared to other two locations, i.e., at the outer surface and at the center. When placed in the center of the wall the heat gain is the same as that with no VIP installed on the wall.
4. There is approximately 2.8 mW m^{-1} K^{-1} increase in thermal conductivity for AF embedded VIP when the thermal bridge at edges is also considered for heat transfer analysis. The thermal bridge effect is less significant for MF-2 and MF-3 envelope material.
5. It is clearly visible from the analysis that the thermal bridge has a significant effect on the performance of the VIP. It increases the equivalent thermal conductivity of VIP which leads to more heat gain and decrease in the service life.
6. The results show appreciable variation in performance of VIP for up to 50 years, which is the average life of the building. This justifies the consideration of air as a real gas for a more realistic approach.

References

Alotaibi SS, Riffat S (2014) Vacuum insulated panels for sustainable buildings: a review of research and applications. Int J Energy Res. https://doi.org/10.1002/er.3101

Baetens R, Jelle BP, Thue JV, Tenpierik MJ, Grynning S, Uvsløkk S, Gustavsen A (2010) Vacuum insulation panels for building applications: a review and beyond. Energy Build 42:147–172. https://doi.org/10.1016/j.enbuild.2009.09.005

Batard A, Duforestel T, Flandin L, Yrieix B (2018a) Prediction method of the long-term thermal performance of Vacuum Insulation Panels installed in building thermal insulation applications. Energy Build 178:1–10. https://doi.org/10.1016/j.enbuild.2018.08.006

Batard A, Duforestel T, Flandin L, Yrieix B (2018b) Modelling of long-term hygro-thermal behaviour of vacuum insulation panels. Energy Build 173:252–267. https://doi.org/10.1016/j.enbuild.2018.04.041

Benson DK, Potter TF, Tracy CE (1994) Design of a variable-conductance vacuum insulation

Brunner S, Simmler H, Brunner S (2014) Thermal properties and service life of vacuum insulation panels (VIP) Service Life Prediction of VIPs-Project on comparing reality with former predictions

View project Thermal properties and service life of vacuum insulation panels (VIP), https://www.researchgate.net/publication/228681787

Capozzoli A, Fantucci S, Favoino F, Perino M (2015) Vacuum insulation panels: analysis of the thermal performance of both single panel and multilayer boards. Energies. https://doi.org/10.3390/en8042528

Fricke J, Schwab H, Heinemann U (2006) Vacuum insulation panels—exciting thermal properties and most challenging applications. Int J Thermophys. https://doi.org/10.1007/s10765-006-0106-6

Handout (n.d.) che31.weebly.com/uploads/3/7/4/3/3743741/handout_e-realgasconstants.pdf. Accessed 19 June 2019

Hans S, Samuel B, Ulrich H, Schwab H, Kumar K, Phalguni M, Daniel Q, Hébert S, Klaus N, Cornelia S, Esra K, Martin T, Hans C, Markus E (2005) Vacuum insulation panels—study on VIP-components and panels for service life prediction of VIP in building applications (Subtask A), HiPTI—High Perform. Therm. Insul.—IEA/ECBCS Annex 39

Herek SJ, Nsofor EC (2014) Performance of vacuum insulation panels in building energy conservation, pp 149–160

Kalnæs SE, Jelle BP (2014) Vacuum insulation panel products: a state-of-the-art review and future research pathways. Appl Energy. https://doi.org/10.1016/j.apenergy.2013.11.032

Kan A, Kang L, Wang C, Cao D (2015) A simple and effective model for prediction of effective thermal conductivity of vacuum insulation panels. Futur Cities Environ. https://doi.org/10.1186/s40984-015-0001-z

Krarti M, Ihm P (2009) Implementation of a building foundation heat transfer model in EnergyPlus. J Build Perform Simul 2:127–142. https://doi.org/10.1080/19401490802613610

Liang Y, Wu H, Huang G, Yang J, Wang H (2017) Thermal performance and service life of vacuum insulation panels with aerogel composite cores. Energy Build 154:606–617. https://doi.org/10.1016/j.enbuild.2017.08.085

Liang Y, Ding Y, Liu Y, Yang J, Zhang H (2019) Modeling microstructure effect on thermal conductivity of aerogel-based vacuum insulation panels. Heat Transf Eng 1–14. https://doi.org/10.1080/01457632.2019.1576443

Molleti S, Lefebvre D, van Reenen D (2018) Long-term in-situ assessment of vacuum insulation panels for integration into roofing systems: Five years of field-performance. Energy Build 168:97–105. https://doi.org/10.1016/j.enbuild.2018.03.010

Nayak JK, Prajapati JA (2006) Handbook on energy conscious buildings. In: Handb. Energy Conscious Build., Solar Energy Centre, Ministry of Non-conventional Energy sources, Government of India, New Delhi, India

Quenard D, Sallee H (2005) Micro-nano porous materials for high performance thermal insulation micro-nano porous materials for high performance. In: 2nd international symposium on nanotechnology in construction, pp 1–10

Ramos JI (1990) Basic heat transfer. Appl Math Model 14:666. https://doi.org/10.1016/0307-904X(90)90027-3

Saikia P, Azad AS, Rakshit D (2018a) Thermodynamic analysis of directionally influenced phase change material embedded building walls. Int J Therm Sci 126:105–117. https://doi.org/10.1016/j.ijthermalsci.2017.12.029

Saikia P, Azad AS, Rakshit D (2018b) Thermal performance evaluation of building roofs embedded PCM for multi-climatic zones. Springer, Singapore, pp 401–423. https://doi.org/10.1007/978-981-10-7188-1_18

Schwab H, Heinemann U, Beck A, Ebert H-P, Fricke J (2005a) Dependence of thermal conductivity on water content in vacuum insulation panels with fumed silica kernels. J Therm Envel Build Sci 28:319–326. https://doi.org/10.1177/1097196305051792

Schwab H, Heinemann U, Wachtel J, Ebert HP, Fricke J (2005b) Predictions for the increase in pressure and water content of Vacuum Insulation Panels (VIPs) integrated into building constructions using model calculations. J Therm Envel Build Sci 28:327–344. https://doi.org/10.1177/1097196305051793

Schwab H, Heinemann U, Beck A, Ebert HP, Fricke J (2005c) Prediction of service life for vacuum insulation panels with fumed silica kernel and foil cover. J Therm Envel Build Sci. https://doi.org/10.1177/1097196305051894

Schwab H, Heinemann U, Beck A, Ebert HP, Fricke J (2005d) Permeation of different gases through foils used as envelopes for vacuum insulation panels. J Therm Envel Build Sci. https://doi.org/10.1177/1097196305051791

Sharma P, Rakshit D (2017) Quantitative assessment of orientation impact on heat gain profile of naturally cooled buildings in India. Adv Build Energy Res 11:208–226. https://doi.org/10.1080/17512549.2016.1215261

Simmler H, Brunner S (2005a) Vacuum insulation panels for building application: basic properties, aging mechanisms and service life. Energy Build. https://doi.org/10.1016/j.enbuild.2005.06.015

Simmler H, Brunner S (2005b) Vacuum insulation panels for building application: Basic properties, aging mechanisms and service life. Energy Build 37:1122–1131. https://doi.org/10.1016/j.enbuild.2005.06.015

Tenpierik MJ, Cauberg JJM (2010) Encapsulated vacuum insulation panels: theoretical thermal optimization. Build Res Inf 38:660–669. https://doi.org/10.1080/09613218.2010.487347

Tenpierik M, Van Der Spoel W (2007) Simplified analytical models for service life prediction of a vacuum insulation panel Double Face 2.0 View project Research through Design for Values View project Analytical VIP Service Life Prediction Model Simplified Analytical Models for Service Life Prediction of a Vacuum Insulation Panel, https://www.researchgate.net/publication/284778729

Venkateshan SP (2009) Heat transfer. Ane Books

Wegger E, Jelle BP, Sveipe E, Grynning S, Gustavsen A, Baetens R, Thue JV (2011) Aging effects on thermal properties and service life of vacuum insulation panels. J Build Phys 35:128–167. https://doi.org/10.1177/1744259111398635

Xu T, Chen Z, Yang Y, Chen Z, Zhang J, Wu C, Liu Y (2018) Correlation between the thermo-physical properties and core material structure of vacuum insulation panel: role of fiber types. Fibers Polym 19:1032–1038. https://doi.org/10.1007/s12221-018-7949-x

Yrieix B, Pons E (2018) New method to assess the air permeance into vacuum insulation panel. Vacuum. https://doi.org/10.1016/j.vacuum.2017.11.033

Solar-Based Electric Vehicle Charging Stations in India: A Perspective

Rajan Kumar, Rabinder Singh Bharj, Jyoti Bharj, Gurkamal Nain Singh, and Monia Sharma

1 Introduction

1.1 Background

Many countries are quickly adopting the technologies related to EVs and slowly eliminating fossil fuel-based vehicles to battle climate alteration and increasing pollution. Most air pollution produced the world over is by burning fossil fuel to produce electricity, heating, transportation, and industries. Fossil fuels accounted for 80.110% of the world's total primary energy supply in 2019 (monthly energy review 2020). It is observed that energy-related use of fossil fuel is a major source of air pollution amongst high- and middle-income countries, while the use of biomass is a prime concern for air pollution resulting from energy generation in the case of low-income countries. The combustion of fossil fuels in vehicles produces harmful gases which have an adverse effect on the environment and human beings. Emissions of acidic sulfur dioxide and nitrogen oxides, mixed with atmospheric moisture, cause acid rain that leads to harmful effects to the community as well as forest area. The most hazardous impact of fossil fuels is carbon dioxide (CO_2) emission into the atmosphere (Perera 2018). CO_2 is one of the major trace gases in the atmosphere, the level of which had increased by 35% in the last 200 years. From 1985 to 2005 along the level of CO_2 had increased from 316 to 375 ppm, resulting in global warming, which results in the overall increase of 1.25 °F (0.7 °C) in the temperature level of the

R. Kumar (✉) · R. S. Bharj · G. N. Singh
Department of Mechanical Engineering, Dr. B. R. Ambedkar National Institute of Technology Jalandhar, Jalandhar, Punjab 144011, India
e-mail: rajank@nitj.ac.in; rajan.rana9008@gmail.com

J. Bharj · M. Sharma
Department of Physics, Dr. B. R. Ambedkar National Institute of Technology Jalandhar, Jalandhar, Punjab 144011, India

H. Tyagi et al. (eds.), *New Research Directions in Solar Energy Technologies*,
Energy, Environment, and Sustainability,
https://doi.org/10.1007/978-981-16-0594-9_10

planet. This may result in rising sea levels which could threaten a large population and may cause extreme environmental changes. The international community has taken action on global warming and has been working on reducing carbon emissions through the Kyoto Protocol. Therefore, a move toward EVs and plug-in hybrid electric vehicles (PHEVs) are much promising and promising alternative. PHEVs have both, an electric motor and an internal combustion (IC) engine. These vehicles are powered by conventional fuel, such as gasoline (petrol) or by an alternative fuel such as electricity and using a battery for charging. The charging can be done by plugging either into an electrical outlet or at a charging station. Due to the good response of EVs worldwide, national and international governments have come out with various policies and allocated funds to support EVs and PHEVs implementation in the market. Long-term planning scenarios express that the EVs will capture the whole of the global vehicle fleet, mostly run by renewable energy sources, by 2050 (https://www.theicct.org/electric-vehicles).

1.2 Worldwide Adaptation of EVs

The occurrence of major events taken place in the European countries that announced to phase out petrol and diesel-based vehicles in support of EVs are as follows (https://www.roadtraffic-technology.com/features/european-countries-banning-fossil-fuel-cars):

- In 2016, Norway announced its proposed ban on fossil fuel vehicles. By 2025, the purchasing of vehicles run on petrol and diesel will be banned. More than half of the new vehicle sold in Norway in January 2017 was either electrical or hybrid. This turned Norway amongst the first nation to sell cars with zero or low emissions. Till 2018, over 135,000 EVs have been registered in the country. In order to continue this hike, the association of EVs in the country is making policies for having more than 4 lakh battery-powered vehicles on highways by 2020.
- Germany's Bundesrat federal council agreed to ban on vehicles based on fossil fuels by 2030. In October 2016, the European Commission announced an EU prohibition on petrol and diesel vehicles. The nation plans to reduce its CO_2 emissions by 95% by 2050, hoping that EVs will help accomplish this objective.
- In July 2017, in addition to phasing out oil and gas production, France proposed to ban all petrol and diesel vehicles by 2040. This will play a major part in the country's objective of becoming carbon neutral by 2050. By 2030, Paris suggested the removal of all fossil fuel cars from the city. The city is implementing a provisional ban to reduce air pollution. In 2018, nearly 150,000 EVs were registered in France and 1.98% of new cars registered in 2017 contributed to electric cars.
- In July 2017, the UK supported banning the sale of new petrol or diesel vehicles with a 2040 target of a full ban. When plans were first announced, plug-in vehicles' contribution is less than 1% of total car sales in the country. This figure increases in the first half of 2018 to 2.2% of the UK's total vehicle sales.

- In September 2017, Scotland plans to remove fossil fuel vehicles by 2032. When compared with the rest of the UK and Scotland is eight years ahead.
- In October 2018, Netherland gives its confirmation to ban gasoline and diesel vehicles by 2030, and all new vehicles are emission free. Plans were initially developed in April 2016. In the Netherlands, 6.4% of the cars were electric when plans were acknowledged.
- Within the framework of the climate action scheme, Ireland plans to completely ban nonzero emission cars by 2045. To accomplish this goal, Ireland plans to have 500,000 EVs on its highways and prevent nonzero emission vehicle sales by 2030.

2 India's Growth in EV Global Market

In 2013, the GoI made a significant shift toward the adoption of EVs by 2030 to resolve the issues of national energy security, vehicle emissions, and the development of domestic built-up capacities, reiterating its pledge to the Paris agreement (https://beeindia.gov.in/content/e-mobility). Indian car manufacturing companies, like Reva Electric Car Company (RECC) and many other companies such as Nissan, TATA, Mahindra, Maruti Suzuki, would launch their EVs by 2020 (Taumar 2019). Niti Aayog's draft proposal which is chaired by the Prime Minister of India is playing a significant part in policy making and has suggested electrifying most motorcycles and scooters over the next six to eight years. It further insisted to electrify the country's famous three-wheeler auto-rickshaws (Shah 2019).

India sold more than 21 million motorcycles and scooters in 2018–19 making it one of the world's largest two-wheeler markets. Over the same time, it sold only 3.3 million cars and utility vehicles. Although the electric scooters accounted for just a fraction of the total sales of two-wheelers in India, the data from the society of manufacturers of electric vehicles reveals that the sales of scooter rose to 126,000, more than doubled, in the last fiscal year, from only 54,800 a year earlier (Shah 2019). If the initiative to electrify vehicles operating with the Ministries of Heavy Industry, Road Transport, and Power were to be accepted, a new market will open up for companies such as Japan's Yamaha Motor and Suzuki Motor, which are developing plans to introduce electric two-wheelers in India. The state now supposes electric cars to make up 15% of all fresh revenues in 5 years from presently less than 1%. India's cabinet endorsed a plan in February 2019 to spend $1.4 billion over three years sponsoring electric and hybrid vehicle sales (https://www.reuters.com/article/us-india-electric-policy-idUSKCN1SS2HS).

2.1 The Faster Adoption and Manufacturing of Hybrid and Electric Vehicles (FAME)

In 2012, the GoI has launched the National Electric Mobility Mission Plan (NEMMP) 2020, to encourage eco-friendly vehicles in the country. The main objectives of this plan are providing economic and financial incentives for the implementation of both hybrid and electric technologies vehicles. Incentives have been imposed on all vehicle segments including two-, three-, and four-wheeler vehicles, light commercial vehicles, and buses. This plan covers hybrid and electrical technologies such as powerful hybrid, plug-in hybrid, and battery EVs (National Electric Mobility Mission Plan 2020).

The government introduced the faster adoption and manufacturing of hybrid and EVs (FAME India) scheme under NEMMP 2020 in the Union Budget for 2015–16 with an initial expenditure of INR 750 million FAME II scheme has been launched by the union cabinet on February 28, 2019, with a total amount of 10,000 crores. The chosen amount will be laid out in the next 3 years with the scheme being implemented from April 1, 2019. The current allocation of 10,000 crores has been utilized to speed up the development of EVs and EVs infrastructure in a bid to meet the target of 100% electrification of vehicles by 2030 (Cabinet approves Scheme for FAME India Phase II 2019). FAME II scheme basically aims to incentivize the purchase of EVs and establishing the required charging infrastructure for EVs. The emphasis will also continue to be on the electrification of public transport vehicles and other shared mobility solutions like last-mile connectivity solutions such as three-wheelers and light commercial vehicles. The government is planning to sell only electric three-wheelers from April 2023 and electric two-wheelers of less than 150 cc from April 2025. Steps will be taken to lessen the level of rising air pollution in India. The incentives will be mainly distributed to the commercial three-wheeler and four-wheeler vehicles along with a private two-wheeler. Incentives will be distributed only to those vehicles which are powered by a lithium-ion battery or work on other advanced technology like fuel cell as an effort to encourage new technologies. This will eliminate the use of EVs powered by lead-acid batteries which form a major chunk of electric two-wheelers in India. The revised FAME II scheme eliminates all the ambiguity and places EVs in the fast track. Mahindra promotes the government's focus on increasing public transportation EVs and is now calling on local authorities to assist promoting the use of EVs on Indian roads. Government support under the FAME II scheme is holistic and also would include focusing on charging EV infrastructure with a clear vision of "Make in India."

The idea is to make at least one charging station in a 3×3 km grid, according to the FAME II scheme. The charging stations will also be set up at a distance of 25 km on either side of major highways which connect major cities of India. The easy availability of charging stations also helps to give a boost to the sales of EVs as well. It will also be influential for automakers like Mahindra and Tata Motors who are selling electric cars in India and also encouraging other automakers to include EVs in their product lineup. The state-run energy efficiency services (EESL) declared its plan to acquire up to 20,000 EVs for government use with an investment of 2,400 crores. The scheme aims at promoting the quicker implementation of electric and hybrid vehicles by proposing excellent incentives for the buying of EVs and likewise by developing the required charging infrastructure for EVs. The scheme will help to deal with the problem of air pollution and fuel security.

3 Electric Charging Infrastructure

Approximately, 200 million vehicles are currently on Indian roads, of which less than 1% are EVs. By 2040, India is expected to be home to 31 million EVs. In order to keep this fleet running, widespread charging infrastructure for EVs is needed. With the government's goal of achieving 30% e-mobility by 2030, we need not only to step up efforts in the manufacturing of EVs but also to ensure that there are adequate charging stations in the country (https://www.eeslindia.org/content/raj/eesl/en/Programmes/EV-Charging-Infrastructure/About-EV-Charging-Infrastructure.html). In this section, different levels/types of EV charging stations and the components of EV chargers are discussed.

3.1 EV Charging Station

A charging station for EVs, also known as an EV charging station, electronic charging station (ECS) is a component of an infrastructure that provides electrical energy to recharge EVs, including electric cars and plug-in hybrids. EV chargers are classified as follows: (i) Level 1 charging stations: home charging, (ii) level 2 charging stations: home and public charging, and (iii) level 3 charging stations: DC fast charging—public charging. Table 1 illustrates the comparison between level 1, level 2, and level 3 charging stations.

3.1.1 Level 1 Charging Stations: Home Charging

Level 1 chargers mean an alternating current (AC) plug that connects the on-board charger and a standard household (120 V/15–20 A). All EVs are equipped with a portable cord set charger. Only a three-pronged plug should be attached to a cord set charger. Therefore, these kinds of chargers do not require any extra equipment to be mounted and are ideal for residential usage. EV will charge at a slow rate by using these chargers. This option may not work great; however, this is good for those commuters whose traveling distance is not more than 40 miles in a day and has to charge the EV throughout the night.

Table 1 Comparison between level 1, level 2, and level 3 (DC fast charger) charging stations (electric vehicle charging stations 2015)

	Level 1	Level 2	Level 3 DC fast charger
Voltage (V)	120 1-Phase AC	208 or 240 1-Phase AC	200–450 DC
Current (A)	12–16	12–80 (Typ. 32)	<200 (Typ. 60)
Useful power (kW)	1.4	7.2	50
Max. output (kW)	1.9	19.2	150
Charging time	12 h	3 h	20 min
Connector	J1772	J1772	J1772 combo, CHAdeMO and supercharger
Charging speed	3–5 miles of range in 1 h	10–20 miles of range in 1 h	24 or 50 miles of range in 20 min
Cost	Generally low	Generally higher	Much higher
Access control	Available	Available	Available
Energy monitoring	Not available, but available on the secondary system	Available	Available

3.1.2 Level 2 Charging Stations: Home and Public Charging

Level 2 charging stations offer charging through a 240 V, AC plug and need a dedicated 40 Amps circuit. These are useful for both housing and public charging stations. Specialized installation equipment is required for the plugging of these chargers. Depending on the electric car and the charger, the charging rate is 5 to 7 times faster than the level 1 charging station (https://chargehub.com/en/how-to-choose-home-charging-station.html). An electric car battery can be charged within three hours by using these chargers. Therefore, it is a good option for the customers seeking fast charging as well as for business purposes providing charging stations as a benefit for the customers when shopping, banking, etc. More efforts are given to boost the charging capacity and reduce the charging time of level 2 chargers. ClipperCreek, JuiceBox, Chargepoint, and Siemens are the popular manufacturers of level 2 chargers. Most producers of EVs, such as Nissan, have their own level 2 charger goods.

Due to the increased demand for electric cars in the market, there are more level 2 home charging stations. Different types of AC vehicle-side connectors used in level 2 charging stations are as follows (https://chargehub.com/en/how-to-choose-home-charging-station.html): Type 1 (SAE J1772), type 2 (Mennekes, IEC 62196), Tesla connector. There should be a universal SAE J1772 connector for all charging stations. This connector can be used with both EVs and PHEVs which is shown in Fig. 1. The SEA J1772 connector has been modified by Tesla chargers and cars. Every Tesla car is fitted with a SAE J1772 connector to Tesla adaptor which is shown in Fig. 2, which helps the driver quickly charge the vehicle in the charging station. Level 2

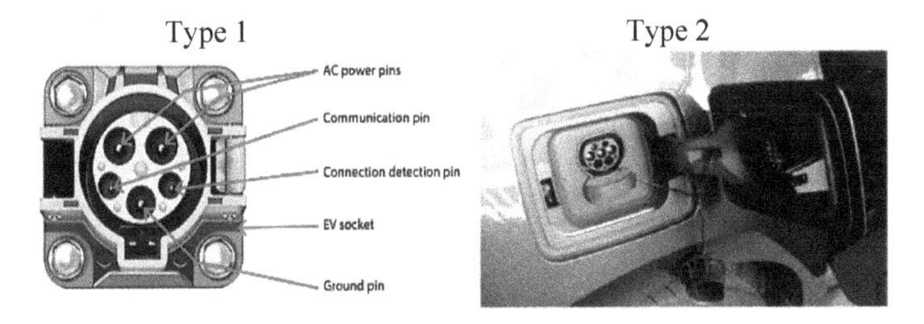

Fig. 1 AC connectors: Type 1 (SAE J1772), type 2 (Mennekes, IEC 62196) (https://www.hydroquebec.com/data/electrification-transport/pdf/technical-guide.pdf; https://www.flickr.com/photos/120167116@N06/14120629806/in/photostream/)

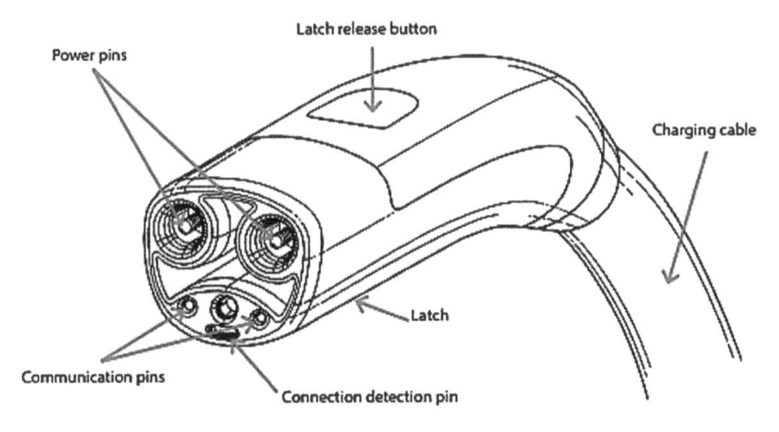

Fig. 2 Tesla connector (https://www.hydroquebec.com/data/electrification-transport/pdf/technical-guide.pdf)

home charging stations operate at 208–240 V voltage range. The factors on which the fast charging will depend are as follows:

- Charger amperage (i.e., 16, 30, 32, or 40 A)

  Powers that can be taken from 16, 30, 32, and 40 A chargers are the follows:

Charger amperage (Amps)	Power (kW)	Voltage (V)
16	3.84	240
30	7.2	
32	7.7	
40	9.6	

- The acceptance rate of the car in kW.

Most EVs have a maximum power acceptance of 7.2 kW or less. It means that mostly EV cannot charge quicker by using a charger of 9.6 kW at 40 Amps. With a 40 Amps charging station, an EV with a power acceptance greater than 7.2 kW can charge quickly (Figs. 1 and 2).

There are two ways for linking level 2 charging station to the electrical panel: a plug or a direct (hardwired) connection. Plugs are further of two kinds: The first one is NEMA 6-50P (welder plug), and the second one is NEMA 14-50P (dryer plug). Length of cable is also an important part of the charging station that charges EVs. Cable length varies between 12 and 25 ft, but the recommended cable length is at least 18 ft long. One can charge EVs with a 25 ft cable whether it is parked backward or forward. Another important factor is the flexibility of the cable. If the cable is more flexible, rolling it up or using it during winter conditions will be easier (https://www.flickr.com/photos/120167116@N06/14120629806/in/photostream/).

3.1.3 Level 3 Charging Stations: DC Fast Charging—Public Charging

Most of the consumer of EVs wants quicker charging in lesser time; thereby, the fast-charging DC chargers were developed. DC fast chargers use three different types of plug and cannot be interchanged. DC fast chargers use three different types of connectors as shown in Fig. 3: combined charging system (CCS), CHAdeMO, Tesla (https://evcharging.enelx.com/news/blog/552-ev-charging-connector-types). The CCS connector uses the J1772 charging inlet and adds two additional pins (high-speed charging pins) below. CHAdeMO connectors do not share part of the connector with the J1772 inlet and therefore require an additional CHadeMO inlet on the car. Usually, Japanese manufacturers use the CHAdeMO model; the CCS is used by many European and American manufacturers; Tesla's supercharging stations use their own vehicle-specific adapter. Tesla uses the same connector for level 1, level 2, and DC fast charging (Fig. 3).

The own vehicle-specific adapter initiative was taken by Japanese when Nissan Leaf and Mitsubishi i-MiEV were introduced. They developed the chargers known as CHAdeMO EV charging stations. This form of charger is identical to a traditional petrol pump-sized unit as shown in Fig. 4. In just 20 min of charging this adapter will deliver 60–100 miles of range. The charging infrastructure of CHAdeMO has not grown very rapidly. The reason is that many manufacturers protested against the introduction of CHAdeMO because the model was not as per the standard approved by SAE. It was standard co-developed by Tokyo electric power company (TEPCO) and the Japanese automakers as an alternative of adopting CHAdeMO, the SAE prepared their own fast-charging standard (combo charging system), and Tesla Motor developed a fast-charging system called as superchargers (2018). Tesla is deemed the leader creator in the EVs sector, taking advantage of the fact that fast-charging does not have a single standard. Tesla superchargers with a maximum power output of 120 kW will charge a battery in about 20 min (Fig. 4).

The international electro-technical commission (IEC) is another standards organization which describes the charging in various modes as follows:

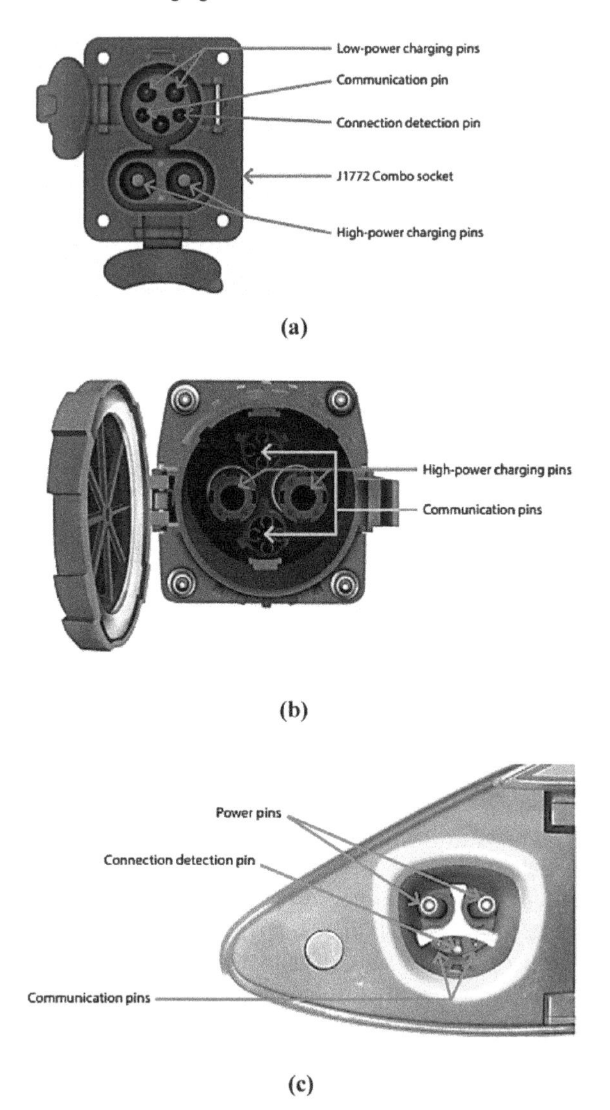

Fig. 3 Three different types of connectors used in DC fast chargers: **a** CCS, **b** CHAdeMO, and **c** Tesla (https://www.hydroquebec.com/data/electrification-transport/pdf/technical-guide.pdf)

Mode 1	Slow charging from a standard single or three-phase electrical socket
Mode 2	Slow charging from a standard socket but with unique safety commitments for some EV (e.g., the Park and Charge or the PARVE systems)
Mode 3	Slow or fast charging with a unique EV multi-pin socket with control and safety features functions (e.g., SAE J1772 and IEC 62196)

(continued)

(continued)

Mode 1	Slow charging from a standard single or three-phase electrical socket
Mode 4	Quick charging with special charger technology like CHAdeMO

There are three connection cases (2019):

- In case A, any adapter connected to the power supply (the power supply cord is attached to the adapter) is typically concerned with modes 1 or 2.
- Case B is an on-board vehicle adapter with a power supply cord, typically mode 3, which can be disconnected from both the supply and the vehicle.
- Case C is a charging station with a DC supply to the vehicle. The cable of the power supply can be permanently attached to the charging station as in mode 4.

3.2 Wireless EV Charging System

Early EV's adopters are highly probable to have easy access to charge EV, either in the home, garage, or on the drive. However, the mass market will bring different charging needs with them. Because of a shortage of dedicated private parking spaces, many EV drivers would have to use street parking, in city apartments. Thus, demand for public charging will rise as ownership of EVs grows. Accordingly, authorities have two options: to implement plug-in charging with a rise in street furniture, which carries the risk of damage, theft, and loosening and oxidizing electrical connections; or to introduce ground-based or underground wireless charging without street congestion. Underground wireless charging also reduces the possibility of tripping and falling over high-voltage cables to the general public about health and safety. EV drivers themselves will also choose the simplicity of wireless charging; no mess and no hassle from bulky, filthy cables that are hard to manage when it is cold and wet. Furthermore, we agree that many drivers should be careful to use plug-in cables in heavy rain. Wireless charging also has a direct economic benefit; keeping the battery charge between 40 and 80 percent will maximize the battery life of the EV (https://www.qualcomm.com/media/documents/files/wireless-charging-for-electric-vehicles-brochure.pdf). Therefore, wireless EV charging systems can provide a possible alternative solution for charging EVs without any plug-in problems. Wireless charging system (WCS) will offer benefits in the context of usability, performance, and user-friendliness relative to plug-in charging systems (Barth et al. 2011). The flaw or disadvantage identified with WCS is that it can only be used when the car is in stationary conditions, such as parking lots, garages, or road signals (Leskarac et al. 2015). Furthermore, stationary WCS has also several challenges, e.g., problems with electromagnetic compatibility, bulky structures, limited power transfer, and shorter range (Covic and Boys 2013, 2013; Moon et al. 2014). The EVs with dynamic (in motion) wireless charging method is explored in order to increase the range and adequate battery storage capacity (Eghtesadi 1990). This method permits battery storage systems to be charged while the EV is in motion. The

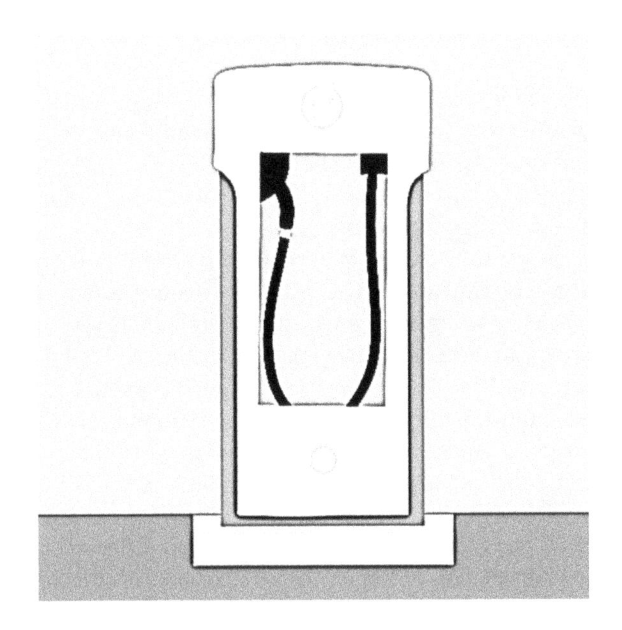

Fig. 4 Level 3 DC charging stations (https://pxhere.com/en/photo/1453945)

EV needs inexpensive battery storage and requires more travel range (Fuller 2016). Nonetheless, before it becomes widely adopted, a dynamic WCS has to meet two major challenges, coil misalignment, and large air-gap. Both coil misalignment and large air-gap affect power transfer efficiency.

There are four methods for the design of wireless EV charging systems are employed: capacitive wireless power transfer (CWPT), inductive power transfer (IPT), magnetic gear wireless power transfer (MGWPT), and resonant inductive power transfer (RIPT).

3.2.1 Inductive Power Transfer (IPT)

Convenience is the main benefit of inductive charging. This facilitates the transfer of electrical energy from the grid to an EV without the help of wires. Energy transfer occurs at the same frequency through magnetic resonance coupling between two copper coils known as primary and secondary coils (embedded in roadways and EVs). The primary coil named the Magne-charge charging paddle (inductive coupler) is placed in the charging port of EV, whereas the power is picked up by the secondary coil that enables charging the EV. The AC power received by the secondary coil is transferred to a standard charger such as the level 1 and level 2 chargers (Musavi and Eberle 2014). For more than 10 years, the system has been in use in shuttle buses that pursue a quick, well-defined path, picking up energy from coils placed on the road at each bus stop. While this method is still not used in light passenger vehicles, researchers are working on appropriate solutions with power transfer levels of up to 10 kW.

3.2.2 Capacitive Power Transfer (CPT)

Due to the relatively directed nature of electrical fields, CWPT systems have potential benefits over inductive systems which eliminate the need for electromagnetic field shielding. These can also be worked at higher frequencies as CWPT systems do not use ferrites, causing them to be smaller and less costly. Therefore, CWPT could allow dynamic charging of EV a reality.

In the CPT, coupling capacitors are used instead of using coils or magnets to transfer power from the transmitter to the recipient. The main AC voltage is transmitted by the power factor adjustment circuitry to an H-bridge converter. The H-bridge produced high-frequency AC passes through the receiver side of the coupling capacitors. In order to reduce the impedance in the resonant system between the transmitter and the receiver side, the extra inductors are mounted in series with the coupling capacitors. The level of power transfer depends entirely on the size of the capacitor and space between two plates. The CWPT offers excellent output for a small air gap and enhanced field constraints formed between two capacitor plates (Kim and Bien 2013). Because of large air differences and high-power demands, the implementation of CWPT for EVs has been restricted to date.

However, due to the extremely limited capacity between road and vehicle surfaces, effective power transmission can only work at very high frequencies, making it very difficult to develop these systems. With the latest development of wide-band (silicon carbide (SiC) and gallium nitride (GaN)) performance semiconductor devices for higher-frequency application, medium-range high-power CWPT systems are becoming feasible (Regensburger et al. 2017; Zhang et al. 2016). Two major challenges of CWPT charging for EVs are: (i) attaining high-power transmission rate at reasonable efficiencies thus meeting the requirements of electromagnetic protection and (ii) maintaining efficient transfer of power even when the relative position of the couplers shifts.

3.2.3 Magnetic Gear Wireless Power Transfer (MGWPT)

Every transmitter and receiver in this method consists of winding armature and synchronized permanent magnets inside the winding. At the transmitter, the side process is alike to motor operation. As AC is supplied to the transmitter winding, it produces a mechanical torque on the transmitter magnet, which causes it to rotate. Due to the change in the transmitter's magnetic contact, the permanent magnet field induces torque on the receiver permanent magnet which results in the transmitter magnet being rotated synchronously. The alteration in the permanent magnetic field of the receiver also produces the AC in the winding, i.e., the receiver works as a generator for the mechanical input of a permanent magnet to the receiver converted into an electrical output at the winding of the receiver. The magnetic gear is considered the coupling of revolving permanent magnets. After rectification and filtering via power converters, the produced AC power on the receiver side fed into the battery (Leskarac et al. 2015).

3.2.4 Resonant Inductive Power Transfer (RIPT)

The RIPT is one of the most common and improved variants of the standard IPT for power electronics and wireless transformer coils. Resonators with high-quality factors transfer energy at a much higher rate, and working at resonance will transfer the same quantity of power as in IPT, even with weaker magnetic fields. The power can be transmitted without wires too long distances. Max. power transmission over the air occurs when the transmitter and receiver coils are matched, i.e., the two coils will suit the resonant frequencies. In order to obtain reasonable resonant frequencies, the transmitter and receiver coils are connected to additional compensation networks in the series and parallel configurations. An expanded network of compensation together with increased resonant frequency further eliminates further losses. The operating frequency of RIPT is between 10 and 150 kHz (Leskarac et al. 2015).

3.2.5 Wireless EV Charging Standards

If each company makes its own specifications for WCS that are not compliant with other systems, then that is not going to be a good thing. Several international organizations such as the society of automotive engineers (SAE), the international electro-Technical Commission (IEC), Underwriters Laboratories (UL) Institute of Electrical and Electronics Engineers (IEEE) are collaborating on standards to make wireless EV charging more users friendly (Vilathgamuwa and Sampath 2019).

- SAE J1772 standard outlines EV/PHEV Conductive Charge coupler.
- SAE J1773 standard describes EV inductively coupled charging.
- SAE J2847/6 standard defines communication between wireless charged vehicles and wireless EV chargers.
- SAE J2836/6 standard outlines use cases for wireless charging communication for PEV. SAE J2954 standard describes WPT for light-duty plug-in EVs and alignment methodology.
- UL subject 2750 defines outline of investigation, for WEVCS.
- IEC 61980–1 Cor.1 Ed.1.0 describes EV WPT systems general requirements.
- IEC 62827–2 Ed.1.0 expresses WPT management: multiple device control management.
- IEC 63028 Ed.1.0 describes WPT air-fuel alliance resonant baseline system specification.

3.3 Solar-Based EV Charging System

Renewable energy comprises approximately 66.7% of worldwide power capacity for roughly 165 gigawatts (GW) in the Renewable 2017 report (Momidi 2017). Solar photovoltaic (SPV) has been noted to be the fastest growing industry in 2016. The SPV electricity offers a possible source of mid-day charging of EVs and PHEVs

Fig. 5 Solar-based EV charging station (Mouli et al. 2015)

(Renewables 2009). EVs charging by using SPV provide a sustainable method for charging the batteries of EVs. Places like factories, official buildings, and warehouse areas are perfect places to use solar EV charging where the space under the rooftops of the building and car parking can be used to mount PV modules as shown in Fig. 5. The power generated in a PV module is used immediately for EV charging without the need for the storage system. Employee vehicles usually stay parked at the parking lot for 6–9 h with a long charge period and also cover the grid support through vehicle-to-grid (V2G) technology (Li and Lopes 2015). Vehicle-to-home (V2H) technology is reachable in the off-grid system.

The advantages of solar-based EV charging station are as follows:

- There is a decrease in load on the grid.
- The voltage issue is eliminated in the distribution system (Cheng et al. 2015).
- There is also a drop in power rates charged to the utility.
- Higher efficiency in direct DC EV-PV interconnection.
- Possibility of V2G and V2H strategies (Mwasilu et al. 2014, 2012; Xu and Chung 2016; Fattori et al. 2014; Shemami et al. 2017).
- Charging cost is lowered and no emissions at the tailpipe (Mouli et al. 2016).
- Due to no moving part, there is no maintenance/noise.
- Easy installation is possible (Fara and Craciunescu 2017).

3.3.1 Solar Rooftop or Ground-Mounted System

In the world, this is mostly used and practice-wise accepted system. In this, the solar modules are installed on the roof of the domestic and commercial buildings, or the land area can be used as a ground-mounted system. For electricity generation, the system includes PV modules, solar inverter, cables, mounting systems, and other

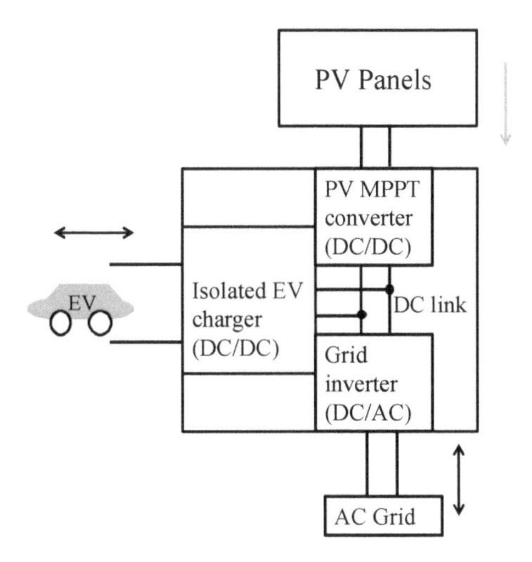

Fig. 6 On-grid solar EV charging system (Mouli et al. 2016)

electrical equipment. The roofing system is small in size 5–20 kW for domestic buildings and 100 kW (or more) for commercial buildings, while the ground-mounted system can be very high in megawatt capacity (Anto and Jose 2014). A research on the installation of a 6 kW_p (kW_p stands for kW peak of the system) grid connection solar system capable of generating 25 units is explained (Ghosh et al. 2016). There are three categories based on grid interface: on-grid, off-grid solar rooftop system, and hybrid systems.

On-Grid Solar EV Charging System

An on-grid solar rooftop system is the system in which the PV modules are mounted on a building's roof catering individual load and are also connected to the grid power (Sharma and Goel 2017). The output of the PV module is connected directly to the delivery system through grid-controlled inverters and cannot work without the grid reference voltage and frequency. There is no requirement for a battery bank, the power supply, and the load fluctuations are balanced by the grid, i.e., the augmented load demand is met by the grid supply, and if the load requirements are small and solar is available, additional energy may be applied to the grid, as shown in Fig. 6. The PV power, EVs and battery energy storage are integrated into a grid-based charging station (Singh et al. 2015, 2016). The charging facilities can be used in the grid-connected during cloudy or rainy seasons. The grid connection also enables transferring electricity from the PV modules to the grid in excess generation conditions. It is suggested that the second-life lithium-ion batteries be included in the system to act as a power reserve and offer immediate support during the grid supply shortage (Khan et al. 2018).

Off-Grid Solar EV Charging System

The off-grid charging system is a stand-alone power system which forms a micro-grid to feed the charge during the day without a battery bank (Sharma et al. 2016, 2017). An off-grid electrical car charger can also be called "EV Autonomous Renewable Charger." PV modules are installed for distributing the loads to the individual proprietor or domestic/commercial EVs and are not linked to any utility grid which is called the off-grid solar power system (Mueller and Mueller 2014). Off-grid electric car chargers can be mounted almost everywhere because there is no need for an electrical grid link. For a continuous nonstop power supply, a battery bank is required to compensate for changes in the input side due to differences in solar radiation on the PV module. For deserted places that are not linked to the grid, an off-grid PV charging system is a good option (Ciric 2017).

Hybrid Solar EV Charging System

Hybridization means that EVs are charged from different sources that work together. An example of one such system is the solar–wind system (Castillo et al. 2015). There is a grid connection of the system along with a battery backup. Without grid reference voltage and frequency, one can work in island mode under unsuitable grid situations (Adithya 2016).

3.4 Vehicle-Integrated PV (ViPV) System

The ViPV system marks an accompanying power source not only to charge the battery during the stoppage of the vehicle but also during the operation of the vehicle. This is a complex source that can provide extra power to electronic control systems, displays, actuators, ventilation, heating, and other electronic equipment such as 110 or 230 V_{AC} power converters, freezers, and microwave ovens (Kronthaler et al. 2014). When the vehicle is in a parking lot, unused generated electrical energy can provide the power to a ventilation/air cooling system in a hot climate (https://www.renewableenergyworld.com/articles/2005/05/from-bipv-to-vehicle-integratedphotovoltaics-31149.html). This could be perfect in India's tropical climate for public buses. The regular air conditioning on buses is not provided adequate ventilation and can danger to the health of passengers during the spread of diseases. The third-generation Prius provides a roof-mounted solar module (optional) for operating an air cooling/ventilation system. Besides, roof-mounted solar modules can be efficient in trickle charging the battery and avoiding self-discharge. The Nissan Leaf is known to pursue these technologies for India. The efficiency of PV modules with multi-crystalline technology is between 11 and 18%, while the usage of mono-crystalline silicon makes it possible to improve the conversion efficiency by about 4%. Gallium arsenide has allowed reaching 40% of cell efficiency. In ViPV systems, solar cells to be implanted into the chassis part exposed to solar radiations like the cover, roof depending on the design of the car. As a result, the vehicle itself looks like a massive solar power plant, and hybrid cars have become much more environmentally friendly. The source of supply is the electric motor and combustion engine to the vehicle.

With other techniques for painting the surface of the car with a light-resistant material, the technology of thin-film photovoltaic solutions is inexpensive and can be painted on the body (Kadar 2013). The efficiency of the system through the use of thin-film technology is almost 4 to 6%. In the research work, the efficiency of the thin film is more than 20% and with organic matter about 10% (https://www.nrel.gov/pv).

3.5 Solar Parking Lots

Workplaces such as commercial spaces or office buildings are ideal places for the photovoltaic electric solar charging of vehicles, where the roof and the parking lot of the building can be equipped with PV modules (Ma and Mohammed 2014). PV parking lot gives onsite energy generation for supplying the EVs batteries (Lee et al. 2016), and therefore, the power is accessible directly for EV charging via an EV-PV charger (Mouli et al. 2015). For large areas and multi-story car parks, this sort of system is becoming particularly fascinating. Vehicles are parked at office parking spaces throughout the day and can be charged via a PV-based charging system at standard charging levels. Despite its impact on the cost of charging, it will help to lessen the grid load and emissions (Tulpule et al. 2011). This PV-based parking model can be used in nations with strong sunlight and good solar potential (Benela et al. 2013). India is a tropical country where sunlight is accessible at high intensity for longer hours a day. Throughout the year, the daily average solar energy remains above 5 kWh/m^2 for a large part of the country; therefore, this PV-based parking model can be promising in India.

4 EV Charging Infrastructure in India

Under the new guidelines, a regulatory agency under the Ministry of Power has been nominated as the Central Nodal Agency, the Bureau of Energy Efficiency (BEE). Besides, the guidelines provide for a provision for the State Nodal Agency for their own states. Such nodal authorities will serve as the main facilitator in the country-wide deployment of EV charging infrastructure. A phase-wise construction of a sufficient network of charging infrastructure across the nation was proposed in the guidelines to solve the range of concerns of the owners of EVs.

Here is a brief about the guidelines issued for charging infrastructure (https://beeindia.gov.in/content/e-mobility):

- At least one charging station should be located in the cities on a 3 km × 3 km grid.
- One charging station on both sides of the highway/road per 25 km.
- In the first phase (i.e., 1–3 years) for all mega-cities with a population of more than 4 million, all current expressways connecting to these mega-cities and largely linked highways may be covered.
- UT headquarters may be covered for spread and demonstrative impact in the second phase (3–5 years) in large cities such as state capitals.
- Fast charging station for long-range and/or heavy-duty EVs, such as busses/trucks, etc., installed every 100 km, one on either side of the highway/road, conveniently located within/along the public charging station (PCS).
- The guidelines explained that private charges at houses or offices are allowed and that distribution company should encourage the same.
- The establishment of PCS is a de-licensed activity, and any person or organization can set up free public charging stations.
- Domestic billing is similar to domestic energy usage and is paid as such.
- In the case of PCS, the tariff for the supply of electricity to PCS is fixed by the relevant commission in compliance with the tariff policy laid down in Sect. 3 of the Electricity Act, 2003.

In support of the "National Electric Mobility Mission" of the Indian Government, Tata Power set up the first set of charging stations for EVs in Mumbai, India's financial capital. Tata Power has plans to set up and extend the network of charging stations for EVs in India. Currently, the company has 85 charging stations in 15 Indian cities like Delhi, Mumbai, Hyderabad, Bengaluru, Vijayawada, Indore, and Hosur (Soni 2019). The aim is to build 500 charging stations across the country's key cities by 2020, as well as another 100 charging stations in Maharashtra in the coming months. The oil marketing companies (OMCs) tie-up also plays an important role in growing to build a nationwide network of EV charging.

There are nearly 500 eV charging stations today covering major metropolitan cities such as Chennai, Mumbai, Bengaluru, and Kolkata. But by 2030, a city like Delhi alone could need around 300,000 fast chargers, assuming an EV penetration of 30% into an approximate 10 million car park. Meeting this need for infrastructure will require an investment of approximately $1–1.5 billion. Therefore, the total investment will rely on how widely EVs enter the automotive market. According to the Maharashtra Electricity Regulatory Commission (MERC) latest tariff orders, for the next two years, the electricity tariff for EV charging will be charged at ₹6 per unit. The commission has proposed a "time-of-the-day" tariff to facilitate charging during non-peak charging hours, thus ensuring that the local power grid is not burdened unnecessarily by the overconsumption from the charging of EV. A benefit of ₹0.75 per kWh is given for charging during low-load night hours, while an additional charge of ₹0.50–1.00 per kWh for morning and evening peak load hours is imposed. Moreover, tariffs for power supply for EV charging stations have also been announced by some other states. In addition to these power supply rates, a service charge may be added to public EV charging stations. To encourage greater use of EVs, though,

Tata Power currently does not offer any service charges to EV users. Tata Power will introduce a mobile app that EV users can use to find chargers, control the charging process, book a charging slot, and finally make payments (Garg 2019).

MG Motor India planned to release the ZS EV in the country in December 2019, and before the launch of the new electric SUV, the car manufacturer had built its first fast-charging station in the county. The new 50 kW DC fast charging station is located at the company's flagship dealer in Gurugram and was built in partnership with the Finnish clean energy company major Fortum (Contractor 2019). The latter has constructed four public 50 kW fast-charging stations in West Delhi, South Delhi, Gurugram, and Noida under the alliance between MG Motor and Fortum. Furthermore, six more public 50 KW DC fast chargers were installed at MG's Mumbai, Hyderabad, Bengaluru, and Ahmedabad dealer locations. Bharat Heavy Electricals Limited (BHEL) is installing on the Delhi-Chandigarh highway a network of solar-based EV chargers. The first in the series was commissioned at Rai (Sonepat) in March 2019. The new five charging stations are situated at Haryana Tourism Corporation Ltd's resorts on Delhi-Chandigarh highway at Ambala, Kurukshetra, Karnal, Panipat, and Samalkha (Sonepat) (Gupta 2019). The project is covered by the FAME scheme of the Indian Government. The establishment of EV chargers at regular intervals over the entire 250 km distance between Delhi and Chandigarh is expected to ease the range anxiety among EV users and enhance their confidence in inter-city travel. Each EV charging station shall be equipped with a roof solar power plant for the supply of renewable energy to EV chargers.

In partnership with Exicom, Magenta, a renewable energy solution company, launched the DC fast charger in Navi Mumbai with a commitment to extending by the end of the year. Mumbai-based Magenta Power is setting up EV charging stations across the Mumbai–Pune expressway. In April, Shah Complex IV, a residential complex located in the Sanpada area of Navi Mumbai, has constructed three solar-powered charging stations for EVs. The installation was completed by Magenta Power, a Mumbai-based firm (Wangchuk 2019). And now, India's first EV charging corridor has been developed by the firm. The charging station network was set up at the Mumbai–Pune Expressway, Lonavala, Hotel Center Point (https://www.financialexpress.com/auto/car-news/after-indias-first-solar-charging-station-magenta-power-sets-up-countrys-first-ev-charging-corridor/1283569/). Finally, the network will be expanded to Bangalore and Mysore. The solar-powered charging station at Navi Mumbai, established by Magenta Power, is already a first in India. Because the fast charger connected to the grid uses solar power to charge the vehicles, the cost was incurred in almost nothing.

5 Challenges and Social Barriers

The transition to electric mobility gives India not only the potential to increase its productivity and change the transport market but also the many problems and social barriers that the country is currently facing.

5.1 Environmental Challenges

IC engine-based vehicles produced harmful gases, which include carbon monoxide and CO_2. HEVs and PHEVs have IC engines, but their emissions of harmful gases are less than fossil fuel-based vehicles. The use of renewable energy sources in conjunction with the EVs can reduce the emission of gases from both the power generation and the transportation sector. EVs cause less emission than fossil fuel-based vehicles. This factor is called as well-to-wheel emission with lesser value for EVs. According to the Paris Agreement, the goal was to reduce the rise in global average temperature below 2 °C above pre-industrial levels (Moriarty and Wang 2017). The adoption of EVs would lead to an increase in CO_2 pollution from the power sector unless the fuel used to charge EVs comes from renewable energy sources.

According to an estimate, CO_2 emissions from thermal plants in India from 2001–2002 to 2009–2010 was 910 g/kWh to 950 g/kWh (https://unfccc.int/paris_agreement/items/9485.php). India's energy consumption and installed electricity capacity have raised 16-fold and 84-fold, respectively, in the last six decades (Mittal 2010). In India, the majority of electricity is generated by coal combustion, generating even more carbon emissions than petrol or diesel. Commonly, coal creates 0.92 kg of CO_2 per kWh and gas, solar energy creates 50 g of CO_2 per kWh, and wind energy creates 10 g of CO_2 per kWh. In 2015, the government set an ultimatum of clean sources generating 175 GW of electricity annually by 2022, including about 100 GW of solar power. Approximately, 50 GW of solar power has already been achieved as per national statistics (Agrawal et al. 2014). Therefore, only when these two components made to act opposite to each other, only then can the actual potential of shifting to EVs will be addressed effectively.

When the price of PV modules decreases, the economic advantage of solar charging can be incredible due to its localized adverse effects (Ba 2017; Almansa Lopez 2016; Dubey et al. 2013). It is very important to transfer to solar energy for the production of electricity. India is highly dependent on foreign companies for technology as there are fewer studies is being conducted in this field. In Indian cars, manufacturers spend almost 1–2% of their turnover on research and development. Lithium-ion battery prices are significantly reduced, and EVs are projected to become as affordable as ICE fuel cars on the automotive market in the upcoming years. Currently, the biggest problem facing in the field of an EV is the price and range of the battery (Khan 2017). Most nations have a common goal of reducing pollution-related problems and global warming. The solution is the worldwide use of renewable energy sources. It is not realistic to think economically of the 100% usage of a renewable-based charging system, one of the key reasons is that fossil fuels are at the heart of the existing grid-connected power network, and constraints on large-scale development and use of PV panels also impede its global adoption. In view of the present situation, there is a need to increase the share of renewable resources. The best way to do this is to provide an on-grid solar charging network with maximum solar energy allocation and the rest is to supply the grid. A stand-alone off-grid system is the other alternative.

5.2 High Price of EVs and Battery

The average cost of electric cars in India is around INR 13 lakh, far higher than the average INR 5 lakh for low-cost cars running on conventional fuel. In India, the price range of electric scooters and bikes is between INR 70 K–INR 1.25 lakh compared to INR 30 K–INR 40 K for the cost range of IC motorcycles and much lower for scooters (Soni 2019).

C V Raman, senior executive director (Engineering) of Maruti Suzuki India, said in his speech that it would be difficult to make a good value proposal quickly under current circumstances unless the cost decreases significantly. This is estimated that the market segment of EVs will also cost two and a half times more than the same type of vehicle powered by a conventional petrol/diesel.

Another challenge is the battery price. Battery price is primarily responsible for the higher prices of EVs. Many EV batteries are designed by combining multiple individual cells; the price of the battery is determined by the cost of each cell. Despite individual cell costs, safety features, packaging costs, mass production, and business strategies are also responsible for driving higher battery prices. The cost of EVs in India is mainly due to Li-ion batteries. Batteries make up around 70% of the cost of the vehicle. That is why the battery packs that are imported cost a lot, about $275/kWh in India. On top of it, the GST component of 28% makes the case even worse. Obviously, the price has dropped in the last five years, but still, it has not reached a value that would allow the general public to opt for EV. Already, annual sales of EVs amount to 0.1 percent of 3 million vehicles a year. The inflection point for such batteries is estimated to be around $100–150/kWh, which is projected to be reached by 2025. When the cost of the battery crosses this level, there will be a big increase in the demand for EVs, and an EV boom will occur in India (Electric Vehicle Industry 2018).

5.3 Charging Infrastructure and Charging Time

India officially had 650 charging stations in 2018, while in the same year China had over 456 K charging points (Kotoky 2019). In addition to charging points, the shortage of private parking spaces is still recognized as a hurdle to the adoption of EVs, and the scarcity of accessible renewable energy indicates that charging EVs will put a strain on the increasingly depleted coal-fired power system. According to research carried out by auto giant Maruti Suzuki, 60% of Indian consumers do not have their individual parking space. "There is no way they can charge the vehicle, therefore they would not adopt it," said C V

Raman, senior executive director (engineering) of Maruti Suzuki India. The shortage of charging stations is also a challenge against the implementation of EVs. Setting up charging infrastructure for fast charging is the most important thing to do. Across India, there are about 222 charging stations for about 187,802 EVs with 353 charging points. For 222 million conventional cars, there are about 56,000 petrol pumps. Unlike in developed countries, people do not have sufficient electrical supplies in their homes to charge or set up charging infrastructure (Chauhan 2018).

Table 2 presents the details of the certified full charge range, battery capacity, maximum power, and charging time of the available EVs in the Indian market

Table 2 Details of the certified full charge range, battery capacity, maximum power, and charging time of the available EVs in the Indian market

EVs		Certified Full Charge range (as per Modified Indian Driving Cycle) (km)	Battery Capacity (kWh)	Max. Power	Charging time		
					Normal Charging Time (0–100%) from any 15 A plug point	AC Charging	Fast/DC Charging Time (0–80%)
Sedans	Tata Tigor EV	213 km	21.5 kWh	30 kW @ 4500 rpm	11.5 h @ 25 ± 2 degrees ambient and battery temperature	Not available	2 h with 15 kW charger @ 25 ± 2 degrees ambient and battery temperature
	Mahindra e-Verito	181 km	21.2 kWh	31 kW @ 4000 rpm	11.5 h (± 15 min) @ 25 °C	Mahindra e-Verito	1 h 30 min @ 25 °C
SUVs	Tata Nexon EV	312 km	30.2 kWh	94.88 kW	State of charge: 10% to 90%: 8.5 h @ ambient temperature	Not available	State of charge: 0%-80%: 60 min @ ambient temperature
	MG ZS EV	340 km	44.5 kWh	104.999 kW @3500 rpm	80% charging in 16–18 h	(0–80%) 6 to 8 h @ Estimated 7 kW	Up to 80% charging within 50 min @ 50 kW
	Hyundai Kona EV	452 km	39.2 kWh	99.92 kW	Not available	(0–100%) approx. 6 h 10 min	Approx. 57 min @50 kW

(https://tigor.tatamotors.com/electric/specification, https://www.mahindraelectric.com/vehicles/everito/, https://nexonev.tatamotors.com/features/, https://s7ap1.scene7.com/is/content/mgmotor/mgmotor/documents/mg-dc-pdf-0060.pdf, https://www.hyundai.com/in/en/find-a-car/kona-electric/specification). The time needed to charge the EV (0 percent–100 percent) is approximately 11.5 to 18 h when normal charging is considered, which creates another time management challenge. It can go well under 90 min or less when talking about the fast charging of EVs, but that depends on the accessibility of charging stations. DC charging requires charging infrastructure, another barrier to overcoming. A petrol pump will normally add about 500 km of range in just 10 min. However, the EVs need to be charged for a longer period, and they give a range of approximately 150–200 km which is very small.

5.4 *Safety*

The installation of charging stations needs various types of permissions from the local electricity board the municipal corporation and central approval as to what ISI standards need to be maintained. Safety is a crucial parameter when there is a high-voltage charging station. A person can bear a shock from a charging station that carries up to 60 V of electricity. The rise in EV infrastructure is hurdled by these kinds of realistic hurdles. The EVs are noise-free, so it is impossible for a pedestrian to realize whether or not an EV is approaching them. The USA has passed a regulation to adjust the minimum sound level for EVs through artificial sounds when approaching a pedestrian. Nevertheless, since EVs promote concepts such as artificial sounds as factors to noise emissions, the regulation is in question.

5.5 *Solar PV Modules*

Solar PV modules with multi-crystalline technology have efficiencies varying from 11 to 18%, whereas the use of mono-crystalline silicone improves the output by about 4%. The use of materials such as gallium arsenide and concentrating technologies has allowed reaching 40% of cell efficiency (Rizzo 2010). But the cost is still too high for a mass application on EVs. A moving PV module was found to raise the solar output from around 46%, at low latitudes, up to 78%, at high latitudes (Ai 2016). Regardless of the energy needed to rotate the module and potential kinematic restrictions preventing perfect orientation, the actual benefits will be smaller than those expected.

5.6 The Need for Public Policy Support

While nobody believes that the GoI is doing everything it can to promote EVs, the industry has in the past opposed the FAME scheme. Originally, the government concentrated on automotive standardization with FAME, which was abandoned to prioritize manufacturing. At the moment, the government is currently preparing the infrastructure framework for charging EVs. The government still aims to tax more aggressively on non-electric vehicles, even though sales of EVs may not warrant such a forced move and that put undue strain on OEMs in the automotive industry.

EV market growth is projected to keep on increasing quickly. Despite numerous emerging regulatory and economic opportunities for EV implementation and infrastructure growth, there are no specific planning recommendations for selecting a site for the development of an EV infrastructure. The development of EV infrastructure was mainly driven by market factors such as prices, characteristics of the electricity grid, and the choice of interested parties. Present EV programs that prioritize domestic and workplace charging as core places for EV chargers establish exclusivity and inadequacy in the operation of facilities as well as barriers to more EV adoption. Through the continuous promotion of private sector investment (including companies and households) is beneficial, a more active involvement for the public sector is required (Deaton 2019).

5.7 Range Anxiety

Range anxiety is when consumers are afraid to think that the EV may not have enough range to take them to their destination. This is directly related to the country's shortage of charging facilities, and although traditional vehicles may be refueled at petrol stations, there is still no such regularized network accessible for EVs. Many studies have shown that consumers remain away from EVs because they are worried about the shortage of charging stations. The studies also indicate that when they see stations around the area, customers are more likely to buy an electric car. Although fears of range anxiety are generally baseless—even the cheapest EVs have sufficient range to meet almost all the driver's needs–the lack of charging stations is a real worry for longer trips, and this discourages consumers from going all-electric (Rahman et al. 2016). The certified full charge ranges (as per modified Indian driving cycle) of Tata Tigor EV, Mahindra e-Verito, Tata Nexon EV, MG ZS EV, and Hyundai Kona EV are 213 km,181 km, 312 km, 340 km, and 452 km, respectively, as given in Table 2. It can be observed that fears of range anxiety are largely baseless, as the lowest certified full charge range is 181 km which is sufficient range except for long trips. The battery capacity of Tata Tigor EV, Mahindra e-Verito, Tata Nexon EV, MG ZS EV, and Hyundai Kona EV are 21.5 kWh, 21.2 kWh, 30.2 kWh, 44.5 kWh, and 39.2 kWh, respectively.

BMW group India's president and CEO, Rudratej Singh, also said earlier that the EV infrastructure is still unclear and unpredictable, impacting Indian consumer's quality and car acceptability. Recently, Toyota has stopped producing electric and hybrid cars for the Indian market, claiming inadequate charging facilities (Soni 2019).

"The problem is that the charging infrastructure does not have a viable business model yet," says David Greene, a professor of Civil and Environmental Engineering at the University of Tennessee. "Although, there are some companies who are working on it hard." Private firms such as EvBox and ChargePoint are aiming to rapidly increase the number of available charging stations, but these plans are focused on an exponential rise in EV purchases. ChargePoint expects to install 2.5 million charging stations on its global network of only 50,000, a goal that it says is based on a "conservative view" of future EV sales. Furthermore, EvBox is planning 1 million new charging stations (Deaton 2019). Analysts expect to see a drastic rise in EV sales in the coming years, but significant roadblocks stand in the way of future adoption. Even if EV sales take off and charging stations increase, hurdles will remain. Making EVs more practical means that not only installing more chargers but also faster chargers that allow drivers to travel longer. There is also the fact that there is no standardization of the technology. Different EVs use different plugs. Ford and GM use one type, and Tesla uses another. Not every station meets the needs of every driver (Deaton 2019). The Ministry of Power has released recommendations and standards for India's EV charging system.

6 Financial Benefits

From the point of view of the EV customer, EV's operating cost is lower due to its superior efficiency (Rahman et al. 2016), and it can be up to 70% relative to IC engine vehicles with efficiencies ranging from 25 to 50%. The customer has earned a financial gain from their EV by delivering grid service using the grid voltage method. The power service provider profits immensely from EV deployment, primarily through the implementation of synchronized charging and grid voltage. EV provides \$200–\$300 savings in cost per vehicle per year. Due to a reduction in the cost of PV modules, there is a major shift to renewable energy through the PV system. The integration of EV and PV creates the potential for efficient EV charging. According to the German Renewable Energy Act, the share of renewable energy will rise by 2025 to 40–45% and by 2035 to 55–60% (Weida et al. 2016). China is the leading contributor to PV and EV production in the number one spot for the PV manufacturing country, with about 58% of global market shares (Li et al. 2017).

7 Conclusions

There are two main fields, solar PV and EVs, which are used for the new mode of transport and can reduce CO_2 pollution and use of fossil fuel, and unfortunately, both have large-scale obstacles. Because of its intermittent existence and concentration of energy generation during the day, PV restricts its commitment to satisfying a large fraction of the average demand for electricity. Compared to the fossil-fuel-based vehicle, EVs offer a clean, effective, and noise-free shifting way.

The future benefits from solar-based EV charging technologies have been outlined. More practical models create social obstacles and bear in mind the robust application of this technology, especially in developing countries where there is a resistance to the adoption of new technologies. Therefore, policy-making choices will be the most effective way. In light of these challenges, if both technologies are used simultaneously, it may be possible to improve the economic performance of EVs by daytime charging, which improves the distance traveled by using low-cost electricity, and carbon emissions will also be controlled. The best option is to install multiple solar PV-EV charging stations globally dispersed. Therefore, the use of renewable energy is growing, and EVs can boost economic growth, generating new job opportunities and ensuring an environmentally sustainable future.

References

Adithya SN (2016) Large-scale implementation of gridconnected rooftop solar photovoltaic system in India—potential, challenges, outlook, and technical impact. In: 2016 International Symposium on Electrical Engineering (ISEE), pp 1–7

After India's first solar charging station, Magenta Power sets up country's first EV charging corridor, Express Drives Desk (2018). https://www.financialexpress.com/auto/car-news/after-indias-first-solar-charging-station-magenta-power-sets-up-countrys-first-ev-charging-corridor/1283569/

Agrawal KK, Jain S, Jain AK, Dahiya S (2014) Assessment of greenhouse gas emissions from coal and natural gas thermal power plants using life cycle approach. Int J Environ Sci Technol 11(4):1157–1164

Ai N (2016) Integrated approaches to EV charging infrastructure and transit system planning. National Center for Transit Research (NCTR), University of South Florida Joel Volinski, Project Manager

Almansa Lopez E (2016) Environmental Impacts of Solar Energy Systems 1(1):75–79

Anto R, Jose J (2014) Performance analysis of a 100kW solar photovoltaic power plant. In: 2014 Annual International Conference on Emerging Research Areas: Magnetics, Machines and Drives (AICERA/iCMMD), p 1–4

Ba I (2017) The future of energy systems in cities. Aalborg Univ Tech Fac IT Des Dep Plan, p 81

Barth H, Jung M, Braun M, Schmülling B, Reker U (2011) Concept evaluation of an inductive charging system for electric vehicles. In: 3rd European Conference Smart Grids and E-Mobility

Benela RA, Jamuna K (2013) Design of charging unit for electric vehicles using solar power. In: 2013 International Conference on Information Communication and Embedded Systems, ICICES, pp 919–924

Cabinet approves Scheme for FAME India Phase II, Press Information Bureau, Government of India Cabinet, Scheme for FAME India Phase II. 28 Feb 2019. https://pib.gov.in/newsite/PrintRelease.aspx?relid=189081

Castillo JP, Mafiolis CD, Escobar EC, Barrientos AG, Segura RV (2015) Design, construction and implementation of a low cost solar-wind hybrid energy system. IEEE Lat Am Trans 13(10):3304–3309

Charging infrastructure for Electric Vehicles(EVs)-Revised Guidelines and Standards, issued on 01.10.2019. https://beeindia.gov.in/content/e-mobility

Chauhan A (2018) Electric vehicles in India: the trends, challenges and future. https://medium.com/@an223c/trends-challenges-and-future-for-electric-vehicles-in-india-b6191f4a70b6

Cheng L, Chang Y, Huang R (2015) Mitigating voltage problem in distribution system with distributed solar generation using electric vehicles. IEEE Trans Sustain Energy 6(4):1475–1484

Ciric RM (2017) Off-grid photovoltaic system as a solution for sustainability of remote farms—an application in Engineering education. In: 2017 International conference on Optimization of Electrical and Electronic Equipment (OPTIM) & 2017 Intl Aegean Conference on Electrical Machines and Power Electronics (ACEMP), pp 159–164

Contractor S (2019) MG Motor India installs first fast charging station for EVs in Gurugram, carandbike. https://www.carandbike.com/news/mg-motor-india-installs-first-fast-charging-station-for-evs-in-gurugram-2135346

Covic GA, Boys JT (2013) Inductive power transfer. Proc IEEE 101(6):1276–1289

Covic GA, Boys JT (2013) Modern trends in inductive power transfer for transportation applications. IEEE J Emerging Selected Topics Power Electronics 1(1):28–41

Deaton J (2019) Everybody wants EV charging stations, but barely anyone is building them. https://www.fastcompany.com/90321889/everybody-wants-ev-charging-stations-but-barely-anyone-is-building-them

Diwan P (2018) Slow, Fast and Super: EV Chargers Conundrum. https://medium.com/@pdiwan/slow-fast-and-super-ev-chargers-conundrum-d35ea0da5a87

Dubey S, Jadhav NY, Zakirova B (2013) Socio-economic and environmental impacts of silicon based photovoltaic (PV) technologies. Energy Proc 33:322–334

Eghtesadi M (1990) Inductive power transfer to an electric vehicle-analytical model. In: 40th IEEE Conference on Vehicular Technology, pp 100–104

Electric vehicle charging stations. Technical installation guide, 2nd edn, August 2015, Hydro-Québec. https://www.hydroquebec.com/data/electrification-transport/pdf/technical-guide.pdf

Electric Vehicle Industry 2018: Emerging Trends and Challenges (2018) https://www.engineersgarage.com/egblog/electric-vehicle-industry-2018-emerging-trends-and-challenges/

Fara L, Craciunescu D (2017) Output analysis of stand-alone PV systems: modeling, simulation and control. Energy Proc 112:595–605

Fattori F, Anglani N, Muliere G (2014) Combining photovoltaic energy with electric vehicles, smart charging and vehicle-to-grid. Sol Energy 110:438–451

"From BIPV to Vehicle-Integrated Photovoltaics—Renewable Energy World." [Online]. Accessed 2017 Nov 18. Available from: https://www.renewableenergyworld.com/articles/2005/05/from-bipv-to-vehicle-integratedphotovoltaics-31149.html

Fuller M (2016) Wireless charging in California: range, recharge, and vehicle electrification. Transp Res Part C: Emerging Technol 67:343–356

Garg A (2019) Tata Power to Set Up 500 EV Charging Stations in India by 2020, Says Ramesh Subramanyam, NEWS18.COM. https://www.news18.com/news/auto/tata-power-to-set-up-500-electric-vehicle-charging-stations-by-2020-in-india-interview-2288637.html

Ghosh D, Karmakar SM, Roy A, Manna S, Samanta S, Mitra S (2016) Installation of 6kWp grid tie rooftop solar system for generation of 25 units of energy per day. In: 2016 IEEE 7th Annual Information Technology, Electronics and Mobile Communication Conference (IEMCON), pp 1–3

Gupta U (2019) BHEL launches 5 solar EV charging stations on Delhi-Chandigarh Highway, pv magazine. https://www.pv-magazine-india.com/2019/09/02/bhel-launches-5-solar-ev-charging-stations-on-delhi-chandigarh-highway/
How to Choose the Right Home Charging Station? ChargeHub. https://chargehub.com/en/how-to-choose-home-charging-station.html
https://beeindia.gov.in/content/e-mobility
https://nexonev.tatamotors.com/features/
https://pxhere.com/en/photo/1453945
https://s7ap1.scene7.com/is/content/mgmotor/mgmotor/documents/mg-dc-pdf-0060.pdf
https://tigor.tatamotors.com/electric/specification
https://www.eeslindia.org/content/raj/eesl/en/Programmes/EV-Charging-Infrastructure/About-EV-Charging-Infrastructure.html
https://www.flickr.com/photos/120167116@N06/14120629806/in/photostream/
https://www.flickr.com/photos/departmentofenergy/28749539888/in/photostream/
https://www.hyundai.com/in/en/find-a-car/kona-electric/specification
https://www.mahindraelectric.com/vehicles/everito/
https://www.reuters.com/article/us-india-electric-policy-idUSKCN1SS2HS
https://www.roadtraffic-technology.com/features/european-countries-banning-fossil-fuel-cars
https://www.theicct.org/electric-vehicles
Kadar P, Varga A (2013) Photovoltaic EV charge station. In: 2013 IEEE 11th international Symposium on Applied Machine Intelligence and Informatics (SAMI), pp 57–60
Khan S, Ahmad A, Ahmad F, Shafaati Shemami M, Saad Alam M, Khateeb S (2018) A comprehensive review on solar powered electric vehicle charging system. Smart Sci 6(1):54–79
Khan NA (2017) Electric vehicle landscape and future forward: bringing clarity in commotion, ETAuto. https://auto.economictimes.indiatimes.com/news/industry/ev-landscape-and-future-forward-bringing-clarity-in-commotion/61178731
Kim J, Bien F (2013) Electric field coupling technique of wireless power transfer for electric vehicles. In: IEEE 2013 Tencon-Spring, pp 267–271
Kotoky A (2019) India has 150 million drivers and only 8000 want electric cars. https://www.bloomberg.com/news/articles/2019-10-06/india-has-150-million-drivers-and-only-8-000-want-electric-cars
Kronthaler L, Maturi L, Moser D, Alberti L (2014) Vehicle-integrated Photovoltaic (ViPV) systems: energy production, diesel equivalent, payback time; an assessment screening for trucks and busses. In: 2014 ninth international conference on Ecological Vehicles and Renewable Energies (EVER), pp 1–8
Lee S, Iyengar S, Irwin D, Shenoy P (2016) Shared solar-powered EV charging stations: Feasibility and benefits. In: 2016 seventh International Green and Sustainable Computing Conference (IGSC), pp. 1–8
Leskarac D, Panchal C, Stegen S, Lu J (2015) PEV charging technologies and v2g on distributed system and utility interfaces. In: Vehicle-to-Grid: Linking Electric Vehicles to the Smart Grid, vol 79, pp 157–209
Li Y, Cai W, Wang C (2017) Economic impacts of wind and solar photovoltaic power development in China. Energy Proc 105:3440–3448
Li X, Lopes LA, Williamson SS (2009) On the suitability of plug-in hybrid electric vehicle (PHEV) charging infrastructures based on wind and solar energy. In: 2009 IEEE Power & Energy Society General Meeting, vol 26, pp 1–8
Ma T, Mohammed OA (2014) Optimal charging of plug-in electric vehicles for a car-park infrastructure. IEEE Trans Ind 50(4):2323–2330
Mittal ML (2010) Estimates of emissions from coal fired thermal power plants in India. Dep Environ Occup Heal 39:1–22
Momidi K. Wireless Electric Vehicle Charging System (WEVCS), Jul 12, 2019, https://circuitdigest.com/article/wireless-electric-vehicle-charging-systems

Monthly Energy Review, June 2020, DOE/EIA-0035(2020/6), U.S. Energy Information Administration, 1000 Independence Ave., SW Washington, DC 20585

Moon S, Kim BC, Cho SY, Ahn CH, Moon GW (2014) Analysis and design of a wireless power transfer system with an intermediate coil for high efficiency. IEEE Trans Industr Electron 61(11):5861–5870

Moriarty P, Wang SJ (2017) Can electric vehicles deliver energy and carbon reductions? Energy Proc 105:2983–2988

Mouli GC, Bauer P, Zeman M (2016) System design for a solar powered electric vehicle charging station for workplaces. Appl Energy 168:434–443

Mouli GRC, Mark L, Venugopal P et al (2016) IEEE Transportation Electrification Conference and Expo (ITEC). 2016:1–7

Mouli GC, Bauer P, Zeman M (2015) Comparison of system architecture and converter topology for a solar powered electric vehicle charging station. In: 2015 9th International Conference on Power Electronics and ECCE Asia (ICPE-ECCE Asia), pp 1908–1915

Mouli GRC, Bauer P, Zeman M (2015) Comparison of system architecture and converter topology for a solar-powered electric vehicle charging station. In: 2015 9th International Conference on Power Electronics and ECCE Asia (ICPE-ECCE Asia), pp 1908–1915

Mueller OM, Mueller EK (2014) Off-grid, low-cost, electrical sun-car system for developing countries. In: IEEE Global Humanitarian Technology Conference (GHTC 2014), pp 14–17

Musavi F, Eberle W (2014) Overview of wireless power transfer technologies for electric vehicle battery charging. IET Power Electronics 7(1):60–66

Mwasilu F, Justo JJ, Kim EK, Do TD, Jung JW (2014) Electric vehicles and smart grid interaction: a review on vehicle to grid and renewable energy sources integration. Renew Sustain Energy Rev 34:501–516

National Electric Mobility Mission Plan (NEMMP) 2020. Department of Heavy Industry, Ministry of Heavy Industries & Public Enterprises, Government of India. https://dhi.nic.in/writereaddata/Content/NEMMP2020.pdf

“Off-grid or hybrid?—Solar off-grid / hybrid.”[Online]. Accessed 2017 Oct 02. Available from: https://offgridhybrid.com/off-grid-or-hybrid-battery-storage/

Perera F (2018) Pollution from fossil-fuel combustion is the leading environmental threat to global pediatric health and equity: solutions exist. Int J Environ Res Public Health 15(1):16

“Photovoltaic Research | NREL.” [Online]. Accessed 2017 Nov 16. Available from: https://www.nrel.gov/pv

Qualcomm Halo Wireless Electric Vehicle Charging | No fuss, just wireless. https://www.qualcomm.com/media/documents/files/wireless-charging-for-electric-vehicles-brochure.pdf

Rahman MM, Baky MA, Islam AS, Al-Matin MA (2016) A techno-economic assessment for charging easy bikes using solar energy in Bangladesh. In: 2016 4th International Conference on the Development in the in Renewable Energy Technology (ICDRET), pp. 1–5

Regensburger B, Kumar A, Sinha S, Doubleday K, Pervaiz S, Popovic Z, Afridi K (2017) High-performance large air-gap capacitive wireless power transfer system for electric vehicle charging. In: 2017 IEEE Transportation Electrification Conference and Expo (ITEC), pp 638–643

Renewables 2017. Analysis and forecasts to 2022. https://www.iea.org/reports/renewables-2017

Rizzo G, Arsie I, Sorrentino M, Solar energy for cars: perspectives, opportunities and problems. InGTAA Meeting 2010 May 26, pp 1–6

Schweber B (2019) Charging electric vehicles, Part 2: The connections. https://www.powerelectronictips.com/charging-electric-vehicles-part-2-connections/

Shah A (2019) India proposes electrifying motorbikes, scooters in six-eight years: source, Reuters. https://www.reuters.com/article/us-india-electric-policy/india-proposes-electrifying-motorbikes-scooters-in-six-eight-years-source-idUSKCN1SS2HS

Sharma R, Goel S (2017) Performance analysis of a 11.2 kWp roof top grid-connected PV system in Eastern India. Energy Rep 3:76–84

Sharma P, Bojja H, Yemula P (2016) Techno-economic analysis of off-grid rooftop solar PV system. In: IEEE 6th International Conference on Power Systems (ICPS), pp 1–5

Shemami MS, Alam MS, Asghar MSJ (2017) Load shedding mitigation through plug-in electric vehicle-to-home (V2H) system. IEEE Transp Electrification Conf Expo (ITEC) 2017:799–804
Singh SA, Azeez NA, Williamson SS (2015) A new single-stage high-efficiency photovoltaic(PV)/grid-interconnected dc charging system for transportation electrification. In: IECON 2015—41st Annual Conference of the IEEE Industrial Electronics Society, pp 005374–005380
Singh SA, Carli G, Azeez NA, Ramy A, Williamson SS (2016) Modeling and power flow control of a single phase photovoltaic/grid interconnected modified Z-source topology based inverter/charger for electric vehicle charging infrastructure. In: IECON 2016–42nd Annual Conference of the IEEE Industrial Electronics Society, pp 7190–7196, IEEE
Soni Y (2019) Tata boosts EV plans: 500 charging stations and EV App by 2020, Inc42 Staff. https://inc42.com/buzz/tata-boosts-ev-plans-500-charging-stations-and-ev-app-by-2020/
Soni Y (2019) What are the challenges for the EV market in India? Inc42 Staff. https://inc42.com/features/what-are-the-challenges-for-the-ev-market-in-india/
Taumar D (2019) Here are 11 electric cars that will be launched in the next two years, ETAuto. https://auto.economictimes.indiatimes.com/news/passenger-vehicle/cars/here-are-11-electric-cars-that-will-be-launched-in-the-next-two-years/68705794
"The Paris Agreement—main page." [Online]. Accessed 2017 Nov 01. Available from: https://unfccc.int/paris_ agreement/items/9485.php
The different EV charging connector types. https://evcharging.enelx.com/news/blog/552-ev-charging-connector-types
Tulpule P, Marano V, Yurkovich S, Rizzoni G (2011) Energy economic analysis of PV based charging station at workplace parking garage. In: IEEE 2011 EnergyTech, pp 1–6
Turker H, Bacha S, Chatroux D, Hably A (2012) Modelling of system components for Vehicle-to-Grid (V2G) and Vehicle-to-Home (V2H) applications with Plug-in Hybrid Electric Vehicles (PHEVs). In: 2012 IEEE PES Innovative Smart Grid Technologies (ISGT), pp 1–8
Vilathgamuwa DM, Sampath JPK (2015) Wireless Power Transfer (WPT) for Electric Vehicles (EVs)—present and future trends. In: Rajakaruna S, Shahnia F, Ghosh A (eds) Plug in electric vehicles in smart grids. Power systems. Springer, Singapore
Wangchuk RN (2019) Zero-energy mumbai society uses sun to charge cars, cuts power bills to 0! The Better India. https://www.thebetterindia.com/200185/mumbai-electricity-bill-discount-solar-power-electric-vehicle-charging/
Weida S, Kumar S, Madlener R (2016) Financial viability of grid-connected solar PV and wind power systems in Germany. Energy Proc 106:35–45
Xu NZ, Chung CY (2016) Reliability evaluation of distribution systems including vehicle-to-home and vehicle-to-grid. IEEE Trans Power Syst 31(1):759–768
Zhang H, Lu F, Hofmann H, Liu W, Mi CC (2016) A four-plate compact capacitive coupler design andlcl-compensated topology for capacitive power transfer in electric vehicle charging application. IEEE Trans Power Electron 31(12):8541–8551

Use of Phase Change Materials for Energy-Efficient Buildings in India

Parth Patil, K. V. S. Teja, and Himanshu Tyagi

Nomenclature

A	Altitude (m)
C_p	Specific heat (J/kg K)
G_{sc}	Global solar irradiation (W/m^2)
h_0	Convective heat transfer coefficient (W/m^2 K)
I_0	Solar insolation (kWh/m^2)
I_r	Realistic solar insolation (kWh/m^2)
k	Thermal conductivity (W/m K)
L	Length of wall (m)
L_f	Latent heat of fusion (J/kg)
m	Mass of material (kg)
n	nth day of the year
Nu	Nusselt number
Pr	Prandtl number
Q	Heat stored or extracted per unit area (W/m^2)
Q''	Net heat flux (W/m^2)
Re	Reynolds number
Ste	Stefan number
T	Temperature (°C)
t	Time variable (s)
V	Wind velocity (m/s)
X	Phase front location (Cartesian coordinates) (m)
x	Spatial variable (m

P. Patil (✉) · K. V. S. Teja · H. Tyagi
Department of Mechanical Engineering, Indian Institute of Technology Ropar, Rupnagar, Punjab 140001, India
e-mail: 2015med1004@iitrpr.ac.in

H. Tyagi et al. (eds.), *New Research Directions in Solar Energy Technologies*,
Energy, Environment, and Sustainability,
https://doi.org/10.1007/978-981-16-0594-9_11

Greek Symbols

α Thermal diffusivity (m^2/s)
δ Declination (°)
ε Emissivity
θ Angle (°)
λ Longitude (°)
μ Dynamic viscosity (Pa s)
ρ Density (kg/m^3)
τ Transmittivity
φ Latitude (°)
ω Solar hour angle (°)

Subscripts

1 Initial
2 Final
a Ambient
b Beam radiation
eff Effective
f Fusion
in Inside room temperature
lat Latent
m Melting
max Maximum
min Minimum
r Resultant
sol Solar
z Zenith

1 Introduction

The level of greenhouse gas emissions is continuously increasing due to heavy worldwide uses of hydrocarbon fuels. Also, there is a drastic climb in fuel prices as the fuel sources are limited. These are the main driving forces behind the efforts of scientists all over the world to discover other alternative renewable energy sources. Another option is to utilize the available energy efficiently, keeping the living standard of people high within any country. This can be achieved by decreasing the use of energy wherever possible.

Direct solar radiation is considered to be one of the best renewable sources of energy other than wind, tides, and geothermal heat. The scientists are trying to develop new energy storage devices that can store solar energy in a suitable form by conventionally converting them in the required form (Kenisarin and Mahkamov 2007). Among different solar energy storage techniques, latent thermal energy storage is the most attractive due to its ability to store heat at a constant temperature, which is the phase transition temperature of the corresponding phase change material (Sharma et al. 2009). Also, latent heat energy storage provides a high-energy storage density than other thermal heat storage methods. In latent heat storage (LHS), the heat is released or absorbed when material changes its phase from solid to liquid or liquid to gas and vice versa (Zalba et al. 2003; Hu et al. 2017).

In a tropical country like India, buildings consume more than 40% of total electricity consumption. So, our main focus is to reduce the power consumption within buildings. During summers, a considerable amount of electricity is used for powering the air conditioners, which helps to maintain the ambient temperature conditions inside the premises. It would be beneficial if we could find other passive, cost-effective, and durable methods to replace this power need. Also, in winters, heaters used inside buildings can be replaced by the same passive systems, which can further help to reduce the heating loads. PCM incorporation within premises can prove to be a long-term solution on all these problems as it can help to reduce the cooling as well as heating loads (Khudhair and Farid 2004). Heat cannot be efficiently passed through them due to their poor thermal conductivities. Also, solid-liquid phase transforming PCMs have low volume changes, eliminating the containment issues (Sarbu and Sebarchievici 2018; Cook et al. 2010).

Khudhair and Farid (2004) reviewed some methods for energy conservation in building applications with thermal energy storage using the phase change materials. Kumar et al. (2016) designed a PCM thermal energy storage system for the comfort cooling in which he studied a model of a west-facing room. The selection criteria of different PCMs and their potential applications are studied by Sharma et al. (2009). Kumar et al. (2011) studied the use of PCMs for defense applications like hot jackets for high altitudes and heat sink for instruments.

PCM encapsulation can also be used to avoid the disadvantages of using conventional PCMs like the irregularities during the phase change, leakage of liquid PCM, and corrosion of storage container due to PCM interaction. These problems can be prevented by using the PCM encapsulation method in which the PCM core is coated with a continuous polymer or inorganic shell. Macroencapsulation is widely used for encapsulation as it can increase the rate of heat transfer in the PCM encapsulating matrix and can also avoid the phase separation which can maintain the uniformity of the mixture. The PCM encapsulation is thoroughly analyzed in Cabeza et al. (2011).

Saxena et al. (2020) experimentally studied the incorporation of PCM inside the bricks for the passive conditioning of the building. The PCM is chosen by studying their thermal properties through differential scanning calorimetry and the peak summer climate conditions of the place. He observed the heat transfer reduction up to 60% in a PCM incorporated brick than a normal brick. Different retrofit layers can be implemented along with PCM to improve the energy performance and reduce the heat gain of the buildings which is experimentally and numerically done by Saikia et al. (2020).

In this study, a PCM incorporated model on the roof of a building is analyzed and then compared with the model without PCM. The PCM material is stored between two insulating layers, which is placed over the roof of any typical building in an Indian city. The cities are selected from the three weather zones of India. The appropriate PCM is chosen for each case, taking into account the different parameters like thermal conductivity, latent heat of transformation, the melting point of the material, and weather conditions of the location. Phase change materials must exhibit specific desirable chemical, thermodynamic, and kinetic properties before their implementation in the buildings, as well as their easy availability and economic factors, should also be taken into consideration (Fleischer 2015). In summer, at the daytime, when the outside temperature is more than the melting point of PCM, it is in a liquid state while at night, it is solid. In the regions with extremely hot summers and freezing winters, we can change the type of PCM according to the season because of its easy accessibility.

Weather plays a vital role in the selection of PCM. Different wind speeds can affect the amount of energy stored in the PCM model. Places having high average wind speeds generally have low heat transfer coefficients, which decreases the heat transfer rate. There are many PCMs available for integration into building applications, having different heat-storing capacities. A suitable phase change material is selected by taking into account its availability and any adverse effects it can cause during its containment. Numerous research and development work is going on in this field to discover better energy-efficient methods to tackle the impact of climate change.

2 Mathematical Modeling

For this analysis, consider a typical model of PCM incorporated between two insulating layers of steatite and polyurethane over the horizontal roof. The aim is to analyze the temperature on the inner surface of the roof and then compare it with the model without PCM. The temperature distribution inside the model is plotted using numerical analysis on multiphysics software. There are some assumptions we have to consider before the calculations. Such as, only heat transfer through the roof is discussed in the analysis and not through the walls, windows, and doors of the buildings, which require the roof to be ideally infinite. This makes the temperature distribution one dimensional along with the thickness of the roof.

The energy storage potential of any material when it transforms its phase from solid to liquid at constant pressure can be found by multiplying the latent heat of fusion of material (L_f) with the mass of material (m) that undergoes phase transformation (Soni et al. 2019).

$$E_{\text{lat}} = mL_{\text{f}} \tag{1}$$

The phase change process involves the melting phase front, and thus, Stefan problems are often used to solve the moving boundary problems. If ΔT is the temperature difference between the heat source or sink and the PCM material, the Stefan number (Ste) can be derived as the ratio of sensible heat to the latent heat of fusion. As most PCMs have large values of latent heat of fusion, they often have low Stefan numbers, as shown in Eq. (2),

$$\mathrm{Ste} = \frac{C_p \Delta T}{L_\mathrm{f}} \tag{2}$$

The governing equations for both the solid and liquid phases of PCM can be developed from the heat diffusion equation in one-dimension in x. Considering no heat generation inside of the phase change material, the final governing equation will be,

$$(\rho C_p)\frac{\mathrm{d}T}{\mathrm{d}t} = \frac{\partial}{\partial x}\left(k\frac{\partial T}{\partial x}\right) \tag{3}$$

This governing Eq. (3) is applied separately to the solid and liquid phases of PCM material, assuming constant thermal conductivity and specific heat at constant pressure in different phases. PCMs generally have high values of latent energy storage densities and thus can be used to store thermal energy. The above equations are required to solve the moving boundary problems by the analytical method. In this study, the results are obtained by simulating the models on COMSOL multiphysics software, where temperature values and other boundary conditions are provided to get the final results.

To calculate the heat transferred between the outer and inner surfaces of the models, the temperature difference needs to be obtained. Hence, the sol–air temperature needs to be computed as the outer surface temperature. It is defined as the effective outer surface temperature that takes into account both the convection and radiation (Duffie et al. 1985). Sol–air temperature (T_sol) is used to calculate the total heat gained through external surfaces. (T_sol) is given by the following Eq. (4):

$$T_\mathrm{sol} = T_\mathrm{a} + \left(\frac{\alpha I_0 - \varepsilon \Delta R}{h_0}\right) \tag{4}$$

where α and ε are the absorptivity and emissivity of the surface, which is steatite in this case. Hence, these values are approximately taken as $\alpha = 0.65$, $\varepsilon = 0.87$ (Cengel and Ghajar 1998).

$\varepsilon * \Delta R \approx 60$ W/m^2 for horizontal surfaces, while it is 0 for vertical surfaces. Here, the setup is taken as horizontal. Hence, $\varepsilon * \Delta R \approx 60$ W/m^2 (Tyagi et al. 2020). There are three different variables in Eq. (4). The variables are calculated for a specific location and weather conditions, which after putting in the above equation, we get the value of T_sol. T_a is calculated by taking average temperatures of a place in a respective season by temperature variation charts. In order to get an understanding of different geographical zones and climatic conditions, four cities (Nagpur, Chennai, Amritsar, and Srinagar) have been considered. Nagpur, Chennai, and Amritsar have been chosen for the summer calculations; Amritsar and Srinagar for winter calculations. For all these conditions, both day and night were considered because it gives an estimate of the minimum and maximum for the temperatures, which helps in the selection of appropriate PCM.

Figure 1 shows the PCM implemented model, where a PCM layer is located between two insulating layers of steatite and polyurethane. This setup is placed over the horizontal roof. The model without PCM is shown in Fig. 2, where the thickness of the brick roof is extended to make the same overall thickness of both the models for the comparison. All the boundary conditions and temperatures are provided to the multiphysics software and the results are obtained.

Calculation of h_0

In the above formula of T_{sol}, h_0 is one of the unknown variables. Since forced convection is the dominant mode of heat transfer; here, Nusselt number needs to be calculated to obtain h_0. For a horizontal flat plate which is exposed to outside winds, the Nusselt number is given by (Duffie et al. 1985):

$$\mathrm{Nu} = 0.86 * \mathrm{Re}^{1/2} * \mathrm{Pr}^{1/3} \tag{5}$$

Local seasonal wind speeds have been taken into consideration for the calculation of Reynolds number (Re).

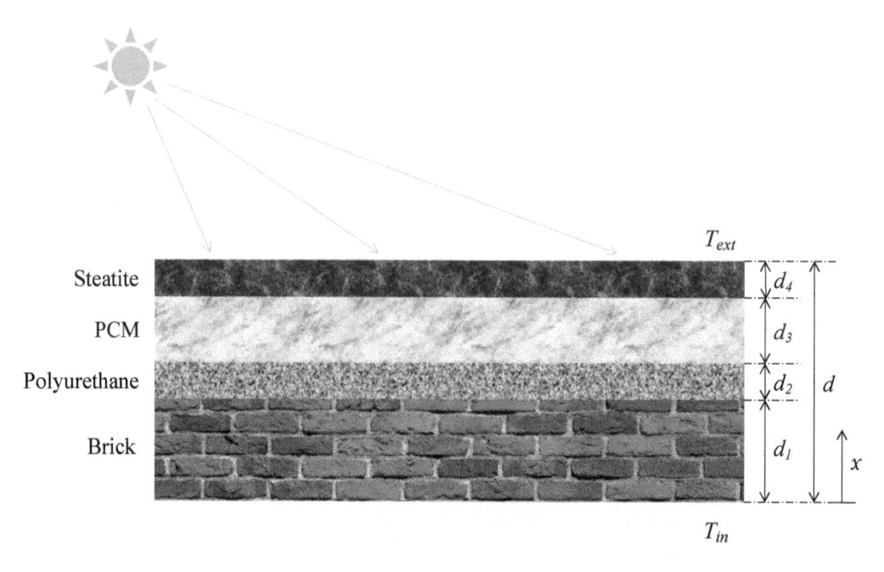

Fig. 1 Schematic of the PCM incorporated model (dimensions: d_1 = 12 cm; d_2 = 1 cm; d_3 = 4 cm; d_4 = 1 cm)

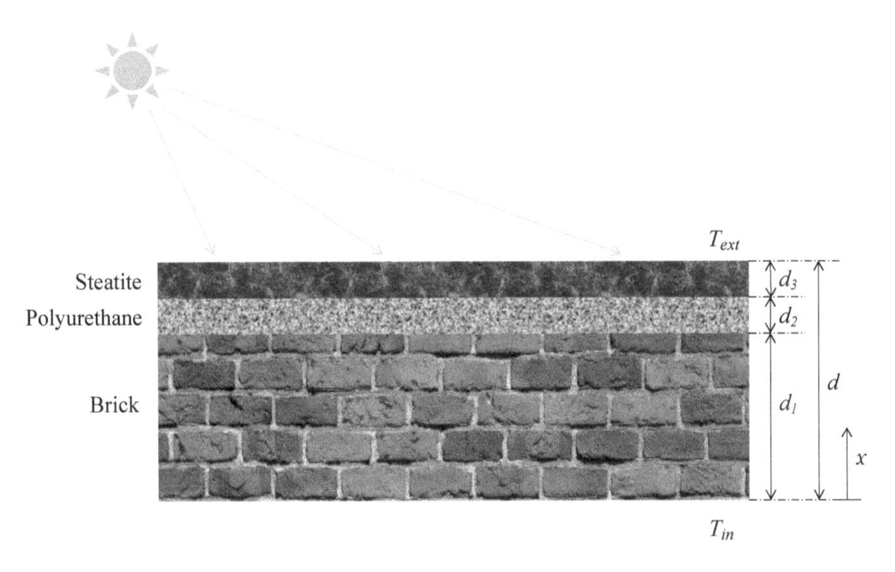

Fig. 2 Schematic of the model without PCM (dimensions: $d_1 = 16$ cm; $d_2 = d_3 = 1$ cm)

$$\mathrm{Re} = \frac{\rho v L}{\mu} \tag{6}$$

Prandtl number is taken as constant throughout for the ease of calculations.

$$\mathrm{Pr} = 0.645 \tag{7}$$

After obtaining the Nusselt number, h_0 can be calculated.

$$h_0 = \frac{\mathrm{Nu} * k}{L} \tag{8}$$

where the characteristic length of the plate (L) is assumed to be 1 m and k is the thermal conductivity of the air.

Calculation of I_0

In Eq. (4), global solar insolation is also an unknown variable. Here, two cases arise to get the value of I_0: (A) Neglecting the atmosphere, and (B) taking the atmosphere into account.

(A) **Case-1 (neglecting the atmosphere):**

To calculate the value of I for case-1, latitude (φ) for each for these locations is necessary. But here the calculations are done by taking the average for the whole season. Summer months are taken as May, June, July, and winter months are December, January, February. The declination is calculated on June 15th for

Table 1 Solar hour angle (ω) for different hours of the day (Duffie et al. 1985)

Time	10 a.m.	11 a.m.	12 p.m.	1 p.m.	2 p.m.	3 p.m.	4 p.m.
Hour angle (°)	−30	−15	0	15	30	45	60

summer and January 15th for winter. G_{sc} is the solar constant which is 1367 W/m^2, while n is the nth day of the year.

Declination is given by Eq. (9) (Duffie et al. 1985)

$$\delta = 23.45 * \sin\left(360 * \frac{284 + n}{365}\right) \tag{9}$$

Hourly incident solar insolation from ω_2 to ω_1 hour angles is given by Eq. (10) (Duffie et al. 1985)

$$I_0 = \frac{12 * 3600 * G_{sc}}{\pi} * \left(1 + 0.033 * \cos\left(\frac{360 * n}{365}\right)\right) * \left[\cos\phi * \cos\delta * (\sin\omega_2 - \sin\omega_1) + \frac{\pi * (\omega_2 - \omega_1)}{180} * \sin\phi * \sin\delta\right] \tag{10}$$

From 10 a.m. to 5 p.m., values of I_0 within each hour are calculated, average I_0 is taken here for further calculations. By substituting the values of h_0 and I_0 in Eq. (4), T_{sol} is calculated. During the night, I_0 becomes 0. Thus, for all the locations, T_{sol} is calculated during the day. T_a is the ambient air temperature. For each location, the minimum and maximum values for ambient temperatures are considered, as this helps in the selection of PCM. The solar hour angles (ω) corresponding to different hours of the day are given in Table 1.

(B) **Case-2 (considering the atmosphere):**

It is to be noted that in case-1, the effect of the atmosphere is neglected. Hence, the T_{sol} values during the day are higher than expected realistic values. Therefore, case-2 arises where a correction factor is multiplied to I_0 to get the real value for solar insolation, which is termed as I_r. This correction factor is the transmittance of the atmosphere. Transmittance depends on the altitude of the location (A) and the zenith angle (θ_z).

Using the values of declination (δ), hour angle (ω), and latitude (φ), zenith angle (θ_z) is calculated using the following relation (Duffie et al. 1985):

$$\theta_z = \cos^{-1}(\cos\delta * \cos\varphi * \cos\omega + \sin\delta * \sin\varphi) \tag{11}$$

The transmittance of the atmosphere for beam radiation is obtained using Eq. (12) (Duffie et al. 1985):

$$\tau_b = a_0 + a_1 * e^{(-k_0/\cos\theta_z)} \tag{12}$$

Table 2 Correction factors for different climate types (Duffie et al. 1985)

Climate type	r_0	r_1	r_k
Tropical	0.95	0.98	1.02
Midlatitude summer	0.97	0.99	1.02
Subarctic summer	0.99	0.99	1.01
Midlatitude winter	1.03	1.01	1.00

where a_0, a_1, k_0 are given by (Duffie et al. 1985)

$$a_0^* = 0.4237 - 0.00821 * (6 - A)^2 \tag{13}$$

$$a_1^* = 0.5055 + 0.00593 * (6.25 - A)^2 \tag{14}$$

$$k_0^* = 0.2711 + 0.01858 * (2.5 - A)^2 \tag{15}$$

$$(a_0 = a_0^* * r_0), (a_1 = a_1 * r_1), (k_0 = k_0^* * r_k) \tag{16}$$

The values of these correction factors for different climatic conditions are given in Table 2 (Duffie et al. 1985).

I_r is given by (Duffie et al. 1985)

$$I_r = I_0 * \tau_b * \cos\theta_z \tag{17}$$

It is to be noted that this gives only the beam component on a horizontal surface during a clear sky. The effect of cloud coverage, fog, etc., is not taken into account. Again, in Eq. (4), replace I_0 with I_r. This gives the corresponding values of T_{sol} by considering the effects of the atmosphere. This T_{sol} value is given to the outer surface of the models and the respective results for both the models with and without PCMs are compared.

The effective heat transfer coefficient of the model (U_{eff}) is (Heat and Mass Transfer 1998),

$$\frac{1}{U_{eff}} = \frac{d_1}{k_1} + \frac{d_2}{k_2} + \frac{d_3}{k_3} + \frac{d_4}{k_4} + \frac{1}{h_0} \tag{18}$$

where d_1, d_2, d_3, d_4, k_1, k_2, k_3, k_4 are the thickness and the thermal conductivity of brick, polyurethane, PCM, and steatite, respectively, and h_0 is the heat transfer coefficient of the outside air.

Heat transfer into the model per unit area (Q) is calculated using,

$$Q = U_{eff}(T_{sol} - T_{in}) \tag{19}$$

Table 3 Altitude (A), latitude (φ), longitude (λ), average wind speeds (V), maximum ambient temperature ($T_{a,max}$), and minimum ambient temperature ($T_{a,min}$) of four selected cities in different seasons (https://weatherspark.com/)

City	Nagpur (summer)	Chennai (summer)	Srinagar (winter)	Amritsar (summer)	Amritsar (winter)
A (m)	310	6.7	1585	232	232
φ (°)	21.1° N	13.1° N	34.1° N	31.6° N	31.6° N
λ (°)	79.1° E	80.3° E	74.8° E	74.9° E	74.9° E
V (mph)	10	11.8	4.8	5.3	5.3
$T_{a,max}$ (°C)	38	36.7	7	38	20.3
$T_{a,min}$ (°C)	26	27	-2	24.7	5.7

where T_{in} is the temperature of the innermost layer of the roof toward the room-side.

Cities are selected considering three climate zones in India; they are: (1) hot zone, (2) cold zone, and (3) the composite zone.

The average wind speeds, maximum ambient temperatures, and minimum ambient temperatures of selected cities in respective seasons are given in Table 3. All the average values are calculated from the wind speed and temperature variation charts which are easily available online (https://weatherspark.com/).

2.1 Hot Zone

This zone includes cities in India having extremely hot weather in summer, which includes Nagpur, Jaipur, Chennai, Bikaner, etc. Among these cities, Nagpur and Chennai are selected to analyze the effect of PCM. Phase change material is chosen according to the respective weather conditions of the place.

2.1.1 Nagpur

In summers, the average maximum temperature of Nagpur can reach over 38 °C in the daytime, while the average minimum temperature of 26 °C at night. Two cases are studied here for analyzing the effectivity of PCM implemented model over the model without PCM: (1) neglecting the atmosphere and (2) taking the atmosphere into account.

Case-1 (Neglecting the Atmosphere)

Neglecting the atmosphere in this case and taking the average temperature values, the maximum value of T_{sol} is 53.8 °C, while the minimum value of T_{sol} is 24.5 °C. The

Table 4 Thermophysical properties of octadecane ($C_{18}H_{38}$) (Salvi and Tyagi 2019)

Properties	Solid	Liquid
Melting temperature (T_m)	27.5 °C	
Latent heat of fusion (L)	244 kJ/kg	
Density (ρ)	814 kg/m^3	774 kg/m^3
Thermal conductivity (k)	0.358 W/m K	0.152 W/m K
Specific heat (C_p)	2150 J/kg K	2180 J/kg K

average solar insolation is calculated based on mean hourly solar insolation between 10 a.m. and 5 p.m. is 1068.8 W/m^2. The mean wind velocity for Nagpur is taken as 10 mph (4.47 m/s). Appropriate PCM is chosen, whose melting point lies between these two extreme temperatures. In this case, octadecane ($C_{18}H_{38}$) is chosen as PCM, whose thermal properties are given in Table 4 (Salvi and Tyagi 2019).

T_{sol} is given to the outer surface of the PCM implemented model and the model without PCM, and the results are compared by analyzing the final temperatures of the inside surface. The results are obtained by running the simulations for 7 h.

Case-2 (Considering the Atmosphere)

Considering the atmosphere in this case, the maximum value of T_{sol} reduced to 46.8 °C from 53.8 °C, while the minimum value to T_{sol} remains the same at 24.5 °C. In this case, the average solar insolation between 10 a.m. and 5 p.m. is reduced to 633.4 W/m^2. Mean wind velocity is taken as 9.9 mph (4.43 m/s). In this case, octadecane is taken as PCM whose melting point lies between extreme T_{sol} temperatures. As in the last case, T_{sol} is given to the outer surface of the models, and both results are compared for the total simulation time of 7 h.

2.1.2 Chennai

In summers, the average maximum temperature of Chennai can reach over 36 °C in the daytime, while the average minimum temperature of 27 °C at night. For Chennai, only the second case is studied where the atmosphere is considered while calculating the temperature values.

After taking into account the effect of the atmosphere on solar radiation, the maximum value of T_{sol} is reduced to 44 °C from 51 °C, while the minimum value to T_{sol} remains at 24.5 °C. In this case, the average solar insolation between 10 a.m. and 5 p.m. is reduced to 583 W/m^2 from 1053 W/m^2. Mean wind velocity is taken as 12 mph (5.36 m/s). In this case also, octadecane ($C_{18}H_{38}$) is taken as PCM as its melting point lies between the two extreme T_{sol} temperatures. T_{sol} is the temperature just outside the outer surface of the models, and both results are compared for the total simulation time of 7 h.

Table 5 Thermophysical properties of OM03 (organic mixture) (Polymers 2016)

Properties	Solid	Liquid
Melting temperature (T_m)	4 °C	
Latent heat of fusion (L)	229 kJ/kg	
Density (ρ)	912 kg/m^3	835 kg/m^3
Thermal conductivity (k)	0.224 W/m K	0.146 W/m K
Specific heat (C_p)	1760 J/kg K	1910 J/kg K

2.2 *Cold Zone*

2.2.1 Srinagar

This zone includes cities in India having freezing weather in winter, which includes Srinagar, Manali, etc. This study aims to decrease the heating load of the places where a considerable amount of energy is required for heating purposes inside buildings. PCM incorporation in buildings can be proven to be an excellent option to reduce the energy demand of the premises during winters. Phase change material is chosen according to the respective weather conditions of the place, which depends on the maximum and minimum average temperatures of the place. Among these cities, Srinagar is selected to analyze the effect of PCM on the realistic values by considering the impact of the atmosphere.

In winters, the average maximum temperature of Srinagar is at 7 °C in the daytime, while the average minimum temperature can reach as low as −2 °C at night. Taking the atmosphere into account and taking the average temperature values, the maximum value of T_{sol} reduced to 9.6 °C, while the minimum value to T_{sol} is the same at − 4 °C. The average solar insolation between 10 a.m. and 5 p.m. is 204.7 W/m^2. The mean wind velocity of Srinagar is taken as 4.8 mph (2.15 m/s). Appropriate PCM is chosen whose melting point lies between these two extreme T_{sol} temperatures. For this case, an organic mixture (OM03) is chosen as PCM, whose thermal properties are given in Table 5. T_{sol} is given to the outer surface of the models, and both results are compared for the total simulation time of 7 h (Polymers 2016).

2.3 *Composite Zone*

2.3.1 Amritsar

This zone includes cities having scorching weather in summer as well as very cold in winter, which includes cities such as Amritsar, Jalandhar, etc. Among these cities, Amritsar is selected for the study. PCM implemented models can be used to decrease the cooling load in summers and also the heating load in winters. Phase change material is chosen according to the weather conditions of the place for any specific

Table 6 Thermophysical properties of OM18 (organic mixture) (Polymers 2016)

Properties	Solid	Liquid
Melting temperature (T_m)	18 °C	
Latent heat of fusion (L)	167 kJ/kg	
Density (ρ)	906 kg/m^3	870 kg/m^3
Thermal conductivity (k)	0.182 W/m K	0.175 W/m K
Specific heat (C_p)	2920 J/kg K	2690 J/kg K

season. Two different phase change materials are chosen for the hot and cold weather conditions. For this reason, the PCM implemented model should be easily accessible to change different PCMs accordingly. Here, the impact of the PCM implemented model in both seasons is analyzed after taking the effect of the atmosphere into account.

During Summer

The average maximum temperature of Amritsar can be reached over 38 °C in summer, while the average minimum temperature of 24.6 °C at night. Two cases of neglecting and considering the atmosphere are studied here.

Considering the effect of the atmosphere on solar insolation, the maximum value of T_{sol} reduced to 49.6 °C from 59 °C, while the minimum value to T_{sol} is at 22.6 °C. The average solar insolation during peak temperature hours is 614.8 W/m^2. The mean wind velocity for Amritsar is 5.3 mph (2.37 m/s). In this case also, octadecane is taken as PCM whose melting point lies between extreme T_{sol} temperatures. T_{sol} is given to the outer surface of the PCM implemented model, and the results obtained by running the simulations for 7 h are compared with the simple model by analyzing the final inside temperatures.

During Winter

In winter, the average maximum temperature of Amritsar is over 20 °C during the daytime, while the average minimum temperature can be as low as 5.6 °C at night.

With the atmosphere affecting the solar insolation, the maximum value of T_{sol} is 22.5 °C, while the minimum value is at 3.6 °C. The average solar insolation during peak hours is 189 W/m^2. The mean wind velocity of Amritsar is 5.3 mph (2.37 m/s). In this case, OM18 (an organic mixture) is taken as PCM, whose thermophysical properties are given in Table 6 (Polymers 2016).

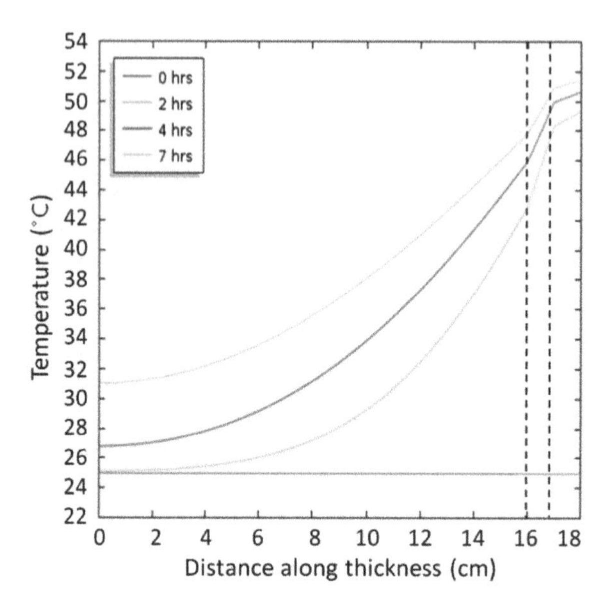

Fig. 3 Temperature distribution within the model without PCM in Nagpur (neglecting the atmosphere). The dashed lines show the model dimensions (roof up to 16 cm; polyurethane from 16 to 17 cm; steatite from 17 cm up to 18 cm)

3 Results and Discussion

3.1 Hot Zone

3.1.1 Nagpur (Altitude: 310 m, Lattitude 21.1° N, Longitude 79.1° E)

Case-1 (Neglecting the Atmosphere)

The temperature variation inside the simple roof model has been plotted. As can be seen in Fig. 3, the inside temperature of the roof reached over 31 °C after 7 h. The temperature distribution inside the PCM implemented model has been plotted without considering the atmosphere along with the thickness of the roof. As we can see in Fig. 4, the temperature inside the model is increasing with time, and the melting front inside the PCM is moving from top to the bottom along its thickness. After almost 4 h, half of the PCM melted, and after 7 h, 3 cm PCM thickness converted into liquid. The temperature inside the roof is about 26 °C after 7 h while it is at 31 °C in the roof model without using PCM.

As a large amount of energy is required for the melting of the PCM, the inside temperature in the PCM implemented model is much less than compared to the simple roof model, which can sufficiently reduce the cooling load of the building. The thermal conductivity of PCM is higher in the solid phase compared to the liquid

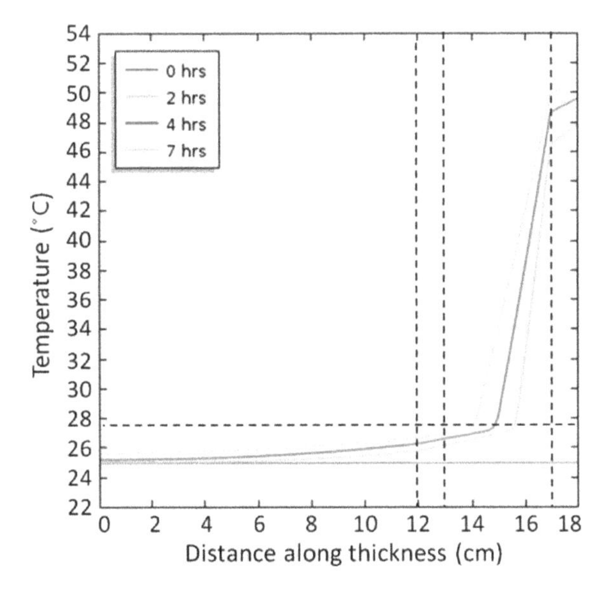

Fig. 4 Temperature distribution within the model with PCM in Nagpur (neglecting the atmosphere). Vertical dashed lines show the model dimensions (roof of 12 cm; polyurethane of 1 cm; PCM from 13 to 17 cm; steatite up to 18 cm); horizontal dashed line shows the melting point of PCM

phase, and thus the energy transfer rate is different in both the phases. For this reason, there is a sharp change in the slope at the melting point of the PCM, which is 27.5 °C for this case.

Case-2 (Considering the Atmosphere)

The atmosphere has been considered, and the temperature distribution inside both models, using PCM and without using PCM, has been compared. In the PCM model, the inside temperature after 7 h is just below 26 °C, while it is below 30 °C in the simple model. So it can be assumed that the implementation of the PCM model above the roof can result in a temperature decrease of approximately 4 °C, and thus the cost of cooling inside the premises.

As we can see in Fig. 6, most of the solar thermal energy is getting stored inside the PCM while changing its phase, so there is a minimal temperature rise inside the room compared to the model without PCM in Fig. 5.

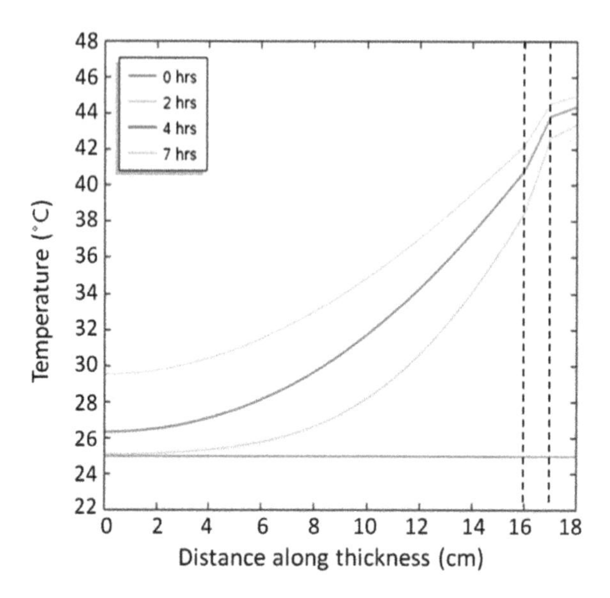

Fig. 5 Temperature distribution within the model without PCM in Nagpur (considering atmosphere). The dashed lines show the model dimensions (roof up to 16 cm; polyurethane from 16 to 17 cm; steatite from 17 cm up to 18 cm)

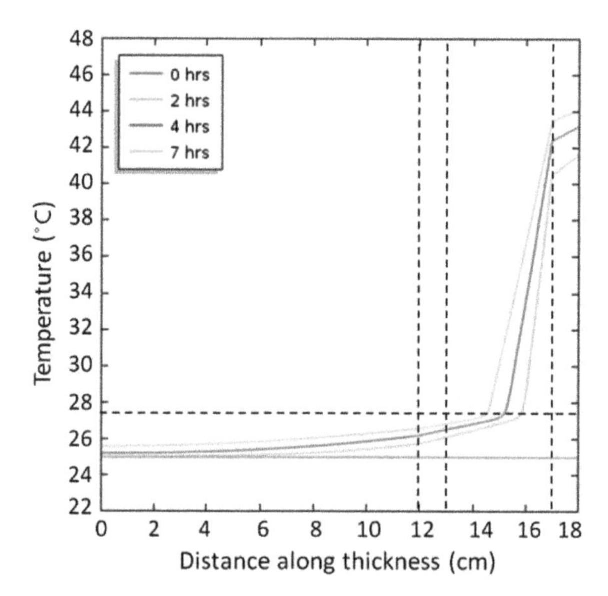

Fig. 6 Temperature distribution within the model with PCM in Nagpur (considering atmosphere). Vertical dashed lines show the model dimensions (roof of 12 cm; insulation of 1 cm; PCM from 13 to 17 cm; steatite up to 18 cm); horizontal dashed line shows the melting point of PCM

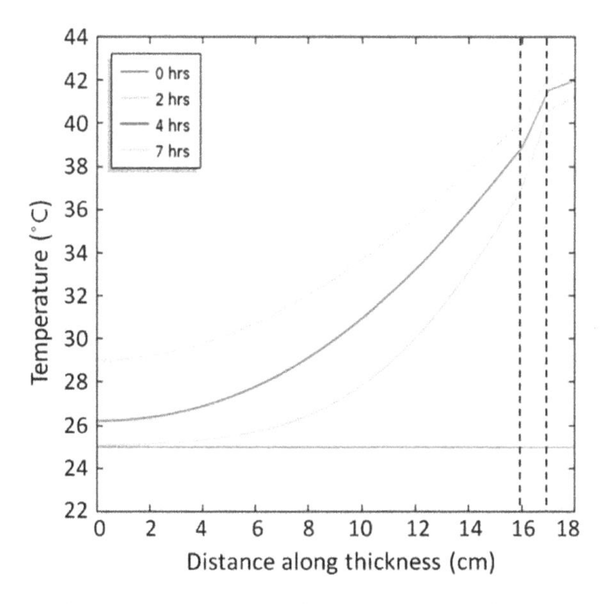

Fig. 7 Temperature distribution within the model without PCM in Chennai (considering atmosphere). The dashed lines show the model dimensions (roof up to 16 cm; polyurethane from 16 to 17 cm; steatite from 17 cm up to 18 cm)

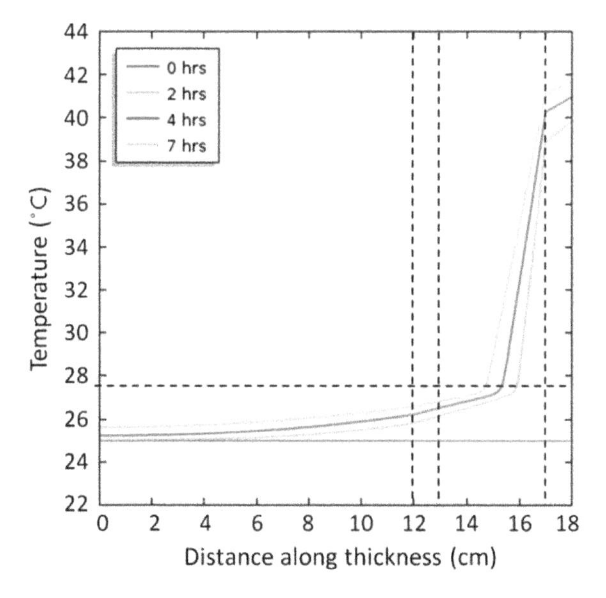

Fig. 8 Temperature distribution within the model with PCM in Chennai (considering atmosphere). Vertical dashed lines show the model dimensions (roof of 12 cm; insulation of 1 cm; PCM from 13 to 17 cm; steatite up to 18 cm); horizontal dashed line shows the melting point of PCM

3.1.2 Chennai (Altitude: 6.7 m, Lattitude 13.1° N, Longitude 80.3° E)

For Chennai, only the second case with the atmosphere is considered. As can be seen from Figs. 7 and 8, the implementation of PCM results in almost 3.5 degrees decrease in the inside temperature after 7 h.

3.2 Cold Zone

3.2.1 Srinagar (Altitude: 1585 m, Lattitude 34.1° N, Longitude 74.8° E)

Figure 9 shows the temperature variation inside the simple roof model in Srinagar. The initial temperature of the model is taken as 25 °C while the outside temperature is 4 °C. After 7 h, the final inside temperature inside the room is just above 19 °C. After using the model with PCM, the final interior temperature is increased to approximately 24 °C. So, PCM implemented models can also be used to decrease the heating loads inside the buildings efficiently.

In Fig. 10, there is a slight change in slope in the plots at 4 °C (melting point of PCM OM03) because PCMs have different values of thermal conductivity values in different phases. As the initial temperature of the model is higher than the phase change temperature of the PCM, it is in a liquid state initially. PCM started

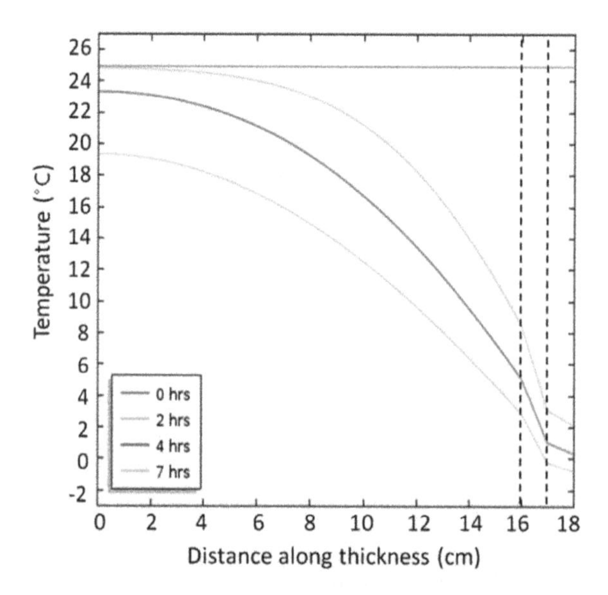

Fig. 9 Temperature distribution within the model without PCM in Srinagar (considering atmosphere). The dashed lines show the model dimensions (roof up to 16 cm; polyurethane from 16 to 17 cm; steatite from 17 cm up to 18 cm)

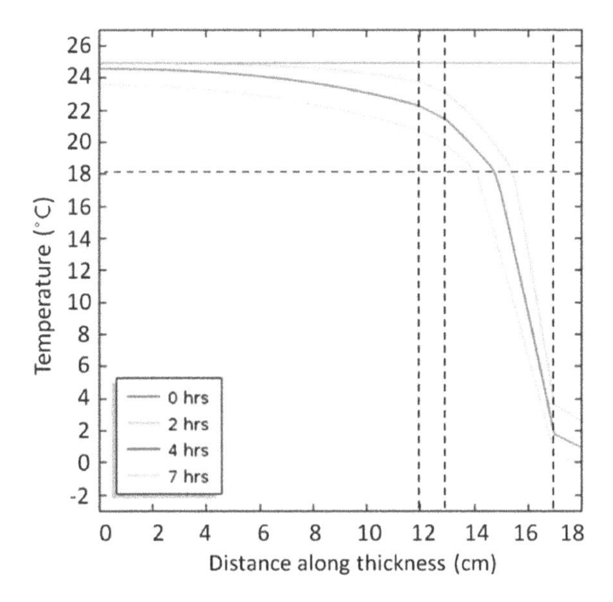

Fig. 10 Temperature distribution within the model with PCM in Srinagar (considering atmosphere). Vertical dashed lines show the model dimensions (roof of 12 cm; insulation of 1 cm; PCM from 13 to 17 cm; steatite up to 18 cm); horizontal dashed line shows the melting point of PCM

converting into solid-state throughout the cold night and again started melting when the surrounding temperature is above its melting point during the day. Whole PCM must be converted into a liquid state during the day for this model to work.

3.3 Composite Zone

3.3.1 Amritsar (Altitude: 232 m, Lattitude 31.6 °N, Longitude 74.9 °E)

During Summer

In Fig. 11, the initial temperature inside the room is maintained at 25 °C, which increased to above 30 °C after 7 h. While in Fig. 12, we can see a minimal temperature rise inside the room because of the latent heat storage of phase change material.

As can be seen in Figs. 11 and 12, PCM implementation over the roof can result in a temperature decrease of almost 5 °C, which can result in decreasing the cooling load and thus the cost of electricity during summers.

The next case is using the PCM model over the roof in winter. For this, the PCM material should be easily accessible and can be easily changed according to the season.

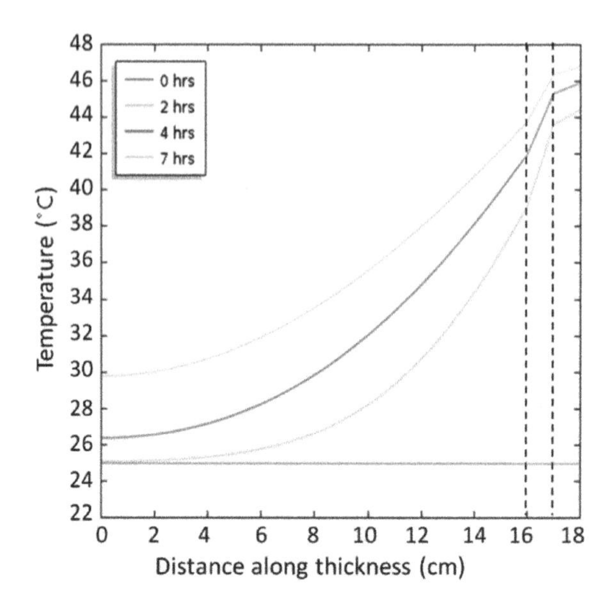

Fig. 11 Temperature distribution in summer within the model without PCM in Amritsar (considering atmosphere). The dashed lines show the model dimensions (roof up to 16 cm; polyurethane from 16 to 17 cm; steatite from 17 cm up to 18 cm)

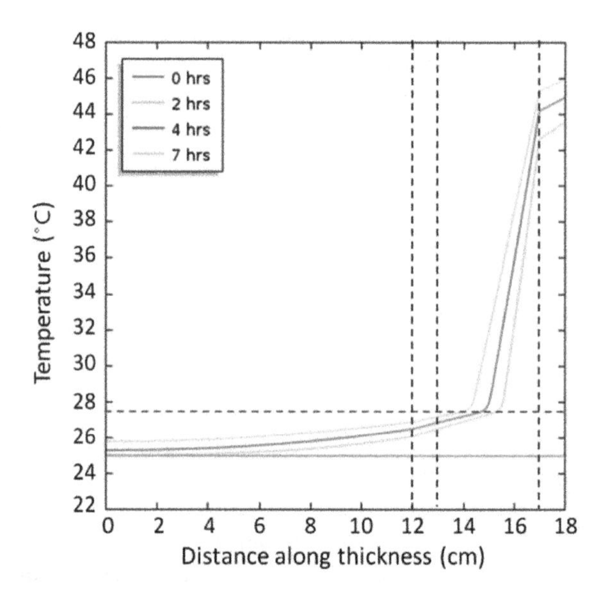

Fig. 12 Temperature distribution in summer within the model with PCM in Amritsar (considering atmosphere). Vertical dashed lines show the model dimensions (roof of 12 cm; insulation of 1 cm; PCM from 13 to 17 cm; steatite up to 18 cm); horizontal dashed line shows the melting point of PCM

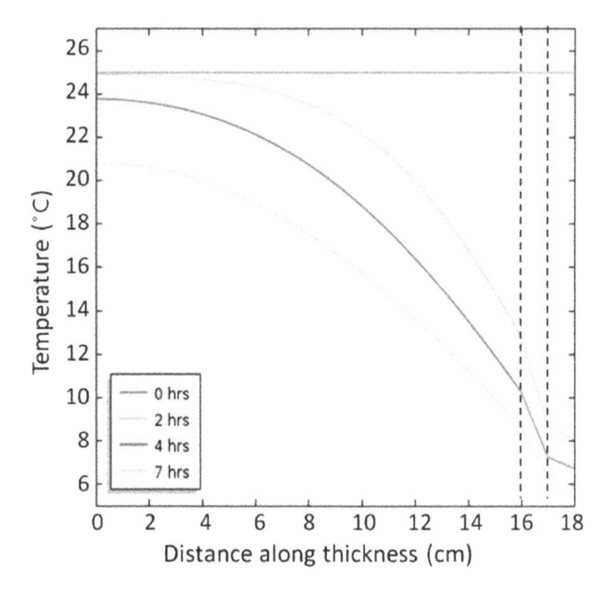

Fig. 13 Temperature distribution in winter within the model without PCM in Amritsar (considering atmosphere). The dashed lines show the model dimensions (roof up to 16 cm; polyurethane from 16 to 17 cm; steatite from 17 cm up to 18 cm)

During Winter

During winters, the average temperature in Amritsar varies from 3 °C to over 22 °C. OM18, an organic mixture, whose freezing point is 18 °C is chosen as PCM. The PCM is in solid-state during the cold nights, while it is liquid during the daytime. As we can compare from Figs. 13 and 14, this model can significantly decrease the heating load during winters as it lowers the temperature decrease by almost 3 °C inside the buildings after comparing it with the model without PCM.

4 Conclusions

Mathematical modeling of a hypothetical PCM model over the roofs of various climate zones in India is done. Numerical modeling by taking into consideration the realistic weather conditions and dimensions are considered for analyzing the effect of PCM implementation. From the above analysis, it can be inferred that

1. As we can see for the hot climate zone, the PCM model can be effectively used to decrease the temperature rise inside the building by about 4 to 5 degrees. This reduces the cost of the cooling load of the building during the summer.
2. As can be seen from the cold zone, this model can also be useful during winters as it decreases the temperature reduction inside the buildings by approximately

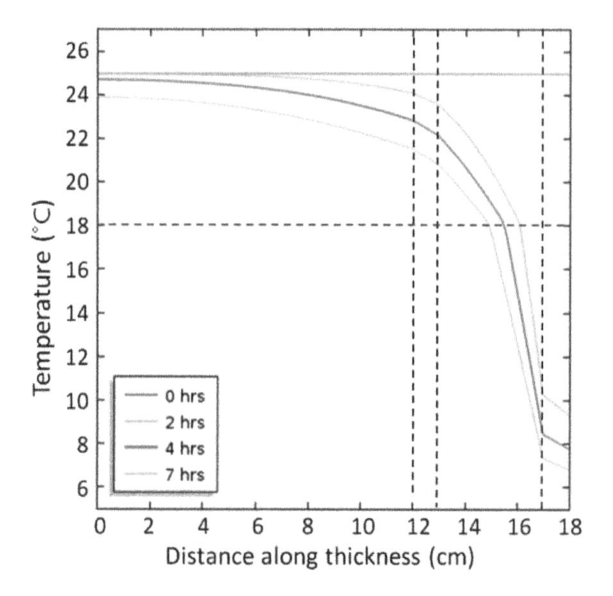

Fig. 14 Temperature distribution in winter within the model with PCM in Amritsar (considering atmosphere). Vertical dashed lines show the model dimensions (roof of 12 cm; insulation of 1 cm; PCM from 13 to 17 cm; steatite up to 18 cm); horizontal dashed line shows the melting point of PCM

3 to 4 degrees, which can lower the heating cost sufficiently. Due to its easy accessibility, different PCMs can be used according to the season.

3. The wind has a significant effect on the amount of thermal energy that can be stored in the model and thus the room temperature inside the building. For example, the average wind speed in Chennai is more than in Nagpur. Therefore, the average heat transfer coefficient in Chennai is more than that of Nagpur, which relatively lowers the thermal energy stored in Chennai compared to Nagpur.

Acknowledgements The authors gratefully acknowledge the support received from the Department of Mechanical engineering at the Indian Institute of Technology Ropar.

References

Cabeza LF, Castell A, Barreneche C, De Gracia A, Fernández AI (2011) Materials used as PCM in Thermal energy storage in buildings: a review. Renew Sustain Energy Rev 15(3):1675–1695

Cengel YA, Ghajar AJ (1998) Heat and mass transfer, fundamentals and applications. McGraw Hill, New York

Cook TR, Dogutan DK, Reece SY, Surendranath Y, Teets TS, Nocera DG (2010) Solar energy supply and storage for the legacy and nonlegacy worlds. Chem Rev 110(11):6474–6502

Duffie JA, Beckman WA, Mcgowan J (1985) Solar engineering of thermal processes. Wiley, Hoboken

Fleischer AS (2015) Springer briefs in applied sciences and technology thermal engineering and applied science. In: Thermal energy storage using phase change materials fundamentals and applications

Hu WJ, Chang MN, Gao Y, Zhang QL, Yang LY, Li DY (2017) Experimental study on the cooling charge and discharge characteristics of a PCM based fin-tube thermal energy storage exchanger. Procedia Eng 205:3088–3095

Kenisarin M, Mahkamov K (2007) Solar energy storage using phase change materials. Renew Sustain Energy Rev 11(9):1913–1965

Khudhair AM, Farid MM (2004) A review on energy conservation in building applications with thermal storage by latent heat using phase change materials. Energy Convers Manage 45(2):263–275

Kumar R, Misra MK, Kumar R, Gupta D, Sharma PK, Tak BB, Meena SR (2011) Phase change materials: technology status and potential defence applications. Defence Sci J 61(6):576–582

Kumar R, Soni V, Kumar A (2016) Design of thermal energy storage system using phase change material for comfort cooling. In: National conference on smart cities, Nov

Polymers (2016) Technical data sheet of save ® om11. Technical Data Sheet, 122016, pp 1–2

Saikia P, Pancholi M, Sood D, Rakshit D (2020) Dynamic optimization of multi-retrofit building envelope for enhanced energy performance with a case study in hot Indian climate. Energy 197:117263

Salvi SS, Tyagi H (2019) Numerical analysis of phase change materials for use in energy efficient buildings. In: Ting D, Carriveau R (eds) Energy generation and efficiency technologies for green residential buildings. The Institution of Engineering and Technology (IET)

Sarbu I, Sebarchievici C (2018) A comprehensive review of thermal energy storage. Sustainability 10(1)

Saxena R, Rakshit D, Kaushik SC (2020) Experimental assessment of phase change material (PCM) embedded bricks for passive conditioning in buildings. Renew Energy 149:587–599

Sharma A, Tyagi VV, Chen CR, Buddhi D (2009) Review on thermal energy storage with phase change materials and applications. Renew Sustain Energy Rev 13(2):318–345

Soni V, Kumar A, Kumar A, Jain VK (2019) Real-time experimental study and numerical simulation of phase change material during the discharge stage: thermofluidic behavior, solidification morphology, and energy content. Energy Storage 1(1):E51

The typical weather anywhere on earth—weather spark [Online]. Available: https://weatherspark.com/

Tyagi H, Chakraborty P, Powar S, Agarwal A (eds) (2020) Solar energy: systems, challenges, and opportunities. Springer, Berlin. https://doi.org/10.1007/978-981-15-0675-8

Zalba B, Marín JM, Cabeza LF, Mehling H (2003) Review on thermal energy storage with phase change: materials, heat transfer analysis, and applications. Appl Therm Eng 23(3):251–283

Energy Storage

Parabolic Dish Solar Cooker: An Alternative Design Approach Toward Achieving High-Grade Thermal Energy Storage Solution

Sairaj Gaunekar, Amit Shrivastava, and Prodyut Ranjan Chakraborty

1 Introduction

Thermal energy demands associated with cooking is majorly catered by fossil fuels. The ever-increasing global demand for energy over the last century led to extensive usage of fossil fuels, causing the depletion of natural reserves at an alarming rate. It is not only the severe depletion of these natural reserves, but also the adverse effects on the environment due to the burning of fossil fuels, that led the scientific community to explore alternative renewable resources of energy with lesser environmental impact. Solar energy has a tremendous potential to substitute fossil fuels for many energy extensive applications, particularly in those areas having an abundance of solar irradiation. According to NREL (https://www.nrel.gov/gis/data_solar.html), the average Direct Normal Irradiance (DNI) at Jodhpur India is significantly high (~5 $kWh/m^2/day$). The abundance of solar irradiation in and around Jodhpur (aptly named as Sun-city) along with the entire northwestern belt of India (consisting of Rajasthan, Gujrat, Haryana Punjab and part of Madhya Pradesh) makes this region ideal for harvesting solar energy as a renewable alternative. Among various solar photovoltaic and solar thermal applications, the usage of solar thermal energy for household and commercial cooking has a huge potential to replace conventional fossil fuels to a large extent. Barring the monsoon season and few odd cloudy days over the year solar cooking can reliably cater to energy requirement for cooking with zero carbon footprint. However, to make this application acceptable and more appealing at large, we need to improve the technology in terms of flexibility in cooking schedule. Almost all the commercially available solar cookers do not offer this flexibility of cooking schedule. And the handful of those offer this flexibility are based on sensible

S. Gaunekar · A. Shrivastava · P. R. Chakraborty (✉)
Department of Mechanical Engineering, Indian Institute of Technology Jodhpur, Jodhpur 342037, India
e-mail: pchakraborty@iitj.ac.in

H. Tyagi et al. (eds.), *New Research Directions in Solar Energy Technologies*,
Energy, Environment, and Sustainability,
https://doi.org/10.1007/978-981-16-0594-9_12

heat storage with limited storage capacity at comparatively low temperature (~100 to 150 °C) (Cuce 2018). The energy storage density being small for sensible heat storage, large volume of the storage unit poses a big challenge. Large temperature gradient within the storage medium also hinders the optimal utilization of stored heat. Since the energy storage density of latent heat storage is significantly larger than sensible heat storage, the volume of the storage unit can be reduced to a large extent. Latent heat storage mechanism also ensures heat supply during cooking to be maintained within a narrow temperature range. All these special features associated with latent heat thermal energy storage (LHTES) makes this technology an ideal thermal storage solution for providing the flexibility of cooking schedule with concentrated solar cooker.

The latent thermal energy storage unit consists of a metallic storage container filled with PCM. Different ways of cooking require heat supply at different ranges of temperatures, namely boiling (100–150 °C), simmering (120–200 °C), frying (180–250 °C), grilling (200–300 °C), and so on. Therefore, an ideal PCM for the LHTES should have melting temperature above 250 °C to cater to all these cooking requirements. Conclusive details on various types of PCMs along with their thermophysical properties are summarized by Sharma et al. (2009), Zalba et al. (2003), Iverson et al. (2012), and Kenisarin (2010). From these data on thermophysical properties, it is evident that suitable PCMs having melting temperatures in the range of 250–300 °C are mostly pure inorganic salts or salt compositions with thermal conductivity raging between 0.25 and 0.5 W/mK. Therefore, thermal conductivity enhancement of the PCM is one of the key challenges in designing the LHTS for solar cooking. The charging process of the LHTES involves complete melting of the PCM inside the storage container during the on-sun period, thus storing the thermal energy in the form of latent heat slightly above the melting temperature of the PCM. On the other hand, discharging process involves heat release from the storage device during off-sun hours cooking activities, rendering the PCM to resolidify in the process. For the current analysis, the sizing of the LHTES unit is done on the basis of thermal energy requirement for cooking a typical evening meal for a six-member family, with the menu consisting of roti, dal, rice, and vegetables. The major design challenge is storing sufficient energy above 300 °C within on-sun duration (between 10:00 am and 4:00 pm) to meet the abovementioned cooking requirement.

A brief overview on existing literature reporting the usage of latent heat storage unit integrated with solar cookers is now presented. Domanski et al. (1995) presented a box-type solar cooker with concentric cylindrical vessels having the annulus portion filled with magnesium nitrate hexahydrate as the PCM. The experiments revealed that the effectiveness of this box-type solar cooker strongly depends on solar intensity, cooker medium mass, and the thermophysical properties of the PCM. The heat transfer from the PCM to the cooking pot is also observed to be slow during discharging process. Buddhi and Sahoo (1997) also reported the performance of a box-type solar cooker with stearic acid as the latent heat storage medium. One of the major outcomes of their experimental observation is the possibility of cooking two batches of about 200 gm rice per pot per day. The maximum temperature obtained in this setup was around 100–110 °C. Buddhi et al. (2003) presented a box-type solar cooker with acetanilide as a latent heat storage medium. They used ternary reflectors to increase the solar radiation. Vigneswaran et al. (2017) studied a box-type solar cooker with multiple reflectors where oxalic acid dihydrate is used as the latent heat storage medium. It was observed that time duration to reach melting temperature is reduced when multiple reflector boosters are used. Sharma et al. (2005) experimentally investigated an evacuated tube solar collector coupled with a latent heat storage unit. Commercial-grade erythritol was used as the PCM. Experimental results showed that the system can cook successfully twice, afternoon and evening daily. Temperatures reached more than 110 °C during evening cooking. Hussein et al. (2008) constructed a solar cooker with heat pipes and double flat plate reflectors. It was noted that the ratio of steel wool incorporated with the PCM can be increased to enhance its effective thermal conductivity. Coccia et al. (2018) constructed a double-walled container storage composed of two stainless steel cylindrical pots constructed concentrically. The annular area in between the pots was filled with a PCM consisting of a ternary mixture of nitrite and nitrate salts (53 wt% KNO_3, 40 wt% $NaNO_2$, 7 wt% $NaNO_3$). A significant improvement of load thermal stabilization during the off-sun period is reported when PCM is used as the storage medium. Lecuona et al. (2013) used a parabolic reflector integrated with LHTES having Paraffin and erythritol as PCM. It was reported that it is feasible to cook the three meals for a family during summer and also in winter. Paraffin was recommended to be a better option than erythritol. Kumaresan et al. (2015, 2016) studied a parabolic trough reflector focussing solar radiation onto the absorber tube. Therminol 55 was used as the heat transfer fluid (HTF) and encapsulated of D-mannitol was used as the PCM. It was observed that the tava cooking unit developed could cater heat comparable to an LPG stove running in the simmering mode. Veremachi et al. (2016) designed and developed a double reflector with axis tracking mechanism. A mixture of sodium nitrate and potassium nitrate at a ratio of 60:40 (mol%) was used as the PCM. A hotspot temperature of 277 °C could be attained with moderate temperatures gradient within the storage unit. Bhave and Thakare (2018) proposed a parabolic solar concentrator integrated with LHTES having magnesium nitrate hexahydrate as the storage medium and thermic mineral oil as HTF. It was reported that in about 50 min the PCM temperature reached 135 °C. Fifty grams of rice with 100 ml water were found to be completely cooked in about 30 min. El-Sebaii et al. (2009) studied degradation associated with thermal cycling of commercial-grade acetanilide and magnesium chloride hexahydrate for solar cooker application. Acetanilide was found to be a more promising PCM for this application. Magnesium chloride hexahydrate degrades easily during thermal cycling because of the phase segregation. Foong et al. (2010) performed numerical and experimental study on KNO_3-$NaNO_3$ salt composition (60:40 weight ratio) as the latent heat storage medium. A double reflector-based parabolic dish concentrator is used for this purpose. The storage container is augmented with fins for improving the thermal performance of the storage unit. The temperature range of 227–327 °C was achieved.

The review of the existing literature on solar cookers with integrated LHTES provides a basis for categorizing such solar cookers containing various design aspects with specific salient features. The block diagram presented in Fig. 1 provides a

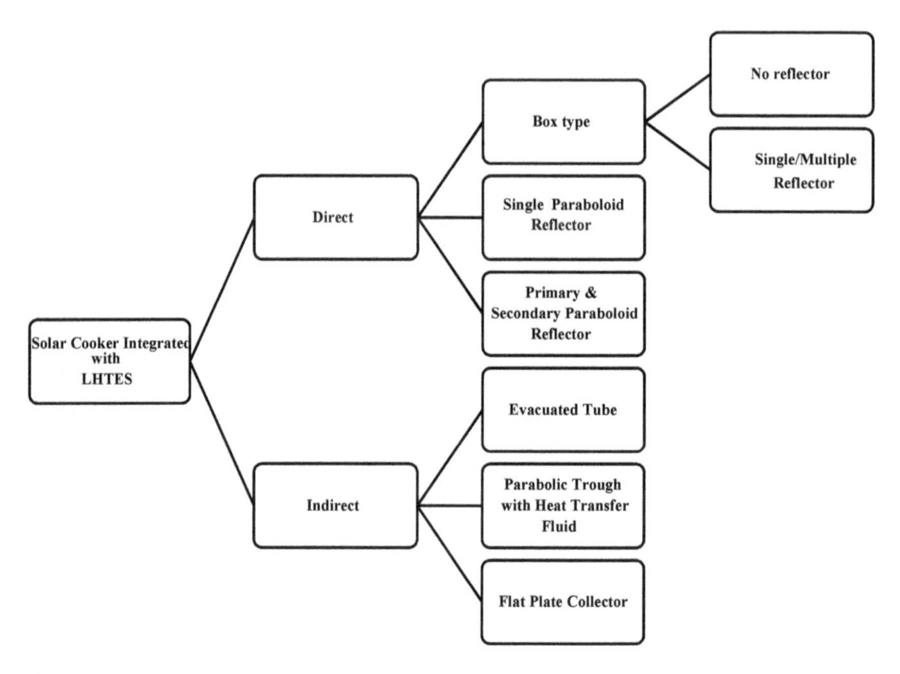

Fig. 1 Classifications of solar cookers with LHTES

consolidated list of existing solar cooker technologies addressing thermal energy storage. The salient features of each of these solar cookers are now discussed in brief.

For direct solar cookers, incident solar radiation is directly used for cooking or storing thermal energy to be dispatched during the hour of need. Solar cookers enlisted in this category are box-type solar cooker, and solar cookers with single or double paraboloid concentrators. Box-type solar cookers can be farther classified as box-cookers without reflector and box-cookers with single or multiple reflectors.

Box-cookers without reflector: Fig. 2 shows a typical design of box-cooker without reflector (Buddhi and Sahoo 1997). A double-walled tray-type container with PCM filled within the space between the walls attributes to the key design feature. The thermal storage tray containing PCM is kept under a glass lid. The upper surface of the tray is covered with absorber material. Sunrays pass through the glass lid, strike the absorber tray, and transfer heat to the PCM underneath it for charging. The tray is surrounded by insulating material. The performance of this type of cooker depends upon the ability of the transparent glass to permit passage of shorter wavelength which forms a significant part of the solar spectrum but opaque to higher wavelength coming out from the box. The temperature inside the box rises because of the green-house effect. However, this setup is found to be inefficient because solar rays are not concentrated by any means, and the charging time is very long.

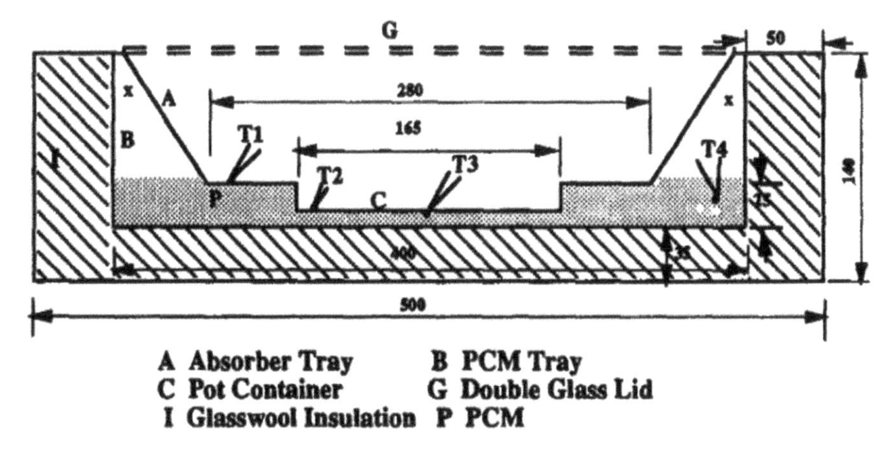

Fig. 2 Schematic of box-cooker without reflector having LHTES (Buddhi and Sahoo 1997)

Box-cookers with single/multiple reflectors: These box-cookers contain single/multiple reflectors (Fig. 3) to concentrate solar radiation on a vessel containing PCM (Vigneswaran et al. 2017). Cooking containers and interior of these box-cookers are made of absorbing material. The box is insulated from sides. The mirrors need to be adjusted with respect to each other to concentrate the solar radiation. Multiple reflectors provide concentrated solar energy to the cooking and storage container enabling the system to achieve temperatures in the range of 100–120 °C.

Concentrated solar cooker with single paraboloid reflector: These solar cookers contain a single paraboloid reflector to concentrate solar irradiation (Fig. 4) on a double-walled container cum cooking pot with PCM filled within the space between the walls Lecuona et al. (2013). The bottom surface of the storage cum cooking container is placed closed to the focal point of the paraboloid reflector. One of the major challenges for this design is associated with continuous solar tracking

Fig. 3 Box-cooker with multiple reflector (Vigneswaran et al. 2017)

Fig. 4 Solar cooker with paraboloid reflector (Lecuona et al. 2013)

during the operation. Hence, integration of an axis tracking mechanism is a must for this design concept. Locating the heavy storage cum cooking container above the reflector renders this tracking even more challenging. However, we can obtain very high temperatures of about 250–300 °C adjacent to the focal point location, enabling the cooker to cater frying and grilling requirements unlike box-cookers adequate only for boiling-type cooking requirements.

Primary paraboloid reflector coupled with secondary reflector PCM solar cooker: Since the positioning of heavy cooking cum storage vessel over the paraboloid reflector poses considerable challenge with respect to the solar tracking requirement, arrangement of a double reflector-based design (Fig. 5) is proposed to circumvent this challenge Veremachi et al. (2016). The key feature of this design is the positioning of a secondary reflector close to the focal point of the primary reflector. The primary reflector is a paraboloid reflector which reflects solar radiation onto the secondary reflector. The secondary reflector concentrates the radiation on the storage container during the charging process. A hole at the center of the primary reflector allows the radiation from the secondary reflector to reach the storage container kept beneath the primary paraboloid reflector. This setup is ergonomically more viable for solar tracking operation as the heavy storage container sits at bottom most location. Another advantage of this mechanism is obtaining a fixed hotspot location on the storage container during charging process which is otherwise very difficult to obtain with single reflector design. However, for this design, simultaneous cooking and charging operation is limited to boiling type of cooking only. Cooking requirements

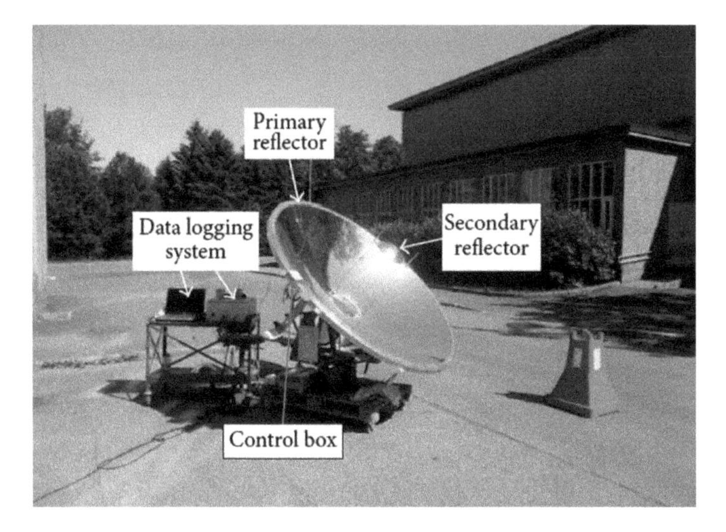

Fig. 5 Primary and secondary reflector solar cooker Veremachi et al. (2016)

like frying or grilling can only be addressed once the charging process of the storage unit is over.

Next, we briefly discuss some of the existing designs of indirect solar cookers. Unlike direct solar cookers solar irradiation is not directly used for cooking or energy storage, rather a heat transfer fluid (HTF) is used as the thermal transport medium to transfer heat from the receiver to the cooking cum storage unit. The thermic fluid is continuously circulated between the receiver and storage cum cooking unit during the cooking and storage operation. Three types of receivers are particularly of interest, namely evacuated tube, parabolic trough, and flat plate collector.

Evacuated tube solar cooker: Fig. 6 depicts the key components of solar cooker with an evacuated tube receiver (Sharma et al. 2005). The evacuated tube consists of an inner metallic pipe encased within a transparent glass tube. The exterior surface of the inner metallic tube is coated with a solar selective coating having high absorptivity and low emissivity. Vacuum is maintained within the gap between the inner metallic tube and an outer glass tube to protect the solar selective coating from degradation due to contact with air. The arrangement also ensures minimum heat loss through convection and radiation mechanism. During on-sun hours, solar radiation is absorbed by the evacuated tube solar collector, and heat is transferred to the HTF circulating through the heated tube. The HTF circulation is achieved using a closed-loop pumping setup. The pump drives the HTF in a close loop through a steel tubing arrangement wrapping around the storage cum cooking container and the evacuated tube receiver. Once again the storage cum cooking container is a double-walled vessel with the gap between the two walls filled with PCM. The HTF circulation system passes through this PCM-filled domain. The HTF circulation loop transports heat from the evacuated receivers to the storage cum cooking container. The flow of hot HTF through the PCM causes the melting of PCM and stores heat in PCM in the

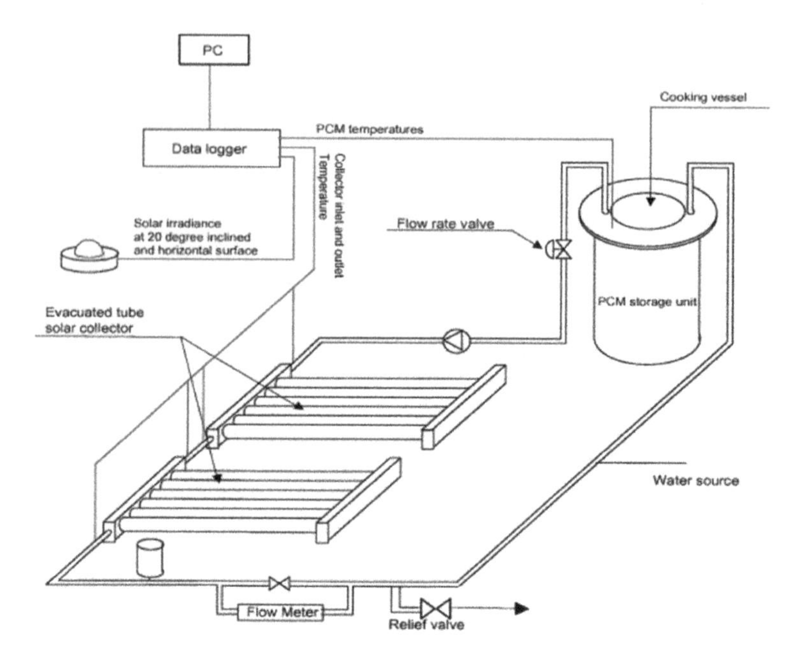

Fig. 6 Schematic diagram of evacuated tube solar cooker with LHTES (Sharma et al. 2005)

form of latent heat, which can be used for cooking at later duration. The absence of solar irradiation concentrator restricts this system to attain high temperature and confines the temperature below 100 °C. Additional pump setup, flow meter, relief valves, and electrical power are required for this closed-loop setup.

Parabolic trough solar cooker: The functionalities of a solar cooker with parabolic trough as the receiver is quite similar to the evacuated tube solar cooker. The only difference between the two being receiver type. However, owing to the usage of a parabolic concentrator, the very high HTF temperature can be achieved by using multiple parabolic concentrators. The performance of this solar cooker can be further improved by using a combination of evacuated tube and parabolic reflector. The schematic of such an arrangement is shown in Fig. 7 (Kumaresan et al. 2015, 2016).

Flat plate collector PCM solar cooker: This setup consists of a flat plate solar collector through which the HTF is circulated (Fig. 8) before delivering the heat to the cooking cum storage container containing PCM as storage material similar to the evacuated tube and parabolic trough solar cooker systems. Once again the absence of concentrator limits the obtainable temperature below 100 °C.

Classification of PCMs used for solar cookers

Several types of PCM have been explored by research groups across the globe as latent heat thermal energy storage medium for solar cooking applications. A consolidated list of PCMs used as thermal energy storage medium for cooking application is shown

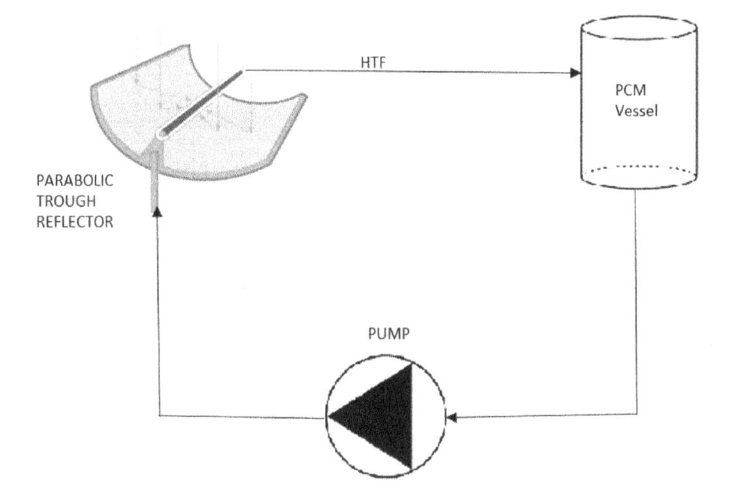

Fig. 7 Schematic diagram of parabolic trough solar cooker with heat transfer fluid

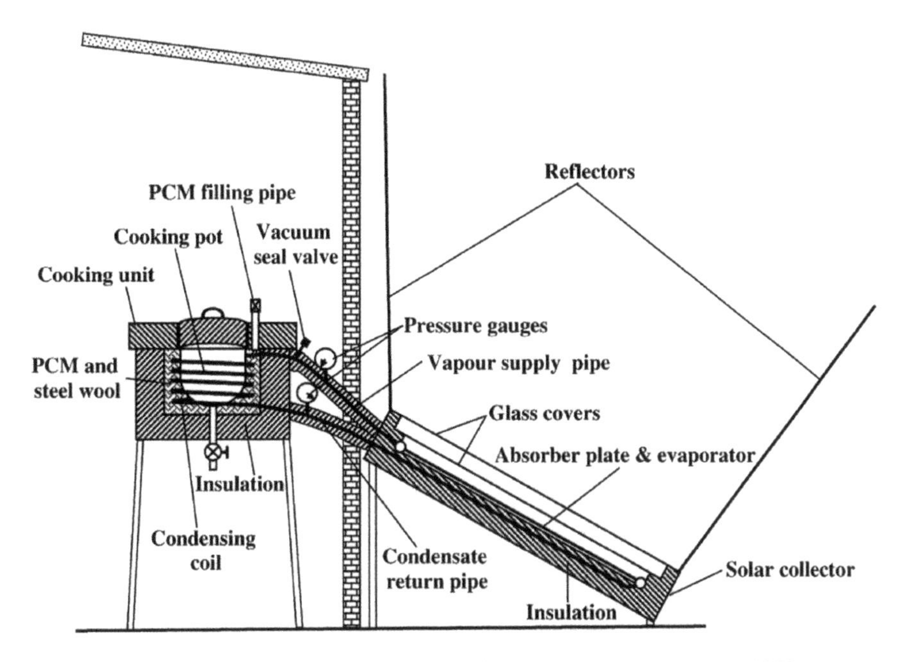

Fig. 8 Schematic diagram of flat plate collector PCM solar cooker (Hussein et al. 2008)

in the chart presented by Figs. 9 and 10. The pros and cons of organic, inorganic, and eutectic PCMs are listed in Table 1.

Selection standards for PCM used in the PCM solar cooking device: There are specific features to be addressed while choosing PCM as thermal storage medium

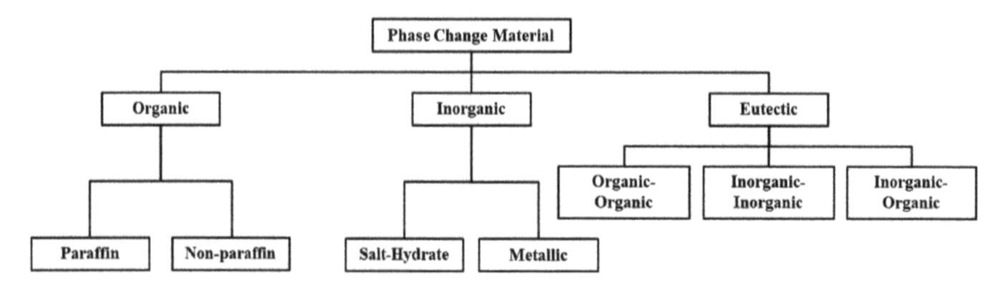

Fig. 9 Classification of PCM (Memon 2014)

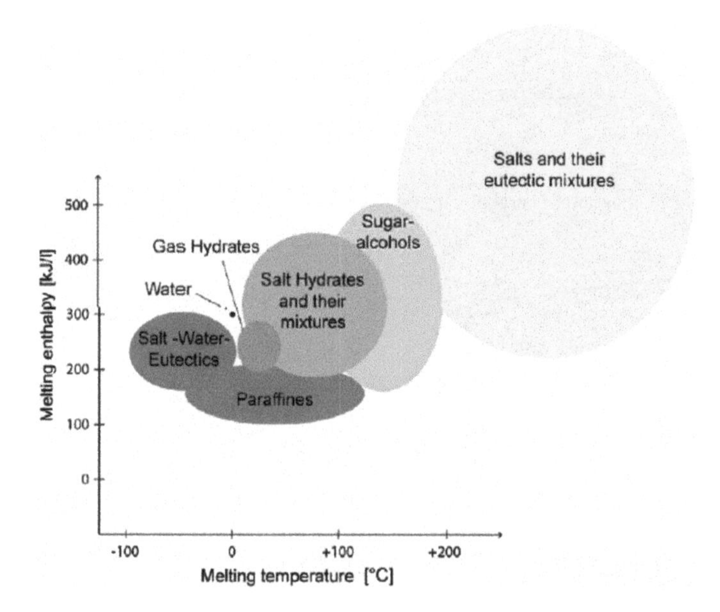

Fig. 10 Fusion enthalpy (kJ/l) versus melting temperature (°C) of various types of PCMs (Kalnæs and Jelle 2015)

for solar cooking application. The preferable features can be summarized as follows (Memon 2014):

A. **Thermophysical features**

1. Large energy storage density to ensure a compact storage unit
2. Reasonably high thermal conductivity to achieve rapid charging process and maintain thermal uniformity and rapid heat flow during cooking or discharging process
3. Large specific heats for solid and liquid phase also ensure higher storage density
4. Small volumetric changes during phase transitions allow simple design requirement for storage container

Table 1 Merits and demerits of organic, inorganic, and eutectic PCMs

PCM type	Merits	Demerits
Organic	• Significant deviation (from 20 to 70 °C) in phase change temperature • Chemically dormant • Better thermal steadiness in a long run melt cycle • Intermediate latent heat of fusion (120–210 J/g) • Superior thermal specific heat in comparison to inorganic PCM • Usually non-corrosive • Economically viable • Compatibility with commonly available container materials excluding plastics • Variation in volume at the time of phase change process is minimal • Minor supercooling during solidification • Usually safe environmentally • Mostly stable at temperatures under 500 °C • Replenishable	• Low melting point • Reasonably small thermal conductivity (about 0.21 W/mK) • Moderately flammable
Inorganic	• Latent heat storage capacity is twice in comparison to organic PCMs • High latent heat of fusion • Thermal conductivity is generally much higher than organic PCM (about 0.5 W/mK) • Economical and readily accessible • Fire resistant • Affinity for plastic container • Small changes in volume on melting • Potentially recyclable • Insignificant solubility in water	• Prone to endure supercooling during freezing • High probability of phase segregation during freezing • Corrosive nature to most metals
Eutectic	• Transformation of phase is at a constant temperature • Energy stored per unit volume is marginally higher than the organic PCM	• Data on the thermophysical properties are not broadly available

5. Thermal stability of PCM and low degradation rate allow a large number of thermal cycles and infrequent replacement requirement for thermal storage material.

B. **Chemical features**

1. The chemical composition must remain stable and uniform over a large number of thermal cycles involving phase change process

2. PCM must be non-toxic, non-corrosive, and non-explosive.

C. **Economic features**

1. The material used for the storage medium must be inexpensive
2. Readily accessible.

D. **Environmental features**

1. The PCM must be non-contaminating
2. The PCM must be easily replenishable.

Thermophysical properties of some commonly used and potential PCMs for solar cooking applications are enlisted in Table 2.

Although discussions regarding design aspects of box-type solar cookers with latent heat storage are frequently encountered (Domanski et al. 1995; Buddhi and Sahoo 1997; Buddhi et al. 2003; Vigneswaran et al. 2017; Coccia et al. 2018), and reports on dish concentrator-based solar cookers with latent heat storage are rarely available (Bhave and Thakare 2018; Foong et al. 2010). In the subsequent section, we provide a detailed modeling description of double reflector-based parabolic dish solar cookers integrated with latent heat storage unit.

2 Design Details of Double Reflector-Based Parabolic Solar Cooker Integrated with Latent Heat Storage Unit

In order to store sufficient thermal energy to cook meal for a family, the total weight requirement of the thermal storage medium is quite high (~10 kg). The weights of the container and insulation material are added over it. Placing this heavy storage unit above the parabolic dish concentrator poses serious structural constraint. For any practical purpose, such top heavy arrangement is not desirable. Continuous solar tracking requirement for the parabolic dish concentrator complicates the matter farther. Locating the bulky storage structure above, the concentrator also restricts a significant amount of sunlight to reach the parabolic dish reflector due to the shadow effect. All these structural and optical challenges can be comprehensively addressed by an alternative design approach of locating the storage unit below the parabolic dish concentrator with double reflector arrangement. Figure 11 shows such an arrangement. The solar irradiation captured by the primary reflector is first focussed on a much smaller secondary reflector (Fig. 11), which directs the focussed solar rays on the storage container located below the primary reflector. A small hole at the center of the primary reflector allows the rays from the secondary reflector to pass through and incident on the latent heat storage container. The large primary reflector is a parabolic dish and the secondary smaller reflector is a hyperbolic dish. The design constraints of these two reflectors are briefed in the subsequent discussions.

Primary and secondary reflector design: Solar energy received from the solar irradiation needs to be concentrated to receive high heat flux at high temperature in

Table 2 Property table for potential PCMs for solar cooking application

PCM	H (kJ/kg)	T_m (°C)	ρ (kg/m^3)		c_p (kJ/kg K)		k (W/mK)		μ (Ns/m^2)	References
			S	L	S	L	S	L		
Paraffin	140	100	880	770	1.8	2.4	0.21	0.2	0.049	Bhave and Thakare (2018)
Erythritol	340	118	1480	1300	1.38	2.76	0.733	0.326	0.01	Bhave and Thakare (2018)
Stearic acid	157	54	940	–	1.76	2.27	0.29	0.17	–	Sarbu and Dorca (2019)
Manganese nitrate hexahydrate	280	89	1640	–	2.5	3.1	0.65	0.5	–	Sarbu and Dorca (2019)
Commercial-grade acetanilide	118.9	222	1318	–	2.00	–	–	–	–	Buddhi et al. (2003)
53 wt% KNO_3,40 wt%$NaNO_2$, 7wt% $NaNO_3$	145.14	101.50	–	–	1.441	1.462	–	–	–	Coccia et al. (2018)
D-mannitol	300	165	1490	–	1.31	2.36	0.19	0.11	–	Sarbu and Dorca (2019)
Oxalic acid	356	105	1900		1.62	2.73	–	–	–	Mukherjee (2018)
Magnesium chloride hexahydrate	117	167	1450	1570	2.61	2.25	0.570	0.704	–	Sarbu and Dorca (2019)
Acetamide	260	82	1160	–	2.00	3.00	0.4	0.25	–	Mukherjee (2018)

H = latent heat of fusion, T_m = melting point temperature, ρ = density, c_p = specific heat capacity, k = thermal conductivity, μ = dynamic viscosity

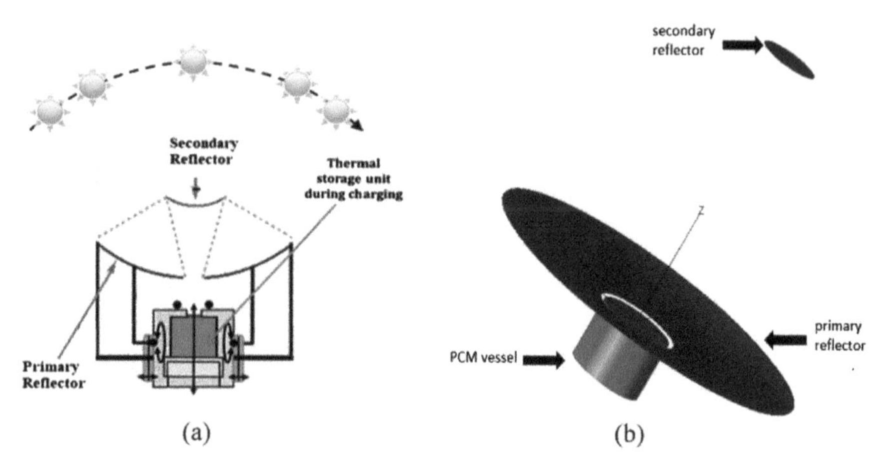

Fig. 11 Schematic diagram of double reflector-based high temperature thermal storage arrangement, **a** 2-D view, **b** 3-D view

the designated receive location of the storage container. The primary and secondary reflector arrangement ensures the concentration of high heat flux at a reasonably high temperature ($\sim$ 350 °C) to be incident on the receiver location of the storage container. To design the reflectors, we require the knowledge of total thermal energy for cooking (E_c) the target meal proportion, target charging duration (Δt_{ch}) of the thermal storage unit, average local DNI (I), and the maximum temperature ($T_{\max} > T_m$) attainable by the PCM without degrading. Once the total requirement of energy for cooking and charging duration is identified, the incident area (A) of the primary reflector can be estimated from the following relation.

$$E_c = I \cdot A \cdot \Delta t_{\mathrm{ch}} \cdot \alpha \tag{1}$$

where α represents the absorptivity of the vessel surface.

Average cooking energy (E_c) requirement above 100 °C for a six-member family is typically 1.2–1.5 kWh (Buddhi et al. 2003). The average DNI observed at Jodhpur is approximately 600 W/m^2. However, to be on the safer side, an average DNI of 450 W/m^2 is considered. Typically, the aperture diameter of the parabolic dish reflector varies within the range of 1.2–2 m. On the other hand, the target charging time should ideally be less than 5 h. A safe value of absorptivity (α) can be assumed to be 85%. Considering all these facts, and taking into account of central hole and shadow effect of the structural frame to support the secondary reflector, Eq. 1 can provide an accurate diameter requirement of the primary reflector. TracePro ray tracing software is used for designing the primary (paraboloid) and secondary (hyperboloid) reflectors (Fig. 12) along with the incident heat flux distribution on the receiver area of the thermal storage container.

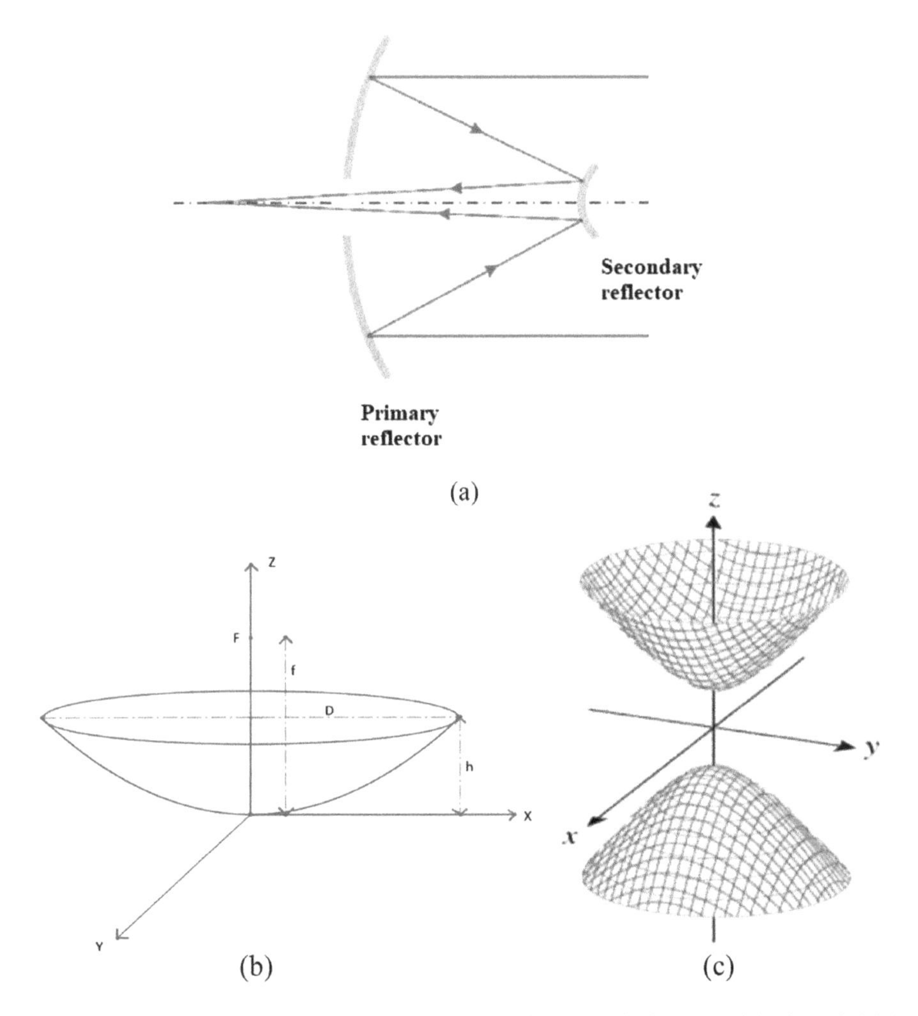

Fig. 12 **a** Light path of Cassegrain model; graphical definition of the **b** paraboloid, **c** hyperboloid reflectors

The 'classic' Cassegrain reflector model is considered for concentrating solar energy (Fig. 12a). A hole is made in the center of the paraboloid reflector to allow the rays to pass through to the receiver location of the thermal storage container. The hyperboloid is kept above the primary reflector and aligned along the same vertical axis of the paraboloid. The sun rays are reflected from the primary paraboloid reflector to the secondary hyperboloid reflector, which in turn concentrates solar radiation upon the cooking vessel. The paraboloid reflector reflects all incoming light rays parallel to its axis of symmetry toward a single point, the focus. The converging light rays reflected by the parabolic concentrator toward its focus are intercepted by a secondary hyperboloid reflector. The hyperboloid reflector surface having two foci reflects all light rays directed toward one of its two foci in the direction of its other focal point (Fig. 12a). The folded optical path in this manner results in a very compact system; hence, hyperboloid geometry is considered for the secondary reflector (Horne 2012). The light rays reflected from the hyperboloid need not be parallel. This novel design helps in decreasing the shadow effect since the aperture of the hyperboloid is considerably smaller than the radial span of the storage container. This bottom heavy arrangement also facilitates structural stability and ergonomics of the storage cum cooking container.

Recalling the basic features of paraboloid geometry, the curved surface formed by rotation of a parabola about its axis is called a paraboloid of revolution. The mathematical equation for the paraboloid (Fig. 12b) in Cartesian coordinate system with the z-axis as the axis of the paraboloid is as follows.

$$x^2 + y^2 = 4fz \tag{2}$$

where the distance f is the focal length.

In cylindrical coordinates, the equation becomes:

$$z = \frac{r^2}{4f} \tag{3}$$

The aperture area of the paraboloid is given by:

$$A_p = \frac{\pi D^2}{4} \tag{4}$$

where D is the aperture diameter.

In terms of aperture diameter and focal length, the height of the dish h is:

$$h = \frac{D^2}{16f} \tag{5}$$

The surface area of the paraboloid is given by:

$$A_s = \frac{8\pi f^2}{3}\left\{\left[\left(\frac{D}{4f}\right)^2 + 1\right]^{3/2} - 1\right\} \tag{6}$$

Concentration ratio is defined as the ratio of aperture area of the concentrator to the focal area of the receiver.

$$\mathrm{CR} = \frac{A_p}{A_r} \tag{7}$$

where A_p denotes the aperture area of the concentrator and A_r denotes the focal area on the receiver surface.

Collector Efficiency
The solar energy collection efficiency η_{col} of thermal collectors is defined as the ratio of the rate of useful thermal energy leaving the collector, to the useable solar irradiance falling on the aperture area. Simply stated, collector efficiency is

$$\eta_{\text{col}} = \frac{Q_{\text{useful}}}{A.I}$$

Optical Efficiency of Collector
Optical efficiency is formulated as

$$\eta_{\text{opt}} = \tau \cdot \alpha$$

where τ is the transmittance of reflector glass and α is the absorptivity of receiver surface.

Recalling the hyperboloid geometry the equation representing a hyperboloid in Cartesian coordinate system is given by:

$$\frac{x^2}{a^2} + \frac{y^2}{b^2} - \frac{z^2}{c^2} = -1 \tag{8}$$

The intercepts of the hyperboloid with z-axis is given by $(0, 0, \pm c)$. There are no intersections with the x, y-axes.

Design of thermal storage container: Dimensions of the storage container are obtained on the basis of the required cooking energy to be stored above 300 °C. We chose the total cooking energy requirement to be 1.5 kWh which is sufficient to cook a sumptuous Indian style meal for a six-member family (Buddhi et al. 2003). A cylindrical steel vessel of wall thickness 3 mm is considered. A ceramic coating of 2 mm thicknesses on the inner surface of the steel container is also considered. Ceramic coating is considered to prevent the corrosive effect of molten PCM at elevated temperatures on the steel container (Moreno et al. 2014). However, steels with anticorrosive properties are available, and thermal storage container made with such corrosion resistant steel may not require this protective ceramic layer.

Two different designs for thermal storage container are considered for the present study (Fig. 13). The first design (Fig. 13a) consists of a cylindrical steel vessel without fin arrangement, while the second design (Fig. 13b) involves a cylindrical steel vessel with circumferentially distributed copper fins. For the finned design of thermal storage vessel, the central copper rod is of diameter 10 mm, and each of the radially extended copper plates from the central rod has a thickness of 2 mm.

Selection of PCM material: In the present study, $NaNO_3$-CEG (sodium nitrate-compressed expanded graphite) composite is used as latent heat storage medium

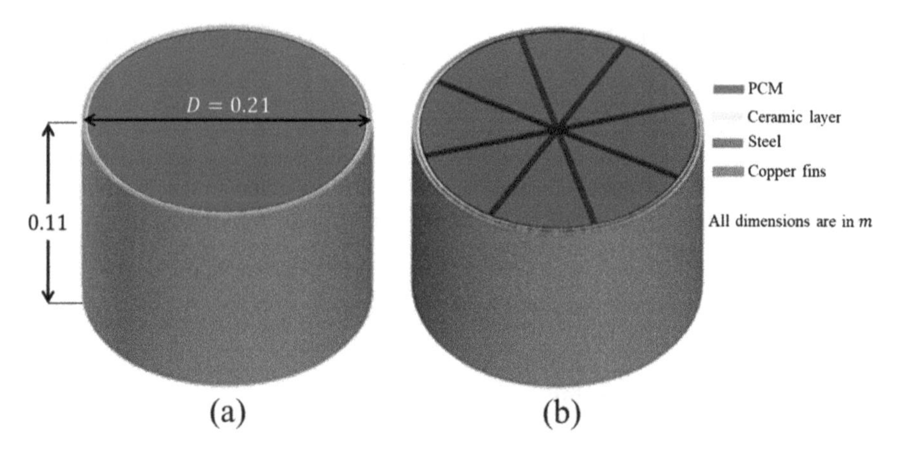

Fig. 13 Schematic diagram of storage container, **a** without fin, **b** with circumferentially distributed fins

with a melting temperature of 308 °C. The major drawback of pure $NaNO_3$ is that its low thermal conductivity (0.5 W/mK) causing accumulation of heat near the receiver area of the storage container. If pure $NaNO_3$ is used as the latent heat storage medium, the low thermal conductivity causes a large temperature gradient within the storage medium with a local hotspot near the receiver region. Very high temperature (~400 °C) at the hotspot may lead to decomposition of nitrate to nitrite and degrade the thermal properties of $NaNO_3$. Therefore, enhancement of thermal conductivity is of absolute necessity for such high-temperature application. The thermal conductivity of $NaNO_3$ is enhanced by incorporating 10% graphite by volume in the form of compressed expanded graphite (CEG). $NaNO_3$-CEG composite consists of a highly porous CEG matrix impregnated with $NaNO_3$. The thermal conductivity of $NaNO_3$-CEG is higher by two orders of magnitude as compared to pure $NaNO_3$ (Kenisarin 2010). Incorporating graphite in PCM results in anisotropic thermal conductivity in the PCM domain. An interesting aspect of PCM-CEG composite is that natural convection does not play a dominant part in the heat transfer and is dominated by heat diffusion only (Py et al. 2001). The small pore size of CEG matrix allows the composite to be analyzed as a homogeneous material. The thermophysical properties have been estimated using volume averaging approach. Homogeneous distribution assumption of very small pore size allowed us to consider this volume averaged estimation of thermal conductivity approach with reasonable accuracy (Mallow et al. 2016). The major limitation of this approach is negating the existence of thermal non-equilibrium between the graphite and PCM due to large difference in thermal conductivities of these two materials. Inclusion of thermal non-equilibrium effect between the graphite and PCM in the numerical model is beyond the scope of the present work and needs to be explored farther providing a future scope for the present study. Based on the volume calculation of the storage container, the total weight of $NaNO_3$ content is approximately 7 kg, which is capable of storing approximately 1.2

Table 3 Thermophysical properties of container material, PCM (Venkateshwar et al. 2017), and CEG (Bodzenta et al. 2011)

Material	c_p(kJ/kg K)	k(W/mK)	$\rho(\mathrm{kg/m^3})$	T_m(°C)	h_{sl}(kj/kg)
PCM	1.820	0.5	2260	308	172
CEG	1.820	11 (axial) 42.3 (radial)	2250	–	–
SS	0.502	16.27	8030	1577	–
Copper	0.381	387.6	8978	1085	–

kWh energy above 100 °C. Thermophysical properties of container material, PCM (Venkateshwar et al. 2017), and CEG (Bodzenta et al. 2011) are provided in Table 3.

3 Numerical Modeling

In this section, we present the numerical model to capture storage dynamics during the charging cycle. The first step in this endeavor is to evaluate incident flux distribution on the receiver region of the thermal storage container. Once this incident flux profile is obtained, it is used as the boundary condition to solve the energy conservation equation in the storage domain.

Incident flux profile: The incident heat flux profile is obtained by using ray tracing TracePro software (Figs. 14 and 15). Average DNI data measured at IIT Jodhpur is used as the input to the software for attaining irradiation flux profile on the receiver surface. The flux profile obtained follows Gaussian distribution.

The incident flux profile $q(r)$ on the PCM vessel is formulated as

$$q(r) = A_1 e^{\left(\frac{r-b_1}{c_1}\right)^2} + A_2 e^{\left(\frac{r-b_2}{c_2}\right)^2} + A_3 e^{\left(\frac{r-b_3}{c_3}\right)^2} + A_4 e^{\left(\frac{r-b_4}{c_4}\right)^2} \tag{9}$$

where $A_1, A_2, A_3, A_4, b_1, b_2, b_3, b_4, c_1, c_2, c_3, c_4$ are discrete functions of time which remain constant for each time intervals of 20 min. MATLAB has been used to obtain the approximate Gaussian distribution (Eq. 9) from the solar irradiance data with uncertainty of less than 5%.

Numerical modeling: The governing energy balance equation for the PCM-CEG domain is derived on the basis of volume averaging formulations proposed by Brent et al. (1988), Bennon and Incropera (1987), and Shrivastava and Chakraborty (2019). Since the PCM is trapped within the porous matrix of CEG in the composite, convection within PCM melt is neglected and the energy balance equation is diffusion dominated. In deriving the energy balance equation, the difference between specific heat capacities of solid and liquid phases of PCM is considered (Shrivastava and

Fig. 14 Ray tracing diagram of the double reflector system

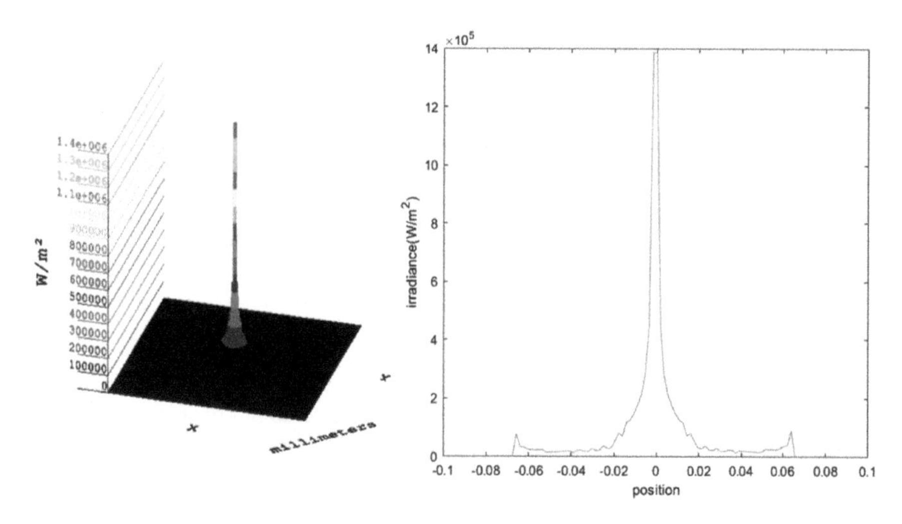

Fig. 15 Flux profile obtained in Ray Tracepro software

Chakraborty 2019). The sensible heating of graphite matrix is also considered (Shrivastava and Chakraborty 2019). The energy conservation equation in the PCM-CEG composite domain using temperature as the primary dependent variable can be formulated as follows (Shrivastava and Chakraborty 2019):

$$\frac{\partial(T)}{\partial t} = \nabla.\left(\frac{k}{\rho c_{ps}}\nabla T\right) - \frac{\partial}{\partial t}\left(\frac{g_l h_{sl}}{c_{ps}}\right) - \frac{\partial}{\partial t}\left[g_l\left(\frac{c_{pl}}{c_{ps}} - 1\right)(T - T_m)\right] - \frac{\partial}{\partial t}\left[g_g\left(\frac{c_{pg}}{c_{ps}} - 1\right)T\right] \quad (10)$$

The second, third, and fourth source terms appearing on the right-hand side of Eq. 10 represent contributions due to phase change, difference between liquid and solid phase specific heats of PCM and difference between graphite and solid phase PCM specific heat, respectively. Since the graphite matrix has anisotropic thermal conductivity, the effective thermal conductivity (k) is defined in the following manner (Shrivastava and Chakraborty 2019).

$$k_i = g_l k_l + g_s k_s + g_{g,i} k_{g,i} \quad (11)$$

where subscript i represents (r, x) direction in cylindrical coordinate system, subscripts l, s represents liquid and solid phases of PCM, and subscript g represents graphite.

The energy equation for ceramic, steel, and copper fin regions is formulated as follows.

$$\frac{\partial}{\partial t}(T) = \nabla.\left(\frac{k}{\rho c_p}\nabla T\right) \quad (12)$$

Heat loss from the circumferential surface and bottom surface is assumed to be zero, i.e., these surfaces are considered to be perfectly insulated. The surface area of the storage container pertaining to the receiver location is provided with the heat flux boundary condition given by Eq. 9.

Volume fraction updating scheme is implemented to estimate the liquid volume fraction using the following updating scheme (Shrivastava and Chakraborty 2019)

$$g_{lp}^{n+1} = g_{lp}^{n} + \lambda\left[\left\{\frac{A_p + D_p + C_p g_{lp}^{n}}{B_p}\right\}(T - T_m)\right] \quad (13)$$

where subscript p represents the nodal point where volume fraction g_l is being updated. A_p represents coefficient of T_p obtained by finite volume discretization of Eq. 10 (Patankar 2018) when contribution of only left-hand side and first term (diffusion term) of the right-hand side of Eq. 10 is considered (Chakraborty 2017). λ is an under-relaxation factor. Superscripts 'n' and '$n + 1$' denote iteration steps during implicit calculation of T and g_l field at a given time step. B_p, C_p and D_p are given as follows:

$$B_p = \frac{h_{sl}}{c_{ps}}; \quad C_p = \frac{c_{pl}}{c_{ps}} - 1; \quad D_p = g_g\left(\frac{c_{pl}}{c_{ps}} - 1\right) \quad (14)$$

The liquid volume fraction g_l ranges within the limit: $0 \leq g_l \leq 1 - g_g$, where g_g represents volume fraction of graphite at an elementary control volume. If the numerical value of g_{lp}^{n+1} from Eq. 13 comes outside the range $0 \leq g_l \leq 1 - g_g$, it is updated as the nearest limit such that:

$$\begin{aligned} g_{lp}^{n+1} &= 0 \quad\quad\;\; \text{for } g_{lp}^{n+1} < 0 \\ g_{lp}^{n+1} &= 1 - g_g \text{ for } g_{lp}^{n+1} > 1 - g_g \end{aligned} \tag{15}$$

Since the energy equation solved by ANSYS-FLUENT solve (Fluent ANSYS 2015) either enthalpy or total energy as primary variable, Eq. 1 could only be solved by treating temperature T as the user-defined scalar (UDS). The enthalpy updating scheme is applied through user-defined function. The anisotropic thermal conductivity (Eq. 11) and the heat flux boundary condition (Eq. 9) are also implemented through UDFs. A grid size of 1 mm has been used for numerical simulations followed by 0.5 s time step.

4 Results and Discussions

The thermal storage dynamics during the charging process is studied for two different sets of storage configurations. The first configuration consists of thermal storage container without any fin arrangement inside the PCM-CEG domain (Fig. 13a). The second configuration contains circumferentially distributed copper fins within the storage medium (Fig. 13b).

Case study 1 (Thermal storage container without fin): This study could have been performed for 2-D axisymmetric geometry. However, the other case study involving circumferentially oriented fin cannot be resolved with 2-D axisymmetric simplification, and a 3-D approach is a must. The 3-D domain can however be reduced conveniently because the fins are oriented with regular angular intervals. In order to have a better comparison with the finned configuration, the case study involving unfinned thermal storage container is carried out in 3-D domain. One-eighth symmetry with an angular span of 45° (Fig. 16) is chosen as the reduced domain for case study 1 in order to reduce total computation time.

The initial temperature of the complete system is assumed to be at ambient temperature 27 °C. The charging process of the PCM is carried out between 10 am to 1:20 pm, i.e., for 3 h and 20 min since the average DNI in this period is maximum. The maximum temperature attained in the system during charging after 12,000 s is 374 °C (647 K) which is tantalizingly close to the temperature 380 °C at which decomposition of $NaNO_3$ in $NaNO_2$ starts. The minimum temperature of the storage medium is found to be 303 °C (5 °C below the melting point temperature $T_m = 308$ °C of $NaNO_3$) indicating incomplete melting of the storage medium.

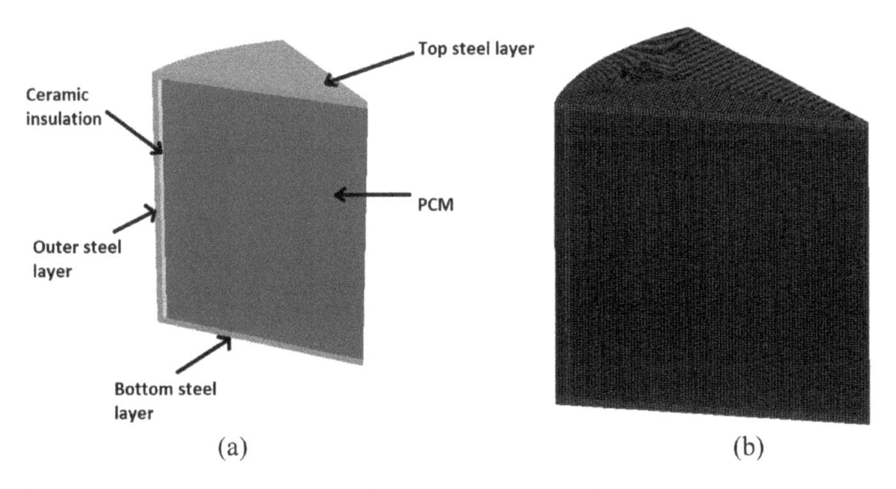

Fig. 16 Computational domain for no-fin storage container configuration, **a** schematic diagram, **b** hex mesh generated in ANSYS

The temperature profile evolution in time for the PCM-CEG domain within the storage unit is presented in Fig. 17 at various time intervals. The solar irradiation flux profile at the receiver region of the CEG-PCM thermal storage container is such that it creates a hotspot at the center of the vessel top surface. The non-uniform temperature distribution (Ranging over a temperature difference of ~30 to 70 °C at different time instants) within the PCM may be attributed to the low thermal conductivity of the PCM. The melting fraction evolution is shown in Fig. 18 at the same time intervals 9000, 10,000, 11,000, and 12,000 s. It is to be noted from Fig. 18, that melting of the PCM starts at around 10,000 s. The complete melting is achieved at 12,100 s, and the time required for complete melting is around 2100 s. One of the interesting non-intuitive features of the melting dynamics is found to be the location where melting process occurs at the end. We found that the location where the melting process defers till the end is adjacent to the central region of the storage container (Fig. 18d) and not the farthest region from the hotspot or receiver area. The reason for this non-intuitive melting feature is attributed to the fact that steel having reasonably larger thermal conductivity ($\sim 16\,\mathrm{W/mK}$) carries the heat from the hotspot along the container wall, promoting faster meltdown of the PCM adjacent to the wall. On the other hand, comparatively low thermal conductivity of CEG in the axial direction due to anisotropy causes slower heat transfer from the hotspot in the axial region, rendering the central region of the storage container not to reach melting temperature till the very end.

Case study 2 (Thermal storage container with fin arrangement): Case study with fin arrangement consists of three subsystem studies with three different fin orientations in the circumferential direction. These three different fin configurations consists of radially extended copper plane plate fins attached to a central copper rod with an angular pitch of 30°, 45°, and 60°. The schematic of 45° fin orientation

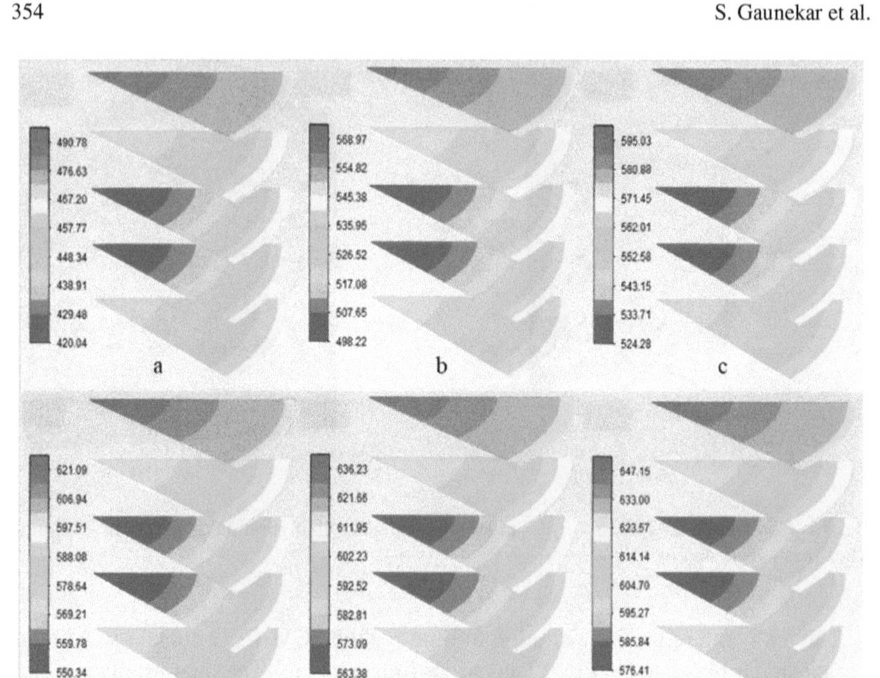

Fig. 17 Temperature (Kelvin) profile of PCM-CEG domain without fin at: **a** 6000 s, **b** 9000 s, **c** 10,000 s, **d** 11,000 s, **e** 11,500 s, and **f** 12,000 s

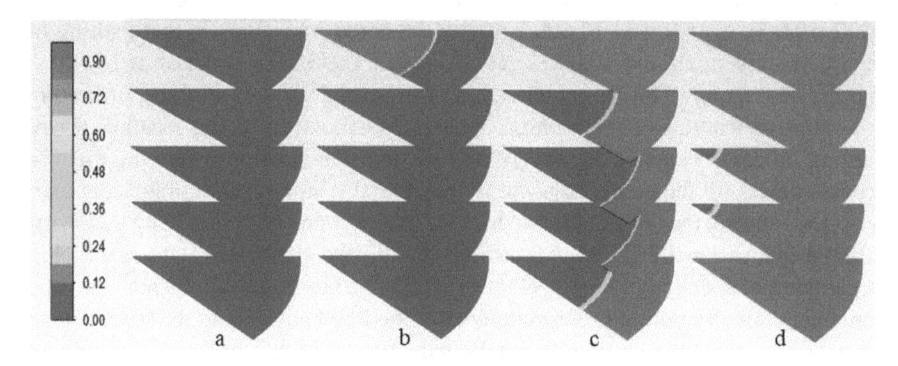

Fig. 18 Melting fraction profile of PCM-CEG domain without fin at: **a** 9000 s, **b** 10,000 s, **c** 11,000 s, **d** 12,000 s

is presented in Fig. 19a. Since the fins are located at regular angular intervals, the computation domain can be reduced comprehensively by considering the volume enclosed by two consecutive fins. Figure 19b shows the computation domain for 45° fin orientation. Figure 20 shows the mesh generated in ANSYS for three different

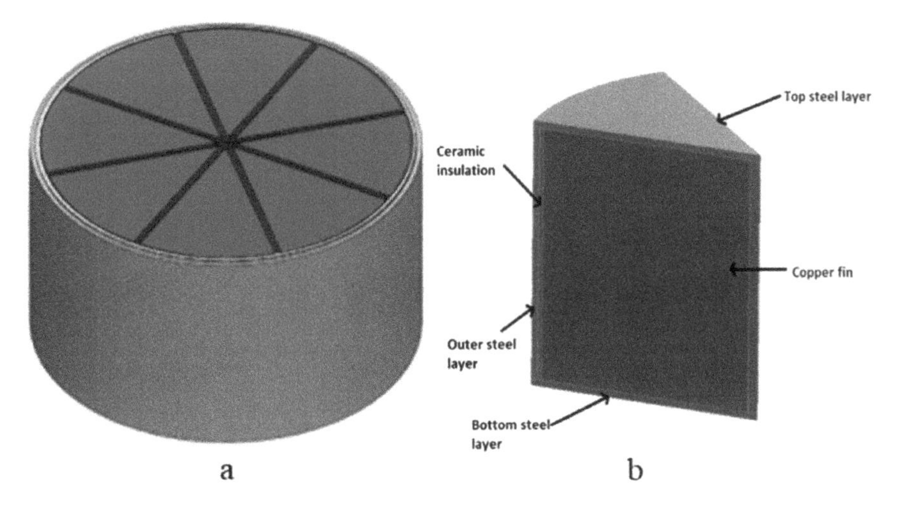

Fig. 19 **a** Schematic diagram of thermal storage container with 45° fin orientation, **b** reduced computational domain for 45° fin orientation

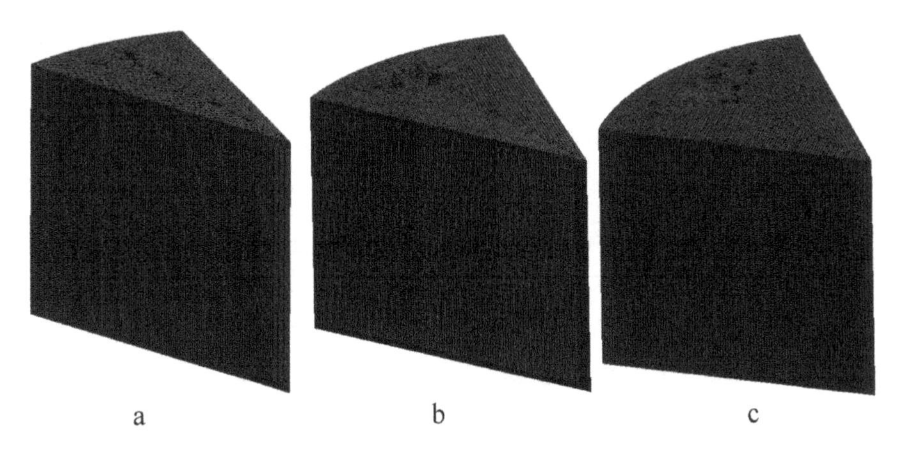

Fig. 20 Hex mesh generated in ANSYS for: **a** 30°, **b** 45°, and **c** 60° fin orientations

fin arrangements. The charging process starts at 10 am and is carried out for 3 h and 20 min from the start time.

Once again, the initial temperature of the complete system for all the cases is assumed to be at ambient temperature 27 °C Figs. 21, 22, 23, 24 and 25 represents temperature profile and melt fraction evolution in time for the PCM-CEG domain within the storage unit for fin orientation of 30°, 45°, and 60°, respectively. When the temperature profiles of finned containers (Figs. 21, 23 and 25) are compared with that of non-finned case study, we find that the difference between the highest and lowest temperature within the PCM-CEG domain is much smaller (~10 °C for 30°, ~13 °C for 45°, and ~16 °C for 60° fin orientations, respectively) indicating much better

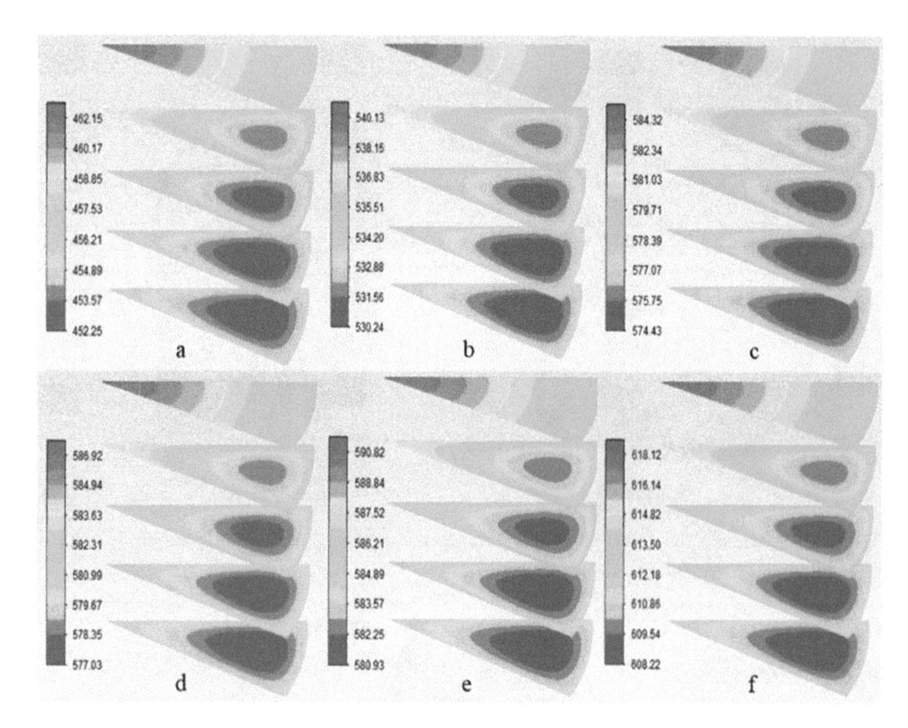

Fig. 21 Temperature (Kelvin) profile of PCM-CEG domain for 30° fin orientation at: **a** 6000 s, **b** 9000 s, **c** 10,700 s, **d** 10,800 s, **e** 10,950 s, and **f** 12,000 s

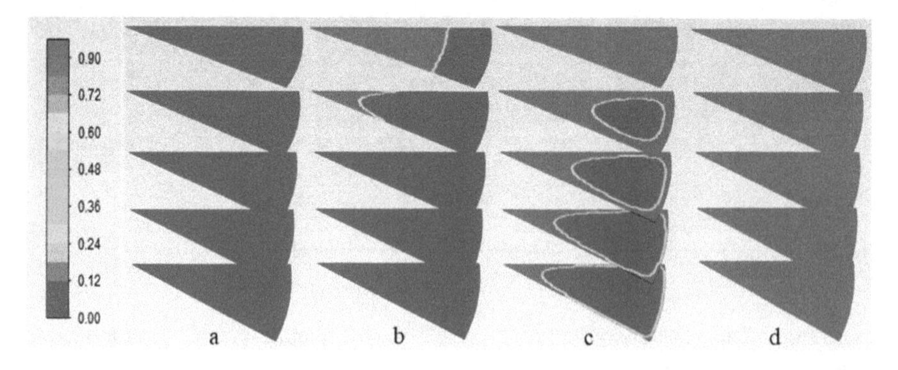

Fig. 22 Melting fraction profile in PCM-CEG domain for 30° fin orientation at: **a** 9000 s, **b** 10,700 s, **c** 10,800 s, and **d** 10,950 s

uniformity of temperature within the storage medium. The temperature maximums at 12,000 s are also found to be significantly lower (345, 339 and 335 °C for 30°, 45°, and 60° fin orientations, respectively) than the decomposition temperature (380 °C) of $NaNO_3$.

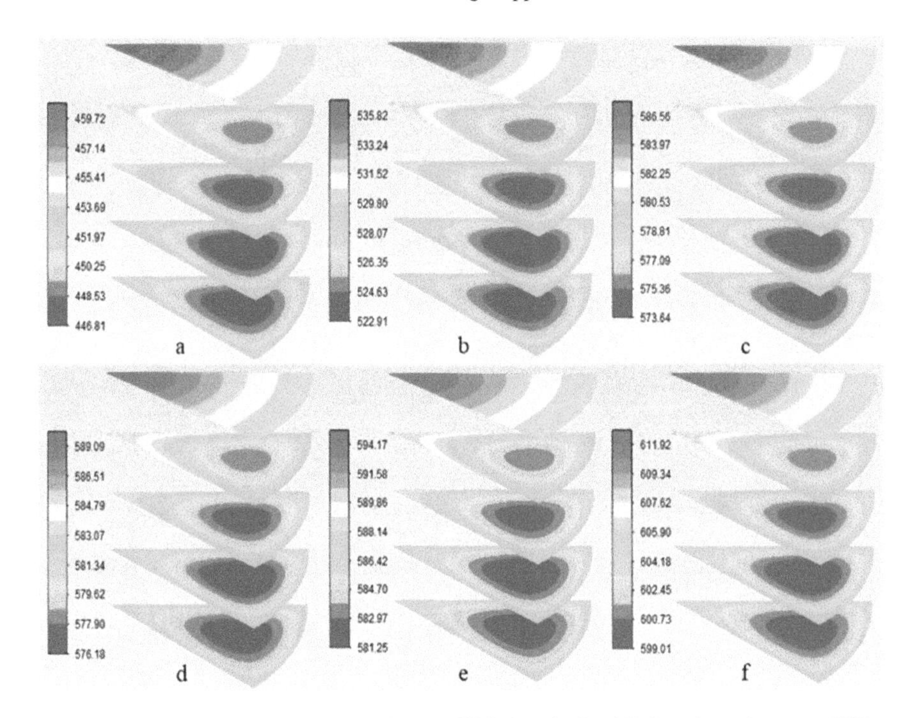

Fig. 23 Temperature (Kelvin) profile of PCM-CEG domain for 45° fin orientation at: **a** 6000 s, **b** 9000 s, **c** 11,000 s, **d** 11,100 s, **e** 11,300 s, and **f** 12,000 s

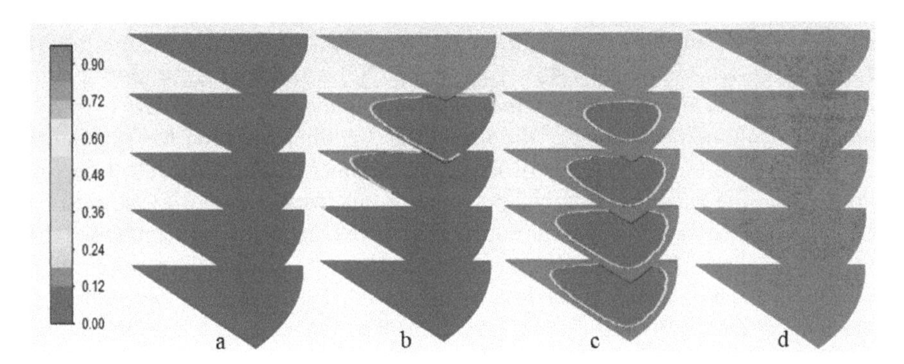

Fig. 24 Melting fraction profile in PCM-CEG domain in 45° fin orientation at: **a** 9000 s, **b** 11,000 s, **c** 11,100 s, and **d** 11,300 s

A close look at maximum temperatures obtained for these three orientations reveals another non-intuitive finding. Since 30° fin orientation offers more thermal uniformity, we expect the hotspot temperature to be the lowest for this configuration among the three chosen orientations. However, the temperature profiles given by Figs. 21f, 23f, and 25f predict the maximum temperatures in an exactly reverse

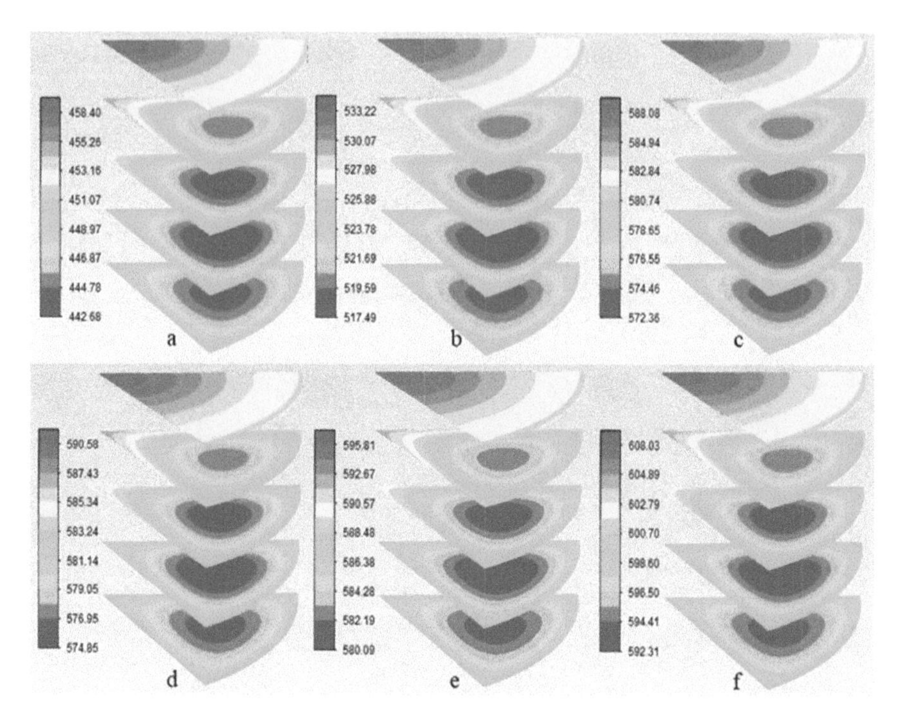

Fig. 25 Temperature (Kelvin) profile of PCM-CEG domain for 60° fin orientation at: **a** 6000 s, **b** 9000 s, **c** 11,200 s, **d** 11,300 s, **e** 11,510 s, and **f** 12,000 s

manner. The maximum temperature for 60° orientation is found to be the lowest (T_{max} = 335 °C), followed by 45° (T_{max} = 339 °C) and 30° (T_{max} = 345 °C) fin orientations. This reverse trend might be attributed to the loss of PCM-CEG volume due to the volume occupancy of the fines. The 30° fin orientation has maximum volume loss due to the presence of maximum number of fins. Equivalent energy pertaining to the latent heat storage potential of this lost volume of PCM must be accounted for and manifested by an equivalent rise in the sensible temperature of the entire storage unit. The excess maximum temperature is essentially the manifestation of the latent heat of fusion of the lost volume of PCM. However, in this context, a question may arise: why is the no-fin arrangement manifesting the lowest maximum temperature? The answer to this question can be comprehensively addressed by temperature larger non-uniformity of the no-fin arrangement. The absence of fins causes a much larger temperature gradient from the hotspot to lowest temperature region. Comparison of minimum temperatures between non-finned and finned orientations shows that the minimum temperature is lowest (~303 °C) for no-fin configuration. We must remember that the total amount of heat supplied to all these orientations is the same. Therefore, larger non-uniformity of temperature leads to larger maximum temperature.

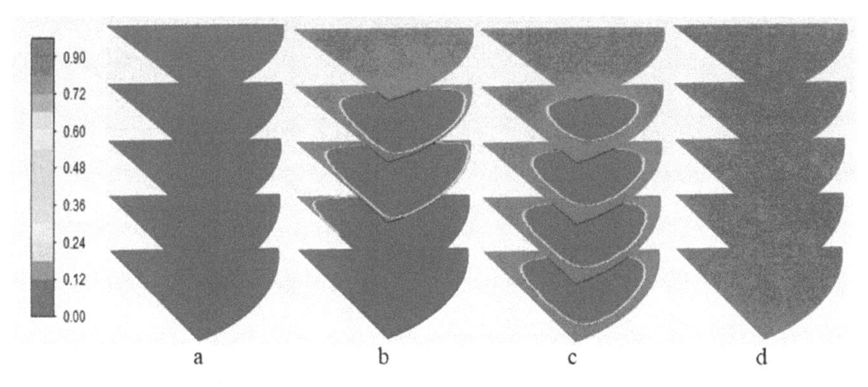

Fig. 26 Melting fraction profile in PCM-CEG domain for 60° fin orientation at: **a** 9000 s, **b** 11,200 s, **c** 11,300 s, and **d** 11,510 s

Next we compare the melt fraction evolution presented by Figs. 22, 24, and 26. Unlike no-fin configuration, the last region to melt is found to be the farthest region from the hotspot. As is evident, 30° fin orientation promotes fastest meltdown of the entire PCM with maximum temperature uniformity (minimum difference between T_{max} and T_{min}). We observe another interesting fact corresponding to the onset and termination of melting process. The earliest onset of melting occurs for the no-fin configuration (Fig. 18). Large temperature non-uniformity leads to accumulation of heat in the hotspot region leading to an early rise in temperature above the melting point (T_m) in this region. However, the same large temperature non-uniformity causes the most delayed completion of melting process. For the finned configurations (Figs. 22, 24 and 26), the onset of melting gets delays as compared to no-fin configuration because the temperature uniformity causes the entire PCM-CEG domain to reach the melting point (T_m) uniformly deferring the hotspot temperature to reach T_m at an early stage. However, the complete melting duration is significantly smaller once onset of melting occurs.

In continuation to the discussion regarding the effect of fin arrangement on maximum temperature (T_{max}) of the domain,, the time evolution of T_{max} and T_{min} is shown in Fig. 27 for different fin configurations. The case study involving no fin predicts maximum non-uniformity of the temperature field with $(T_{max} - T_{min}) \sim$ 70 K. As the fin numbers are progressively increased (60°, 45°, and 30° fin orientations), $T_{max} - T_{min}$ values reduce subsequently indicating better uniformity of the thermal field. It is to be noted here, that the higher values of T_{max} associated with higher number of fins ($T_{max\,@30^\circ} > T_{max\,@45^\circ} > T_{max\,@60^\circ}$) are also subjected to higher values of T_{min} evolution, i.e. $T_{min\,@30^\circ} > T_{min\,@45^\circ} > T_{min\,@60^\circ}$, which is physically consistent.

Figure 28 shows the variation of overall melt fraction with respect to time. No-fin configuration promotes earliest onset of melting and most delayed completion of the same. The 30° fin configuration promotes next earliest onset of melting followed by the 45° and 60° fin configurations. The 30° fin configuration also has the steepest melt

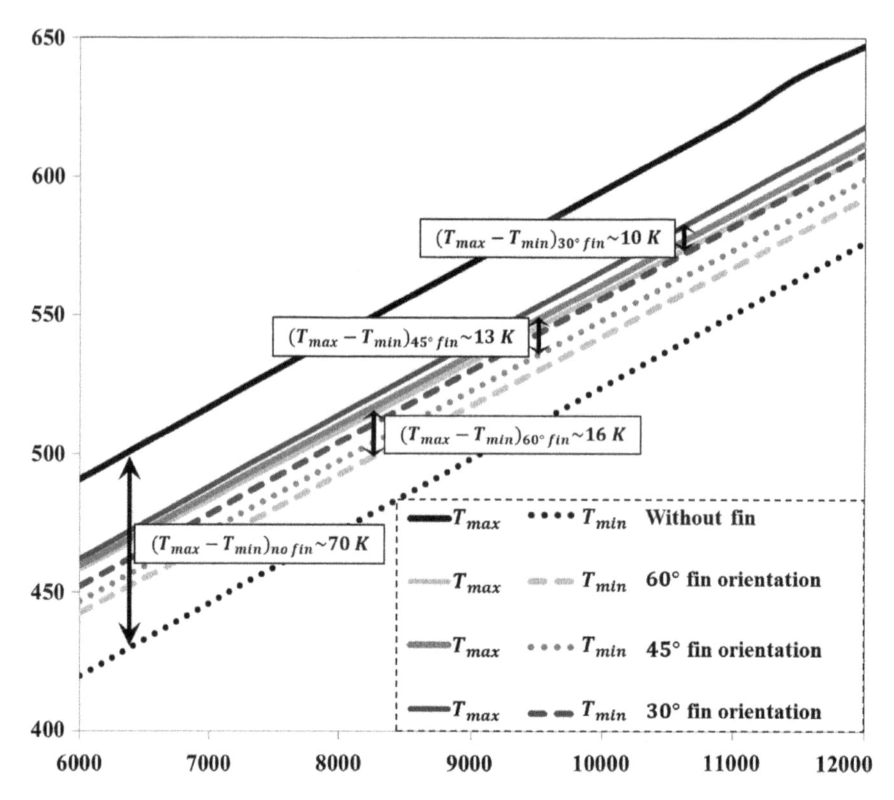

Fig. 27 Variation of maximum and minimum temperatures with time for different fin configurations; smaller difference between maximum and minimum temperature denotes more uniformity of temperature field

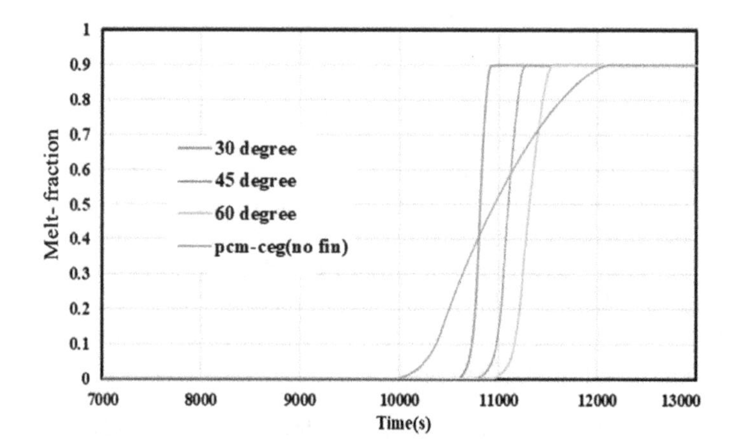

Fig. 28 Evolution of overall melt fraction with time

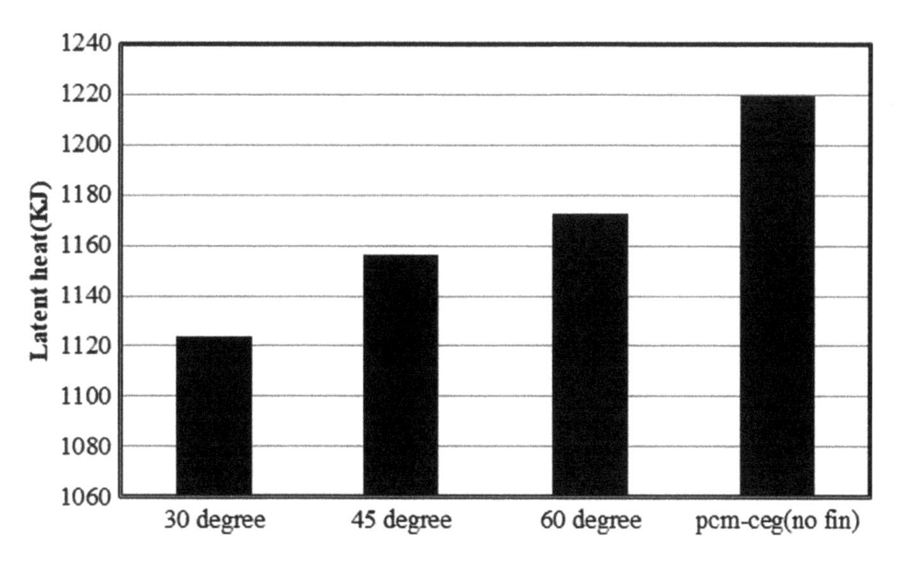

Fig. 29 Heat storage capacity (latent. heat) versus fin angle

fraction versus time plot, depicting fastest completion of melting process followed by 45° and 60° fin configurations.

Figure 29 shows latent heat storage capacity of four different configurations under consideration. Although increasing the total number of fins reduces the charging time significantly, the price we pay is the reduction in total latent heat capacity of the storage. Therefore, a tradeoff must be put into place to optimize between the charging duration and latent heat storage capacity.

5 Conclusion

An alternative design of parabolic dish solar cooker with double reflector arrangement integrated with latent heat storage unit located below the parabolic dish (primary reflector) is explored in terms of its thermal performance during the charging process. A diffusion dominated numerical model to address the melting of PCM in PCM-CEG composite is described. Effect of plate fins extended in radial direction and distributed circumferentially with prescribed angular pitch on the overall charging dynamics is studied. If fins are not incorporated, the maximum temperature (hotspot temperature) attained by the storage unit after completion of the charging process is found to be very close to the decomposition temperature of the PCM $NaNO_3$. Also, the temperature distribution in the storage medium is found to be highly non-uniform in the absence of fins. Addition of circumferentially distributed radial fins not only reduces the hotspot temperature to a safer margin, but it also establishes a much uniform thermal field within the storage medium. Three different fin orientations are studied for angular pitch of 30°, 45°, and 60° with progressively lesser number of fins. Larger number of fins resulted in earlier attainment of complete melting. Although the onset of melting occurs at a much earlier time when fins are not added, the completion of melting can only be obtained after significantly large time duration due to large non-uniformity of the thermal field. On the other hand, addition of fins delays the onset of melting process, but once melting process starts completion of melting is attained at an incredibly short time span. When the maximum temperature evolution is compared for 30°, 45°, and 60° fin orientations, surprisingly higher values of maximum temperature is obtained for progressively higher number of fins ($T_{\max @30^\circ} > T_{\max @45^\circ} > T_{\max @60^\circ}$). However, the same trend is also observed when minimum temperatures are compared ($T_{\min @30^\circ} > T_{\min @45^\circ} > T_{\min @60^\circ}$) with the difference between maximum and minimum temperatures progressively reducing for higher number of fines ($[T_{\max} - T_{\text{mean}}]_{@30^\circ} < [T_{\max} - T_{\text{mean}}]_{@45^\circ} < [T_{\max} - T_{\text{mean}}]_{@60^\circ}$) indicating attainment of better uniformity in the temperature field with larger number of fins, which is physically consistent.

References

Bennon WD, Incropera FP (1987) A continuum model for momentum, heat and species transport in binary solid-liquid phase change systems—I. Model formulation. Int J Heat Mass Transf 30(10):2161–2170

Bhave AG, Thakare KA (2018) Development of a solar thermal storage cum cooking device using salt hydrate. Sol Energy 171:784–789

Bodzenta J, Mazur J, Kaźmierczak-Bałata A (2011) Thermal properties of compressed expanded graphite: photothermal measurements. Appl Phys B 105(3):623–630

Brent AD, Voller VR, Reid KJ (1988) The enthalpy porosity technique for modelling convection diffusion phase change: application to the melting of a pure metal. Numer Heat Transf 13:297–318

Buddhi D, Sahoo LK (1997) Solar cooker with latent heat storage: design and experimental testing. Energy Convers Manag 38(5):493–498

Buddhi D, Sharma SD, Sharma A (2003) Thermal performance evaluation of a latent heat storage unit for late evening cooking in a solar cooker having three reflectors. Energy Convers Manag 44(6):809–817

Chakraborty PR (2017) Enthalpy porosity model for melting and solidification of pure substances with large difference in phase specific heats. Inte Commun Heat Mass Transf 81:183–189

Coccia G, Di Nicola G, Tomassetti S, Pierantozzi M, Chieruzzi M, Torre L (2018) Experimental validation of a high-temperature solar box cooker with a solar-salt-based thermal storage unit. Sol Energy 170:1016–1025

Cuce PM (2018) Box type solar cookers with sensible thermal energy storage medium: a comparative experimental investigation and thermodynamic analysis. Sol Energy 166:432–440

Domanski R, El-Sebaii AA, Jaworski M (1995) Cooking during off-sunshine hours using PCMs as storage media. Energy 20(7):607–616

El-Sebaii AA, Al-Amir S, Al-Marzouki FM, Faidah AS, Al-Ghamdi A, Al-Heniti S (2009) Fast thermal cycling of acetanilide and magnesium chloride hexahydrate for indoor solar cooking. Energy Convers Manag 50(12):3104–3111

Fluent ANSYS (2015) Ansys fluent. Academic Research. Release, 14

Foong CW, Hustad JE, Løvseth J, Nydal OJ (2010) Numerical study of a high temperature latent heat storage (200–300 °C) using eutectic nitrate salt of sodium nitrate and potassium nitrate. In: Proceedings of the COMSOL conference

Horne S (2012) Concentrating photovoltaic (CPV) systems and applications. In: Concentrating solar power technology. Woodhead Publishing, pp 323–361
https://www.nrel.gov/gis/data_solar.html
Hussein HMS, El-Ghetany HH, Nada SA (2008) Experimental investigation of novel indirect solar cooker with indoor PCM thermal storage and cooking unit. Energy Convers Manage 49(8):2237–2246
Iverson BD, Broome ST, Kruizenga AM, Cordaro JG (2012) Thermal and mechanical properties of nitrate thermal storage salts in the solid-phase. Sol Energy 86(10):2897–2911
Kalnæs SE, Jelle BP (2015) Phase change materials and products for building applications: a state-of-the-art review and future research opportunities. Energy Build 94:150–176
Kenisarin MM (2010) High-temperature phase change materials for thermal energy storage. Renew Sustain Energy Rev 14(3):955–970
Kumaresan G, Raju G, Iniyan S, Velraj R (2015) CFD analysis of flow and geometric parameter for a double walled solar cooking unit. Appl Math Model 39(1):137–146
Kumaresan G, Vigneswaran VS, Esakkimuthu S, Velraj R (2016) Performance assessment of a solar domestic cooking unit integrated with thermal energy storage system. J Energy Storage 6:70–79
Lecuona A, Nogueira JI, Ventas R, Legrand M (2013) Solar cooker of the portable parabolic type incorporating heat storage based on PCM. Appl Energy 111:1136–1146
Mallow A, Abdelaziz O, Graham S Jr (2016) Thermal charging study of compressed expanded natural graphite/phase change material composites. Carbon 109:495–504
Memon SA (2014) Phase change materials integrated in building walls: a state of the art review. Renew Sustain Energy Rev 31:870–906
Moreno P, Miró L, Solé A, Barreneche C, Solé C, Martorell I, Cabeza LF (2014) Corrosion of metal and metal alloy containers in contact with phase change materials (PCM) for potential heating and cooling applications. Appl Energy 125:238–245
Mukherjee D (2018) A review study on the thermo physical properties and storage applications of phase change materials. World Sci News 98:185–198
Patankar S (2018) Numerical heat transfer and fluid flow. CRC Press
Py X, Olives R, Mauran S (2001) Paraffin/porous-graphite-matrix composite as a high and constant power thermal storage material. Int J Heat Mass Transf 44(14):2727–2737
Sarbu I, Dorca A (2019) Review on heat transfer analysis in thermal energy storage using latent heat storage systems and phase change materials. Int J Energy Res 43(1):29–64
Sharma SD, Iwata T, Kitano H, Sagara K (2005) Thermal performance of a solar cooker based on an evacuated tube solar collector with a PCM storage unit. Sol Energy 78(3):416–426
Sharma A, Tyagi VV, Chen CR, Buddhi D (2009) Review on thermal energy storage with phase change materials and applica-tions. Renew Sustain Energy Rev 13(2):318–345
Shrivastava A, Chakraborty PR (2019) Shell-and-tube latent heat thermal energy storage (ST-LHTES). In: Advances in solar energy research. Springer, Singapore, pp 395-441
Venkateshwar K, Mathur V, Chakraborty PR (2017) Feasibility of using phase change material for thermal storage at high temperature for concentrated solar cooker: a numerical approach. In: Proceeding of SEEC2017-121
Veremachi A, Cuamba BC, Zia A, Lovseth J, Nydal OJ (2016) PCM heat storage charged with a double-reflector solar system. J Solar Energy 2016
Vigneswaran VS, Kumaresan G, Sudhakar P, Santosh R (2017) Performance evaluation of solar box cooker assisted with latent heat energy storage system for cooking application. In: IOP conference series: earth and environmental science, vol 67, no 1, p 012017. IOP Publishing
Zalba B, Marın JM, Cabeza LF, Mehling H (2003) Review on thermal energy storage with phase change: materials, heat transfer analysis and applications. Appl Therm Eng 23(3):251–283

Experimental Investigation of a Sensible Thermal Energy Storage System

Vishwa Deepak Kumar, Yudhisther Surolia, Sudipto Mukhopadhyay, and Laltu Chandra

1 Introduction

Solar energy is a promising renewable source to meet the growing energy demand. Using concentrated solar thermal (CST) technologies, solar energy can be harnessed to meet thermal as well as electrical demand Häberle (2012); Blanco and Miller (2017). CST technologies need to concentrate sunlight. Optical concentration is achieved using the direct beam irradiance component. Direct beam irradiance is generally quantified by the amount of solar radiation received per unit area by a surface normal to the sun and is called direct normal irradiance (DNI). The DNI distribution across the globe varies with the highest values around the subtropics. For example, DNI map of India NREL (2021) shows the abundance of solar radiation in Rajasthan and Gujarat, with an availability of more than 5.5 $kWh/m^2/day$. In CST technologies, the DNI is concentrated onto a receiver using curved reflectors, like parabolic trough, linear Fresnel, heliostat, and dish, etc. In heliostat-based CST system, DNI is concentrated onto a receiver, viz. open or closed volumetric air receiver. The open volumetric air receiver (OVAR)-based solar towers use ambient air as heat transfer fluid (HTF) and are capable to heat air beyond a temperature of 800°C Pitz-Paal et al. (1997); Hoffschmidt et al. (2003); Ávila Marín (2011); Sharma et al. (2014). Thus, OVAR-based central tower technology can be used for metals processing operations, such as heat treatment and soaking. Sharma et al. Sharma et al. (2015) proposed the novel concept of a solar convective furnace (SCF) which is capable of heat treatment of metal via forced convection. In this system, hot air from OVAR is transported to a SCF through thermal energy storage (TES).

V. D. Kumar · Y. Surolia · S. Mukhopadhyay
Department of Mechanical Engineering, Indian Institute of Technology Jodhpur, Jodhpur, Rajasthan 342037, India

L. Chandra (✉)
Department of Mechanical Engineering, IIT BHU Varanasi, Varanasi 221 005, India
e-mail: chandra.mec@iitbhu.ac.in

H. Tyagi et al. (eds.), *New Research Directions in Solar Energy Technologies*, Energy, Environment, and Sustainability,
https://doi.org/10.1007/978-981-16-0594-9_13

A solar air tower simulator (SATS) facility is installed at IIT Jodhpur for assessment of this integrated concept. It consists of all the necessary sub-systems, such as OVAR, two TES viz. primary and secondary, cross-flow heat exchanger, and SCF. This facility is used for the experimental investigation of sub-systems. The mounted SCF is a part scaled-down model of retrofitted solar convective furnace Patidar et al. (2015, 2017). In the SATS facility, hot air is generated in OVAR through Joule heating and transported to TES or SCF through insulated pipes.

Due to intermittent nature of solar energy, TES is essential for a plant to run continuously without interruptions Fath (1998); Powell and Edgar (2012). Depending upon the duration of the heat storage, TES can have capacity to cater from diurnal to annual requirements. Some of the criteria for selecting a storage material are as follows:

- High thermal conductivity and high thermal diffusivity.
- Abundant and cheap.
- Non-explosive and non-corrosive in nature.
- Environment friendly.
- Stable under various loading conditions.

Two most commonly used thermal energy storage methods are discussed below briefly.

Latent heat thermal energy storage (LHTES) also called phase change storage as they have the capability to absorb or release energy with change in physical state. The energy storage density is inversely proportional to volume. A large amount of heat is generally stored during the phase change process nearly at constant temperature and this directly related to latent heat of a storage medium. The main advantage of using this over sensible storage system is that it has a high energy storage density. Initially, it behaves as sensible heat storage system when temperature rises linearly, and then, it stores heat at constant temperature with phase change Sarbu and Sebarchievici (2018). Some commonly used phase change materials (PCM) are salt hydrates, fluorides, nitrides, chlorides, paraffins, and fatty acids Pielichowska and Pielichowski (2014). The LHTES system can be mathematically described as below:

$$Q_s = \int_{T_i}^{T_m} mc_{ps}dt + mf\Delta q + \int_{T_m}^{T_f} mc_{pl}dT \tag{1}$$

$$Q_s = mc_{ps}(T_m - T_i) + mf\Delta q + mc_{pl}(T_f - T_m) \tag{2}$$

where
m = mass of the phase change medium (kg)
T_i = initial temperature of the storage medium (°C)
T_f = freezing temperature of the storage medium (°C)
T_m= melting emperature of the storage medium (°C)
f = melt fraction
δq = latent heat of fusion (J/kg)

Table 1 Characteristics of some common sensible heat storage solid materials Tian and Zhao (2013)

Storage Material	Working Temperature (°C)	Density (kg/m^3)	Specific heat (kJ/kg °C)	Thermal Conductivity (W/m K)
Sand–rock minerals	200–300	1700	1.30	1.0
Reinforced concrete	200–400	2200	0.85	1.5
Cast iron	200–400	7200	0.56	37.0
NaCl	200–500	2160	0.85	7.0
Cast steel	200–700	7800	0.60	40.0
Silica fire bricks	200–700	1820	1.00	1.5
Magnesia fire bricks	200–1200	3000	1.15	5.0

C_{ps} = average specific heat of solid phase between T_i and T_m (J/kg K)
C_{pl}= average specific heat of liquid phase between T_m and T_f (J/kg K)

Sensible heat thermal energy storage (SHTES) is one of the simplest methods in which energy is stored by heating or cooling a storage medium which may be solid or liquid. TES systems for sensible heat are commonly inexpensive as they consist of a simple enclosure for the storage medium and the equipment to charge/discharge. Storage media (e.g., water, soil, rocks, concrete, or molten salts) are generally commonly available and relatively cheap. Table 1 shows some of the most widely used solid SHTES materials and their properties. However, the container of the storage material requires effective thermal insulation, which may increase the TES cost Sarbu and Sebarchievici (2018).

For large-scale applications, underground storage of sensible heat in solid and liquid media is used. The two main advantages of SHTES are that firstly, it is cheap and secondly, it is environment friendly. It utilizes the heat capacity of storage medium and temperature of the medium changes during charging and discharging. The working principle of SHTES system can be mathematically described as below:

$$Q_s = \int_{T_i}^{T_f} mc_p \mathrm{d}T = mc_p(T_f - T_i) \tag{3}$$

In above equation, the amount of heat stored by storage medium depends upon the quantity of the storage medium, specific heat of the medium, and the temperature change of the medium.

Q_s = amount of heat stored in Joule (J)
m = mass of the storage medium (kg)
C_p= specific heat of the storage medium (J/kg K)

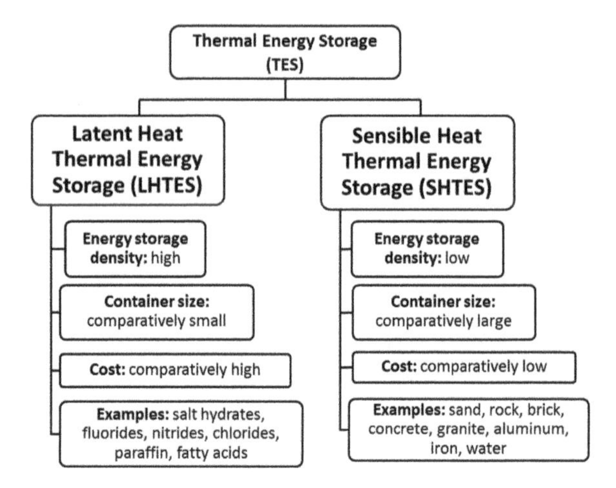

Fig. 1 Comparison of LHTES and SHTES systems

T_i = initial temperature of the storage medium (°C)
T_f = final temperature of the storage medium (°C)

The main differences between LHTES and SHTES system are depicted in Fig. 1.

SHTES is widely deployed in practical applications and thus of present interest. In the SATS system studied here, both the TES are of SHTES type. The energy storage material used in this system is pebbles. The primary TES stores the heat generated in OVAR and supplies to SCF. The secondary storage recovers waste heat of air leaving from the furnace. The TES is very crucial in the overall operation of the SCF. The energy storage capacity of the TES and their charging/discharging capacity have to be measured over the time. In this work, secondary TES which is used for recovering waste heat is experimentally investigated. First, the experimental setup along with the construction of the secondary TES is described. Next, the experimental methodology for investigating the charging and discharging of the secondary TES is presented. This is followed by presentation of the experimental results obtained during the charging and discharging of the secondary TES. Lastly, the energy balance of the secondary TES, and then, main conclusions are summarized.

2 Experimental Setup

Solar air tower simulator (SATS) experiment setup of 4 kW is installed at IIT Jodhpur. Experimental setup is shown below in Fig. 2. The main components of the setup are:

- Open volumetric air receiver (OVAR).
- Primary TES system.
- Solar convective furnace.
- Secondary TES system.

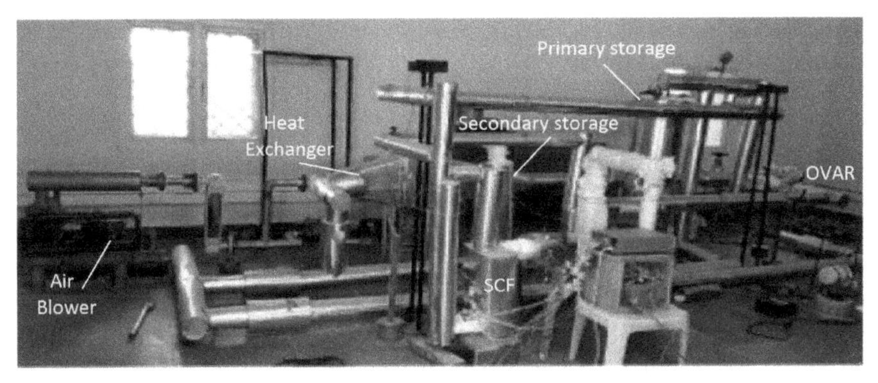

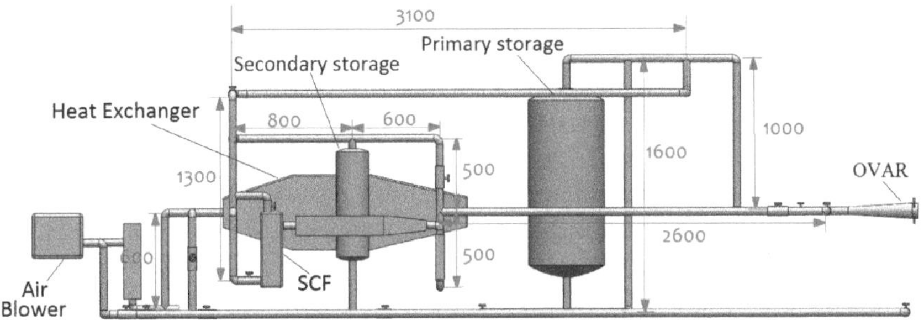

Fig. 2 Solar air tower simulator (SATS) setup at IIT Jodhpur (top) and its schematic with dimensions in mm (bottom)

- Air blower.
- Heat exchanger.

OVAR is the main component of the SATS where concentrated solar irradiation is incident and the thermal energy is absorbed for operation of the system. The OVAR is comprised of straight-pore-based absorbers and provides air at a high temperature. In order to perform experiments under controlled energy input, instead of concentrated irradiation, the receiver is heated by Joule heating. Nichrome wire is wrapped on the absorber circumference, for the Joule heating and wrapped with Kapton tape for electrical insulation to avoid short circuiting. The heat transfer fluid (HTF), which is air in this case, is sucked by the help of variable frequency drive twin lobe blower. It is installed to draw in air from the receiver to the entire system. Variable frequency drive controls the speed of motor to set a particular mass of air to be drawn in the system. To measure mass flow rate, rotameter is mounted before the air gets heated in order to prevent heat damage.

A cross-flow heat exchanger is also incorporated in the SATS to exchange heat with other fluids which can be used for other heating applications. With help of pipe valve arrangement, direction of HTF flow can be changed. HTF flow from OVAR to primary storage where heat can be stored and further utilized or it pass to the solar convective furnace. Solar convective furnace (SCF) is used for metal processing like tempering, annealing, soaking. From solar convective furnace, HTF flows to the atmosphere through secondary thermal energy storage (TES) system. Schematic of SATS setup with the various dimensions are also shown in Fig. 2.

2.1 Thermal Energy Storage

As shown in Fig. 2, there are two heat storage systems: primary and secondary thermal energy storage. Both the TES are of SHTES type. The storage is built using pebbles enclosed in metal structure with insulation cover. The hot air can be directly introduced to the furnace or through pebble-bed primary thermal energy storage. The primary, high capacity, thermal energy storage is charged by introducing the generated hot air from receiver. The function of the primary TES is to supply hot air to SCF through its discharging. Thus, the primary storage is use to store heat for the metal processing operation with a sustained heating.

The secondary storage is typically used for waste heat recovery. The size of secondary storage is half of the primary storage, and heat storage capacity of secondary storage is 1/8 of the primary storage. The enthalpy of the fluid at outlet of the SCF is less than that at the outlet of the heater. This is because as the HTF passes through the SCF, it loses enthalpy during the metal processing as well heat is transferred to the system. However, the HTF still has sufficient energy which will be lost to ambient on exit from the SATS. This can be significant depending on the enthalpy of the HTF at outlet of solar convective furnace. The secondary TES serves to extract this enthalpy before it exits into the atmosphere, and this in turn improves the efficiency of the system.

The exploded view of the secondary TES is shown in Fig. 3. It consists of four cast iron cylindrical compartments for easy handling and packing of material connected

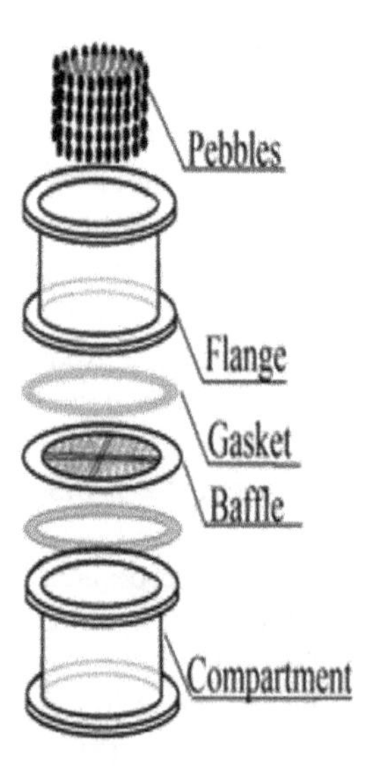

Fig. 3 TES: Exploded view (left) and assembly before enclosing (right)

Fig. 4 Components of TES: **a** Baffle **b** empty compartment with resin bond **c** compartment filled with pebbles

Table 2 Secondary thermal energy storage design parameters

Parameter	Value
Length of compartment (mm)	125
Diameter of compartment (mm)	125
Number of compartment	4
Gap between two compartment (mm)	11
No of holes in baffle	192
Diameter of pebble (mm)	20
Inlet and outlet pipe diameter (mm)	50

with each other with the help of flange. For uniform distribution of HTF in every compartment, a baffle (porous plate) of thickness 10 mm is used. Gasket is also used to ensure no leakage of air. To ensure perfect insulation, storage system is surrounded by silicone resin bond sheet from inside and glass wool on the outside. The baffle, empty compartment with silicon resin bond and the compartment filled with pebbles, is presented in Fig. 4. The insulation is protected by 0.5 mm aluminum sheet on the outside. Dimensions of TES and the material properties of the components play important role for determining the energy storage capacity. The specification of the secondary TES is presented in Table 2.

3 Experimental Methodology

The experimental methodology comprises mainly of providing a measured Joule heating, measuring the mass flow rate of the air, and assessing the energy stored in secondary TES through temperature measurements. The schematic of the experimental layout is shown in Fig. 5. Joule heating is carried out by maintaining the current and voltage with variacs. A rotameter is used to measure the volumetric flow rate of the air. Temperatures at several locations are measured by using K-type

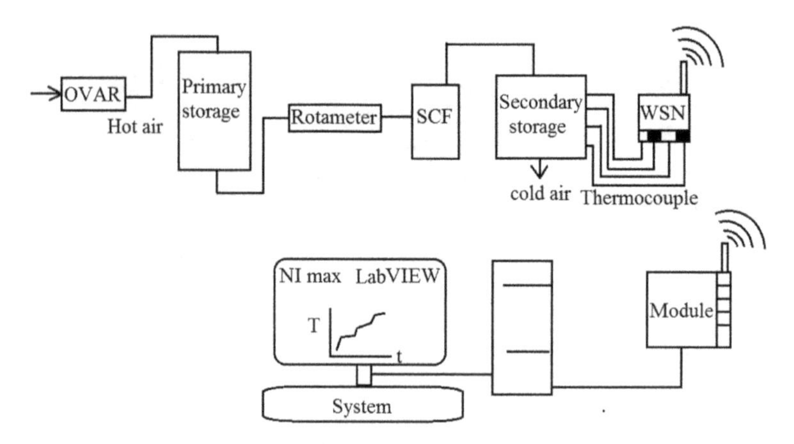

Fig. 5 Schematic of the experimental layout

thermocouples. National instrument wireless sensor board (WSN3212) is used to transmit thermocouple reading, and LABVIEW is used to monitor it online. The layout of the thermocouples will be described in detail next.

3.1 Temperature Measurement in TES

The temperature distribution within the secondary storage has been measured with respect to time while charging and discharging of the TES. This is carried out by using K-type thermocouples fixed at different locations. The schematic view of the secondary TES with layout of various thermocouples is shown in Fig. 6 with centerline indicated by dashed line. The four compartments of the secondary TES are denoted by C1, C2, C3, and C4. The thermocouples are placed inside the various component materials of the secondary TES to study the heat distribution.

The axial direction is indicated by z and radial direction by r. The axial and radial coordinates of the thermocouples in the r and z plane withe respect to the centerline are summarized in Table 3. In order to measure the temperature of the pebbles which is the sensible thermal energy storage medium, two thermocouples are placed diametrically opposite inside each compartment. The average temperature reading of the two thermocouples is taken as pebble temperature for that compartment. Thus, in order to monitor the pebble temperature inside each compartment, eight thermocouples are used. To measure the metal baffles temperature, four thermocouples are placed at the surface of the metal, one in each compartment. Four thermocouples are placed in contact with the glass wool outer surface, one in each compartment. For measuring the aluminum casing temperature, only one thermocouple is used at the outer surface and mid-height of the TES ($z = 250$ mm). Thermocouples are also placed at the inlet and outlet of TES to record inlet and outlet air temperature.

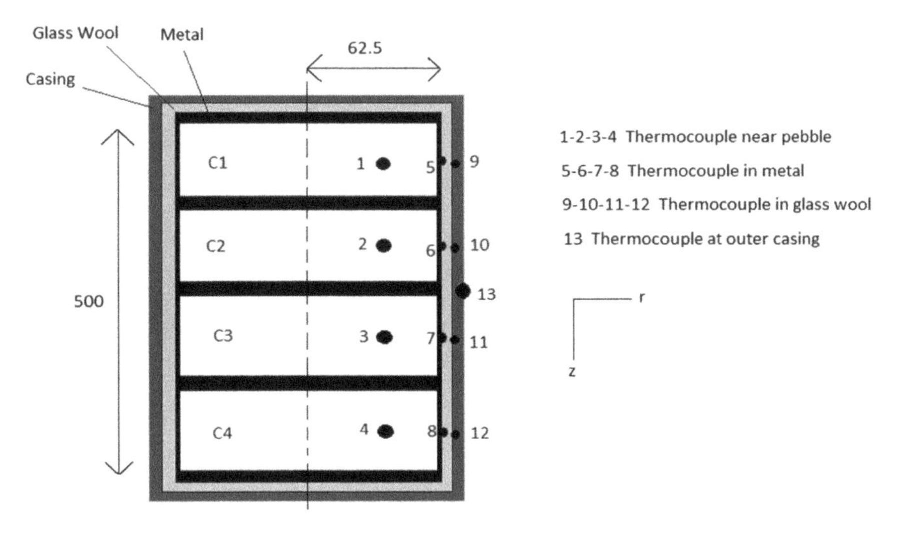

Fig. 6 Schematic view of secondary TES with thermocouple location

Table 3 Co-ordinates of thermocouple placement in the secondary TES

	Material	Pebble	Metal	Glass wool
Compartment	Axial location (z mm)	Radial location (r mm)		
C1	62.5	35	65.5	70.5
C2	187.5	35	65.5	70.5
C3	312.5	35	65.5	70.5
C4	437.5	35	65.5	70.5

3.2 *Procedure*

Blower is use to suck the ambient air, and the mass flow rate is set to a particular flow rate using the rotameter by adjusting the variable frequency drive blower. For performing the charging experiments, the air is heated by forced convection and supplied measured thermal energy in the form of Joule heating. Hot air enters the TES from the top during charging, and readings from all the thermocouples are recorded over the time. The flow path during the charging experiment is shown in Fig. 7.

After charging for a significant duration the thermal energy stored in TES is extracted by passing cold air flow through it. During discharging, with the help of the blower, ambient air is introduced from the bottom inside the charged (heated)

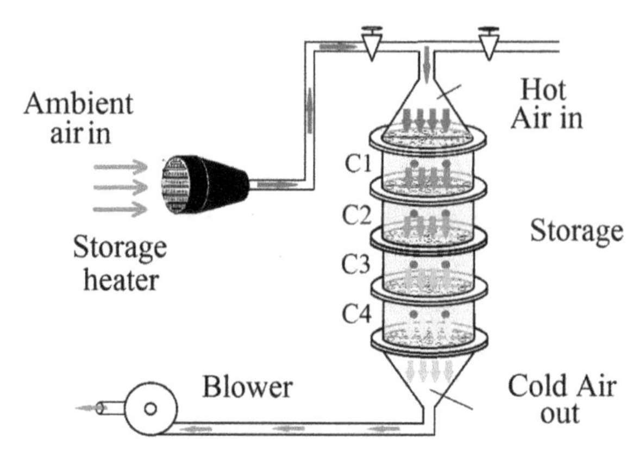

Fig. 7 Air flow path during charging

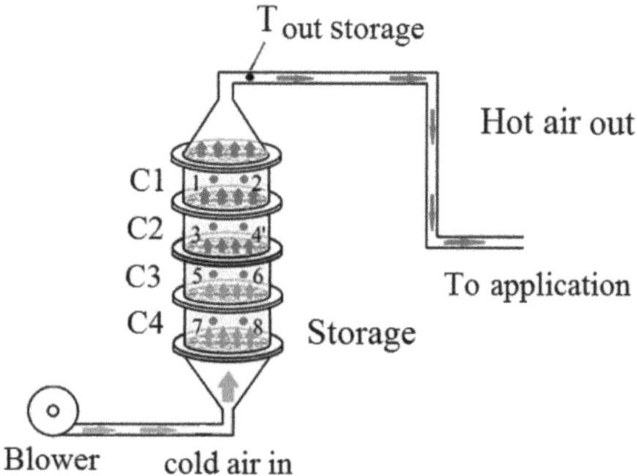

Fig. 8 Air flow path during discharging

Table 4 Duration and mass flow rate during charging and discharging experiments

Parameter	Charging	Discharging
Time (sec)	12000	7000
Mass flow rate (kg/s)	0.0042	0.0042

TES. The data from the thermocouples are recorded till a significant amount of heat is taken away from the TES. The flow path during the discharging experiment is shown in Fig. 8. The mass flow rates during charging and discharging experimentation along with the time duration are provided in Table 4.

4 Experimental Results

4.1 Heat Loss in Pipe Flow

Even though the SATS is insulated with glass wool completely, there is heat loss. In order to measure the temperature drop, thermocouples are also set up at different locations. The temperature at outlet of heater is 260 °C, and at the inlet of storage is 150 °C. So there is around 110 °C drop of temperature while flowing from heater outlet to secondary TES inlet. This drop is observed over a length of 1.5 m of insulated pipe line which indicates a drop of 73 °C per meter. Thus, for improving the efficiency of the system, the heat loss needs to be reduced.

Next, the details of the charging and discharging of secondary storage are presented. Charging of the secondary TES is carried out for 11,900 s (three and half hours) followed by discharging for 7000 s (two hours). Thus, the total duration of experiments is approximately 7 h including initial preparation and running.

4.2 Pressure Drop in Secondary TES

The pressure drop in the secondary TES is also calculated using the pressure drop characteristics of rock beds with air as the heat transfer medium. At first, the porosity is calculated by filling a cylinder of known volume with pebbles and then filling up the cylinder with water. The volume of the cylinder and the volume of the water are used to calculate the porosity. Diameter of the cylinder is 0.125 m and length of the cylinder is also 0.125 m which results in volume of cylinder as 0.001533 m^3. The amount of water required to fill the cylinder is thus 1.533 L. The amount of water required to fill the void of cylinder after filling the pebbles in cylinder is 0.6 L. The void fraction or porosity is calculated as 0.6/1.533 = 0.3913.

The pressure drop across the bed is given by Hänchen et al. (2011)

$$\Delta p = \frac{LG^2}{\rho_f d}\left(150\frac{(1-\epsilon)}{\epsilon^3}\frac{\mu_f}{Gd} + 1.75\frac{(1-\epsilon)}{\epsilon^3}\right) \tag{4}$$

where

Δp = pressure drop (Pa)
L = height of the storage (m)
ρ_f = density of fluid (kg/m^3)
G= mass flow per unit cross section (kg/(m^{2}s))
ϵ = porosity
δq = latent heat of fusion (J/kg)
μ_f = dynamic viscosity of fluid (Pa-s)

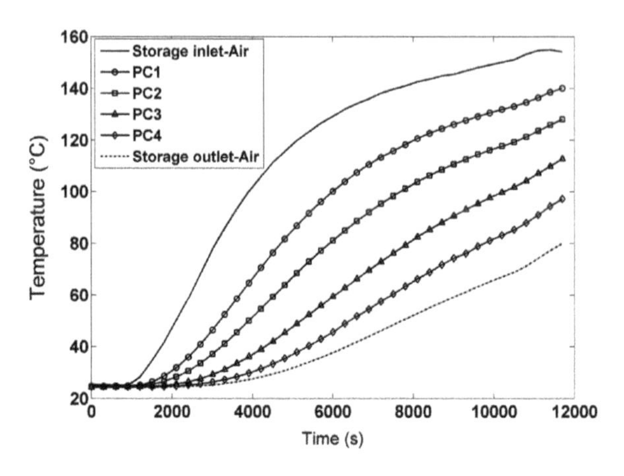

Fig. 9 Secondary storage charging: temperature rise of pebbles in the compartment with time

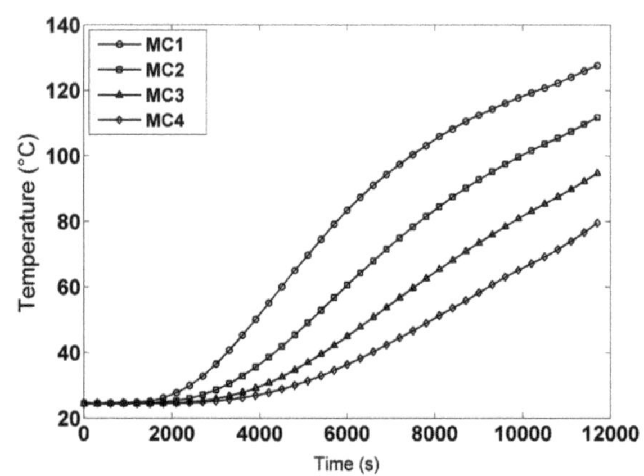

Fig. 10 Secondary storage charging: metal compartment wall temperature

For the mass flow rate of 0.0042 kg/s and considering the parameters of secondary storage, the pressure drop in secondary storage is approximately 62 Pa.

4.3 Charging of Secondary TES

Ambient air is sucked in through the blower and heated by Joule heating. A constant mass flow rate of 0.0042 kg/s is used to charge the secondary TES. Hot air enters from top of the storage, and heat is transferred from air to the pebble by forced convection. The temperature distribution which is indicator of the energy stored is measured with help of thermocouples placed inside as described earlier.

The temperature recorded by the thermocouples at different locations is presented in Fig. 9. It is observed that the temperature at the inlet of the secondary storage is at the ambient temperature initially up to 1000 s and then increases rapidly. This is

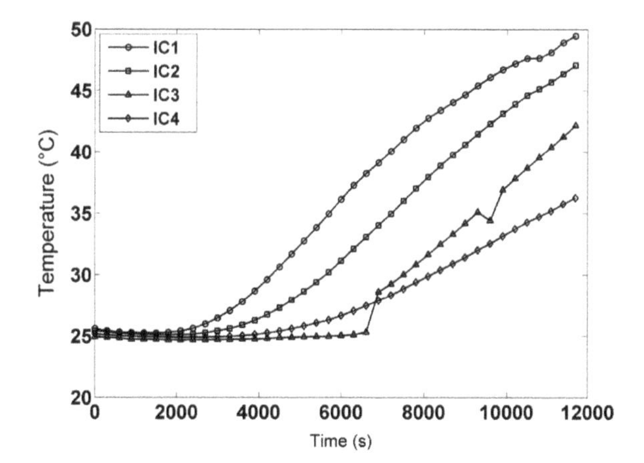

Fig. 11 Secondary storage charging: insulation temperature

due to the fact that initially, the SATS piping is at ambient temperature so the heat is transferred from the hot air to the pipe causing temperature at inlet of the storage to be almost same as ambient air. As the time passes, the difference in temperature between fluid and pipe decreases so temperature at the inlet of storage increase. With time the pebble temperature in all the four compartments indicated by PC1, PC2, PC3, and PC4 are seen to be increasing in Fig. 9. In compartment C1, temperature increases from 25 °C to 140 °C, in C2, temperature increases from 25 °C to 130 °C, in C3, temperature increases from 25 °C to 110 °C, and in C4, temperature increase from 25 °C to 100 °C. Temperature in compartment 1 and 2 increases rapidly up to 6000 s because the hot fluid has more enthalpy with it compared to the pebbles. As the hot air flows down, it's enthalpy decreases so rise in temperature in compartment 3 and 4 is lesser.

The temperature rise over time in the metal compartments of TES is shown in Fig. 10. It follows the same trend as the pebbles in different compartments. It can be again observed that there is no appreciable temperature rise in the first 2000 s. Figure 11 shows the temperature inside the insulation of secondary TES for different compartments. Temperature rise in the thermal insulation is observed to be quite less as compared to the metal walls of the TES.

One thermocouple is placed at the outer surface of casing (at mid-height of the storage system), and one thermocouple is left exposed in ambient air to measure the casing and ambient temperature, respectively. The two temperatures are shown in Fig. 12. The casing temperature is observed to increase from 25 °C to 33 °C and ambient temperature increases from 25 °C to 27 °C during the entire duration of the charging experiment. The outer casing temperature is not significantly higher than ambient temperature at all times during the experiment. Thus, convective and radiative heat losses can be neglected. However, from all the temperature graphs, it can be observed that steady state of the secondary TES is not attained even after 3.5 h. The experiments could not be carried further due to malfunction of Joule heating.

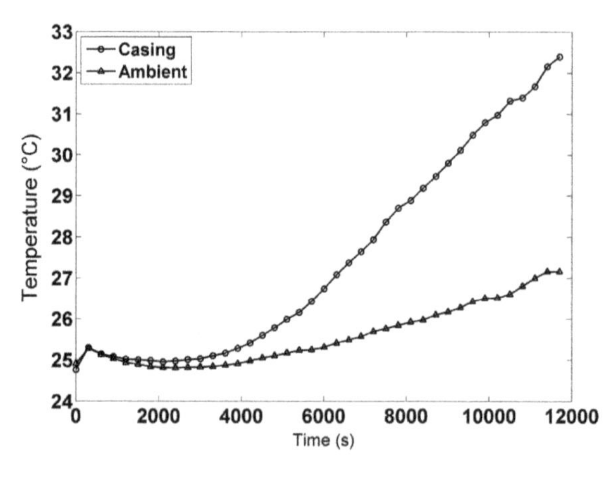

Fig. 12 Secondary storage charging: casing and ambient temperature

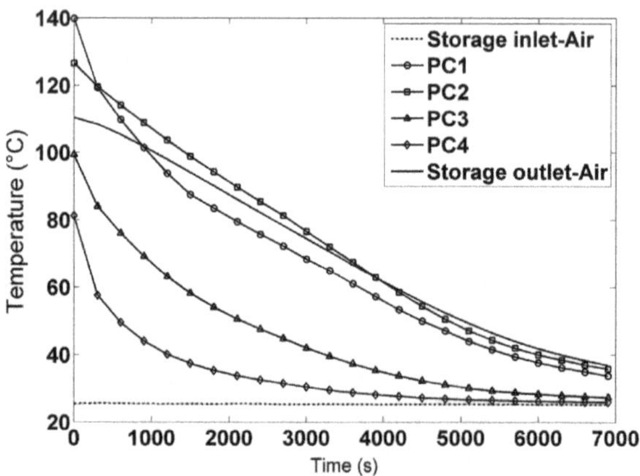

Fig. 13 Secondary storage discharging: temperature drop of pebbles in the compartment with time

4.3.1 Discharging of Secondary TES

After charging for a period of about 11900s, discharging of TES is started by reversing air flow. Ambient air is sucked by blower and is made to enter into the secondary TES from bottom. Discharging is carried out for a period of about 7000 s, and mass flow rate is kept same as in charging, i.e., 0.0042 kg/s.

Figure 13 shows the thermocouple readings at the inlet of the TES (bottom of TES), exit of TES (top of TES), and the pebbles in different compartments. The temperature of pebbles in compartment 3 and 4 decreases more rapidly in comparison to the pebble temperature in compartment 1 and 2 because the ambient air first comes in contact with pebble of C4 and C3. As the air heats up, the temperature difference between air and pebble of C1 and C2 decreases so heat transfer rate decreases. It is also observed that the temperature of the hot air from the secondary TES outlet decreases with time from approximately 110 °C to 50 °C. Steady state is achieved for pebbles in compartment 3 and 4 at around 6500 s. The temperature of pebbles in

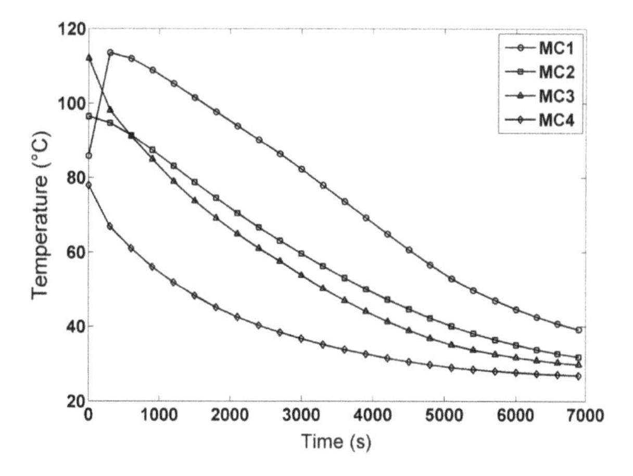

Fig. 14 Secondary storage discharging: temperature drop of metal walls in the compartment with time

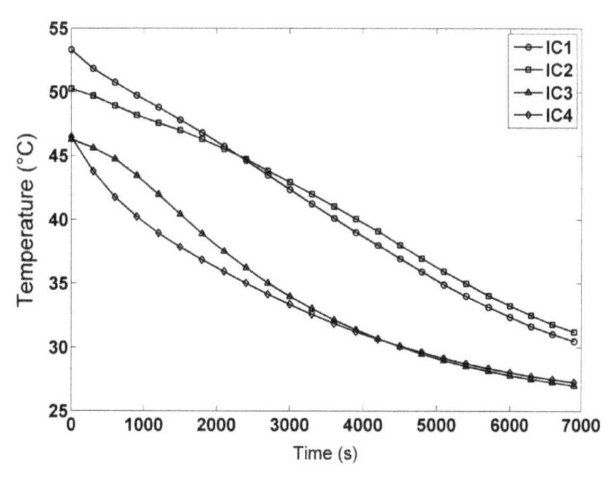

Fig. 15 Secondary storage discharging: temperature drop in insulation in the compartment with time

C1 and C2 still continues to drop indicating that these compartments are not fully discharged. Due to decreasing temperature difference between the ambient air and pebbles in compartments (C1 and C2), complete discharge will take excessive time, and the experiments are stopped around 7000 s.

The temperature of the metal walls of the compartments during discharging is shown in Fig. 14. It also follows the same trend as the pebble but in the compartment, 1 rise in temperature is observed up to 1000 s. Figure 15 shows the temperature drop with time in the insulation of the compartments. It is observed that insulation temperature in compartments 3 and 4 are almost similar. The insulation temperature in compartment 1 and 2 are also similar and approximately 10 °C higher than the temperature in the compartments 3 and 4.

5 Energy Balance

In this section, an approximate of the energy stored in the secondary TES is estimated using the measured temperatures and the material properties available from the literature. The properties are taken at a constant temperature in order to simplify the calculations and are presented in Table 5.

The geometrical dimensions of the various components of the secondary TES are known. Using the dimensions and density, the approximate mass of each TES constituent such as pebbles, metal structure, thermal insulation (glass wool), and outer metal casing is computed. Using the approximate mass and temperature rise of the material, an approximate energy balance is arrived at assuming steady state, i.e., at the end of the charging 7000 s). It is to be noted here that neither steady state is attained nor the mass of the materials very precise. However, this helps to have a qualitative idea of the heat storage distribution. Table 6 gives a qualitative overview of the energy balance of the secondary TES. It is observed that approximately 60% of the inlet energy leaves out of the TES with the outgoing fluid. Only 12% of the incoming thermal energy is stored in the pebbles. The metal components such as the metal compartment, baffles, and nut bolts store almost around 17% of the incoming energy. The energy stored by metal components is higher as the secondary TES is small in size and is mainly used for recovering waste heat. As expected, the energy stored in insulating material is negligible.

Table 5 Properties at average temperature 80°C

Parameter	Notation	Value
Density of pebble(kg/m^3)	ρ_s	2200
Specific heat of pebble(J/kgK)	c_s	710
Pebble diameter(m)	d	0.02
Thermal conductivity of pebble(W/mK)	k_s	1.83
Porosity	ε	0.39
Density of metal casing(kg/m^3)	ρ_m	7200
Specific heat of metal casing(J/kgK)	c_{pm}	500
Thermal conductivity of metal casing (W/mK)	k_m	50
Density of air(kg/m^3)	ρ_f	1
Specific heat of air (J/kgK)	c_{pf}	1009
Thermal conductivity of air (W/mK)	k_f	0.0299
Thermal conductivity of insulation (W/mK)	k_{in}	0.04
Mass flow rate (kg/s)	$\dot{m}$	0.0042
Velocity (m/s)	v	0.276
Prandtl no.	Pr	0.7

Table 6 Energy balance of the secondary TES

Enthalpy of fluid at inlet of storage	630 W
Enthalpy of fluid at outlet of storage	378 W
Charging power (enthalpy lost by fluid)	252 W
Energy stored by pebble in C1	23.12 W
Energy stored by pebble in C2	21.11 W
Energy stored by pebble in C3	17.09 W
Energy stored by pebble in C4	15.08 W
Energy stored by metal in C1	22 W
Energy stored by metal in C2	17.85 W
Energy stored by metal in C3	15.75 W
Energy stored by metal in C4	11.55 W
Energy stored by nut bolt	25 W
Energy stored by baffle	17 W
Energy stored by glass wool	1 W
Heat loss in convection and radiation	3 W
Difference in energy balance	62 W

6 Conclusion

The experimental results of a secondary TES are presented in this work. Temperature readings of the TES components are presented while charging and discharging of the TES using air as working fluid. It is found that the SHTES using pebbles enclosed within metal container performs well. The system can be used to store waste heat after metal processing, and the stored heat can be extracted while discharging. The heat losses from the setup have to be minimized. Further, improvements have to be carried out in the setup to enable to attain steady state while charging. This work is a first demonstration of the use of locally available material to construct a SHTES system and can be considered as proof of the concept.

References

Ávila Marín AL (2011) Volumetric receivers in solar thermal power plants with central receiver system technology: a review. Solar Energy 85(5):891–910

Blanco M, Miller S (2017) Introduction to concentrating solar thermal (CST) technologies. In: Blanco MJ, Santigosa LR (eds) Advances in concentrating solar thermal research and technology. Woodhead Publishing Series in, Energy. Woodhead Publishing, pp 3–25

Fath HE (1998) Technical assessment of solar thermal energy storage technologies. Renew Energy 14(1):35–40 (6th Arab International Solar Energy Conference: Bringing Solar Energy into the Daylight)

Häberle A (2012) Concentrating solar technologies for industrial process heat and cooling. In: Lovegrove K, Stein W (eds) Concentrating solar power technology. Woodhead Publishing Series in, Energy. Woodhead Publishing, pp 602–619
Hänchen M, Brückner S, Steinfeld A (2011) High-temperature thermal storage using a packed bed of rocks-heat transfer analysis and experimental validation. Appl Thermal Eng 31(10):1798–1806
Hoffschmidt B, Te'llez FM, Valverde A, Ferna'ndez J, Ferna'ndez V (2003) Performance evaluation of the 200-kWth HiTRec-II open volumetric air receiver. J Solar Energy Eng 125(1)01:87–94
NREL. DNI Map of India
Patidar D, Pardeshi R, Chandra L, Shekhar R (2017) Solar convective furnace for heat treatment of aluminium. Lecture Notes in Mechanical Engineering, pp 1531–1541
Patidar D, Tiwari S, Sharma P, Chandra L, Shekhar R (2015) Open volumetric air receiver based solar convective aluminum heat treatment furnace system. In: Energy procedia, vol 69, pp 506–517. International Conference on Concentrating Solar Power and Chemical Energy Systems, SolarPACES 2014
Pielichowska K, Pielichowski K (2014) Phase change materials for thermal energy storage. Progress Mater Sci 65:67–123
Pitz-Paal R, Hoffschmidt B, Böhmer M, Becker M (1997) Experimental and numerical evaluation of the performance and flow stability of different types of open volumetric absorbers under non-homogeneous irradiation. Solar Energy 60(3):135–150
Powell KM, Edgar TF (2012) Modeling and control of a solar thermal power plant with thermal energy storage. Chem Eng Sci 71:138–145
Sarbu I, Sebarchievici C (2018) A comprehensive review of thermal energy storage. Sustainability 10(1):191
Sharma P, Sarma R, Chandra L, Shekhar R, Ghoshdastidar P (2015) Solar tower based aluminum heat treatment system: part i. Design and evaluation of an open volumetric air receiver. Solar Energy 111:135–150
Sharma P, Sarma R, Chandra L, Shekhar R, Ghoshdastidar P (2014) On the design and evaluation of open volumetric air receiver for process heat applications. In: Energy Procedia, vol 57, pp 2994–3003. 2013 ISES Solar World Congress
Tian Y, Zhao C-Y (2013) A review of solar collectors and thermal energy storage in solar thermal applications. Appl Energy 104:538–553

PCM-Based Energy Storage Systems for Solar Water Heating

Akshay Sharma, Prasenjit Rath, and Anirban Bhattacharya

1 Introduction

In recent years, there has been a prominent shift toward the use of renewable energy sources instead of using fossil fuel for energy generation. Use of renewable energy sources has multiple benefits such as the reduction of greenhouse gas emission and long-term sustainability. Among different renewable energy sources, solar energy is the most widely available and has been considered for large-scale energy production. Although solar energy has lot of potential for electricity generation, one of the most widely used applications of solar energy is domestic water heating. In solar water heating, cold water is directly heated using incident solar radiation without the necessity of any external power source.

1.1 Solar Domestic Water Heating

Solar domestic water heaters have been used for many decades due to their simplicity. A typical solar water heater consists of a solar energy collector which heats a heat transfer fluid or water directly. The heating is done by concentrating the incident solar radiation using reflector plates which focus the energy on a cylindrical channel carrying the heat transfer fluid. The hot fluid is then used for heating water stored in a storage tank. Depending on the design, the water in the storage tank can also be directly circulated through the collector and directly heated. Whenever, hot water is necessary for domestic applications, the hot water from the storage tank is provided.

A. Sharma · P. Rath · A. Bhattacharya (✉)
School of Mechanical Sciences, IIT Bhubaneswar, Bhubaneswar, Odisha 752050, India
e-mail: anirban@iitbbs.ac.in

H. Tyagi et al. (eds.), *New Research Directions in Solar Energy Technologies*, Energy, Environment, and Sustainability,
https://doi.org/10.1007/978-981-16-0594-9_14

The main advantage of this type of water heating system is that it directly converts solar energy to thermal energy and thus has high efficiency of conversion. In comparison, an electric water heater will convert electricity to thermal energy and the electricity is previously obtained by conversion of thermal or chemical energy. Also, solar energy is widely available and thus can support the design and installation of small independent units for each user. However, installation of solar domestic heaters requires significant capital costs although the high initial cost is offset by the subsequent lower energy cost. The other major disadvantage of solar water heaters is that they can only provide hot water intermittently during daytime depending on weather conditions and cannot provide hot water during night. It is necessary to use an effective storage method to provide hot water as and when required and particularly during night time.

1.2 Energy Storage for Solar Water Heater

There are two main ways to store energy for solar water heaters (Kee et al. 2018). The traditional designs use a storage tank which is well insulated and stores heated water for future use. The water in the storage tank can be directly heated by supplying it to the collector or it can be heated using a separate heat transfer fluid which transfers thermal energy as sensible heat to the stored water. Various designs have been proposed for the storage tank to increase its storage effectiveness. However, in spite of improved designs, the stored water loses the stored heat gradually and cannot retain it for long durations. This can be overcome by using a large storage tank with high thermal inertia. However, such a design will increase the weight of the system significantly and will require a more robust design for domestic roof top installation.

To limit the weight of the system, storage tanks incorporating phase change materials (PCM) have been developed (Kee et al. 2018; Seddegh et al. 2015; Abokersh et al. 2018). PCM store energy as latent heat by melting during the charging process. When energy is required to be extracted, the PCM is solidified and the stored latent energy is released. Since latent heat of melting is considerably higher as compared to sensible energy change for a substance, significantly lower mass of storage material is required for storing the same quantity of energy.

1.3 PCM-Based Solar Water Heater

A number of designs for PCM-based solar water heaters have been proposed over the years due to the high energy storage density of PCM (Cabeza et al. 2002; Mehling et al. 2003; Al-Hinti et al. 2010; Oró et al. 2012; Kousksou et al. 2011; Fazilati and Alemrajabi 2013; Bouadila et al. 2014). Another advantage of using PCM is that the energy can be extracted at a constant temperature. Thus, a relatively constant temperature supply of hot water can be maintained for domestic applications. Also, by incorporating multiple PCMs with different melting temperatures, water can be extracted at different temperatures for different applications.

For PCM-based solar water heaters, the PCM is either stored in separate chambers to avoid mixing with water (Al-Hinti et al. 2010; Kousksou et al. 2011; Prakash et al. 1985; Bansal and Buddhi 1992) or it is encapsulated in spherical shells which are placed in the storage tank (Nallusamy et al. 2007; Reddy et al. 2012). Melting takes place by supplying hot water from the solar collector which passes over the encapsulated PCM or in between the PCM chambers. For energy extraction, cold water is circulated around the PCM chambers or capsules which removes energy from the PCM and heats the water. The hot water is supplied for domestic applications. During this discharge process, the PCM converts back to the solid phase. For solar water heaters, the PCM undergoes repeated cycles of charging and discharging during which it melts and solidifies alternately. Low thermal conductivity of PCM restricts the total quantity of energy that can be stored in a limited time although the total storage capacity may be higher. Different methods have been proposed for increasing the effective thermal conductivity of PCM-based energy storage systems such as the use of metal fins (Zhang and Faghri 1996; Ismail et al. 2001; Nayak et al. 2006; Gharebaghi and Sezai 2007), metal or graphite foams (Zhou and Zhao 2011; Alshaer et al. 2015; Meng and Zhang 2017; Dinesh and Bhattacharya 2019, 2020; Joshi and Rathod 2019), nanoparticles (Shaikh et al. 2008; Fan and Khodadadi 2012; Sahoo et al. 2019), and encapsulated PCM (Hawlader et al. 2003; Fukahori et al. 2016).

One of the important aspects of designing a suitable PCM-based solar water heater is proper selection of PCM. The important properties that need to be considered for selection of PCM are as follows.

1. High latent heat of phase change which leads to large energy storage density.
2. Suitable temperature of melting and solidification so that heat transfer can occur at the desired temperature depending on the application.
3. Less shrinkage and expansion during solidification and melting.
4. High specific heat in both liquid and solid phases so that significant sensible energy storage can occur. However, this will lead to longer time for the system to reach melting temperature and may be detrimental if heat transfer is required at a specific temperature.
5. Relatively large thermal conductivity for higher heat transfer rates.
6. Chemical stability of PCM over large number of charging and discharging cycles.
7. Low cost and easy availability of PCM.
8. Non-corrosive and non-toxic nature for long-term safety of the system.

Among these properties the most important property for selection of PCM is high energy storage density and suitable temperature of melting and solidification depending on the requirement.

Another important aspect of PCM-based solar water heater is storage tank design. Among different types of storage tank designs one of the most popular is to use a shell and tube heat exchanger with a cylindrical storage tank and cylindrical pipes passing through the tank for carrying charging and discharging water (Nagano et al.

2004; Adine and Qarnia 2009; Hosseini et al. 2012; Mat et al. 2013; Mahfuz et al. 2014; Shrivastava and Chakraborty 2019). Typically, it consists of two concentric cylinders. The outer cylinder contains the PCM while the inner cylinder acts as the flow channel for charging and discharging. The aim of this chapter is to present a study on the effect of important parameters on the performance of shell- and tube-type PCM-based solar water heater. To do this, a two-channel energy storage system is considered with PCM in the outer chamber and water flowing through the inner channel. An enthalpy-porosity-based model is developed to simulate both melting and solidification of PCM. The model is capable of simulating flow of water during charging and discharging, heat transfer in the water, PCM and the separating aluminum wall, phase change and buoyancy driven natural convection in the PCM.

2 Problem Description

For the simulations, a two-dimensional problem domain is considered, as shown in Fig. 1. The problem domain consists of an inner channel through which water flows during charging and discharging. PCM is contained in the outer chamber. The two chambers are separated by a thin wall of aluminum. Only half of the system is considered for the simulations. It is assumed that the effect of natural convection is not strong enough to significantly alter the melting pattern of PCM at the top and bottom of the flow channel. Similar observations have been made in Morales-Ruiz et al. (2016). During charging, hot water enters the inner channel from the left side of the domain and exits from the right. During discharging, the flow direction is reversed and cold water enters from the right side.

It is assumed that the PCM is in solid phase initially. The temperature in the entire domain is uniform initially with a value less than the melting temperature of PCM. For the flow, velocity boundary conditions consist of specified uniform velocity at the inlet and outflow conditions on the opposite side of the flow channel. As mentioned previously, the flow direction changes depending on the charging or discharging process. All the other boundaries have zero velocity condition except the bottom boundary which has symmetry conditions. For the heat transfer, uniform

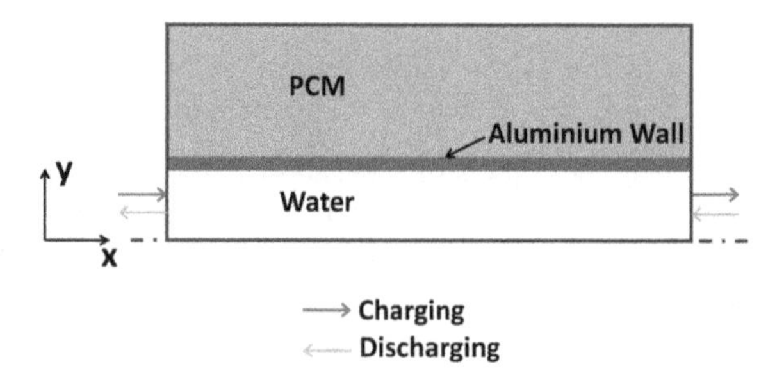

Fig. 1 Schematic representation of the problem domain

temperature boundary condition is applied at the inlet of the flow channel. All the other boundaries are kept insulated.

3 Mathematical Model

The developed model consists of two coupled parts: The flow model and the phase change model. The flow model consists of the continuity and momentum equations which are formulated using volume averaging of the two phases. It should be noted here that the two-phase model is only required for the PCM domain. For calculating the flow in the water channel, only liquid phase momentum equations are considered as described later in Sect. 3.2. However, the numerical code for implementing the flow equations is developed in a generalized form with the capability of solving for both solid and liquid phases. Depending on the conditions, the model appropriately reduces to single phase liquid, single phase solid, or two-phase form. The phase change model is developed using the enthalpy-porosity technique (Voller et al. 1989; Chakraborty and Dutta 2001) which requires the solution of the enthalpy-based energy conservation equation.

All the governing equations are formulated and solved in non-dimensional form. Although for the PCM, water channel, and solid metal wall, the governing equations are different as given in Sects. 3.1, 3.2 and 3.3, the equations are implemented in the CFD code in a generalized form where a single formulation is sufficient to describe the governing equations for all the regions. Depending on different material parameters, the equations reduce to the appropriate equations for the PCM, water and solid metal regions while solving. Hence, the non-dimensionalization of the governing equations for all the three regions is done in the same way by using the same set of parameters. The non-dimensional parameters are taken as $x^* = x/L$, $y^* = y/L$, $T^* = (T - T_{\min})/(T_{\max} - T_{\min})$, $u^* = uL/\alpha_{\text{pcm}}$, $v^* = vL/\alpha_{\text{pcm}}$, $t^* = t\alpha_{\text{pcm}}/L^2$, $\rho^* = \rho/\rho_{\text{pcm}}$, $K^* = K/K_{\text{pcm}}$, $C^* = C/C_{\text{avg}}$, $(\rho C)^* = \rho C/(\rho C)_{\text{avg}}$, $\alpha_{\text{pcm}} = K_{\text{pcm}}/(\rho_{\text{pcm}} C_{\text{avg}})$. x and y denote the distance along the two coordinate axes, L is the length of the domain, T is the temperature and $T_{\min}$ and $T_{\max}$ are the minimum and maximum temperatures equal to the cold water temperature and hot water temperature, respectively. u and v are the velocity components along the x and y directions, α is the thermal diffusivity, t denotes time, and ρ denotes density. K and C are the thermal conductivity and specific heat, respectively. The subscripts 'pcm' and 'avg' denote the properties of PCM and average of two phases, respectively. The non-dimensional values are represented by the superscript '*'. The governing equations assume different forms for the three regions: PCM, water, and metal wall as described in the following sections.

3.1 PCM Volume

For the PCM, both flow and energy conservation equations need to be solved. The flow equations need to be solved as the effect of natural convection is considered in the model. To solve for flow in the domain, the following continuity and momentum equations are required.

$$\frac{\partial \rho^*}{\partial t^*} + \frac{\partial(\rho^* u^*)}{\partial x^*} + \frac{\partial(\rho^* v^*)}{\partial y^*} = 0 \tag{1}$$

$$\rho^*_{\text{pcm}}\left[\frac{\partial u^*}{\partial t^*} + u^*\frac{\partial u^*}{\partial x^*} + v^*\frac{\partial u^*}{\partial y^*}\right] = -\frac{\partial p^*}{\partial x^*} + \text{Pr}\left[\frac{\partial^2 u^*}{\partial x^{*2}} + \frac{\partial^2 u^*}{\partial y^{*2}}\right] + Au^* \tag{2}$$

$$\rho^*_{\text{pcm}}\left[\frac{\partial v^*}{\partial t^*} + u^*\frac{\partial v^*}{\partial x^*} + v^*\frac{\partial v^*}{\partial y^*}\right] = -\frac{\partial p^*}{\partial y^*} + \text{Pr}\left[\frac{\partial^2 v^*}{\partial x^{*2}} + \frac{\partial^2 v^*}{\partial y^{*2}}\right] + Av^* + \text{Pr}\,.\text{Ra}\left(T^* - T^*_m\right) \tag{3}$$

In Eqs. (1–3), Pr and Ra denote the Prandtl number and Rayleigh number. Prandtl number is defined as $\text{Pr} = \mu_{\text{pcm}} / \rho_{\text{pcm}}\alpha_{\text{pcm}}$ while Rayleigh number is defined as $\text{Ra} = \rho_{\text{pcm}} g \beta_{\text{pcm}} L^3 (T_{\max} - T_{\min}) / \mu_{\text{pcm}}\alpha_{\text{pcm}}$ where β_{pcm} is the coefficient of volume expansion of PCM and g is the acceleration due to gravity. The term $\text{Pr}\,.\text{Ra}\left(T^* - T^*_m\right)$ represents the effect of natural buoyancy.

The presence of solid phase is incorporated through the terms Au^* and Av^*. A is defined as $A = C(1-\varepsilon)^2 / \varepsilon^3 + b$ where ε is the liquid fraction. C is taken as a very large valued constant and b is a small number required to avoid division by zero in the numerical solution. These two terms approximate the increased resistance for flow in the two-phase region and drive the velocity to zero in the solid phase.

The phase change and heat transfer in the PCM are modeled using the following volume averaged energy equation.

$$\rho^*_{\text{pcm}} C^*_L\left[\frac{\partial T^*}{\partial t^*} + u^*\frac{\partial T^*}{\partial x^*} + v^*\frac{\partial T^*}{\partial y^*}\right] = K^*_{\text{pcm}}\left[\frac{\partial^2 T^*}{\partial x^{*2}} + \frac{\partial^2 T^*}{\partial y^{*2}}\right] + S_{h1} + S_{h2} \tag{4}$$

In Eq. (4), S_{h1} is a source term which represents the effect of latent heat of phase change. S_{h1} is defined as $S_{h1} = -\frac{1}{\text{Ste}}\left[\rho^*_{\text{pcm}}\frac{\partial \varepsilon}{\partial t^*} - \nabla\left(\rho^*_{\text{pcm}}\vec{u}^*\varepsilon\right)\right]$ where Ste is the Stefan number. The Stefan number is defined as $\text{Ste} = C_{\text{avg}}(T_{\max} - T_{\min}) / L_f$ where L_f is the latent heat of phase change. The second source term, S_{h2}, can be written as $S_{h2} = -\rho^*_{\text{pcm}}\left[\left(C^*_{\text{avg}} - C^*_L\right)\frac{\partial T^*}{\partial t} + T^*\frac{\partial C^*_{\text{avg}}}{\partial t}\right] - \frac{\rho^*_{\text{pcm}} T_{\min}}{(T_{\max} - T_{\min})}\frac{\partial C^*_{\text{avg}}}{\partial t}$ and arises due to the modification of the volume averaged equation in the given form.

3.2 Water Channel

For the water channel, the continuity and momentum equations are formulated as follows.

$$\frac{\partial \rho^*_{\text{water}}}{\partial t^*} + \frac{\partial\left(\rho^*_{\text{water}} u^*\right)}{\partial x^*} + \frac{\partial\left(\rho^*_{\text{water}} v^*\right)}{\partial y^*} = 0 \tag{5}$$

$$\rho^*_{\text{water}}\left[\frac{\partial u^*}{\partial t^*} + u^*\frac{\partial u^*}{\partial x^*} + v^*\frac{\partial u^*}{\partial y^*}\right] = -\frac{\partial p^*}{\partial x^*} + \Pr\frac{\mu_{\text{water}}}{\mu_{\text{pcm}}}\left[\frac{\partial^2 u^*}{\partial x^{*2}} + \frac{\partial^2 u^*}{\partial y^{*2}}\right] \tag{6}$$

$$\rho^*_{\text{water}}\left[\frac{\partial v^*}{\partial t^*} + u^*\frac{\partial v^*}{\partial x^*} + v^*\frac{\partial v^*}{\partial y^*}\right] = -\frac{\partial p^*}{\partial y^*} + \Pr\frac{\mu_{\text{water}}}{\mu_{\text{pcm}}}\left[\frac{\partial^2 v^*}{\partial x^{*2}} + \frac{\partial^2 v^*}{\partial y^{*2}}\right] \tag{7}$$

As the entire water channel contains only liquid water, the flow resistance source terms to represent solid phase are not present in Eqs. (6 and 7). Natural buoyancy effects are also not considered for the water flow as the velocity due to forced flow is drastically higher in this case.

The heat transfer in the water is governed by the following energy conservation equation.

$$\rho^*_{\text{water}} C^*_{\text{water}}\left[\frac{\partial T^*}{\partial t^*} + u^*\frac{\partial T^*}{\partial x^*} + v^*\frac{\partial T^*}{\partial y^*}\right] = K^*_{\text{water}}\left[\frac{\partial^2 T^*}{\partial x^{*2}} + \frac{\partial^2 T^*}{\partial y^{*2}}\right] \tag{8}$$

3.3 Solid Metal Wall

For the metal wall separating the PCM from the heat transfer fluid, only the energy transfer needs to be considered. The temperature evolution in this region is governed by the following energy equation.

$$\rho^*_{\text{Solid}} C^*_{\text{Solid}}\left(\frac{\partial T^*}{\partial t^*}\right) = K^*_{\text{Solid}}\left[\frac{\partial^2 T^*}{\partial x^{*2}} + \frac{\partial^2 T^*}{\partial y^{*2}}\right] \tag{9}$$

In Eqs. (5–9), the subscripts ‘water’ and ‘solid’ denote the water and solid metal regions, respectively.

4 Numerical Solution

The governing equations are discretized in a Cartesian coordinate framework using a finite volume approach. For solving the equations, the SIMPLER algorithm given in

Patankar (2018) is employed. The convection–diffusion terms are discretized using the power law scheme. The discretized equations are solved using line-by-line tri-diagonal matrix algorithm (TDMA) with multiple sweeping from all possible directions. The numerical model is implemented using an in-house code developed using Fortran 90. The developed code is based on the previous numerical program described in Sahoo et al. (2016). It has been validated with analytical solution for phase change and with experimental results for melting of PCM in Sahoo et al. (2016). The solution algorithm consists of the following steps.

1. At first the domain is defined and the numerical grid is generated.
2. The initial and boundary conditions are specified for the entire domain.
3. The liquid fraction values are set equal to zero for the PCM and metal and equal to one for the water. All the other properties are defined for all the regions.
4. The different source terms are calculated.
5. The flow field is calculated by solving the continuity and momentum equations.
6. The energy equation is solved for each region to obtain the temperature field.
7. The liquid fraction is calculated using the enthalpy update scheme given in Sahoo et al. (2016).
8. The solution is checked for convergence. If the solution has not converged, steps 4–7 are repeated until convergence is achieved.
9. The current values are stored for use in the next time step.
10. The calculation for the next time is started and the steps 4–9 are repeated until the time reaches the end of charging phase.
11. The discharging phase is started and the steps 4–9 are repeated until the end of simulation.

5 Results for a Single Charging and Discharging Cycle

5.1 Simulation Parameters

All the simulations considered in this chapter consist of a single charging–discharging cycle. It is assumed that initially the PCM is in solid state kept at a low temperature. High-temperature water is passed through the inner channel which transfers heat to the PCM through the metal wall. At first, sensible heating of PCM occurs until it reaches the melting temperature. Subsequently, the PCM starts melting which continues until the entire PCM gets melted or the charging end time is reached. This is followed by the discharging cycle in which water at low temperature is passed through the channel from the opposite direction. Energy stored in the PCM is transferred to the water through the metal wall. The water goes out of the channel at a relatively high temperature. The PCM loses energy and solidifies back by latent heat removal. After complete solidification, the PCM supplies energy by losing sensible heat and thus undergoing reduction in temperature. This continues until the discharging cycle is stopped. For all the simulations presented in this chapter, paraffin wax is taken as

Table 1 Thermophysical properties used for the simulations

Property	Paraffin	Water	Aluminum
Latent heat (kJ/kg)	169	–	–
Thermal conductivity (W/m K)	0.2	0.591	205
Specific heat (J/kg K)	2100	4187	910
Density (kg/m^3)	880	1000	2830
Viscosity (Ns/m^2)	0.007	0.001	–
Melting temperature (°C)	60–64	–	–
Thermal expansion coefficient (K^{-1})	0.001	–	–

the PCM and the metal wall is assumed to be made of aluminum. The thermophysical properties of the different materials are specified in Table 1.

5.2 *Simulation Prediction for a Single Charging and Discharging Cycle*

At first a single charging cycle for 20,000 s is considered which is immediately followed by a discharging cycle for 10,000 s. The initial temperature for the entire domain is taken as 20 °C. During the charging cycle, hot water enters the flow channel with a temperature of 80 °C and a velocity of 0.002 m/s. During the discharging cycle, cold water enters the flow channel with a temperature of 20 °C and a velocity of 0.001 m/s. The length of the channel is taken as 0.5 m. The thickness of the water channel (for the half domain as specified in the problem due to its symmetric nature) is taken as 3 cm, the thickness of the metal plate is 3 mm while that of the PCM region is 4 cm.

The evolution of temperature and liquid fraction contours is presented in Figs. 2 and 3. It is seen that initially the temperature rises quickly in the PCM and metal due to sensible heating. Subsequently, the rate of increase of temperature slows down as the energy is absorbed as latent heat due to melting of PCM. From the liquid fraction contours, it is observed that melting starts quickly over the entire metal plate initially. The rate of melting is highest near the inlet section of water. During the discharging cycle, the temperature of PCM near the metal wall decreases quickly and the PCM solidifies. PCM present at larger distances from the metal wall takes considerably higher time to solidify and maintains high temperature for a longer duration. The average temperature of water at the outlet is higher than that at the inlet.

To quantify the melting rate of PCM and temperature of water, the variation of liquid fraction of PCM during the entire charging–discharging cycle and the variation of outlet water temperature with time during charging and during discharging are shown in Fig. 4. It is seen that at the end of the charging cycle the liquid fraction of PCM is about 0.6 and it almost reaches 0 at the end of the discharge cycle. During

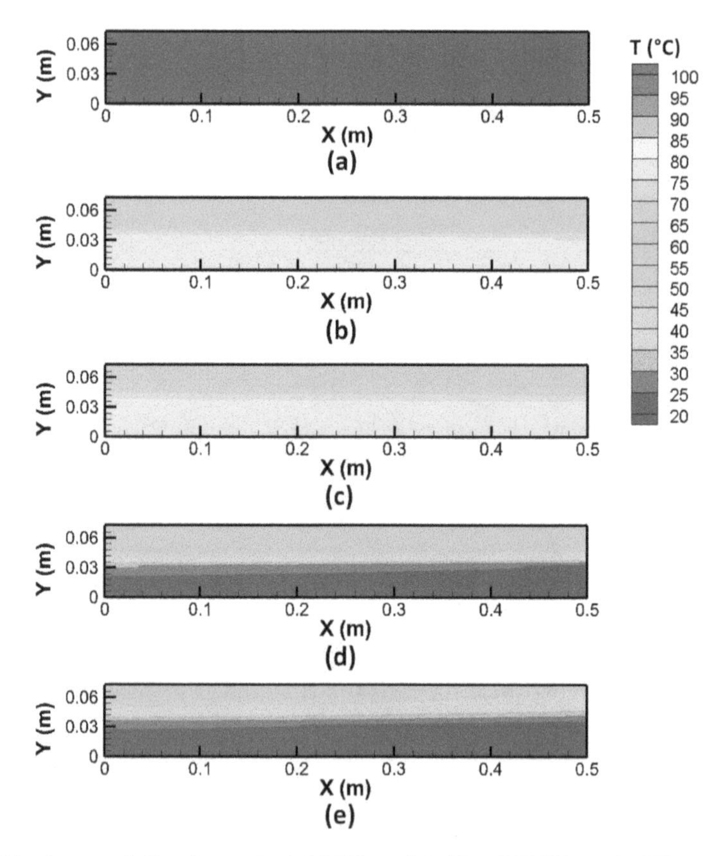

Fig. 2 Temperature field at time. **a** 0 s, **b** 10,000 s, **c** 20,000 s, **d** 25,000 s, **e** 30,000 s

the charging cycle, the water outlet temperature is very low initially as most of its energy is transferred to the PCM due to the large temperature difference with the PCM. Another reason for the initial low outlet temperature is that the water channel is initially kept at 20 °C. However, the outlet temperature quickly increases which signifies that the rate of heat transfer to the PCM decreases. The outlet temperature remains slightly below 80 °C which results in reduced heat transfer to the PCM. During the discharge cycle, the average water temperature reaches a value of 31 °C but gradually drops to around 22.5 °C at the end of the discharge cycle.

6 Effect of Design Parameters

In this section, the effect of important design parameters such as the inlet flow conditions, the charging time, and the dimensions of the system are studied. Seven different parameters are considered: inlet temperature of water during charging, flow

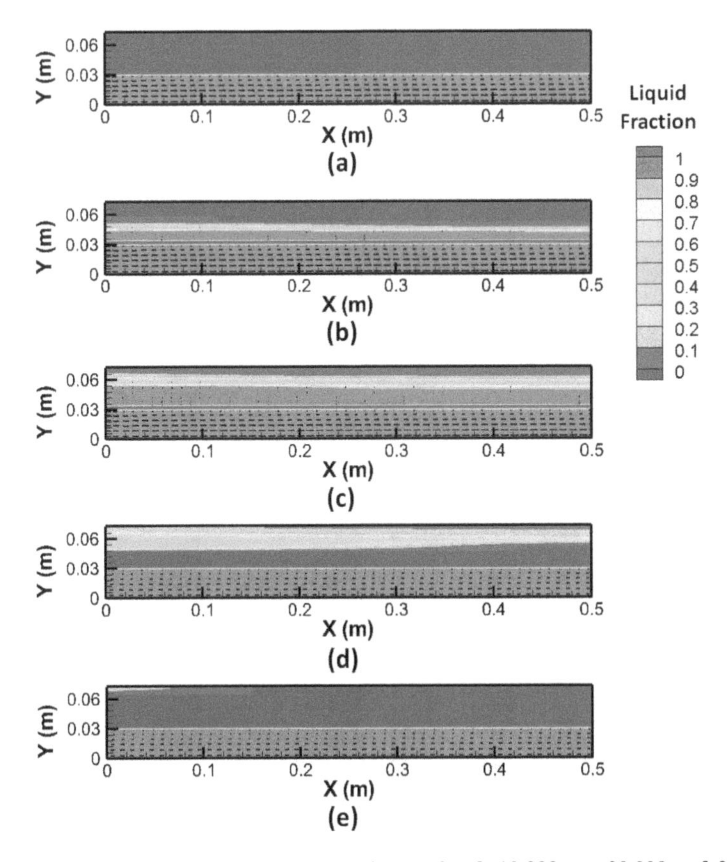

Fig. 3 Liquid fraction and velocity vectors at time. **a** 0 s, **b** 10,000 s, **c** 20,000 s, **d** 25,000 s, **e** 30,000 s

velocity of water during charging, flow velocity of water during discharging, charging duration, length of the domain, width of the flow channel, and width of the PCM region. For each case, the values of all the parameters except the chosen parameter are kept constant and equal to that given in Sect. 5.2. For each study, the rate of melting and the temperature variation is compared.

6.1 *Effect of Inlet Temperature of Water During Charging*

At first the effect of water inlet temperature is studied. Three cases are considered with inlet temperatures of 70 °C, 80 °C, and 90 °C. Charging is continued for 20,000 s which is followed by discharging for 10,000 s. Figures 5 and 6 present the comparison of temperature and liquid fraction for the three cases after the end of the charging cycle. It can be observed that the increase in temperature is very less for the 70 °C

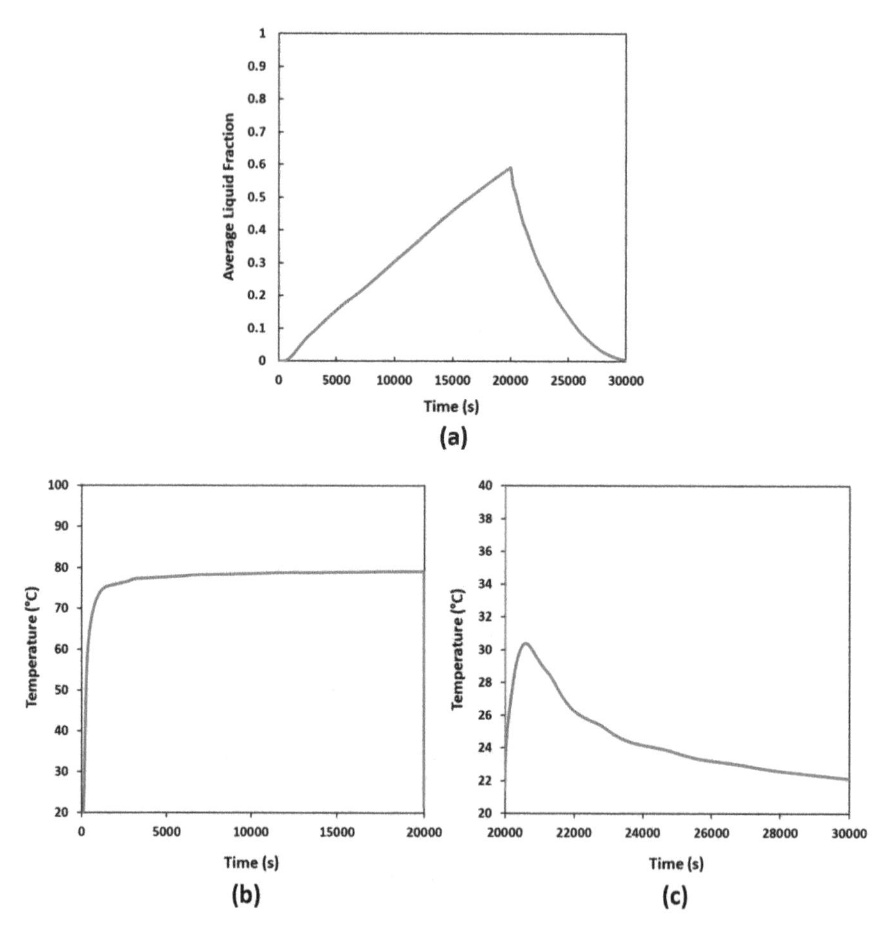

Fig. 4 Variation of **a** liquid fraction with time during the entire cycle, **b** water outlet temperature during charging, and **c** water outlet temperature during discharging

case and the corresponding melting is also small. On the other hand, for high inlet temperature, the rate of melting is significantly higher. The temperature of the PCM near the wall also shows a similar trend although the temperature of PCM far away from the wall is relatively similar because of the effect of phase change between this region and the water.

Figure 7 presents a quantitative comparison of melting and temperature evolution for the three cases. From Fig. 7a, it is observed that the liquid fraction is significantly higher both during the charging and discharging cycles for higher inlet temperature. The water outlet temperature (Fig. 7b) during charging shows similar trends for all the cases as described previously in Sect. 5.2. The water outlet temperature during the discharge is improved significantly for the two high-temperature cases (Fig. 7c). As the water outlet temperature during discharge is the most important parameter for domestic heating applications, the time durations for which water is obtained at

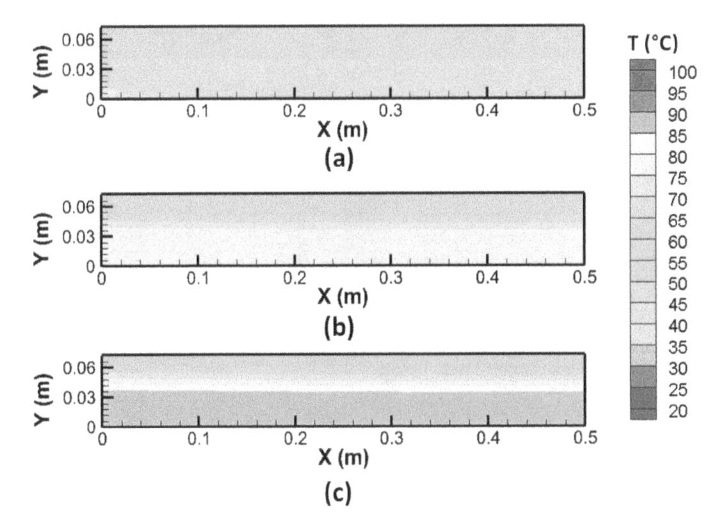

Fig. 5 Temperature field at the end of charging cycle (t = 20,000 s) with water inlet temperature of **a** 70 °C, **b** 80 °C, **c** 90 °C

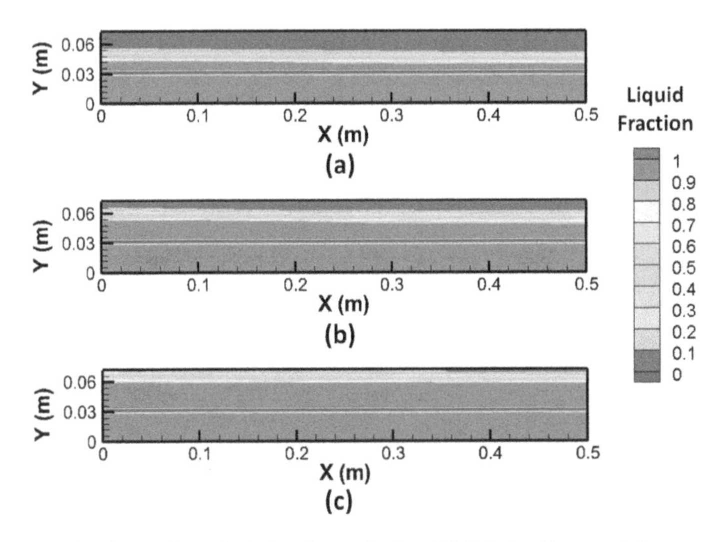

Fig. 6 Liquid fraction at the end of charging cycle (t = 20,000 s) with water inlet temperature of **a** 70 °C, **b** 80 °C, **c** 90 °C

temperatures above 30 °C and 25 °C are compared in Fig. 7d. It is seen that water at more than 30 °C is obtained from the high-temperature system for about 650 s while the low-temperature system cannot provide water at this temperature threshold. For the high-temperature case, water temperature during the discharge is above 25 °C for about 3300 s which is considerably higher than that for the low-temperature case.

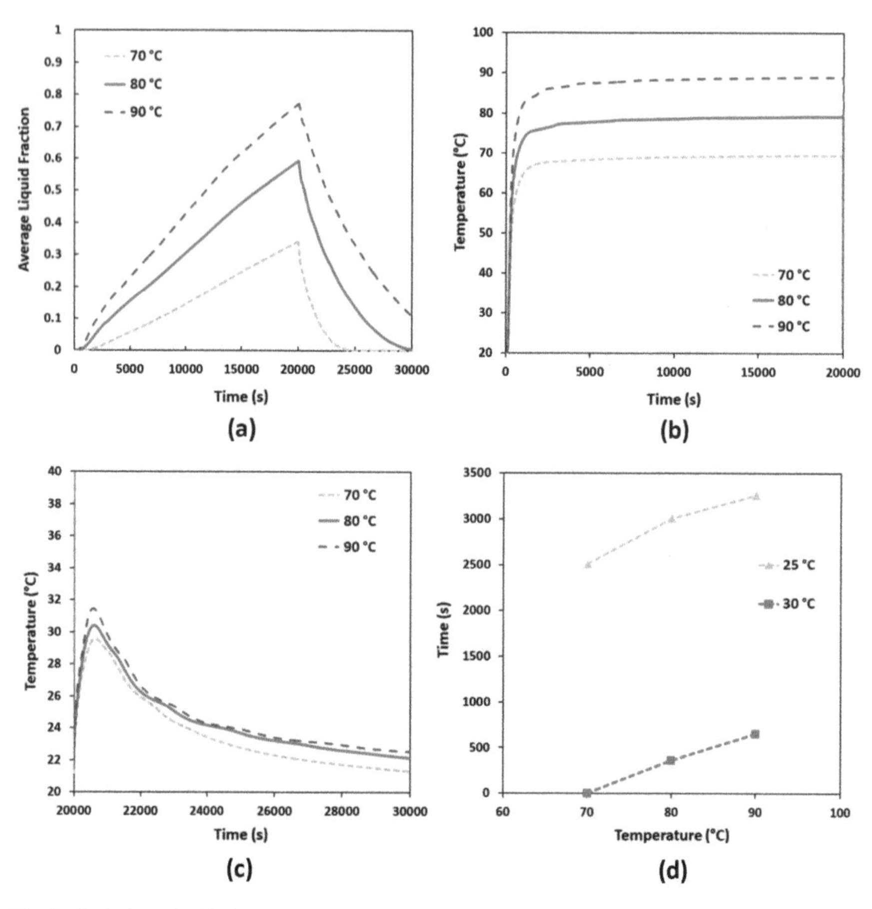

Fig. 7 Variation of **a** liquid fraction with time, **b** water outlet temperature during charging, **c** water outlet temperature during discharging, **d** time for obtaining water at specified temperature

6.2 *Effect of Inlet Flow Velocity During Charging*

In this section, the effect of inlet flow velocity during the charging process is considered. Three different inlet velocities are considered: 0.001 m/s, 0.002 m/s, and 0.004 m/s. Figures 8 and 9 present the comparison of temperature and liquid fraction for the three cases after the end of the charging cycle. Although not much difference can be seen from the temperature and liquid fraction contours, careful observation shows that the liquid fraction increases with increase in flow rate. This can be confirmed from Fig. 10a which shows that the rate of melting increases slightly with increase in flow velocity. However, the variation is not as significant as that seen with different inlet flow temperatures. The temperature variation during charging (Fig. 10b) shows that the rate of temperature increase is slightly higher for the high flow rate case initially although subsequently it becomes almost equal for all the

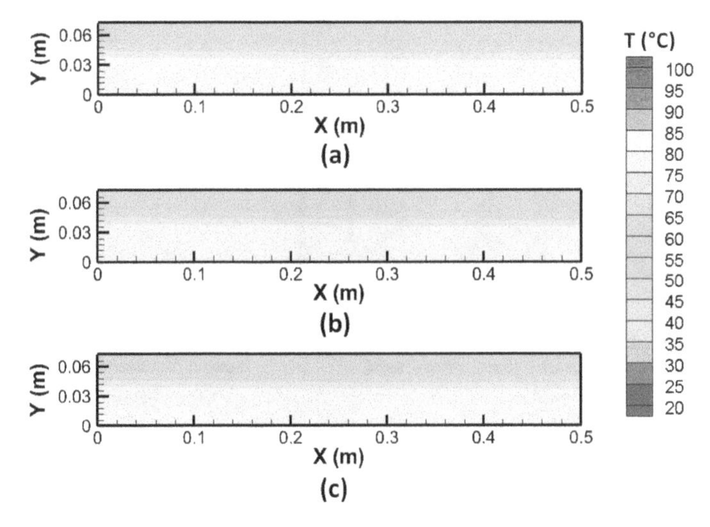

Fig. 8 Temperature field at the end of charging cycle (t = 20,000 s) with inlet flow velocity of **a** 0.001 m/s, **b** 0.002 m/s, and **c** 0.004 m/s

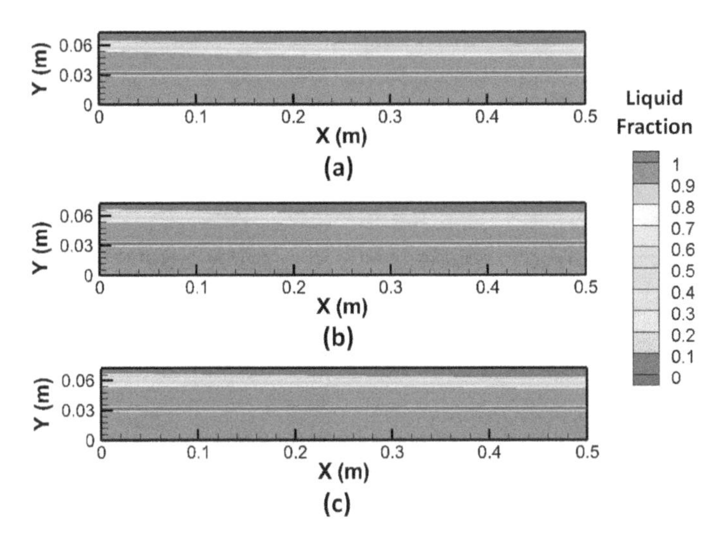

Fig. 9 Liquid fraction at the end of charging cycle (t = 20,000 s) with inlet flow velocity of **a** 0.001 m/s, **b** 0.002 m/s, and **c** 0.004 m/s

cases. The rate of temperature variation for all the three cases is similar during the discharging period (Fig. 10c). This is also confirmed from Fig. 10d which shows that the availability of water at a particular temperature is similar for all the cases considered. This study shows that the inlet flow velocity is not as important as the inlet flow temperature for improving the performance of PCM-based energy storage systems of this type.

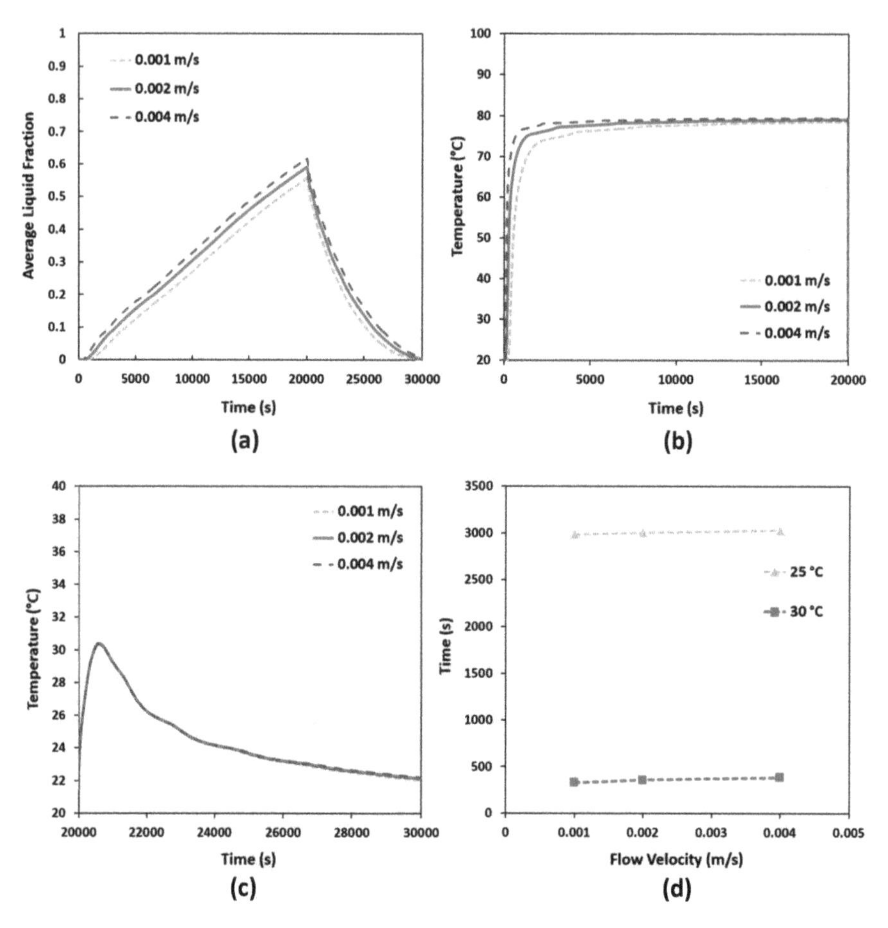

Fig. 10 Variation of **a** liquid fraction with time, **b** water outlet temperature during charging, **c** water outlet temperature during discharging, and **d** time for obtaining water at specified temperature

6.3 Effect of Inlet Flow Velocity During Discharging

The effect of discharge velocity is studied in this Sect. 3 different cases are considered with discharge flow velocity of 0.001 m/s, 0.002 m/s and 0.004 m/s. For all the 3 cases charging is done with hot water at 80 °C with a flow velocity of 0.002 m/s for 20,000 s. Subsequently, the discharging cycle takes place for 10,000 s. As the charging cycles are exactly similar for all the cases, the temperature distributions and liquid fraction distributions after charging are similar to that shown in Figs. 2 and 3. The temperature variation after the discharging cycle is shown in Fig. 11. It is seen that the temperature variation is almost similar in all the three cases. The variation of liquid fraction and temperature are analyzed in Fig. 12. Figure 12a shows that the rate of solidification is slightly higher for the high flow rate case. As discussed previously, the temperature during the charging cycle is exactly similar (Fig. 12b).

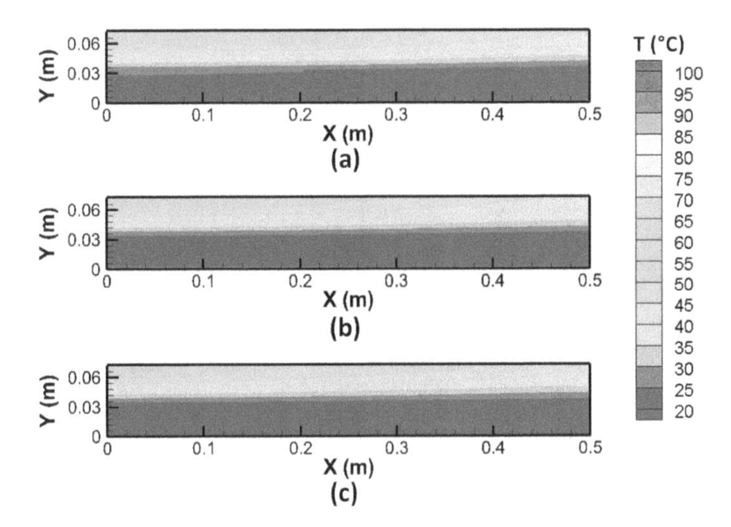

Fig. 11 Temperature field at the end of discharging cycle ($t = 30{,}000$ s) with inlet flow velocity of **a** 0.001 m/s, **b** 0.002 m/s, and **c** 0.004 m/s

However, during the discharge cycle, lower flow rate results in significantly higher temperature for the entire discharge duration (Fig. 12c). Figure 12d shows that at higher flow rates, the outlet temperature is always less than 30 °C and the duration for which water is available at more than 25 °C is drastically reduced. If the results are compared with the predictions presented in Sect. 6.2, it can be observed that the discharge flow velocity has considerably larger effect on the available water temperature and thus is more important.

6.4 *Effect of Charging Time*

Another important parameter which affects the charging process is the total duration of charging. Different charging times result in different quantities of energy stored and different temperature distribution after charging and thus affect the discharging process. For this analysis, three different charging times are considered: 10,000 s, 20,000 s and 30,000 s. The temperature and liquid fraction contours for the three cases after charging are compared in Figs. 13 and 14. It can be observed that although there is not much temperature variation due to the phase change process, the extent of melting is significantly higher after 30,000 s. This is clearly seen from Fig. 15a which shows that the liquid fraction is about 80% after 30,000 s and about 30% after 10,000 s. As a result, the entire PCM quickly solidifies during the discharging process for the 10,000 s case. On the other hand, large quantity of PCM remains in molten state even after the discharge cycle is complete for the 30,000 s case. The outlet water temperature during charging and discharging are presented in Fig. 15b,

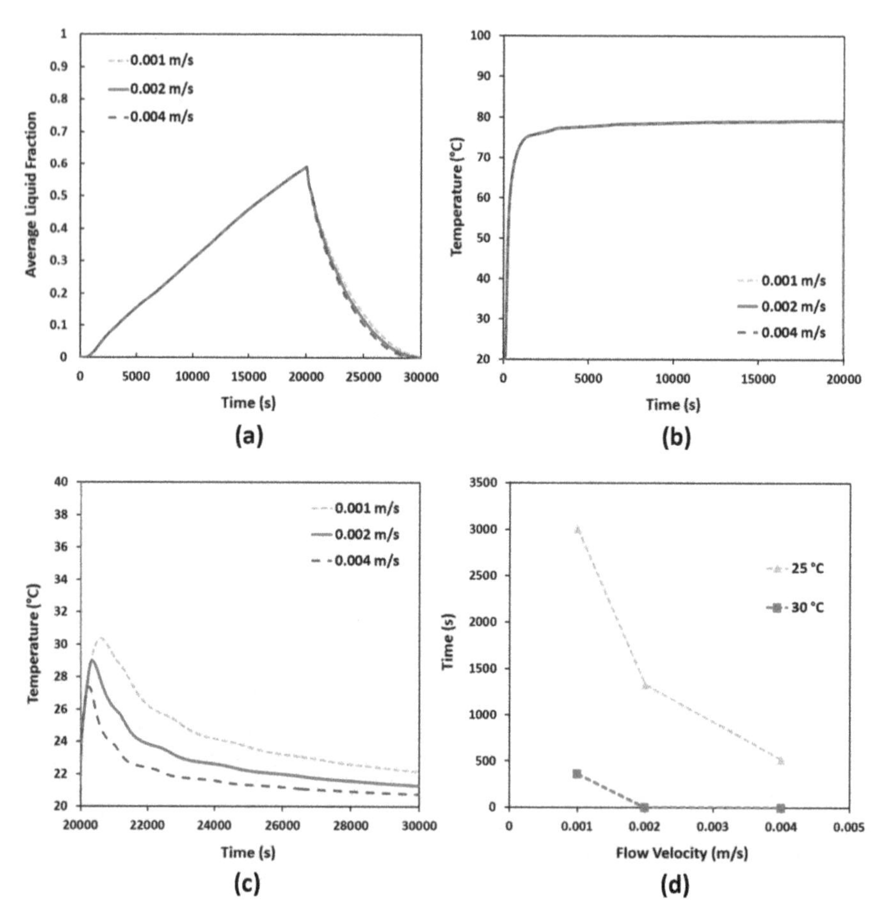

Fig. 12 Variation of **a** liquid fraction with time, **b** water outlet temperature during charging, **c** water outlet temperature during discharging, and **d** time for obtaining water at specified temperature

c. It is seen that for the discharge cycle, the water outlet temperatures are similar for the 20,000 s and 30,000 s cases. For the low charging time case, the water outlet temperature is considerably lower. Figure 15d shows that the effect of charging time is more than that of charging flow velocity but less than that of charging temperature and discharging flow velocity.

6.5 *Effect of Domain Length*

The effect of flow channel length is analyzed in this section. Three different domain lengths are considered: 0.25 m, 0.5 m, and 1.0 m. The temperature and liquid fraction contours for the three cases are compared in Figs. 16 and 17. It should be noted that

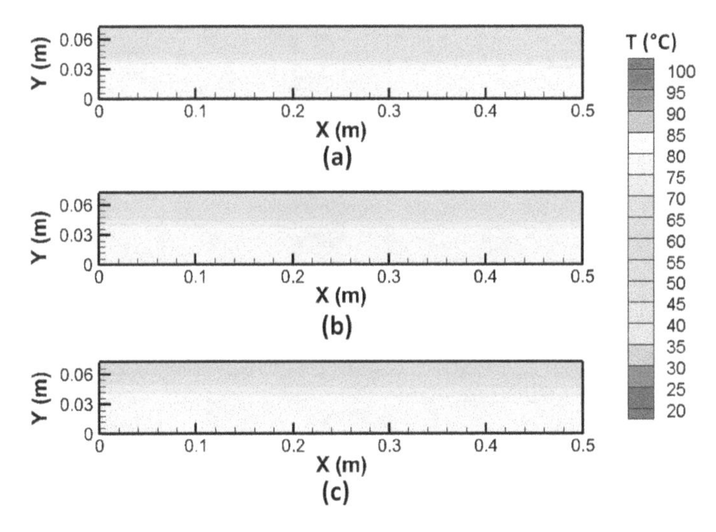

Fig. 13 Temperature field at **a** $t = 10{,}000$ s, **b** $t = 20{,}000$ s, and **c** $t = 30{,}000$ s

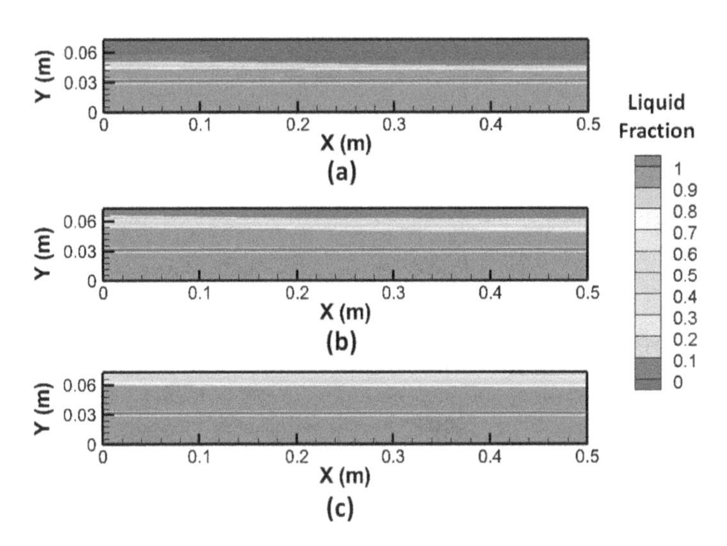

Fig. 14 Liquid fraction variation at **a** $t = 10{,}000$ s, **b** $t = 20{,}000$ s, and **c** $t = 30{,}000$ s

all the three cases are shown by scaling them to the same length which changes the plotted width of the domain accordingly. It is seen that there is not much difference in temperature and melting rate for the three cases. Figure 18a, b shows that the liquid fraction and outlet temperature are slightly higher for the smaller domain. For smaller domain length, the reduction of water temperature is less during the charging cycle as the time of heat transfer to the PCM is less. As a result, the average liquid fraction is higher. However, because the lengths are different, the actual quantity of PCM melted is considerably higher for the longer domains. Figure 18c, d shows that

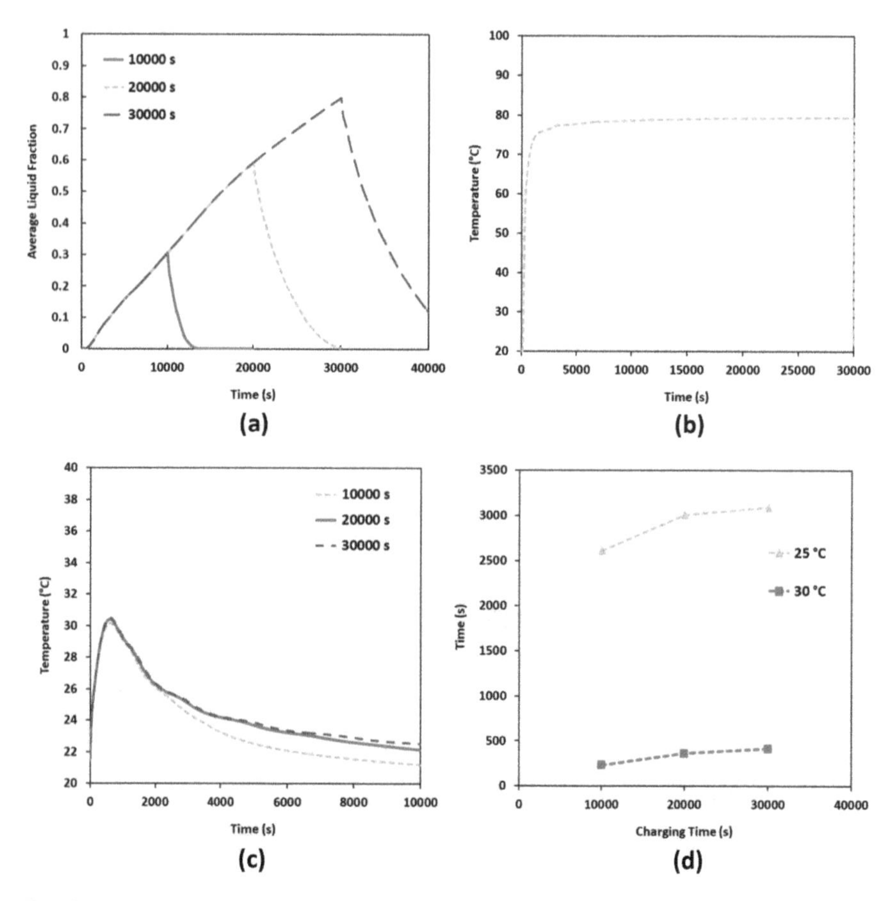

Fig. 15 Variation of **a** liquid fraction with time, **b** water outlet temperature during charging, **c** water outlet temperature during discharging, and **d** time for obtaining water at specified temperature

the water outlet characteristics are drastically improved for longer domain sizes. It is seen that the domain length is the most important factor governing the discharge flow temperature.

6.6 Effect of Height of Flow Channel

The effect of flow channel height is analyzed in this Sect. 3 and different channel heights are considered: 0.015 m, 0.03 m, and 0.06 m. The channel length is equal to 0.5 m for all the three cases. The inlet flow velocities are also kept constant and equal to 0.002 m/s during the charging process and 0.001 m/s during the discharging process. Figures 19 and 20 show the temperature and liquid fraction contours after

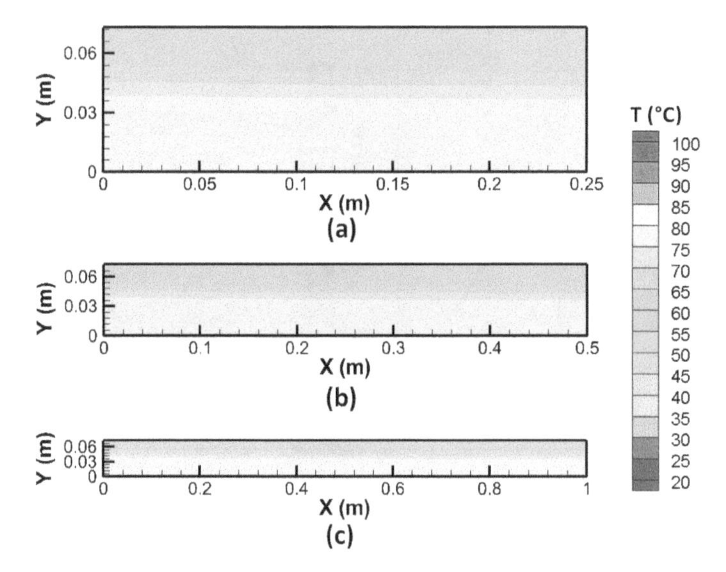

Fig. 16 Temperature field at the end of charging cycle (t = 20,000 s) with domain length equal to **a** 0.25 m, **b** 0.5 m, and **c** 1.0 m

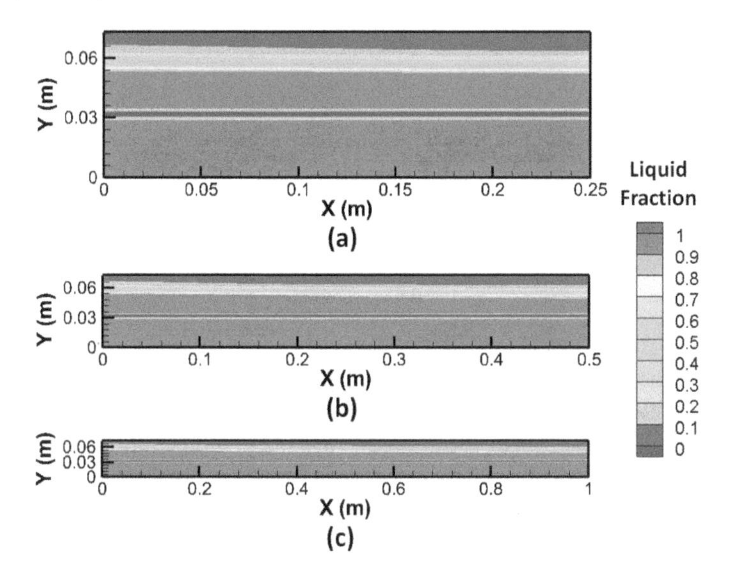

Fig. 17 Liquid fraction at the end of charging cycle (t = 20,000 s) with domain length equal to **a** 0.25 m, **b** 0.5 m, and **c** 1.0 m

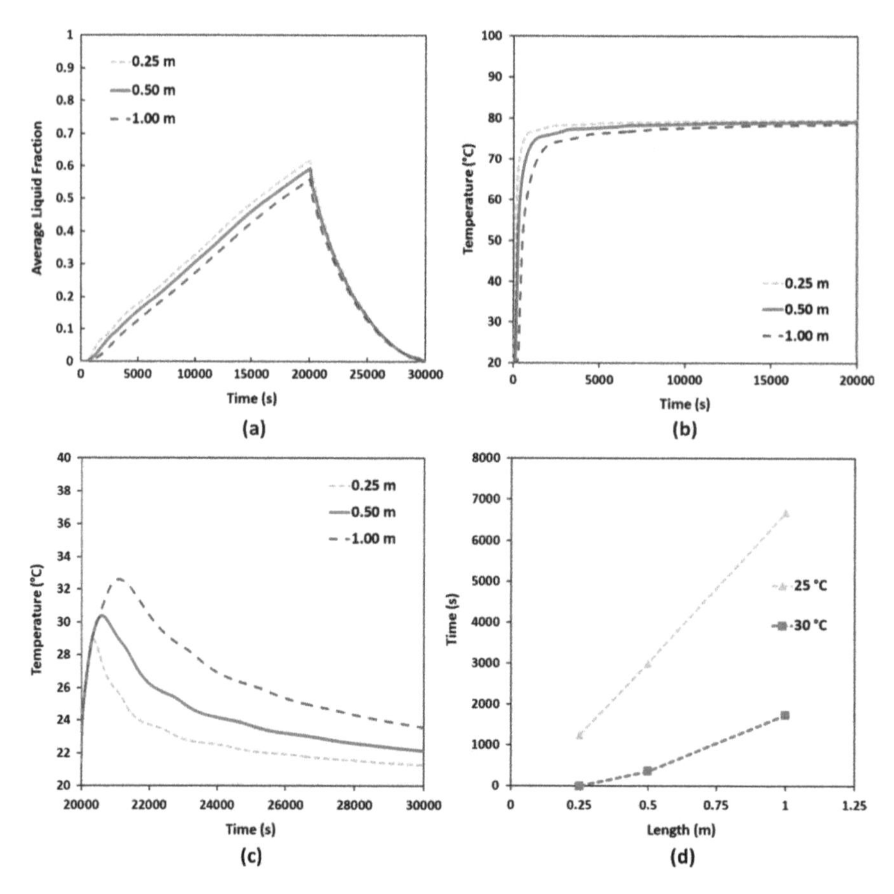

Fig. 18 Variation of **a** liquid fraction with time, **b** water outlet temperature during charging, **c** water outlet temperature during discharging, and **d** time for obtaining water at specified temperature

charging for the three cases. It is seen that there is not much difference in temperature. The variation of liquid fraction is also similar. This is confirmed by the comparisons shown in Fig. 21a, b. However, the average outlet temperature is considerably different during the discharging process, as seen from Fig. 21c. Lower channel width results in faster heating of the water. This results in considerably longer duration at which water is obtained above prescribed thresholds of 25 °C and 30 °C (Fig. 21d).

6.7 Effect of Height of PCM Chamber

The effect of PCM layer height is studied in this Sect. 3 and different PCM chambers are considered with PCM height of 0.02 m, 0.04 m, and 0.06 m. All the other parameters have the same value as given in Sect. 5.2. The variation of temperature and liquid

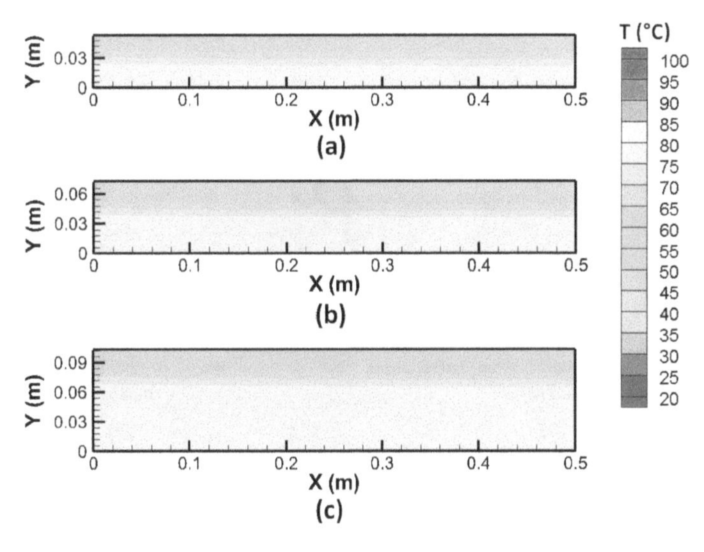

Fig. 19 Temperature field at the end of charging cycle (t = 20,000 s) with flow channel height equal to **a** 0.015 m, **b** 0.03 m, and **c** 0.06 m

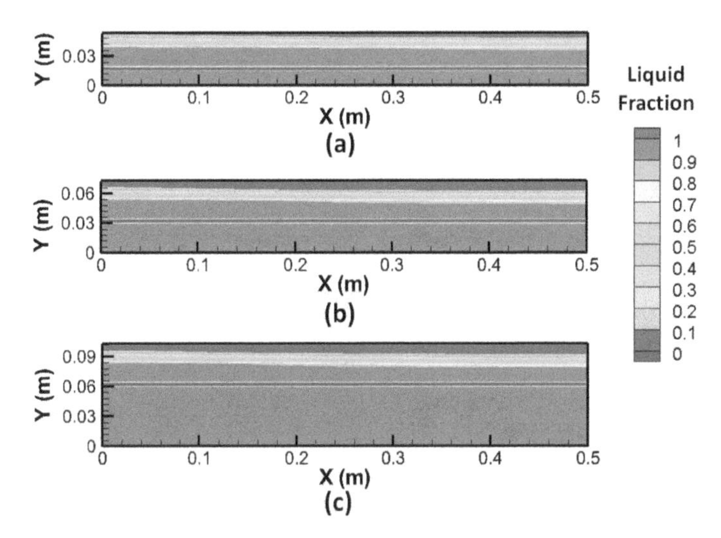

Fig. 20 Liquid fraction at the end of charging cycle (t = 20,000 s) with flow channel height equal to **a** 0.015 m, **b** 0.03 m, and **c** 0.06 m

fraction after the charging process for the three cases is compared in Figs. 22 and 23. It is seen that for lower height of PCM complete melting has occurred and subsequently sensible heating has raised the temperature of PCM significantly beyond the melting temperature. For the larger PCM domain, melting is partially complete and

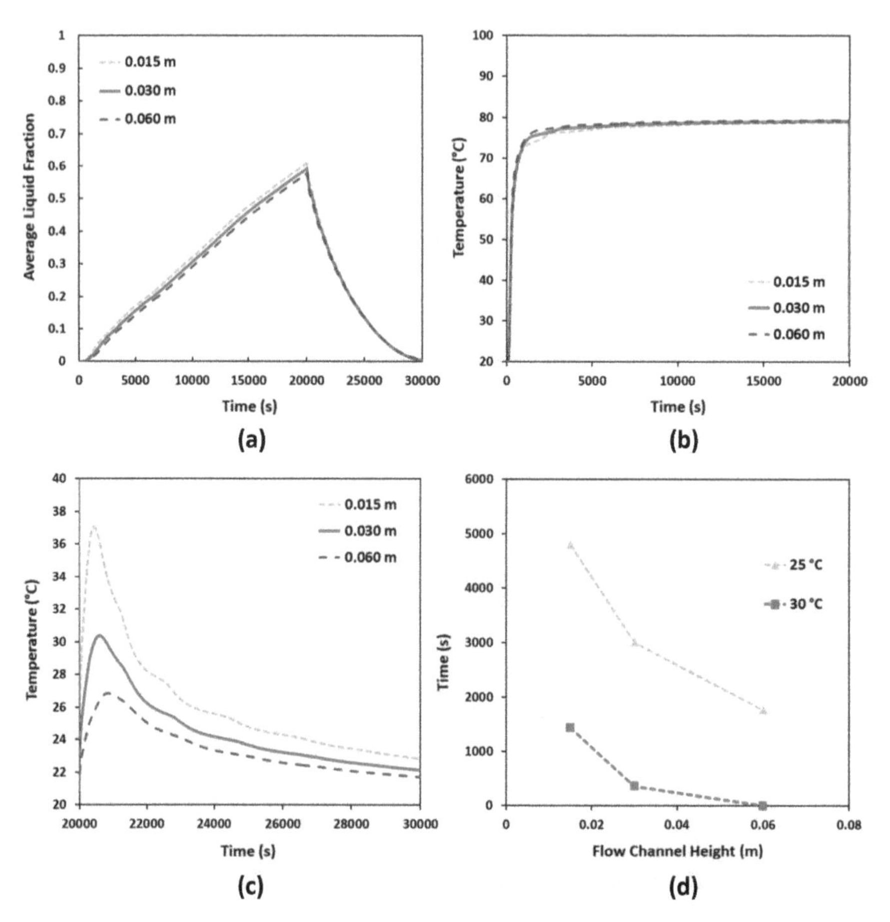

Fig. 21 Variation of **a** liquid fraction with time, **b** water outlet temperature during charging, **c** water outlet temperature during discharging, and **d** time for obtaining water at specified temperature

the temperature is restricted to near melting temperature due to the phase change process.

The variation of liquid fraction with time for the three cases is compared in Fig. 24a. It is seen that the rate of increase of liquid fraction is considerably faster for the smaller PCM domain, and complete melting occurs after about 14,000 s. On the other hand, the maximum liquid fraction for the larger PCM domain is only about 30%. The water outlet temperatures are similar during charging (Fig. 24b) and discharging (Fig. 24c). The smaller PCM domain results in slightly higher water discharge temperature initially. Subsequently, the entire PCM solidifies and the outlet temperature drops below that for the other cases. Figure 24d shows that although the PCM domain height has significant influence on the discharge process, its effect is relatively less as compared to the effects of discharge flow velocity and domain length.

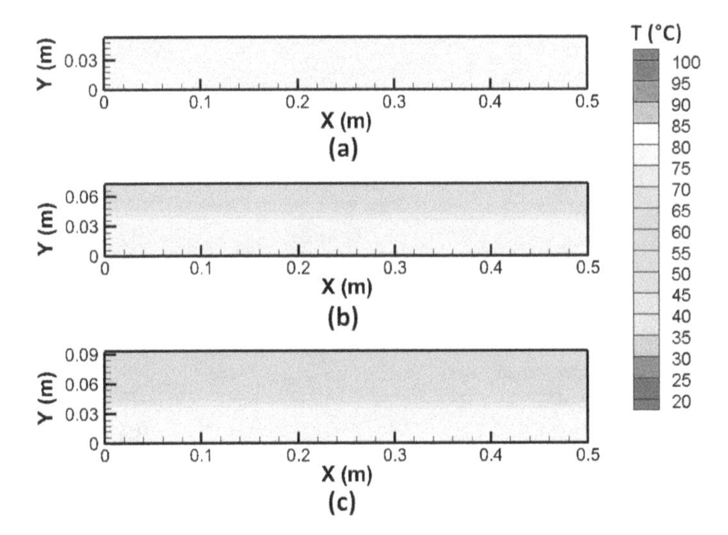

Fig. 22 Temperature field at the end of charging cycle (t = 20,000 s) with PCM height equal to **a** 0.02 m, **b** 0.04 m, **c** 0.06 m

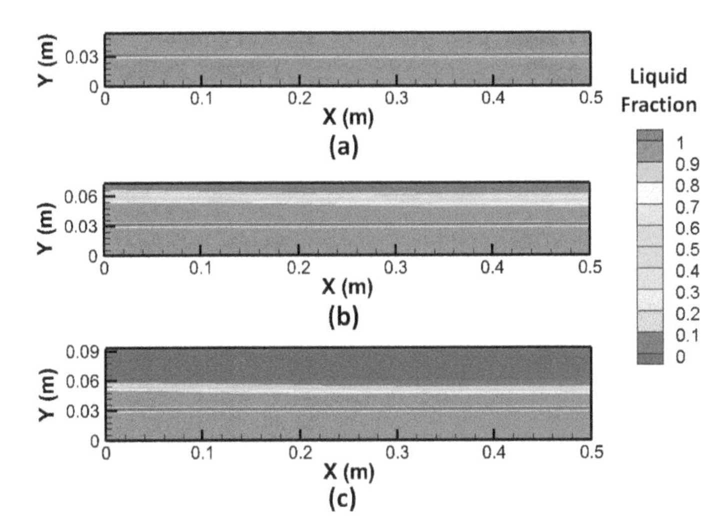

Fig. 23 Liquid fraction at the end of charging cycle (t = 20,000 s) with PCM height equal to **a** 0.02 m, **b** 0.04 m, **c** 0.06 m

7 Conclusion

An enthalpy-porosity-based model has been presented in this chapter for simulating the melting and solidification of PCM in a PCM-based solar water heater. The model captures the heat transfer to the PCM due to the flow of the heat transfer fluid. The effect of important flow parameters such as the flow inlet temperature and velocity

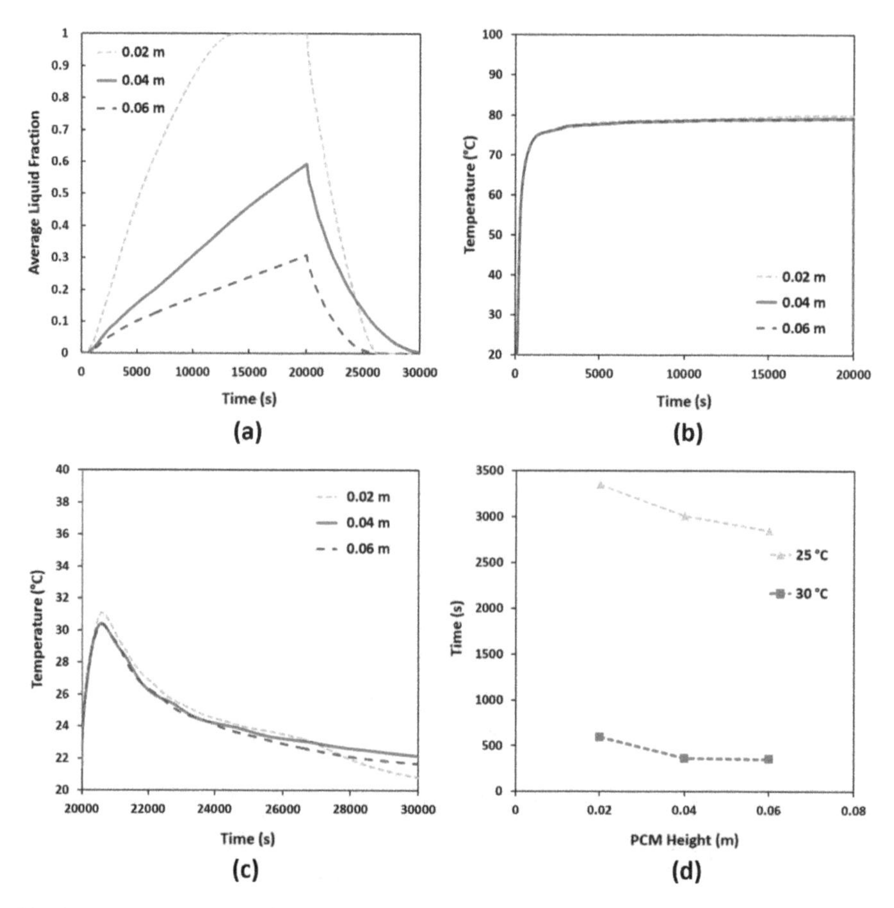

Fig. 24 Variation of **a** liquid fraction with time, **b** water outlet temperature during charging, **c** water outlet temperature during discharging, and **d** time for obtaining water at specified temperature

and geometrical parameters such as the domain length, PCM chamber height and flow channel height on the melting pattern and discharge water temperature are analyzed. It is found that the discharge water temperature is strongly affected by the discharge flow velocity, and the length and height of the flow channel. The initial temperature of water during the charging cycle, the overall charging duration, and height of the PCM domain moderately affects the outlet water temperature during the discharging process. The effect of flow velocity during the charging process was found to be negligible. It can be inferred that to improve the performance of shell- and tube-type PCM-based solar water heaters, the geometrical parameters of the flow channel and the flow rate during the discharge process should be considered carefully.

References

Abokersh MH, Osman M, El-Baz O, El-Morsi M, Sharaf O (2018) Review of the phase change material (PCM) usage for solar domestic water heating systems (SDWHS). Int J Energy Res 42(2):329–357

Adine HA, El Qarnia H (2009) Numerical analysis of the thermal behaviour of a shell-and-tube heat storage unit using phase change materials. Appl Math Model 33(4):2132–2144

Al-Hinti I, Al-Ghandoor A, Maaly A, Naqeera IA, Al-Khateeb Z, Al-Sheikh O (2010) Experimental investigation on the use of water-phase change material storage in conventional solar water heating systems. Energy Convers Manage 51(8):1735–1740

Alshaer WG, Nada SA, Rady MA, Le Bot C, Del Barrio EP (2015) Numerical investigations of using carbon foam/PCM/nano carbon tubes composites in thermal management of electronic equipment. Energy Convers Manage 89:873–884

Bansal NK, Buddhi D (1992) An analytical study of a latent heat storage system in a cylinder. Energy Convers Manage 33(4):235–242

Bouadila S, Fteïti M, Oueslati MM, Guizani A, Farhat A (2014) Enhancement of latent heat storage in a rectangular cavity: solar water heater case study. Energy Convers Manage 78:904–912

Cabeza LF, Mehling H, Hiebler S, Ziegler F (2002) Heat transfer enhancement in water when used as PCM in thermal energy storage. Appl Therm Eng 22(10):1141–1151

Chakraborty S, Dutta P (2001) A generalized formulation for evaluation of latent heat functions in enthalpy-based macroscopic models for convection-diffusion phase change processes. Metall Mater Trans 32(3):562

Dinesh BVS, Bhattacharya A (2019) Effect of foam geometry on heat absorption characteristics of PCM-metal foam composite thermal energy storage systems. Int J Heat Mass Transf 134:866–883

Dinesh BVS, Bhattacharya A (2020) Comparison of energy absorption characteristics of PCM-metal foam systems with different pore size distributions. J Energy Storage 28:101190

Fan L, Khodadadi JM (2012) A theoretical and experimental investigation of unidirectional freezing of nanoparticle-enhanced phase change materials. J Heat Transf 134(9):092301

Fazilati MA, Alemrajabi AA (2013) Phase change material for enhancing solar water heater, an experimental approach. Energy Convers Manage 71:138–145

Fukahori R, Nomura T, Zhu C, Sheng N, Okinaka N, Akiyama T (2016) Macro-encapsulation of metallic phase change material using cylindrical-type ceramic containers for high-temperature thermal energy storage. Appl Energy 170:324–328

Gharebaghi M, Sezai I (2007) Enhancement of heat transfer in latent heat storage modules with internal fins. Numer Heat Transf Part A Appl 53(7):749–765

Hawlader MNA, Uddin MS, Khin MM (2003) Microencapsulated PCM thermal-energy storage system. Appl Energy 74(1–2):195–202

Hosseini MJ, Ranjbar AA, Sedighi K, Rahimi M (2012) A combined experimental and computational study on the melting behavior of a medium temperature phase change storage material inside shell and tube heat exchanger. Int Commun Heat Mass Transf 39(9):1416–1424

Ismail KAR, Alves CLF, Modesto MS (2001) Numerical and experimental study on the solidification of PCM around a vertical axially finned isothermal cylinder. Appl Therm Eng 21(1):53–77

Joshi V, Rathod MK (2019) Thermal performance augmentation of metal foam infused phase change material using a partial filling strategy: an evaluation for fill height ratio and porosity. Appl Energy 253:113621

Kee SY, Munusamy Y, Ong KS (2018) Review of solar water heaters incorporating solid-liquid organic phase change materials as thermal storage. Appl Therm Eng 131:455–471

Kousksou T, Bruel P, Cherreau G, Leoussoff V, El Rhafiki T (2011) PCM storage for solar DHW: from an unfulfilled promise to a real benefit. Sol Energy 85(9):2033–2040

Mahfuz MH, Anisur MR, Kibria MA, Saidur R, Metselaar IHSC (2014) Performance investigation of thermal energy storage system with Phase Change Material (PCM) for solar water heating application. Int Commun Heat Mass Transf 57:132–139

Mat S, Al-Abidi AA, Sopian K, Sulaiman MY, Mohammad AT (2013) Enhance heat transfer for PCM melting in triplex tube with internal–external fins. Energy Convers Manage 74:223–236
Mehling H, Cabeza LF, Hippeli S, Hiebler S (2003) PCM-module to improve hot water heat stores with stratification. Renew Energy 28(5):699–711
Meng ZN, Zhang P (2017) Experimental and numerical investigation of a tube-in-tank latent thermal energy storage unit using composite PCM. Appl Energy 190:524–539
Morales-Ruiz S, Rigola J, Oliet C, Oliva A (2016) Analysis and design of a drain water heat recovery storage unit based on PCM plates. Appl Energy 179:1006–1019
Nagano K, Ogawa K, Mochida T, Hayashi K, Ogoshi H (2004) Performance of heat charge/discharge of magnesium nitrate hexahydrate and magnesium chloride hexahydrate mixture to a single vertical tube for a latent heat storage system. Appl Therm Eng 24(2–3):209–220
Nallusamy N, Sampath S, Velraj R (2007) Experimental investigation on a combined sensible and latent heat storage system integrated with constant/varying (solar) heat sources. Renew Energy 32(7):1206–1227
Nayak KC, Saha SK, Srinivasan K, Dutta P (2006) A numerical model for heat sinks with phase change materials and thermal conductivity enhancers. Int J Heat Mass Transf 49(11–12):1833–1844
Oró E, De Gracia A, Castell A, Farid MM, Cabeza LF (2012) Review on phase change materials (PCMs) for cold thermal energy storage applications. Appl Energy 99:513–533
Patankar S (2018) Numerical heat transfer and fluid flow. CRC Press, Boca Raton
Prakash J, Garg HP, Datta G (1985) A solar water heater with a built-in latent heat storage. Energy Convers Manage 25(1):51–56
Reddy RM, Nallusamy N, Reddy KH (2012) Experimental studies on phase change material-based thermal energy storage system for solar water heating applications. J Fundam Renew Energy Appl 2
Sahoo SK, Rath P, Das MK (2016) Numerical study of phase change material based orthotropic heat sink for thermal management of electronics components. Int J Heat Mass Transf 103:855–867
Sahoo SK, Rath P, Das MK (2019) Solidification of phase change material nanocomposite inside a finned heat sink: a macro scale model of nanoparticles distribution. J Therm Sci Eng Appl 11(4):041005
Seddegh S, Wang X, Henderson AD, Xing Z (2015) Solar domestic hot water systems using latent heat energy storage medium: a review. Renew Sustain Energy Rev 49:517–533
Shaikh S, Lafdi K, Hallinan K (2008) Carbon nanoadditives to enhance latent energy storage of phase change materials. J Appl Phys 103(9):094302
Shrivastava A, Chakraborty PR (2019) Shell-and-tube latent heat thermal energy storage (ST-LHTES). In: Advances in solar energy research. Springer, Singapore, pp 395–441
Voller VR, Brent AD, Prakash C (1989) The modelling of heat, mass and solute transport in solidification systems. Int J Heat Mass Transf 32(9):1719–1731
Zhang Y, Faghri A (1996) Heat transfer enhancement in latent heat thermal energy storage system by using the internally finned tube. Int J Heat Mass Transf 39(15):3165–3173
Zhou D, Zhao CY (2011) Experimental investigations on heat transfer in phase change materials (PCMs) embedded in porous materials. Appl Therm Eng 31(5):970–977

Review on Thermal Performance Enhancement Techniques of Latent Heat Thermal Energy Storage (LHTES) System for Solar and Waste Heat Recovery Applications

Abhishek Agrawal and Dibakar Rakshit

1 Introduction

In the recent past, rising dependency on fossil fuels and rapid depletion of natural resources are the major concerns for sustainable development. However, utilization and storage of solar energy and waste heat recovery (WHR) seem to be the potential solutions to curb the usage of natural resources. Solar energy systems require thermal energy storage (TES) to fill the gap between energy supply and demand. Thermal energy can be stored in the form of thermochemical energy, sensible heat, or latent heat. Apart from the mentioned energy storage techniques, latent heat storage is possibly most effective heat storage technique due to high energy density during phase change (Jose 2016). However, materials can have latent heat in solid–liquid phase change, liquid–vapor phase change, and solid–solid phase change. Out of these, solid–liquid phase change is considered more efficient because liquid–vapor transitions undergo a high-volume variation while the solid–solid transitions have a very low latent heat of phase change (An article on "Test of Two Phase Change Materials for Thermal Energy Storage: Determination of the Global Heat Transfer Coefficient"). Hence, latent heat thermal energy storage (LHTES) using solid–liquid PCMs is a worthwhile option to store certain amount of energy at much smaller volume of material. However, most of the commonly used PCMs suffer low thermal conductivities and poor rates of thermal diffusivity. Thermal performance of a PCM can be improved by various heat transfer enhancement techniques such as addition of high thermal conductivity nanoparticles, insertion of metallic foams, addition of expanded graphite, and encapsulation of PCMs. Other than this, PCMs can also be

A. Agrawal
The Energy and Resources Institute, New Delhi, India

A. Agrawal · D. Rakshit (✉)
Indian Institute of Technology Delhi, New Delhi, India
e-mail: dibakar@iitd.ac.in

H. Tyagi et al. (eds.), *New Research Directions in Solar Energy Technologies*,
Energy, Environment, and Sustainability,
https://doi.org/10.1007/978-981-16-0594-9_15

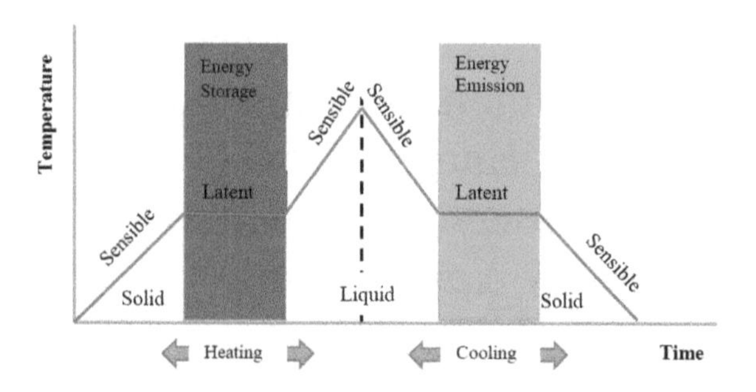

Fig. 1 Temperature versus time curve of PCM melting and solidification

used effectively by efficient design and operational parameters of heat exchangers such as the geometry of the thermal storage, effective heat exchanger design, selection of HTF, HTF inlet temperature, and mass flow rate. The common applications reviewed in the temperature less than 250 °C are solar air heaters, solar stills and solar domestic hot water systems, solar absorption cooling, waste heat recovery, and solar thermal electricity generation. This paper reviews and summarizes the various applications of PCM and common thermal performance enhancement techniques. Figure 1 shows the phase change transition of PCM.

2 Literature Review of PCMs

2.1 *Classifications of PCMs*

PCMs can broadly be classified as organic, inorganic, and eutectic PCMs based on their material properties. Figure 2 shows the different categories and types of PCMs. Thermal conductivity of organic PCMs is very low (0.1–0.7 W/m-K) having melting temperature in the range of 8–172 °C. Inorganic salts consist of salt hydrates and metallic. Thermal conductivity of organic salts is in the range of 0.43–1.6 W/m-K, and their melting temperature is in the range of 30–117 °C. Eutectic PCMs are

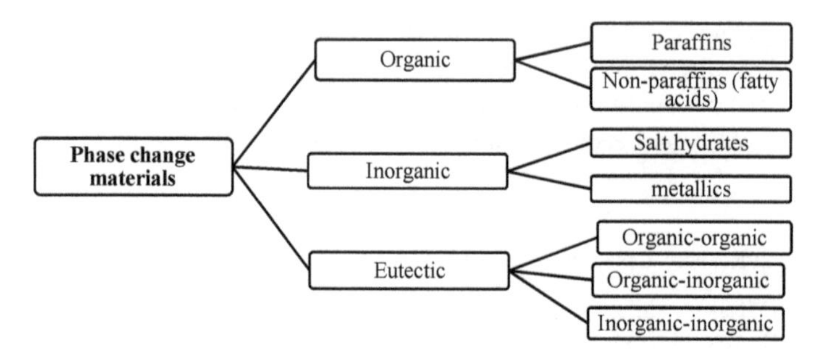

Fig. 2 Different types of PCMs

binary and ternary mixtures of inorganic salts. Melting temperature of eutectic PCMs is in the range of 25–250 °C while their thermal conductivity is in the range of 0.23–1.37 W/m-K (Cunha and Eames 2016). It can be easily observed that thermal conductivity of all type of PCMs is very low and should be improved for maximum heat storage and utilization.

2.2 *Desired Properties of Phase Change Heat Storage Materials*

- **Thermodynamic properties**: The PCM should possess high density and high latent heat of fusion per unit mass in order to store more energy in smaller volume of PCM. Specific heat and thermal conductivity of PCM should also be high for high sensible heat storage effects and high heat transfer rate, respectively. PCM should also possess congruent melting and small volume change during phase change in order to avoid complex heat exchanger designs.
- **Kinetic properties**: PCM should exhibit little or no super cooling during freezing;
- **Chemical properties**: PCM should be chemically stable, non-corrosive, non-poisonous, non-inflammable, and non-explosive. Moreover, PCM should not be chemically disposed to ensure more life of LHTES system (Fig. 3).

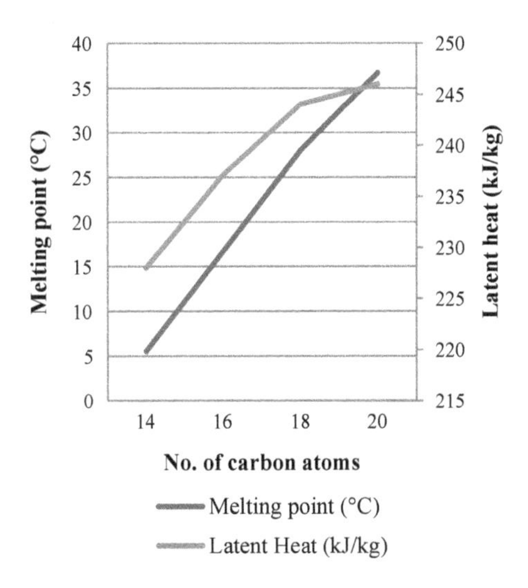

Fig. 3 Trend of latent heat and melting point of paraffin wax of different number of carbon atoms

2.3 Numerical Analysis of LHTES Systems

Numerical analysis of LHTES systems involves commonly used governing equations such as continuity, momentum and energy equations, energy balance equation, and heat generation equation. However, common assumptions should also be taken into consideration to apply governing equations during heat transfer analysis of LHTES systems. Designing of PCM storage tank (heat exchanger) plays very important role in order to charge and discharge the PCM efficiently. Therefore, PCM storage tanks should also be analyzed using mathematical equations.

2.3.1 Common Assumptions to Be Applying Governing Equations While Heat Transfer Analysis of LHTES Systems

- PCM is homogeneous and isotropic.
- HTF flow is laminar and incompressible.
- Inlet temperature and velocity of the HTF is constant.
- While melting and solidification, PCM should be solid phase and liquid phase, respectively.
- Thermophysical properties of the HTF, the tube walls of heat exchanger, and the PCM are constant.

2.3.2 Commonly Used Governing Equations

Continuity equation:

$$\frac{\partial \rho}{\partial t} + \nabla.(\rho V) = 0 \tag{1}$$

Momentum equation:

$$\rho \frac{\partial V}{\partial t} + \rho(V.\nabla)V = -\nabla P + \mu \nabla^2 V + \rho\alpha(T - T_{\text{ref}})g + S \tag{2}$$

where ρ is the PCM density, V is the velocity vector of PCM, P is the pressure, α thermal expansion coefficient, and μ is the dynamic viscosity of PCM.

Energy equation:

$$\frac{\partial \rho H}{\partial t} + \nabla.\rho V H = \nabla.\left(K_{\text{pcm}} \nabla T\right) \tag{3}$$

$$H = H_{\text{ref}} + \beta L + \int_{T_{\text{ref}}}^{T} C_p \text{d}T \tag{4}$$

where β = fraction of solid–liquid quantity present in PCM

$$\beta = \begin{cases} 0, & T < T_{\text{solidus}} \\ \frac{T - T_{\text{solidus}}}{T_{\text{liquidus}} - T_{\text{Solidus}}}, & T_{\text{solidus}} \leq T \leq T_{\text{liquidus}} \\ 1, & T > T_{\text{solidus}} \end{cases} \quad (5)$$

H enthalpy of PCM (sum of sensible enthalpy and latent heat)
H_{ref} reference enthalpy at reference temperature.

Energy balance equation:

$$\frac{\mathrm{d}E}{\mathrm{d}t} = q_{\text{gen}} - q_{\text{loss}} \quad (6)$$

Heat generation equation:

$$q_{\text{gen}} = mc_p \frac{\mathrm{d}T}{\mathrm{d}t} \quad (7)$$

$$q_{\text{conv}} = h_{\text{conv}} \times S \times (T_{\text{amb}} - T) \quad (8)$$

$$q_{\text{cond}} = \frac{k.A.\mathrm{d}T}{e} \quad (9)$$

where

S area exposed,
e is thickness.

Design equations for PCM storage tank

$$\Delta Q = m_{\text{PCM}} \times c_{\text{PCM}} \times \frac{\Delta T}{\Delta t} \quad \text{(Sensible cooling of PCM)}$$

$$\Delta Q = m_{\text{PCM}} \times L_{\text{PCM}} \times \frac{\Delta X_{\text{PCM}}}{\Delta t} \quad \text{(Solidifying of PCM);}$$

where

ΔX_{PCM} is liquid fraction of PCM.

$$\Delta Q = m_{\text{HTF}} \times L_{\text{HTF}} \times \Delta T \quad \text{(Sensible heating of HTF)}$$

$$\Delta Q = m_{\text{HTF}} \times L_{\text{HTF}} \times \frac{\Delta X_{\text{HTF}}}{\Delta t} \quad \text{(Boiling of HTF);}$$

ΔX_{HTF} is vapor fraction (quality) of water/steam.

$$\Delta Q = U \times \Delta A \times (T_{\mathrm{PCM}} - T_{\mathrm{HTF}})$$

$$\mathrm{NTU} = \frac{UA}{m_{\mathrm{HTF}} \times c_{\mathrm{HTF}}} \tag{10}$$

Heat loss coefficient of storage tank (h_L)

$$\frac{m_w \times C_{pw} \times (T_1 - T_2)}{t_1} = h_L \times A_s \times \mathrm{LMTD} \tag{11}$$

$$\mathrm{LMTD} = \frac{(T_1 - T_\infty) - (T_2 - T_\infty)}{\ln \frac{(T_1 - T_\infty)}{(T_2 - T_\infty)}} \tag{12}$$

Heat extraction rate from exhaust gases (Q_e)

$$Q_e = m_g c_{pg} \left(T_{g1} - T_{g2}\right) \tag{13}$$

Charging rate: **Ratio of total heat sored in a tank per unit time**

$$Q_c = \frac{m_w C_{pw} \times \left(T_f - T_i\right) + m_P C_{PP} \times \left(T_f - T_i\right) + m_P L}{t_1} \tag{14}$$

Heat transfer rate per unit length

$$\dot{Q} = \frac{2\pi k (T_A - T_B)}{\frac{k}{r_1 h_1} + \frac{k}{r_2 h_2} + \ln\left(\frac{r_1}{r_2}\right)} \tag{15}$$

Pumping power (P_p)

$$P_p = \frac{V \Delta p}{\eta} \tag{16}$$

where

V Volumetric flow rate (m^3/s),
Δp pressure drop,
η efficiency of pump.

Liquid volume fraction: The liquid volume fraction in the PCM, α, is a function of temperature,

$$\alpha = \begin{cases} 0, & T < T_{\text{solidus}} \\ \frac{T - T_{\text{solidus}}}{T_{\text{liquidus}} - T_{\text{Solidus}}}, & T_{\text{solidus}} \le T \le T_{\text{liquidus}} \\ 1, & T > T_{\text{solidus}} \end{cases}$$

Charging efficiency

$$\eta_c = \frac{Q_c}{Q_e} \tag{17}$$

Percentage energy saved

$$E_s = \frac{Q_c}{m_f \times CV} \tag{18}$$

2.4 Applications of PCMs in Low Temperature Range

Most of the PCMs in the low temperature range (<250 °C) are used for solar-centric applications such as solar air and water heating, solar absorption cooling, solar stills and solar thermal power generation. However, low temperature waste process heat from industries and exhaust waste heat from IC engines can also be recovered. The suitable PCMs in this temperature range are organic compounds (paraffin wax and fatty acids), salt hydrates, and eutectic mixtures. Figure 4 shows the classification of various reviewed low-grade heat recovery applications. Table 1 shows the list of low-temperature application-based industrial sectors where PCM can be used to store and release solar thermal energy.

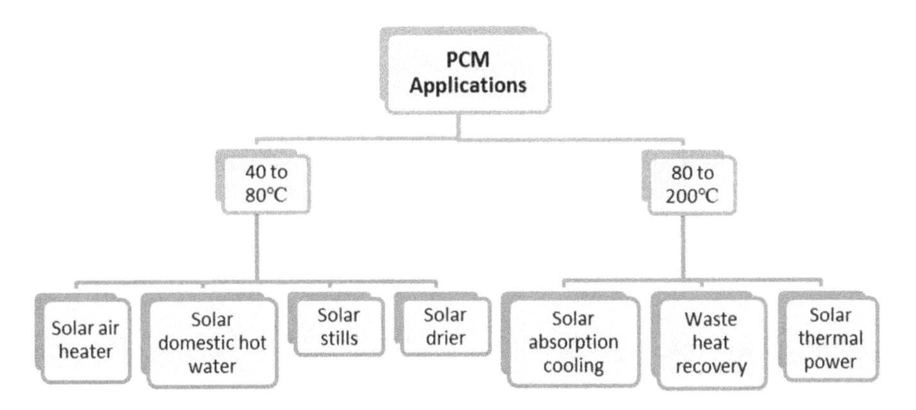

Fig. 4 Classification of various reviewed applications in the specified temperature range

Table 1 Type of industry and various processes with their respective temperature range (https://www.pluss.co.in/upload/application/plus10d82e_Pluss%20PCM%20in%20Solar%20Application.pdf)

Industry	Process	Temperature level [°C]
Food and beverages	Drying, washing, heat treatment	30–90
	Pasteurizing, boiling	80–110
	Sterilizing	140–150
Textile industry	Washing, bleaching, dyeing	40–160
Chemical industry	Boiling	95–105
	Distilling	110–300
Production industry	Preheating of boiler feed water, heating of production halls	30–100

2.4.1 Applications of PCMs in the Temperature Range of 40–80 °C

It is cleared from literature that, organic PCMs such as paraffin wax and fatty acids are suitable for this temperature range (Du et al. 2018). Key applications in this temperature range are discussed below.

Solar Air Heaters

In case of solar air heaters, heat energy stored in PCM during daytime can be utilized after sunset. Kabeel et al. investigated thermal performance of solar air heaters using paraffin wax with melting point (m.p.) 54 °C as PCM. It was found that, the outlet temperature of the v-corrugated plate solar air heater was higher than flat plate solar air heater (Kabeel et al. 2016).

Solar Stills

Faegh et al. used paraffin wax (m.p. 56 °C) as a PCM to experimentally investigate the solar still. Continuous desalination process even after the sunset with 86% increase in yield was noticed (Faegh and Shafii 2017). Kabeel et al. investigated that by introducing PCM in solar stills increases the productivity of fresh water by 1.4 times as compared to conventional solar still (Kabeel and Abdelgaied 2017).

Solar Water Heaters

Mahfuz et al. used paraffin wax (m.p. 56 °C) as a PCM to examine the performance of a solar water heater using shell and tube heat exchanger to exchange heat b/w saline water and PCM. It was investigated that, energy efficiency of solar heater increases with increase in mass flow rate of water as HTF (Mahfuz et al. 2014).

Solar Drier

Solar driers find potential use to dry agriculture/food products to make value added products and long-term preservation of food products. The suitable temperature range for drying such products is 40–75 °C. Hence, organic PCMs with melting temperatures (paraffin wax) are best suitable to store latent heat (Shalaby et al. 2014). Cakmak et al. observed the drying kinetics of seeded grapes in a solar dryer with PCM-based solar integrated collector and analyzed that drying time decrease with increase in air velocity of dryer (Çakmak and Yildiz 2011). Esakkimuthu et al. found that collector efficiency of solar drier increases with increase in mass flow rate of dry air (Esakkimuthu et al. 2013).

2.4.2 Applications of PCMs in the Temperature Range of 80–200 °C

In this temperature range, suitable PCMs are organic compounds, salts hydrates, and eutectic mixtures for LHTES systems. The key applications in this range are waste heat recovery, solar-based absorption chilling, and solar thermal power generation. Commonly used organic PCMs in this temperature range are erythritol, mannitol, and hydroquinone.

Solar Absorption Cooling

Agyenim et al. experimentally investigated that 100 L of erythritol (m.p. 117.7 °C) is enough to store thermal energy to operate LiBr/H_2O solar absorption cooling system (COP 0.7) for about 4.4 h in the space of 82 m^3 (Agyenim et al. 2011). Gil et al. used hydroquinone as a PCM to successfully drive a double-effect absorption chiller (working temperature range between 150 and 200 °C) (Gil et al. 2014). Fan et al. numerically investigated that 12.55 m^3 of hydroquinone as a PCM (m.p. 166–173 °C) is sufficient to provide 100 kW cooling load from double-effect solar absorption system without external energy supply (Fan et al. 2014).

Off-Site Waste Heat Recovery

Nomura et al. found that, with NaOH as PCM, 2.76 times more industrial waste heat can be stored and transported up to 35 km to a distillation tower of benzene, toluene, and xylene (BTX) as compared to sensible heat transportation system (Nomura et al. 2010).

I. C. Engine Exhausts WHR System

Around 35% of combustion heat is wasted out from exhaust of an IC engine which can be utilized for various purposes such as cold start of IC engine at low temperature, cabin heating, efficient combustion during warm up and to control high fuel consumption, etc. Stored heat of stationary diesel power plants can be utilized for various purposes such as air preheating, process water heating, and vapor absorption unit.

Pandiyarajan et al. designed, fabricated, and tested an integrated 7.4 kW diesel engine with PCM-based finned shell and tube tank of 20 MJ capacity. The system recovers 10–15% of exhaust waste heat with around 99% effectiveness of the heat recovery heat exchanger (HRHE) at the end of charging process at all load conditions (Pandiyarajan et al. 2011).

Subramanian et al. developed an experimental setup of diesel engine (80 mm bore, 110 mm stroke, and 3.7 kW rated power at 1500 rpm), coupled to hydraulic dynamometer, heat recovery heat exchanger (HRHE), and a thermal storage system. The concept of combined sensible and latent heat storage system is used. For automobiles, this heat can be utilized during cold start, cabin heating (Subramanian et al. 2004). HRHE fitted into exhaust pipe of engine. Water is taken as sensible heat storage medium and Paraffin (filled in LDPE spherical capsules) as a latent heat storage medium (PCM). Thermal storage tank consists of 5.2 kg paraffin and 8.5 kg of water.

Properties of Paraffin used

- Latent heat of fusion: 214 kJ/kg
- Specific heat capacity: 2.9 kJ/kg-K
- Thermal conductivity: 0.2 W/m K
- Density (solid): 850 kg/m^3
- Density (liquid): 775 kg/m^3
- Melting temperature range: 60–70 °C.

Observations recorded (at no load, 30%, 60%, and at full load)

- Exhaust gas temperature at inlet and outlet of HRHE
- Temperature of water at the inlet and outlet of storage tank
- Temperature at various locations inside the storage tank
- Airflow rate and fuel flow rate of the engine.

Solar Thermal Electricity Generation

Pirasaci and Goswami used NaCl–$MgCl_2$ eutectic mixture as LHTES for direct steam generation (DSG). The important parameters to evaluate the effectiveness of storage system were flow rate of HTF, length and diameter of storage tube (Pirasaci and Goswami 2016).

Tamme et al. established design concepts for solar thermal power generation system using PCM as latent thermal energy storage medium. It was found that thermal conductivity of PCM can be increased from 0.5 to 20 W/m-K while adding graphite as composite. Researchers observed that, PCM for DSG technologies is more advantageous as compared to sensible heat storage systems because of high storage density and compact storage size of PCM storage systems (Tamme et al. 2008).

2.4.3 Market Survey of PCM in Solar Applications (https://www.pluss.co.in/upload/application/plus10d82e_Pluss%20PCM%20in%20Solar%20Application.pdf)

It has been observed that, manufacturing companies are working on development of effective, efficient, and cost-effective phase change materials. This section describes some of the PCM-based solar appliances whose commercialization is still in a nascent stage.

1. **Solar water heaters**: A PCM manufacturing company "Pluss Polymers" claims their own developed hydrated salt "savE HS58" as PCM to use solar water heater effectively. The design helps to warm water even during night. The product is yet to come in the market. Figure 5 shows the schematic of PCM-based solar water heater
2. **Solar air conditioning**: "Solar Energy Centre" in collaboration with "Thermax Limited, Pune" proposing highly efficient and cost-effective 100 kW solar air conditioning systems to meet the increasing air conditioning demand. The present system will cater to air conditioning needs of 13 rooms of Solar Energy Centre with the help of 288 m^2 of solar collector area which generates nearly 60 kW of 210 °C pressurized hot water. This heat is used in vapor absorption machine to generate 7 °C chilled water which in turn circulates through the fan coil installed in all rooms.
3. **Fruits and Vegetable drying**: Drying of various fruits and vegetables for their long-term use is a promising application which can be fulfilled efficiently by the use of PCM-based solar driers which can cater continuous supply of heat energy.
4. **Concentrated Solar Power for generation of electricity**: A European international company "Abengoa solar" is trying to develop a PCM-based system to effectively utilize the solar energy for electricity generation during night time.

3 Different Methods to Effectively Utilize the PCMs

PCMs can be effectively used in two ways

- By improving their thermal conductivity through different methods

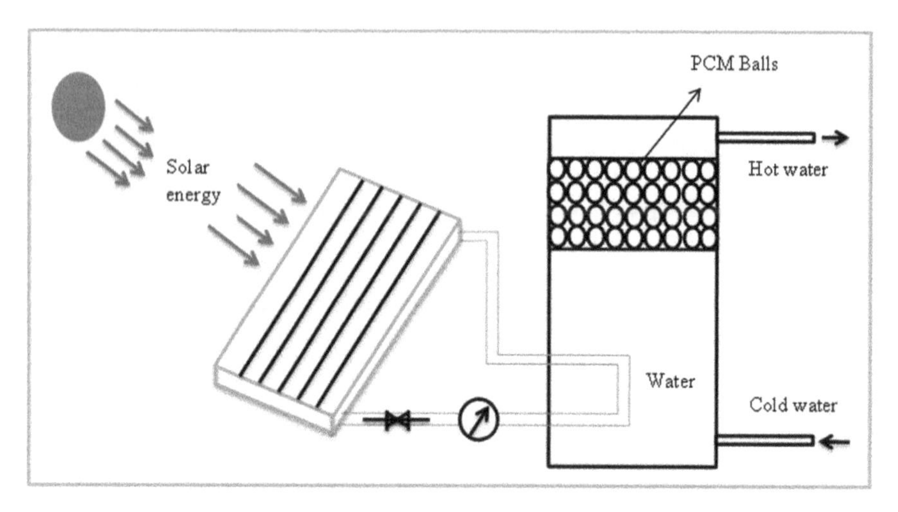

Fig. 5 PCM-based solar water heater

- By efficient design and operational parameters such as selection of HTF, HTF inlet temperature, mass flow rate of HTF, geometry/shape of the thermal storage unit, and effective heat exchanger design.

This section describes both the ways to efficiently use the PCMs.

3.1 Heat Transfer Enhancement Techniques to Improve the Thermal Performance of PCMs

The major drawback of most of the PCMs is their low conductivity which results in sluggish heat exchange and low charging and discharging rate. This limits the use of PCMs for practical applications. To overcome this problem, thermal conductivity of PCM can be enhanced by many ways such as insertion of highly thermally conductive nanoparticles, metallic foams, expanded graphite and encapsulation of PCM.

3.1.1 Addition of High Thermal Conductivity Nanoparticles

Some of the ordinarily used nanoparticles are carbon fibers, graphene nanoplatelets (GnPs), and carbon nanotubes. These particles are highly suitable to enhance the thermal conductivity of PCMs because of their low density, better stability, and easy dispersion in PCM.

Cui et al. added 10% by weight of carbon nanofiber (CNF) and carbon nanotube (CNT) as nanoparticles to enhance the thermal conductivity of Soy wax (m.p. 52–54 °C) from 0.324 W/m-K to 0.469 and 0.403 W/m-K respectively (Cui et al. 2011).

Harish et al. found that addition of 1 vol% of graphene nanoplatelets (GnP) in lauric acid increases the thermal conductivity of CPCM by 2.3 times. The thermal conductivity of PCM is measured by transient hot wire method (Harish et al. 2015). Min Li et al. inserted 10 wt% of nanographite (NG) in paraffin and investigated the thermal conductivity through electron microscope. It was observed that thermal conductivity of paraffin increased significantly up to 7.41 times (Li 2013). Srinivasan et al. added 3.5% by volume of graphite particles in pure eicosane, thermal conductivity of CPCM increased by 4.5 times while measured using KD2 pro thermal properties analyzer (Srinivasan et al. 2017). Mehrali et al. added 5 wt% nitrogen-doped graphene (NDG) as a nanoparticle in palmitic acid ($k = 0.29$ W/m-K) and measured thermal conductivity using laser flash method and observed five times increase in thermal conductivity of composite as compared to pure PCM (Mehrali et al. 2014). Yang et al. prepared composite by adding 4% graphene oxide (GO) and 30% boron nitride (BN) in polyethylene glycol (PEG) as PCM and observed nine times increase in thermal conductivity as compared to pure PEG (Yang et al. 2018). Mehrali et al. used palmitic acid as PCM ($k = 0.29$ W/m-K) and inserted graphene nanoplatelets (GNPs) having specific surface area of 750 m^2/g to absorb 91.94% of palmitic acid by weight. CPCM was prepared with the help of vacuum impregnation method and eight times increase in thermal conductivity has been observed by inserting CPCM (Mehrali et al. 2013). Singh et al. experimentally and numerically investigated the solidification behavior of medium temperature (160–200 °C) binary eutectic PCM (equal proportion of lithium nitrate and potassium chloride) in vertical finned thermal storage system. Solidification time reduces to 49% and thermal conductivity increased by 1.97 times with finned storage system using 5% GnP as compared to conventional storage system. Properties were evaluated using rotational rheometer and DSC (Singh et al. 2018a).

3.1.2 Thermal Conductivity Enhancement Using Metallic Foams

Due to high thermal conductivity of metallic foams, these are widely used to enhance the thermal conductivity of different PCMs. Commonly used metallic foams are nickel, copper, and graphite foams.

Xiao et al. inserted metallic foams in paraffin by "Vacuum impregnation method" and found that thermal conductivity of CPCM increased by 3 times and 15 times when nickel foam and copper foams with 5 PPI and 97% porosity are inserted, respectively (Xiao et al. 2013). Huang et al. recorded 2.8 times and 8.5 times increase in thermal conductivity of CPCM when nickel and copper foams, respectively, are inserted in myristyl alcohol (Huang et al. 2017). Tao et al. prepared CPCM by inserting porous graphite foam into liquid paraffin ($k = 0.22$ W/m-k) and result showed that thermal conductivity of CPCM increase 22.6 times as compared to pure paraffin (Tao et al. 2017). Wang et al. embedded paraffin ($k = 0.065$ W/m-k) into copper foam having pore size 8–12 PPI and 97% porosity. Thermal conductivity of composite formed was increase by around 48 times at temperature of 48 °C (Wang et al. 2016). Hussain et al. prepared the composite of paraffin ($k = 0.2$ W/m-k) and nickel foam with different pore sizes and porosities. Result depicted that at pore size of 20 PPI and

97% porosity, thermal conductivity of CPCM magnified 5.8 times as compared to pure PCM (Hussain et al. 2016).

3.1.3 Thermal Conductivity Enhancement by Adding Expanded Graphite (EG)

In this section, a review of selected research studies has been made to understand the effect of addition of expanded graphite for thermal conductivity enhancement.

Chang et al. added 3 wt% of EG into paraffin ($k = 1.04$ W/m-k) and thermal conductivity of CPCM formed was increased by 2 times (Cheng et al. 2018). Wu et al. prepared CPCM by mixing 20% expanded graphite with paraffin ($k = 0.268$ W/m-K) and it was found that thermal conductivity of CPCM enhanced 28.5 times as compared to pure paraffin (Wu et al. 2015).

3.1.4 Encapsulation of PCM

Encapsulation of PCM in stable organic material having high thermal conductivity can raise the thermal conductivity of PCM and problem of PCM leakage during solid-to-liquid phase change can also be solved. Commonly used shells for encapsulation of PCMs can be polyurea, urea formaldehyde, and melamine formaldehyde resin.

Qureshi et al. summarized the thermal conductivity enhancement technique through encapsulation of PCM. Thermal conductivity of paraffin ($k = 0.3$ W/m-K) magnified 3.87 times when encapsulated in a shell of SiO_2 and GO. Similarly, thermal conductivity of n-octadecane ($k = 0.15$ W/m-K) enhanced by 4.14 times and 11.2 times when encapsulated in a shell of SiO_2 and $CaCO_3$, respectively (Qureshi et al. 2018).

4 Review of PCM-Based Heat Exchangers

4.1 Introduction

A basic heat exchanger consists of tube with one fluid running through it and another fluid flowing by on the outside. Heat transfer through any heat exchanger consists of three typical processes. Initially, heat transfer from the fluid takes place to the inner wall of tube by convection. After that, tube walls of heat exchanger transfer the heat through conduction and finally, convective heat transfer takes place from tube walls to outer fluid.

As we know, melting of PCM causes density changes and hence create buoyant forces. These forces result in natural convection in the melted PCM, which affects the heat transfer. NTU method ignores the effect of natural convection while designing the heat exchanger. Further, ignoring the effect of natural convection causes the under estimation of effectiveness values during charging and discharging of PCM. PCM-based heat transfer is a combination of convection and conduction. Heat is dominantly transferred through conduction during melting and through convection after melting (Hosseini et al. 2014). Natural convection takes place during solid–liquid phase change because of buoyancy forces created due to density difference. In ε-NTU method of heat exchanger design, effect of natural convection is overlooked, which results in imprecise effectiveness values during charging/discharging processes (Tay et al. 2012).

This section reviews the commonly used PCM heat exchangers suitable for low temperature range of less than 250 °C such as shell and tube, triplex tube heat exchanger (TTHX) and polymeric hollow fiber heat exchangers (PHFHE).

4.2 Types of PCM-Based Heat Exchangers

4.2.1 Shell and Tube Heat Exchanger

Literature review reveals that most common geometry of PCM containers is shell and tube heat exchanger employed with cylindrical pipes. This is due to the fact that heat loss from such system is minimal (Agyenim et al. 2010). Shell and tube heat exchanger is suitable to exchange heat between two fluids having extreme temperature difference. These heat exchangers are also best suited to the condition, where low pressure loss is required (https://www.thermaxxjackets.com/plate-and-frame-heat-exchangers-explained/). Hosseini et al. reported that an increase in fins' height of longitudinal fins of a shell and tube heat exchanger enhances the melting rate of PCM (Hosseini et al. 2015). Esapour et al. reported that the increase in the number of tubes from 1 to 4 reduces the melting time of the PCM (RT35) by 29% (Esapour et al. 2016). Hakeem et al. experimentally investigated phase change in pipe, cylindrical and shell and tube heat exchanger. It was found that, for the same mass of PCM and surface area of heat transfer, shell and tube model takes the minimum time for melting the PCM (Niyas and Muthukumar 2013).

4.2.2 Triplex Tube Heat Exchanger (TTHX)

This type of heat exchanger consists of three concentric tubes. In general, for latent heat storage applications, the center tube is filled with PCM while the internal and external tubes carry HTFs. TTHX can be used for solar water heating and waste heat recovery applications. The TTHX has high storage density and large heat transfer area in order to reduce heat transfer resistance of the solid PCM zone. However, simultaneous charging and discharging (SCD) of a PCM is possible in TTHX (Al-Abidi et al. 2013). Mat et al. investigated the heat flux of HHTX and found that increase in heat flux increases melting rate (Mat et al. 2013). Jian et al. investigated that melting rate of PCM in TTHX increase with increase mass flow rate of HTF (Jian 2008).

4.2.3 Polymeric Hollow Fiber Heat Exchangers (PHFHEs)

PHFHE employs thin-wall polymeric fibers for separation of heat transfer fluids. Low thermal conductivity of PCMs can be compensated by large surface area density (around 1400 m^2/m^3) of PHFHEs (Zarkadas and Sirkar 2004).

Due to lighter in weight as compared to traditional metal heat exchanger, cost of manufacturing of PHFHEs is low. PHFHE is well suited for temperature applications below 200 °C (Krasny et al. 2016). Due to very low inner diameter (in the range of 10^{-3} m) of polymeric hollow fibers (mini channels), fluid flow is characterized by low Reynolds numbers. Therefore, heat discharge rate of PCM is reduced with reduction in fiber diameter of heat exchanger. However, very small fiber diameters may lead to higher pressure drop. Therefore, more pumping power will be required for same water flow rate. Hence, fiber diameter should be in optimal range in order to achieve high heat transfer rate (Hejčík et al. 2016).

4.2.4 Spiral Coil Heat Exchangers

The main features of spiral coil heat exchangers are its compactness, less fouling, easier cleaning, high heat transfer coefficient. Heat transfer in case of spiral coil heat exchanger is better as compared to other straight tube heat exchangers due to "swirling" motion created by tubes of uniform cross section within the HTF. The swirling motion of flowing HTF creates turbulence in the fluid at much lower velocity as compared to other heat exchangers such as shell and tube, cylindrical, and triplex tube heat exchangers. However, fluid travels with constant velocity throughout the heat exchanger unit. In spiral coil heat exchanger, heat transfer rate is directly proportional to the coil diameter. With the help of k-ε model for handling turbulence, Kumar et al. analyzed that in spiral-coiled heat exchanger, Reynolds number and heat transfer rate increases with increase in mass flow rate of HTF while Euler number decreases with increase in mass flow rate of HTF (Kumar et al. 2017).

Nusselt number and pressure drop for a spiral coil are higher than the straight tube. Rahimi et al. experimentally analyzed that spiral-coiled heat exchanger is superior in terms of heat transfer as compared to other types of heat exchangers. It has also been observed that, melting process accelerates and by increase in the helical diameter of the HTF tube, PCM absorbs more power (Rahimi et al. 2019).

Thermal design of spiral coil heat exchanger (Khorshidi and Heidari 2016)

$$\dot{m} = \rho \times a \times v \tag{19}$$

$$A = \pi r^2 \tag{20}$$

$$\text{Re} = \rho \times v \times \frac{d}{\mu} \tag{21}$$

$$\text{Eu} = \frac{2 \times \Delta P}{\rho \times v^2} \tag{22}$$

$$Q = U \times A \times F \times \Delta T_{\text{LMTD}} \tag{23}$$

$$U = \frac{1}{\frac{1}{h_h} + \frac{t_w}{k_w} + \frac{1}{h_c} + r_i + r_o} \tag{24}$$

where F is a correction coefficient, U is overall heat transfer coefficient, Eu is Euler number, h_c and h_h are heat transfer coefficients of the cold and hot fluids, respectively, k_w is the conductivity of wall (heat transfer surface), r_i and r_o are the fouling resistances and t_w is the wall thickness.

For laminar flow in radial direction:

$$h = 1.86 * c * G * \text{Re}^{-\frac{2}{3}} * \left[\frac{L}{D_e}\right]^{-\frac{1}{3}} \left[\frac{\mu_f}{\mu_b}\right]^{-0.14} \tag{25}$$

For turbulent flow in radial direction:

$$h = \left(1 + 3.54 * \frac{D_e}{D_h}\right) * 0.023 * c * G * \text{Re}^{-0.2}\,\text{Pr}^{-\frac{2}{3}} \tag{26}$$

where

- D_e equivalent diameter,
- D_h spiral diameter,
- L length of heat transfer surface,
- μ viscosity,
- G mass flow rate.

$$\text{Re} = \frac{G}{\mu} \tag{27}$$

$$G = \frac{m}{H.s} \tag{28}$$

$$D_e = \frac{2Hs}{H + s} \approx 2s \tag{29}$$

As $H \gg s$.

H height of heat exchanger.
s distance between surfaces in each channel.

4.3 *Geometries of PCM-Based Latent Heat Storage Containers*

Cylindrical geometries are the most likely for the devices of commercial heat exchangers. In general, cylindrical latent heat storage containers are of four types.

1. **Pipe model**: PCM fills the shell and the HTF flows through a single pipe.
2. **Cylindrical model**: PCM fills the tube and the HTF flows parallel to the tube.
3. **Shell and tube model**: PCM fills the tube, while HTF flows through shell or PCM fills the shell, while HTF flows through tube.
4. **Triplex tube heat exchanger (TTHX)**: Center tube contains PCM while flow of HTF occurs through the internal and external tubes.

Research shows that, performance of shell and tube heat exchanger in terms of charging time is best as compared to pipe model and cylindrical model. Niyas et al. show that for complete melting of same quantity of paraffin RT 50 (m.p. 48 °C) in shell and tube model takes least time (1860 s) when compared with charging time of pipe model (4425 s) and cylinder model (2319 s). Design software used for thermal modeling is CFD and COMSOL (Niyas and Muthukumar 2013).

Many researchers have investigated different parameters for different configurations of heat exchangers. Findings in terms of improvement in charging/discharging rates, effect of HTF mass flow rate, PCM melting behavior, effect of HTF inlet temperature, and effect of change in various dimensionless numbers such as Reynolds number and Stefan number have been observed and reported. Table 2 shows the parameters investigated and finding of different geometries of PCM-based heat exchangers. Tables 3 and 4 shows the longitudinal and circular fin enhancement techniques, respectively, in PCM-LHTES systems.

4.4 *Effect of Various Dimensionless Numbers on Melting and Solidification of PCM*

1. **Effect of Grashof Number**

As buoyancy forces play major role in flow of PCM in the melt region, therefore, Grashof number is important for natural convection flows (Rahimi et al. 2019).

$$\mathrm{Gr} = \frac{g\beta\left(T_w - T_{\mathrm{PCM},m}\right)L^3}{\nu^2} \tag{30}$$

where

Table 2 Different configuration of cylindrical PCM heat exchangers

Author	Geometry of heat exchanger	Parameters investigated	Type of study[a]	Findings
Hosseini et al. (2014)	Shell and tube	Inlet temperature of HTF	E/N	Heat transfer rate and melt fraction increases with increase in HTF inlet temperature
Akgun et al. (2008)		Mass flow rate and inlet temperature of HTF	N	Improving the melting and solidification rate
Yazici et al. (2014)		Heat flux	E	Enhancing the PCM melting behavior
Jesumathy et al. (2012)	Concentric annulus	Re and HTF mass flow rate	E	Melt fraction and heat transfer rate increases with increase in Reylonds number
Mat et al. (2013)	Triplex tube	Heat flux	N	Increase in heat flux increases melting rate
Jian (2008)		HTF mass flow rate and inlet temperature	E/N	Increase in mass flow rate increases melting rate
Al-Abidi et al. (2013)		Heat flux	N	Amount of PCM melt decreases with decrease in heat transfer rate

[a]*E* experimental, *N* numerical

β thermal expansion coefficient (1/K).
L Characteristic length [gap space between shell and tube, in case of shell and tube heat exchanger].
ν kinematical viscosity (m^2/s).
T_w HTF inlet temperature.
$T_{PCM,m}$ mean melting temperature of PCM.

2. **Effect of Reynolds Number (Re) on Melting Time**

As Reynolds number depends on mass flow rate of HTF, which means, increase in mass flow rate of HTF increases Reynolds number and hence increase in enthalpy flow. However, high mass flow rate of PCM also results in high pumping power. Therefore, optimum mass flow rate and hence optimum Reynolds number should be preferred for efficient thermal energy storage (Akgun et al. 2008). However, Paria

Table 3 Longitudinal fin enhancement techniques in PCM-LHTES system

Authors	Geometry of heat exchanger	Parameters investigated	Type of study[a]	Findings
Al-Abidi et al. (2014)	Triplex tube	HTF mass flow rate and inlet temperature	E	Melting of PCM is influenced more by HTF inlet temperature as compared to HTF mass flow rate
Mat et al. (2013)		Heat flux	N	Melting rate of PCM increase with longitudinal fins and also increase due to increase in length of fins and Stefan number
Velraj et al. (1997)	Cylinder-shaped tube	Heat flux	E	Cylindrical tube increases cooling rate and thus reduces solidification time
Castell et al. (2008)	Shell and tube	Heat flux	E	Improving cooling rate because of fins
Zhao and Tan (2015)		HTF mass flow rate	N	Longitudinal fins improve phase change rate
Wei et al. (2010)	Rounded tube	Heat flux	E	Circular tube maintains uniform temperature distribution of PCM

[a]*E* experimental, *N* numerical

et al. experimentally investigated that solidification and melting time decrease with increase in Reynolds number (Paria et al. 2015).

3. **Effect of Stefan Number (Ste) on Melting Time**

Akgun et al. experimentally investigated that melting time of PCM decreases with increase in Stefan number (Akgun et al. 2008).

$$\mathrm{Ste} = \frac{C_{pl}\Delta T}{\lambda} \tag{31}$$

For a specific PCM, rise in the inlet temperature of the HTF leads to an increase in the Stefan number. However, high HTF inlet temperature means higher energy input. Therefore, HTF inlet temperatures should be in optimum range for efficient thermal energy storage.

Table 4 Circular/annular fin enhancement techniques in PCM-LHTES system

Authors	Geometry of heat exchanger	Parameters investigated	Study type[a]	Outcomes
Mosaffa et al. (2012)	Shell and tube heat exchanger with circular fins	Mass flow rate and inlet temperature of HTF	A	Cylindrical storages solidifies PCM more quickly as compared to rectangular containers
Lacroix (1993)	Shell and tube/annular fins	HTF mass flow rate and inlet temperature	N	Increasing the PCM energy stored
Choi and Kim (1992)			E	Improving cooling rate at all radial PCM points
Seeniraj et al. (2002)		HTF exit temperature	N	Increasing the PCM energy stored
Zhang and Faghri (1996)		Operating temperature	N	Liquid fraction increases with increase in length and number of fins
Singh et al. (2018b)		Melting phenomenon with different concentration of GnP	N	Maximum volume fraction of GnP is 5% and fins should take up to 50% of annular space for enhanced heat transfer

[a]*A* analytical, *E* experimental, *N* numerical

4.5 *Effect of Geometry on Melting and Solidification (Phase Change)*

Singh et al. experimentally and numerically evaluated the melting performance of a multistage solar cooling system consist of finned conical shell in tube storage system along with different concentration of GnP. The system was encapsulated with nanoenhanced phase change material to improve the heat storage performance. 57% reduction in charging time was observed using the conical shell in tube storage system as comparison to conventional cylindrical system. It was also analyzed that effect of inserting fins on heat transfer enhancement is more as compared to dispersion of nanoparticles (Singh et al. 2019).

4.6 Effect of Change in Dimensions on Melting and Solidification (Phase Change) (Singh et al. 2019)

- **Effect of change in tube length:** Melting and solidification of phase change material are proportional to the tube length.
- **Effect of change in Shell diameter:** Shell diameter reduction increase the discharging and charging rate.

5 Concept of Multiple PCM Configurations

Use of multiple PCM configurations is gaining interest among researchers as it increases rate of heat transfer during charging and discharging (Seeniraj and Narasimhan 2008; Tian and Zhao 2013). By arranging the PCMs in a decreasing order of their melting points, nearly a constant temperature difference can be maintained to uphold uniform heat transfer rate. Chiu and Martin (2013) analyzed faster discharging and charging time by using multiple PCM configurations. However, mentioned characteristics features of multiple PCM configurations are mostly proven based on numerical studies.

Peiro et al. experimentally evaluated hydroquinone and d-mannitol with melting temperatures between 150 and 200 °C. Results revealed that uniform HTF temperature difference can be obtained by introducing cascaded PCM configuration. Moreover, cascaded PCM system also enhances the effectiveness by 19.36% when compared with single PCM system (Peiro et al. 2015).

6 Inferences

- Thermal performance of LHTES system can be boosted by experimenting different geometries of heat exchangers, by inserting different geometrical shapes fins and by inserting nanocomposites PCM. Configurations of heat storage systems such as vertical, horizontal, and tilted geometries at different angles also affect the thermal performance of PCM and should be investigated numerically and experimentally.
- The method of heat transfer in the PCM is a combination of convection and conduction, but in the melting process, initially, heat transfer is dominated by conduction and after melting of PCM (even partially), convective heat transfer start dominating.
- Out of all commonly used configurations, performance of cylindrical storage system in terms of solidification rate of PCM is better.

- Amount of nanoparticles to be inserted in PCM should be optimized as very high concentration of nanoparticles causes agglomeration, increases the effective viscosity and convective heat transfer.
- Charging time reduces with increase in HTF inlet temperature and increase in HTF mass flow rate. However, heat transfer inside a PCM is almost unaffected by air inlet pressure.
- Concentric cylinder-based multilayer heat exchangers can be used to fill diverse PCMs with different melting points so that maximum heat can be stored and released uniformly during charging and discharging, respectively.
- Thermal performance enhancement using longitudinal fins is best as compared to other configurations of fins such as circular/annular, plate and pin fins.
- Very limited experimental research has been conducted on use of multiple PCM (Cascade) system.

7 Research Gaps

Heat exchange process in existing storage systems is not uniform. To overcome this, combined latent and sensible heat storage system can be introduced. There is lack of information about performance of PCM after its multiple usages for heat transfer. Very limited research information is there about the phenomenon of simultaneous charging and discharging (SCD) of PCM. It is generally assumed that the PCM start exchanging the heat only after its complete charging/discharging which may not be the case in many real-life applications such as solar water heating for domestic application. Consecutive charging/discharging and simultaneous charging/discharging (Murray and Groulx 2014a) of a vertical cylindrical storage system for domestic water heating were experimentally studied by Murray and Groulx (2014b). However, effect of natural convection is not clearly justified in this study. There exists good scope of research to amplify the thermal performance of LHTES system using multi-tubes instead of single tube with nanocomposites PCM for horizontal, vertical, and various tilted configurations numerically. Different permutation and combinations of different composites can be tried to further improve the thermal conductivity of PCMs. However, ratio of doping of different composites in PCMs also changes the thermal conductivity of PCMs and hence different cases should be analyzed numerically and experimentally. Various applications where there is a huge scope of heat recovery such as biomass gasification should be coupled with PCM-based energy storage systems to increase the overall efficiency of the system. Further research is still required to explore the possible geometrical designs of fins to enhance the heat transfer rate. Very few experimental analyses have been done to validate the numerical studies related to use of multiple PCM (cascade) system to reduce charging and discharging time and to maintain uniform charging and discharging. It has been found that the results of numerical and experimental analyses are varied due to limitations of modeling tools. Hence, more focus should be imposed to carry out experimental analysis of multiple PCM configurations with different storage geometries.

8 Conclusions

An extensive review has been made to study the phase change materials. It includes desirable properties of PCMs and applications of PCM to recover low-grade waste heat having temperature less than 200 °C. Most of the applications of PCM in this temperature range are solar centric such as solar air and water heating, solar absorption cooling, solar stills and solar thermal power generation. However, waste heat recovery from industrial heat and IC engines is also the lucrative applications to utilize waste energy. Most of the PCMs in the temperature of 40–80 °C are organic materials like paraffin and fatty acids. Literature survey reveals that, most of the PCMs suffers from low thermal conductivity which leads to low heat transfer rate and limits their usage in most of the practical applications. However, thermal conductivity of the PCMs can be increased by many techniques such as insertion of high thermal conductivity nanoparticles, addition of metallic foams, adding of expanded graphite, and microencapsulation of PCM. Up to 48 times increase in thermal conductivity is recorded through these thermal conductivity enhancement techniques. Heat transfer performance of PCMs can also be improved by effective design of heat exchanger and its components. Hence, study of commonly used heat exchangers in the defined temperature range has also reviewed. Effect of mass flow rate and inlet temperature of HTF, dimensionless numbers, geometry and dimensions of heat exchanger components are also reviewed. Melting time reduces with increase in HTF inlet temperature and HTF mass flow rate. Very limited information is there about phenomenon of simultaneous charging and discharging (SCD) of PCM and performance of PCM after its multiple usages for heat transfer which are the practical cases of many real-life applications. However, performance of PCM-based LHTES systems can be further improved by experimenting different combinations of CPCMs and different geometries of heat exchangers. Hence, there exists the potential scope for research on these areas.

References

Agyenim F, Hewitt N, Eames P, Smyth M (2010) A review of materials, heat transfer and phase change problems formulation for latent heat thermal energy storage systems (LHTESS). Renew Sustain Rev 14:615–628

Agyenim F, Eames P, Smyth M (2011) Experimental study on the melting and solidification behavior of a medium temperature phase change storage material (Erythritol) system augmented with fins to power a LiBr/H_2O absorption cooling system. Renew Energy 36(1):108–117

Akgun M, Aydin O, Kaygusuz K (2008) Thermal energy storage performance of paraffin in a novel tube-in-shell system. Appl Therm Eng 28:405–413

Al-Abidi AA, Mat S, Sopian K, Sulaiman M, Mohammad AT (2013) Numerical study of PCM solidification in a triplex tube heat exchanger with internal and external fins. Int J Heat Mass Transf 61:684–695

Al-Abidi AA, Mat S, Sopian K, Sulaiman MY, Mohammad AT (2014) Experimental study of melting and solidification of PCM in a triplex tube heat exchanger with fins. Energy Build 68:33–41

An article on "Test of two phase change materials for thermal energy storage: determination of the global heat transfer coefficient"

Çakmak G, Yildiz C (2011) The drying kinetics of seeded grape in solar dryer with PCM-based solar integrated collector. Food Bioprod Process 89(2):103–108

Castell A, Sole C, Medrano M, Roca J, Cabeza LF, Garcia D (2008) Natural convection heat transfer coefficients in phase change material (PCM) modules with external vertical fins. Appl Therm Eng 28(13):1676–1686

Cheng WL, Li WW, Nian YL, Xia W (2018) Study of thermal conductive enhancement mechanism and selection criteria of carbon-additive for composite phase change materials. Int J Heat Mass Transf 116:507–511

Chiu JNW, Martin V (2013) Multistage latent heat cold thermal energy storage design analysis. Appl Energy 112:1438–1445

Choi JC, Kim SD (1992) Heat-transfer characteristics of a latent heat storage system using $MgCl_2{\cdot}6H_2O$. Energy 17(12):1153–1164

Cui Y, Lui C, Hu S, Yu X (2011) The experimental exploration of carbon nanofiber and carbon nanotube additives on thermal behavior of phase change materials. Sol Energy Mater Sol Cells 95(4):1208–1212

Cunha JPD, Eames P (2016) Thermal energy storage for low and medium temperature applications using phase change materials—a review. Appl Energy 177:227–238

Du K, Calautit J, Wang Z, Wu Y, Liu H (2018) A review of the applications of phase change materials in cooling, heating and power generation in different temperature ranges. Appl Energy 220:242–273

Elmeriah A, Nehari D, Aichouni M (2018) Thermo-convective study of a shell and tube thermal energy storage unit. Periodica Polytech Mech Eng 62(2):101–109

Esakkimuthu S, Hassabou A, Palaniappan C, Spinnler M, Blumenberg J, Velraj R (2013) Experimental investigation on phase change material based thermal storage system for solar air heating applications. Sol Energy 88:144–153

Esapour M, Hosseini MJ, Ranjbar AA, Pahamli Y, Bahrampoury R (2016) Phase change in multi-tube heat exchangers. Renew Energy 85:1017–1025

Faegh M, Shafii MB (2017) Experimental investigation of a solar still equipped with an external heat storage system using phase change materials and heat pipes. Desalination 409:128–135

Fan Z, Infante Ferreira CA, Mosaffa AH (2014) Numerical modelling of high temperature latent heat thermal storage for solar application combining with double-effect H_2O/LiBr absorption refrigeration system. Sol Energy 110:398–409

Gil A, Oró E, Miró L, Peiró G, Ruiz Á, Salmerón JM (2014) Experimental analysis of hydroquinone used as phase change material (PCM) to be applied in solar cooling refrigeration. Int J Refrig 39:95–103

Harish S, Orejon D, Takata Y, Kohno M (2015) Thermal conductivity enhancement of lauric acid phase change nanocomposite with graphene nanoplatelets. Appl Therm Eng 80:205–211

Hejčík J, Charvát P, Klimeš L, Astrouski I (2016) A PCM-water heat exchanger with polymeric hollow fibres for latent heat thermal energy storage: a parametric study of discharging stage. J Theor Appl Mech 54(4):1285–1295

Hosseini MJ, Rahimi M, Bahrampoury R (2014) Experimental and computational evolution of a shell and tube heat exchanger as a PCM thermal storage system. Int Commun Heat Mass Transf 50:128–136

Hosseini MJ, Ranjbar AA, Rahimi M, Bahrampouri R (2015) Experimental and numerical evaluation of longitudinally finned latent heat thermal storage systems. Energy Build 99:263–272

https://www.pluss.co.in/upload/application/plus10d82e_Pluss%20PCM%20in%20Solar%20Application.pdf. Last access 16/12/2019

https://www.pluss.co.in/upload/application/plus10d82e_Pluss%20PCM%20in%20Solar%20Application.pdf. Last accessed on 26.12.2019

https://www.thermaxxjackets.com/plate-and-frame-heat-exchangers-explained/. Last accessed 2019/01/21

https://www.world-builders.org/lessons/less/biomes/SunEnergy.html
Huang X, Lin Y, Alva G, Fang G (2017) Thermal properties and thermal conductivity enhancement of composite phase change materials using myristyl alcohol/metal foam for solar thermal storage. Sol Energy Mater Sol Cells 170:68–76
Hussain A, Tso CY, Chao CY (2016) Experimental investigation of a passive thermal management system for high-powered lithium ion batteries using nickel foam-paraffin composite. Energy 115:209–218
Jesumathy SP, Udayakumar M, Suresh S (2012) Heat transfer characteristics in latent heat storage system using paraffin wax. J Mech Sci Technol 26(3):959–965
Jian L (2008) Numerical and experimental investigation for heat transfer in triplex concentric tube with phase change material for thermal energy storage. Sol Energy 82(11):977–985
Jose et al (2016) Thermal energy storage for low and medium temperature applications using phase change materials—a review. Appl Energy 177:227–238
Kabeel AE, Abdelgaied M (2017) Observational study of modified solar still coupled with oil serpentine loop from cylindrical parabolic concentrator and phase changing material under basin. Sol Energy 144:71–78
Kabeel AE, Khalil A, Shalaby SM, Zayed ME (2016) Experimental investigation of thermal performance of flat and v-corrugated plate solar air heaters with and without PCM as thermal energy storage. Energy Convers Manag 113:264–272 (2016)
Khorshidi J, Heidari S (2016) Design and construction of a spiral heat exchanger. Adv Chem Eng Sci 6:201–208
Krasny I, Astrouski I, Raudensky M (2016) Polymeric hollow fiber heat exchanger as an automotive radiator. Appl Therm Eng 108:798–803
Kumar M, Gupta V, Bagri S (2017) CFD analysis of spirally coiled heat exchanger. Int J Sci Res Sci Eng Technol 3(5):157–160
Lacroix M (1993) Study of the heat transfer behavior of a latent heat thermal energy storage unit with a finned tube. Int J Heat Mass Transf 36(8):2083–2092
Li M (2013) A nano-graphite/paraffin phase change material with high thermal conductivity. Appl Energy 106:25–30
Mahfuz MH, Anisur MR, Kibria MA, Saidur R, Metselaar IHSC (2014) Performance investigation of thermal energy storage system with PCM for solar water heating application. Int Commun Heat Mass Transf 57:132–139
Mat S, Al-Abidi AA, Sopian K, Sulaiman MY, Mohammad AT (2013) Enhance heat transfer for PCM melting in triplex tube with internal-external fins. Energy Convers Manage 74:223–236
Mehrali M, Latibari ST, Mahila TMI, Matselaar HSC, Naghavi HS, Sadeghinezhad E, Akhiani AR (2013) Preparation and characterization of palmitic acid/graphene nanoplatelets composite with remarkable thermal conductivity as a novel shape-stabilized phase change material. Appl Therm Eng 61(2):633–640
Mehrali M, Tahan LS, Mehrali M, Mahlia TMI, Sadeghinezhad E, Metselaar HSC (2014) Preparation of nitrogen-doped graphene/palmitic acid shape stabilized composite phase change material with remarkable thermal properties for thermal energy storage. Appl Energy 135:339–349
Mosaffa AH, Talati F, Basirat Tabrizi H, Rosen MA (2012) Analytical modeling of PCM solidification in a shell and tube finned thermal storage for air conditioning systems. Energy Build 49:356–361
Murray RE, Groulx D (2014a) Experimental study of the phase change and energy char-acteristics inside a cylindrical latent heat energy storage system: part 1 consecutive charging and discharging. Renew Energy 62:571–581
Murray RE, Groulx D (2014b) Experimental study of the phase change and energy char-acteristics inside a cylindrical latent heat energy storage system: part 2 simultaneous charging and discharging. Renew Energy 63:724–734
Niyas H, Muthukumar P (2013) Performance analysis of latent heat storage systems. Int J Sci Eng Res 4(12):74–79

Nomura T, Okinaka N, Akiyama T (2010) Waste heat transportation system, using phase change material (PCM) from steelworks to chemical plant. Resour Conserv Recycl 54:1000–1006
Pandiyarajan V, Chinna Pandian M, Malan E, Velraj R, Seeniraj RV (2011) Experimental investigation on heat recovery from diesel engine exhaust using finned shell and tube heat exchanger and thermal storage system. Appl Energy 88:77–87
Paria S, Sarhan AAD, Goodarzi MS, Baradaran S, Rahmanian B, Yarmand H, Alavi MA, Kazi SN, Metselaar HSC (2015) Indoor solar thermal energy saving time with phase change material in a horizontal shell and finned-tube heat exchanger. Sci World J 2015
Peiro G, Gasia J, Miro L, Cabeza LF (2015) Experimental evaluation at pilot plant scale of multiple PCMs (cascaded) vs. single PCM configuration for thermal energy storage. Renew Energy 83:729–736
Pirasaci T, Goswami DY (2016) Influence of design on performance of a latent heat storage system for a direct steam generation power plant. Appl Energy 162:644–652
Qureshi ZA, Ali HM, Khushnood S (2018) Recent advances on thermal conductivity enhancement of phase change materials for energy storage system: a review. Int J Heat Mass Transf 127:838–856
Rahimi M, Hosseini MJ, Gorzin M (2019) Effect of helical diameter on the performance of shell and helical tube heat exchanger: an experimental approach. Sustain Cities Soc 44:691–701
Seeniraj RV, Narasimhan NL (2008) Performance enhancement of a solar dynamic LHTS module having both fins and multiple PCMs. Sol Energy 82:535–542
Seeniraj RV, Velraj R, Narasimhan NL (2002) Thermal analysis of a finned-tube LHTS module for a solar dynamic power system. Heat Mass Transf 38(4–5):409–417
Shalaby SM, Bek MA, El-Sebaii AA (2014) Solar dryers with PCM as energy storage medium: a review. Renew Sustain Energy Rev 33:110–116
Singh RP, Kaushik SC, Rakshit D (2018a) Solidification behaviour of binary eutectic phase change material in a vertical finned thermal storage system dispersed with graphene nano-plates. Energy Convers Manage 171:825–838
Singh RP, Kaushik SC, Rakshit D (2018b) Melting phenomenon in a finned thermal storage system with grapheme nano-plates for medium temperature applications. Energy Convers Manage 163:86–99
Singh RP, Xu H, Kaushik SC, Rakshit D, Ramagnoli A (2019) Charging performance evaluation of finned conical thermal storage system encapsulated with nano-enhanced phase change material. Appl Therm Eng 151:176–190
Srinivasan S, Diallo MS, Saha SK, Abass OA, Sharma A, Balasubramanian G (2017) Effect of temperature and graphite particle fillers on thermal conductivity and viscosity of phase change material n-eicosane. Int J Heat Mass Transf 114:318–323
Subramanian SP, Pandiyarajan V, Velraj R (2004) Experimental analysis of a PCM based I.C. engine exhaust waste heat recovery system. Int Energy J 5(2):81–92
Tamme R, Bauer T, Buschle J, Laing D, Müller-Steinhagen H, Steinmann W-D (2008) Latent heat storage above 120 °C for applications in the industrial process heat sector and solar power generation. Int J Energy Res 32(3):264–271
Tao Z, Wang H, Liu J, Zhao W, Liu Z, Guo Q (2017) Dual-level packaged phase change materials–thermal conductivity and mechanical properties. Sol Energy Mater Sol Cells 169:222–225
Tay N, Bruno F, Belusko M (2012) Experimental validation of a CFD and an ε–NTU model for a large tube-in-tank PCM system. Int J Heat Mass Transf 55:5931–5940
Tian Y, Zhao CY (2013) Thermal and exergetic analysis of metal foam-enhanced cascaded thermal energy storage (MF-CTES). Int J Heat Mass Transf 58:86–96
Velraj R, Seeniraj RV, Hafner B, Faber C, Schwarzer K (1997) Experimental analysis and numerical modeling of inward solidification on a finned vertical tube for a latent heat storage unit. Sol Energy 60(5):281–290
Wang C, Lin T, Li N, Zheng H (2016) Heat transfer enhancement of phase change composite material: copper foam/paraffin. Renew Energy 96:960–965
Wei L, Xinguo L, Jun Z (2010) Experimental study of a finned-tube phase change heat storage system. In: Asia-Pacific power energy engineering conference (APPEEC) 2010

Wu W, Zhang G, Ke X, Yang X, Wang Z, Liu C (2015) Preparation and thermal conductivity enhancement of composite phase change materials for electronic thermal management. Energy Conserv Manag 101:278–284

Xiao X, Zhang P, Li M (2013) Preparation and thermal characterization of paraffin/metal foam composite phase change material. Appl Energy 112:1357–1366

Yang J, Tang L, Bao R, Bai L, Liu Z, Xie B, Yang M, Yang W (2018) Hybrid network structure of boron nitride and graphene oxide in shape-stabilized composite phase change materials with enhanced thermal conductivity and light-to-electric energy conversion capability. Sol Energy Mater Sol Cells 174:56–64

Yazıcı MY, Avci M, Aydin O, Akgun M (2014) Effect of eccentricity on melting behavior of paraffin in a horizontal tube-in-shell storage unit: an experimental study. Sol Energy 101:291–298

Zarkadas D, Sirkar KK (2004) Polymeric hollow fiber heat exchangers: an alternative for lower temperature applications. Ind Eng Chem Res 43(25):8093–8106

Zhang Y, Faghri A (1996) Heat transfer enhancement in latent heat thermal energy storage system by using the internally finned tube. Int J Heat Mass Transf 39(15):3165–3173

Zhao D, Tan G (2015) Numerical analysis of a shell-and-tube latent heat storage unit with fins for air-conditioning application. Appl Energy 138:381–392

Lightning Source UK Ltd.
Milton Keynes UK
UKHW021301130522
402959UK00002B/8